*Ultrafast and
Ultra-parallel
Optoelectronics*

Ultrafast and Ultra-parallel Optoelectronics

Edited by

T. Sueta
Setsunan University, Japan

T. Okoshi
University of Tokyo, Japan

Ohmsha   Tokyo

JOHN WILEY & SONS
Chichester • New York • Brisbane • Toronto • Singapore

Copyright © 1995 by Tadashi Sueta and Takanori Okoshi

Published jointly by Ohmsha, Ltd. and John Wiley & Sons Ltd.

All rights reserved.

No part of this book may be reproduced by any means,
or transmitted, or translated into a machine language
without the written permission of the publisher.

Distributed in:

Japan by
Ohmsha, Ltd.
3-1 Kanda Nishiki-cho, Chiyoda-ku, Tokyo 101, Japan

The rest of the world by

John Wiley & Sons Ltd.
Baffins Lane, Chichester
West Sussex PO19 IUD, England

John Wiley & Sons, Inc., 605 Third Avenue,
New York, NY 10158-0012, USA

Jacaranda Wiley Ltd, 33 Park Road, Milton,
Queensland 4064, Australia

John Wiley & Sons (Canada) Ltd, 22 Worcester Road,
Rexdale, Ontario M9W 1L1, Canada

John Wiley & Sons (SEA) Pte Ltd, 37 Jalan Pemimpin #05-04,
Block B, Union Industrial Building, Singapore 2057

Library of Congress Cataloging-in-Publication Data
Ultrafast and ultra-parallel optoelectronics / edited by T. Sueta,
 T. Okoshi.
    p. cm.
   Includes bibliographical references and index.
   ISBN 0 471 95665 1
   1. Optoelectronics. 2. Optical communications. 3. Picosecond
pulses. 4. Quantum electronics. 5. Integrated optics. I. Sueta,
T. (Tadashi) II. Ōkoshi, Takanori, 1932 —   .
   TA1750.U57   1995   95-9102
   621.381′045 — dc20   CIP

British Library Cataloguing in Publication Data

A catalogue record for this book is available from the British Library

ISBN 4-274-90055-X C3055 (Ohmsha, Ltd)

ISBN 0 471 95665 1 (John Wiley)

Typeset in $10\frac{1}{2}/12\frac{1}{2}$ pt Times by Laser Words, Madras, India
Printed and bound in Great Britain by Bookcraft (Bath) Ltd
This book is printed on acid-free paper responsibly manufactured from sustainable forestation,
for which at least two trees are planted for each one used for paper production.

CONTENTS

Preface	ix
List of Contributors	xi

1 FUNDAMENTALS OF ULTRAFAST AND ULTRA-PARALLEL OPTOELECTRONICS   1
    1.1 Impact of Optical Fiber Amplifiers on Ultrafast Optical Communication Systems   3
    1.2 Historical Review of Ultrafast Optical Devices   10
    1.3 New Concepts in Ultrafast Optoelectronics   16
    1.4 Photonic Integration   21
    1.5 Photonic Devices for Ultra-parallel Lightwave Systems   27
    1.6 Parallel Optical Computing Systems   33
    1.7 Multi-dimensional Optical Information Processing   44

2 BASICS OF ULTRAFAST OPTOELECTRONICS   51
    2.1 Ultrashort Semiconductor Laser Pulse Source   53
    2.2 Ultrashort Pulse Generation Method by Electrooptic Modulation   63
    2.3 Ultrafast Optical Switching by Quantum Well Structures   74
    2.4 Generation of Ultra-wide Band Optical Frequency Grids   84
    2.5 Control of Femtosecond Optical Pulses Using Nonlinear Organic Materials   94

3 NONLINEAR OPTICS: MATERIALS, PROCESSES AND APPLICATIONS   107
    3.1 Image Processing by Phase Conjugation in Organic Dye-doped Polymer Films   109
    3.2 Nonlinear Interactions in Periodic Domain Reversals   117
    3.3 Three-dimensional Optical Memory Using Photorefractive Materials   125
    3.4 Second-harmonic Generation in Multilayered Semiconductors   135

3.5 Second-harmonic Generation and Light Amplification in Organic Waveguides	145
3.6 Multi-photon Diffraction Switching Using a Resonant Two-level System	157

4 FUNCTIONAL PHOTONICS DEVICES AND INTEGRATION — 165

4.1 Mode-locked Fiber Laser Using a Fiber-optic Phase Modulator	169
4.2 Vertical Integration Technology for Fiber-optic Circuits	174
4.3 Three-dimensional Optical Interconnects Using Stacked ARROW Guided Wave Circuits	184
4.4 Design and Modeling of Optical Waveguide Devices	198
4.5 Guided Wave Solid State Laser and Optical Amplifier	210
4.6 Optical Isolator with a Waveguide Structure	227
4.7 Integrated Photonic Functional Devices Using Grating Couplers	238

5 ACTIVE PHOTONIC DEVICES — 247

5.1 Multifunctional Optoelectronic Integrated Devices	249
5.2 Strained Layer Superlattices for High-speed Optical Devices	261
5.3 Functional Integration in Distributed Feedback Lasers	272
5.4 Light-emitting Devices by GaAs-on-Si Technology	283
5.5 Organic Light-emitting Diodes with Microcavity Structures	292

6 HIGH SPEED DEVICE TECHNOLOGIES — 301

6.1 Quantum Wire and Quantum Box Lasers: GaAs Systems	303
6.2 GaInAsP/InP Quantum Well and Quantum Wire Semiconductor Laser Amplifiers	328
6.3 Quantum Well and Wire Lasers: InP Systems	340
6.4 Ultrafast Guided Wave Electrooptic $LiNbO_3$ Modulators	349

7 OPTOELECTRONIC DEVICES FOR PARALLEL PROCESSING AND INTERCONNECTS — 359

7.1 Surface-emitting Laser Diodes and 2D Arrayed Devices	361
7.2 Liquid Crystal Cells for Optical Neural Computing	371
7.3 Realization of High-speed 3D LSIs by Optical Interconnections	385
7.4 Semiconductor-based, Two-dimensional Spatial Light Modulator	400
7.5 Integrated Optoelectronic Neuro Devices	410

8 ULTRA-HIGH CAPACITY OPTICAL COMMUNICATIONS — 421

8.1 Ultrafast Transmission and Processing Using Optical Fiber Nonlinearity	423
8.2 High Precision Simulation of Soliton Problems	433
8.3 Dispersive and Nonlinear Degradation Compensation in Fiber-optic Systems	437

8.4 Ultra-multiplexing and Demultiplexing Scheme of Subcarriers	445
8.5 Ultralong-distance Fiber-optic Transmission and Multiplexing	455
8.6 Control of Fiber Characteristics	465

9 DIGITAL OPTICAL COMPUTING — 475

9.1 Optical Parallel Computing Systems and Parallel Algorithms: Experimental Demonstration — 477
9.2 Parallel Optoelectronic Computing System — 486
9.3 Temporal Coding Logic Array and its Application to Hybrid Computing — 495

10 NEURAL COMPUTING — 507

10.1 Neural Network Learning: Generalization and Over-learning — 509
10.2 Integrated Image Processing System Using an Optical Neural Network — 517
10.3 Real Time Optical Correlator with a $Bi_{12}SiO_{20}$ Crystal — 525
10.4 Dynamics of Complex Neural Fields in a Phase-conjugate Resonator — 536

11 MULTI-DIMENSIONAL OPTICAL SENSING AND PROCESSING — 547

11.1 Display Systems of Autostereoscopic 3D Images — 549
11.2 Selective Image Retrieval by Synthesis of the Coherence Function — 561
11.3 Ultraweak Biophoton Imaging and Information Characterization — 570
11.4 Super-parallel Fourier-transform Spectral Imaging — 581
11.5 Optical Heterodyne Spatiotemporal Polarimetry — 591

Index — 599

PREFACE

The academic superiority of lightwaves over radio waves is in the ultrafast and ultra-parallel properties, or equivalently in the high temporal and spatial resolutions. In the field of information transmission and processing, gigabit technologies are being established by electronic and ordinary optoelectronic means. The demand for speed and capacity, however, is steadily increasing year by year toward the next century. Thorough investigation of the ultrafast and ultra-parallel properties of lightwaves will lead to achievements of higher speed and larger capacity, and to the establishment of the so-called terabit technologies. This book is a comprehensive review of this new field of research in the frontiers of optoelectronics, written by the most active experts of the Japanese universities.

The book begins with a chapter of general review of ultrafast and ultra-parallel optoelectronics. Ultrafast technologies are discussed in the chapters on basic fast phenomena in time and frequency domain, nonlinear optics, high-speed photonic devices, large-capacity optical communications, etc. Ultra-parallel technologies are treated in connection with optical computing, optical interconnection and image processing, etc. Fundamental principles and basic technologies of relevant fields are included for general readers. The book will be useful for researchers/engineers in optoelectronics as well as graduate students who are going to start work in this promising research field.

To pursue the potentialities in ultrafast and ultra-parallel optoelectronics, a cooperative three-year research project entitled 'Ultrafast and Ultra-parallel Optoelectronics' started in April 1991 under the sponsorship of Grant-in-Aid for Scientific Research on Priority Areas from the Ministry of Education, Science and Culture. Selected research laboratories of Japanese universities joined this project, which covered a wide range of problems in this specific field including new materials, devices, and systems. Owing to the interdisciplinary nature of the subject, laboratories of both electronics and optics were chosen. The project ended successfully in March 1994.

At the beginning of the project, 44 university laboratories were organized to form five research groups: (1) basic studies on ultra-high performance optical devices; (2) highly functional optical devices; (3) ultrafast, ultra-multiplex lightwave transmission and networks; (4) ultra-parallel optical computing systems; (5) multi-dimensional information and image processing. At the end of the project the number of laboratories had increased to 51. This book is written by the members of the project based on the achievements of these research laboratories. In the book, the contributions from five research groups are rearranged into 11 chapters in a logical order.

We wish to express our sincere thanks to the Ministry of Education, Science and Culture for sponsorship of the project. Also we are grateful to the External and Internal Advisors for their valuable advice and encouragement and are deeply indebted to Professors T. Kamiya, H. Nishihara, K. Iga, Y. Ichioka and Y. Ohtsuka as Group Leaders, and to Professors K. Hotate and M. Ohtsu as Project Secretaries for the management and accomplishment of the project. Finally much appreciation is owed to all contributors and the members of the editorial board.

Publication of this book is supported by the Grant-in-Aid for Scientific Research from the Ministry of Education, Science and Culture.

T. Okoshi
T. Sueta
October 1994

LIST OF CONTRIBUTORS

Chief Editors:

Sueta, T. — *Setsunan University and Professor Emeritus of Osaka University*

Okoshi, T. — *Tokyo University of Science, National Institute for Advanced Interdisciplinary Research (NAIR) and Professor Emeritus of The University of Tokyo*

Editorial Board:

Ichioka, Y.	*Osaka University*
Iga, K.	*Tokyo Institute of Technology*
Ito, H.	*Tohoku University*
Haruna, N.	*Osaka University*
Hotate, K.	*The University of Tokyo*
Kamiya, T.	*The University of Tokyo*
Kokubun, Y.	*Yokohama National University*
Nishihara, H.	*Osaka University*
Ohtsu, M.	*Tokyo Institute of Technology*
Ohtsuka, Y.	*Hokkaido University*
Takeda, M.	*The University of Electro-Communications*
Yamada, M.	*Kanazawa University*
Yatagai, T.	*University of Tsukuba*

Authors:

Arai, S.	6.8	*Tokyo Institute of Technology*
Arakawa, Y.	6.1	*University of Tokyo*
Asada, M.	6.2	*Tokyo Institute of Technology*

Choi, Y.-K.	8.4	*Fukui University*
Egawa, T.	5.4	*Nagoya Institute of Technology*
Fujii, Y.	8.1	*University of Tokyo*
Fujiwara, H.	3.1	*Muroran Institute of Technology*
Furuya, K.	6.3	*Tokyo Institute of Technology*
Hamasaki, J.	11.1	*Graduate School, University of East Asia*
Hashi, M.	7.2	*Tokyo University of Agriculture and Technology*
Hirota, R.	8.2	*Waseda University*
Honda, T.	10.3	*Chiba University*
Horiuchi, K.	8.2	*Waseda University*
Hotate, K.	11.2	*The University of Tokyo*
Ichioka, Y.	1.6/9.1	*Osaka University*
Iga, K.	1.5/7.1	*Tokyo Institute of Technology*
Iimura, Y.	7.2	*Tokyo University of Agriculture and Technology*
Imai, M.	4.1	*Muroran Institute of Technology*
Inaba, H.	11.3	*Tohoku Institute of Technology*
Ishikawa, M.	9.2	*The University of Tokyo and Professor Emeritus of Tohoku University*
Ito, H.	3.2	*Tokyo University*
Ito, K.	10.3	*Chiba University*
Itoh, K.	11.4	*Osaka University*
Izutsu, M.	6.4	*Osaka University*
Kamiya, T.	1.3/2.1	*The University of Tokyo*
Kawakami, S.	4.2	*Tohoku Institute of Technology*
Kawata, Y.	3.3	*Osaka University*
Kikkawa, H.	7.2	*Tokyo University of Agriculture and Technology*
Kikuchi, K.	8.3	*The University of Tokyo*
Kishigami, T.	10.4	*The University of Electro-Communications*
Kobayashi, S.	7.2	*Tokyo University of Agriculture and Technology*
Kobayashi, T.	2.2	*Osaka University*
Kokubun, Y.	4.3	*Yokohama National University*
Komori, K.	6.2	*Tokyo Institute of Technology*
Konishi, T.	9.1	*Osaka University*
Koshiba, M.	4.4	*Hokkaido University*
Koyama, F.	7.1	*Tokyo Institute of Technology*
Koyanagi, M.	7.3	*Tohoku University*
Kuwamura, Y.	7.4	*Kanazawa University*
Lee, J.-S.	10.2	*Tokyo Institute of Technology*
Lorattanasane, C.	8.3	*The University of Tokyo*
Maeda, K.	7.2	*Tokyo University of Agriculture and Technology*
Masumoto, Y.	2.3	*University of Tsukuba*
Miyamoto, Y.	6.3	*Tokyo Institute of Technology*
Miyazaki, Y.	4.5	*Toyohashi University of Technology*
Morimoto, A.	2.2	*Osaka University*
Morita, R.	2.5	*Hokkaido University*
Naito, Y.	4.6	*Tokyo Institute of Technology*
Nakagawa, K.	3.1	*Muroran Institute of Technology*
Nakajima, M.	8.4	*Kyoto University*
Nakano, Y.	5.3	*University of Tokyo*
Nishihara, H.	1.4/4.7	*Osaka University*
Noda, S.	5.1	*Kyoto University*

List of Contributors

Ogasawara, N.	3.4	The University of Electro-Communications
Ogawa, H.	10.1	Tokyo Institute of Technology
Ohtsu, M.	2.4	Tokyo Institute of Technology
Ohtsuka, Y.	1.7/11.5	Hokkaido University
Oishi, S.	8.2	Waseda University
Okada, K.	10.3	Chiba University
Okoshi, T.	1.1/8.5	Tokyo University of Science
Ohyama, N.	10.2	Tokyo Institute of Technology
Sasaki, A.	5.1	Kyoto University
Sasaki, K.	3.5	Keio University
Sasaki, Y.	8.6	Ibaraki University
Satoh, O.	3.6	Yamagata University
Satoh, S.	4.1	Muroran Institute of Technology
Suematsu, Y.	8.3	Kogakuin University and Professor Emeritus of Tokyo Institute of Technology
Suemune, I.	5.2	Hokkaido University
Sueta, T.	1.2/6.4	Setsunan University
Suhara, T.	4.7	Osaka University
Tada, K.	5.3	University of Tokyo
Takeda, M.	10.4	The University of Electro-Communications
Takeda, T.	3.1	Muroran Institute of Technology
Tan-no	3.6	Yamagata University
Tanida, J.	9.1	Osaka University
Tsuchiya, M.	2.1	University of Tokyo
Tsujimoto, S.	8.2	Waseda University
Tsujiuchi, J.	10.3	Chiba University
Tsutsui, T.	5.5	Kyushu University
Umeno, M.	5.4	Nagoya Institute of Technology
Ura, S.	4.7	Osaka University
Yamada, M.	7.4	Kanazawa University
Yamashita, M.	2.5	Hokkaido University
Yamashita, S.	8.5	The University of Tokyo
Yatagai, T.	9.3	University of Tsukuba
Yonezu, H.	7.5	Toyohashi University of Technology

1
FUNDAMENTALS OF ULTRAFAST AND ULTRA-PARALLEL OPTOELECTRONICS

1.1

Impact of Optical Fiber Amplifiers on Ultrafast Optical Communication Systems

Takanori Okoshi

Abstract

The advent of an erbium-doped fiber amplifier (EDFA) in 1987 impacted significantly on the entire framework of optical communications technology. The influence and the induced new technological trends are widespread and diverse; however, in this section only two aspects of these are discussed: one could be expressed briefly as 'from optoelectronic systems to all-optical systems', and the other 'from linear to nonlinear'.

1 Introduction

Until the end of 1980s a great difference had existed between optical and radio-wave communications; that is, whether the signal could be amplified or not at their carrier (i.e. lightwave or radio-wave) frequency. In the optical frequency region, semiconductor laser amplifiers had been investigated. However, the technology had not reached a practical level mainly because of large coupling losses between the fiber and the device at the input and output ports of the amplifier. The advent of an erbium doped fiber amplifier (EDFA) in 1987 changed the situation dramatically.

Research on optical amplifiers using rare-earth-doped optical fibers has continued for more than 30 years. The success in manufacturing a neodium-doped optical fiber was first reported as early as 1962, and the first success of optical amplification at the 1.06 μm wavelength using such a fiber was reported only two years later, in 1964 [1]. However, another 23 years had to pass before the success of a practical EDFA was reported by Mears and others in 1987 [2].

Since 1987 the research and development (R&D) of the EDFA have been expanded explosively worldwide because of its overwhelming technical advantages of (1) amplification of a 1.5 μm lightwave which has the lowest transmission loss in silica fibers; (2) a low coupling loss at the input and output ports; (3) a high gain (20–30 dB); (4) a relatively low noise figure (3–4.5 dB at medium signal level); and (5) polarization-independent amplification.

The successful R&D of the EDFA in the past 7 years have brought about a great change in the entire aspect of optical communications research. In this section, however, we review this change from only two points of view: 'from optoelectronic systems to all-optical systems', and 'from linear to nonlinear'.

2 From Optoelectronic Systems to All-optical Systems

An example of the practical applications of the EDFA will appear soon, namely the third Trans-Pacific optical fiber cable (the fifth including copper ones), TPC-5, which will be completed in 1995. This is essentially an all-optical trunk line featuring an EDFA/fiber chain structure with about 300 EDFA optical repeaters between North America and Japan.

Other examples at the research stage are (1) the extraction of a timing signal from a received optical signal using an optical phase-locked loop (PLL) based on the all-optical correlation detection technique; (2) demultiplexing of multiplexed optical signals using all-optical switches; and (3) the 'midway optical spectrum inversion' for canceling the dispersion of an optical fiber.

2.1 EDFA/fiber chain systems

Now that the EDFA/fiber chain structure as employed in TPC-5 is becoming more common, the term 'transmission distance' of an optical communication system has come to have two meanings, as shown in Figure 1.1.1; one could be called (tentatively) the 'repeater-free distance', and the other the 'transparent distance' in the sense that the

```
         Since the advent of "transparent"
   optical amplifiers, the "transmission
   distance" has become to have two
   different meanings:

   (1) Repeater-Free Distance: D_RF

   (2) Transparent Distance: D_TR
```

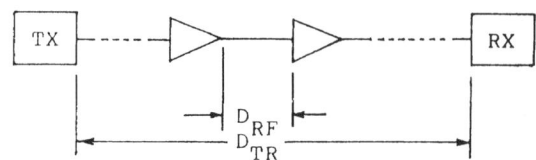

Figure 1.1.1
Two definitions of the transmission distance of an optical fiber communication system

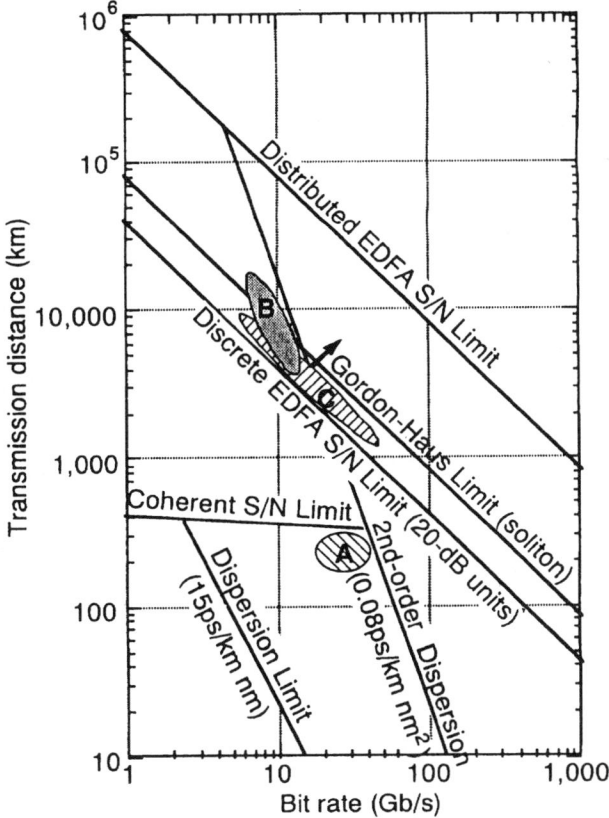

Figure 1.1.2
Estimates of various bitrate-distance limiting factors in optical fiber communications. For details, refer to the text

EDFAs are essentially 'transparent' devices for optical signals. Incidentally, the longest record of a repeater-free distance is 364 km, achieved by a coherent optical communication system, whereas that of the transparent distance is more than 15 000 km with a 'full-length-equipped' system, and much above 1 000 000 km with a 'recirculating experiment' system.

The limiting factors of the repeater-free transmission distance are the signal-to-noise (S/N) limit of the receiver and dispersion limit [3]. The limiting factors of the transparent transmission distance in 'nonsoliton' cases are the (S/N) limit given by the accumulation of the amplified spontaneous emission (ASE) noise, and the dispersion limit. In the case of soliton-type transmission, the limit is given by the accumulation of time jitter of optical pulses, which is called the 'Gordon–Haus limit' [4].

Figure 1.1.2 shows the rough estimates of the above limiting factors in a bitrate-distance graph. The target areas in the near future for the above three cases are depicted as regions A, B, and C. As to the soliton-type transmission, specialists' concern is now directed toward a breakthrough in the Gordon–Haus limit (see the arrow in Figure 1.1.2).

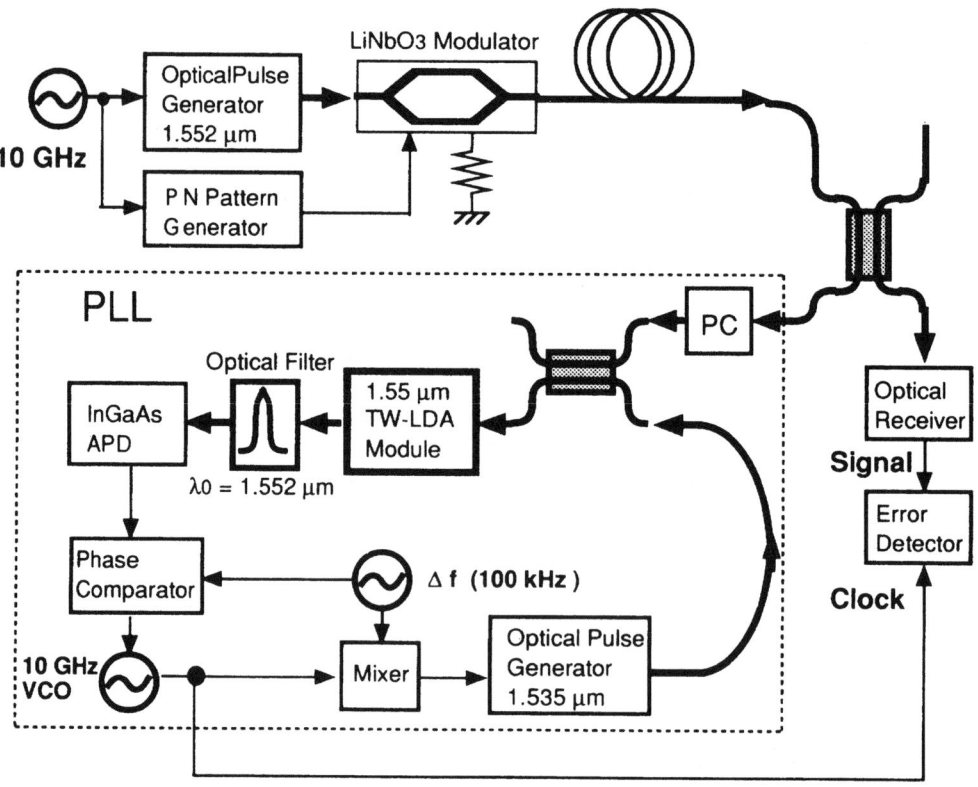

Figure 1.1.3
An experimental setup for the timing signal extraction using an all-optical phase-locked loop. After Kawanishi and Saruwatari [4]

2.2 Timing extraction using an all-optical phase-locked loop

Figure 1.1.3 shows the experimental setup for the timing signal extraction from a 10 Gbit/s optical signal using an all optical PLL [4].

2.3 Demultiplexing of multiplexed optical signal by all-optical switches

When the signal bit-rate exceeds 10 Gbit/s, the direct (or indirect) modulation of laser light and the direct detection (optoelectronic conversion) of the received light all become difficult. The use of all-optical switches has been investigated to overcome this difficulty.

Reference [5] reports an example of such experiments. Four 8 Gbit/s signals are multiplexed to form a 32 Gbit/s light signal, or eight 8 Gbit/s signals are multiplexed to form a 64 Gbit/s light signal. In either case the signal is transmitted via a fiber, and an 8 Gbit/s timing signal is extracted by using a travelling-type laser-diode amplifier (TW-LDA). The extracted timing (clock) signal is then used to drive all optical

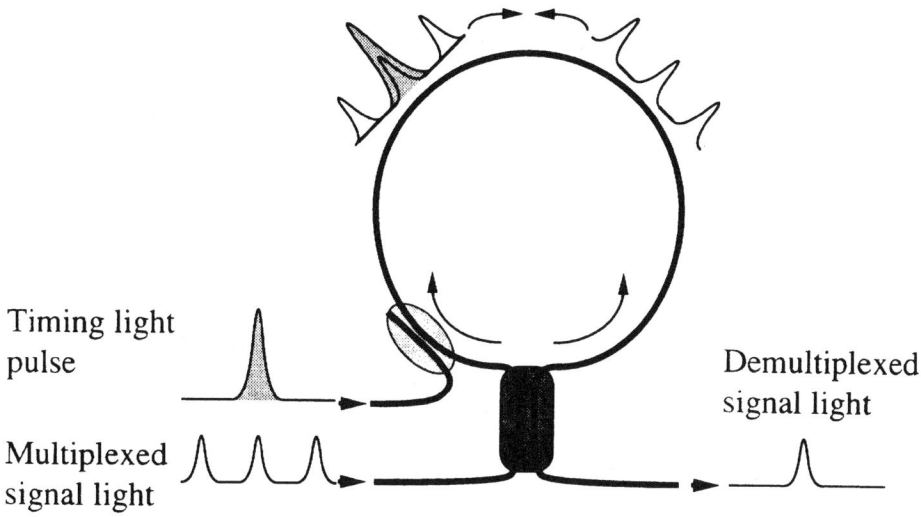

Figure 1.1.4
A nonlinear loop-mirror switch used in the optical demultiplexing experiment. After Kawanishi *et al.* [5]

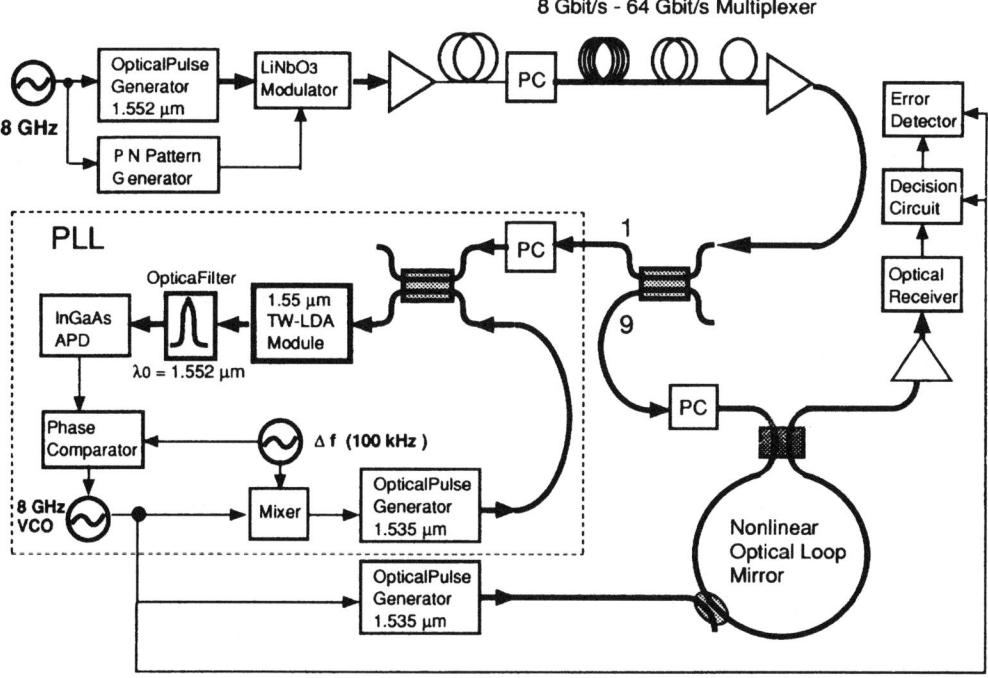

Figure 1.1.5
An experimental setup for demultiplexing 8 Gbit/s signals from a 64 Gbit/s multiplexed signal. After Kawanishi *et al.* [5]

demultiplexing circuits using nonlinear loop-mirror switches, restoring four or eight 8 Gbit/s signals from the 32 Gbit/s or 64 Gbit/s signals, respectively. Figure 1.1.4 shows the construction of the nonlinear loop-mirror switch, whereas Figure 1.1.5 shows the setup for the demultiplexing experiment.

A more recent paper from the same research group reported another experiment of multiplexing and demultiplexing of 6.3 Gbit/s → 100 Gbit/s → 6.3 Gbit/s [6,7].

3 From Linear to Nonlinear

Recently R&D of nonlinear effects and their applications in optical devices and materials have been expanded. Three technological backgrounds seem to exist behind this trend:

(1) the mature of traditional 'linear' fiber technology;
(2) the ease of obtaining a relatively high power lightwave (above 100 mW) achieved by the progress in semiconductor laser technology; and
(3) the ease of lightwave amplification brought about by the EDFA, which is the subject of this section.

The nonlinear effects being investigated are induced Raman scattering, induced Brillouin scattering, the optical Kerr effect, the optical soliton effect, optical pulse compression, and four-wave mixing. Among these, the application of the optical soliton effect is the most significant application and has been pursued most vigorously in the past several years.

In addition to the above, a promising application of the optical nonlinear effect is the 'midway spectral inversion' which aims to cancel the dispersion effect of optical fibers. This was originally proposed by Yariv *et al.* in 1979 [8], and, following the pioneering experiments by Murata *et al.* [9] using four-wave mixing in a semiconductor laser amplifier, as many as four successful 'midway spectral inversion' experiments were reported in 1993 using fiber nonlinearity: from Fujitsu [10], AT&T Bell Laboratories [11], Southampton University [12], and British Telecom Research Laboratories [13].

On the other hand, the techniques to minimize the adverse effects of nonlinearities in fibers and/or devices are becoming equally important as the result of the progress in ultralong distance or ultra-multiplexed optical communication systems.

A typical and very important practical problem is the interaction of the dispersion, nonlinear effect, and the ASE (amplified spontaneous emission) noise in an optical fiber, which limits the maximum transmission distance in EDFA/fiber chains. Intensive research on this combined effect is being performed [14,15].

4 Conclusions

Recent R&D trends in optical communication technologies have been reviewed, with emphasis on the influence of the advent of the EDFA from two points of view: 'from optoelectronic to all optical' and 'from linear to nonlinear'.

References

[1] C.J. Koester and E. Snitzer, *Appl. Opt.* **3**, 1182–1186 (1964).
[2] R.J. Mears *et al.*, *Tech. Digest, OFC'87*, No. WI 2. Reno, Nevada, 1987.
[3] T. Okoshi, *Optical Fibers*, Academic Press, New York, 1982.
[4] S. Kawanishi and M. Saruwatari, *Electron. Lett.* **28**, 510–511 (1992).
[5] S. Kawanishi *et al.*, *Electron. Lett.* **29**, 231–233 (1993).
[6] S. Kawanishi *et al.*, *Tech. Digest, OFC'93*, No. PD-2, San Jose, CA, 1993.
[7] S. Kawanishi *et al.*, *Proc. ECOC'93*, No. ThP-12.1, Montreux, 1993.
[8] A. Yariv, D. Fekete and D.M. Pepper, *Optics Lett.* **4**, (2) 52–54 (1979).
[9] S. Murata *et al.*, *IEEE Photonics Tech. Lett.* **3**, 1021–1023 (1991).
[10] S. Watanabe *et al.*, *Spring National Convention Record*, IEICE Japan, No. B-945, March 1993 [in Japanese].
[11] R.M. Jopson *et al.*, *Tech. Digest, OFC'93*, No. PD3, San Jose, CA, 1993.
[12] R.I. Laming *et al.*, *Proc, ECOC'93*, No. WeC8.2, Montreux, 1993.
[13] M.C. Tatham *et al.*, *Proc, ECOC'93*, No. ThP12.3, Montreux, 1993.
[14] A. Naka and S. Saito, *Tech. Report*, IEICE Japan, No. OQE92-117/OCS92-54, 1992 [in Japanese].
[15] J. Nakagawa and T. Okoshi, *Tech. Report*, IEICE Japan, No. CS93-37/OCS93-13, 1993 [in Japanese].

1.2

Historical Review of Ultrafast Optical Devices

Tadashi Sueta

Abstract

The lightwave has, by nature, an ultrafast property or a high temporal resolution owing to its extremely high frequency. In this section, the developments of ultrafast optical devices are reviewed historically from the ruby laser as a simple pulse source to the recent femtosecond optical pulse generators. The high-speed interaction between light and electric waves is also discussed, putting emphasis on controllability. The importance of the waveguide concept in the design of these control devices is described. Finally, some comments are made on the future prospects of ultrafast optical devices and their applications.

1 Introduction

More than thirty years have passed since the appearance of the laser. During that time many optical devices have been developed for the generation, control, detection, etc. of coherent lightwaves. A new research field, 'optical electronics', which is an interdisciplinary field of optics and electronics, has been established. Also, industries related to optical electronics were born, and are steadily growing for optical components, devices, and systems.

As seen in compact disc players and optical fiber communication systems, some of the applications are taking firm positions in the real world. However, the potential abilities of lightwaves are not fully realized in present-day technologies. The ultrafast property of lightwaves is one kind of potential. Ultrafast optical electronics study the high-speed properties of lightwaves in order to develop high-performance optical devices for advanced applications.

There are many aspects for fast physical phenomena in optics and electronics. The photographic flash lamp of millisecond duration is a rather popular fast optical phenomenon. A typical electronic fast time is of the order of a nanosecond or a subnanosecond. However, lightwaves have potentially fast properties owing to their much higher frequencies and resulting much broader frequency bandwidths, compared with electric signals. The shortest width of light pulses is reported as 6 fs or 6×10^{-15} s. This corresponds to only 2 μm measured in spatial length. If one could control at will such ultrashort pulses, a novel way would be opened toward advanced systems. Although femtosecond technologies are far away, ultrafast optical technologies, even of picosecond order, will contribute much to the development of optical electronics.

2 Ultrafast Properties of Lightwaves

As an electromagnetic wave the lightwave has a much higher frequency than radio waves. This means that the electromagnetic field of a lightwave changes very rapidly. In other words, the lightwave has the possibility of an ultrashort resolvable time comparable to its period of femtoseconds. Actually 6 fs pulses were produced by combining the mode-locked dye laser and the pulse compression technique [1]. If lasers with shorter wavelengths are developed, further shortening would be expected.

While lightwaves have an extremely high time resolution, they also have high spatial resolution owing to their short wavelength. The latter feature was studied extensively for the development of microscopes at a time much earlier than lasers. After lasers became available, the coherent light sources made it possible to realize the high-density optical disc memory.

As can be seen from the fact that the optical disc has attained a diffraction-limited amount of memory, 'the high spatial resolution' of lightwaves was fully studied and widely utilized. On the other hand, the study of 'the high temporal resolution' or ultrafast property of lightwaves is not yet finished. For example, the typical bit rate of commercial optical fiber trunk transmission systems is only 400 Mb/s. The time interval between adjacent bits is 2.5 ns. For even more advanced 10 Gb/s systems, the time interval is 0.1 ns. These values are the so-called electronic times, and are far from the potential resolvable time of femtoseconds for lightwaves.

The above discussion shows that, although an almost limiting pulse width is obtained, the usable resolution time is not short enough compared with the optical resolution. In other words, the potential ultrafast properties of lightwaves have not yet been fully explored.

3 Optical Short Pulse Technology

Short pulse generation is the most basic technology in ultrafast optical electronics. As shown in Figure 1.2.1 [6], the pulse-shortening trace indicates the development of high-speed optical technology.

It is interesting that the output waveform of the first laser, i.e. the ruby laser, is a collection of microsecond pulses. In 1962 the pulse width was shortened to 20 ns by

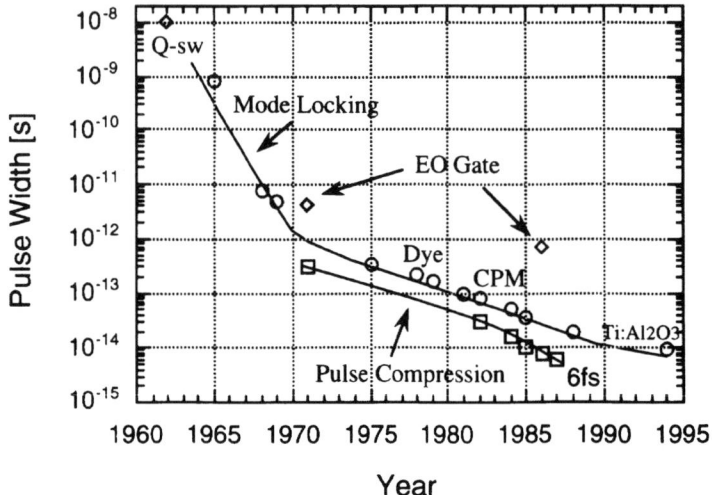

Figure 1.2.1
History of optical pulse shortening [6]

the Q-switching technique. A peak power of megawatts was obtained for the resulting giant pulses. Q-switching was realized by rotating mirrors, Pockels' cells, and saturable absorbers. Giant pulses stimulated the important research field of nonlinear optics, including second harmonic generation (SHG).

The next basic technique in the development of short pulse technology was mode-locking, introduced in 1964. In the case of a He–Ne laser, a train of subnanosecond pulses was produced at a repetition rate of a few hundred megahertz. There are several kinds of mode-locking methods: the first is active, or forced, locking by the internal modulator; the second is self-locking, making use of the nonlinearity of the laser medium; and the third is passive locking by the saturable absorber.

Pulse widths became shorter and shorter as solid-state lasers with wide gain bandwidths, like neodymium YAG and neodymium glass lasers, have been developed. Picosecond pulses became obtainable in about 1970. In 1981 passive mode-locked dye lasers with much wider bandwidths were used to obtain pulses shorter than 0.1 ps by applying the colliding pulse mode-locking (CPM) technique. Recently self-mode-locked titanium sapphire lasers have attracted much attention. Incorporating suitable dispersive optics into the cavity, pulses up to 15 fs were obtained.

Pulse compression techniques further shorten the pulses from the mode-locked lasers. A typical method is to first make use of optical fiber nonlinearity in order to obtain frequency chirping, then a grating pair is used as the dispersive medium to realize pulse compression. As stated earlier, the shortest pulse width of 6 fs was obtained using this type of method.

Short pulse generation from semiconductor lasers is important from the viewpoint of applications. In 1978, 10 ps pulses were obtained either by the mode-locking or direct excitation (gain-switching) methods. Currently even subpicosecond pulses are obtainable.

Soliton transmission in an optical fiber is another interesting problem. Picosecond pulses are transmitted long distances without distortion by balancing the nonlinearity and the dispersion. Optical solitons in the fiber were predicted theoretically in 1973, and later experimentally confirmed. Recently, soliton transmission over a several thousand kilometer fiber was done by using an Er-doped optical fiber laser amplifier. Short pulse generation is also possible by soliton laser using an optical fiber as the feedback loop.

Another interesting pulse generation technique is the electrooptic method. This method is applicable to any kind of laser. Fast electrooptic switches and optical Kerr shutters can be used as a high-speed optical gate to produce short pulses. Deep electrooptic modulation produces side-bands spread over terahertz width. The modulated light may be processed, for example by a dispersive medium, to produce short pulses.

4 *Electronic Control of Optical Signals (EO Control)*

The interaction between microwave sound and lightwaves has been known since as long ago as Brillouin scattering. But it was after lasers that the interaction between lightwaves and radio waves in a certain material became the object of study from the viewpoint of fast electronic control of lightwaves.

The most important technique in high-speed electrooptic control is the microwave modulation of lightwaves. In large-capacity optical transmission systems, for example, high-speed light modulators are the key devices. The basic physical process of modulation is the electrooptic effect, in other words the second-order nonlinear optical effect, of materials like KDP, lithium niobate, lithium tantalate, etc. Although the response time of these materials is fast enough for high-speed control at microwave or even millimeter-wave frequencies, it was difficult to obtain the necessary intensity of the modulating electric field and the necessary interaction length. In early times some light modulators at microwave frequencies were designed and built, but they are bulky, narrow band, and inefficient.

At the end of 1960s the waveguide concept was introduced into optics. Some small-size, guided-wave electrooptical devices were built using the titanium diffused lithium niobate optical waveguides. But the modulation bandwidth was still narrow.

In 1977 the first small-size, broadband modulator was successfully built and tested by adopting travelling-wave operation in a guided-wave modulator. Since that time similar types of modulators have been developed. At the present time, reliable modulators with a 20 GHz bandwidth at reasonable modulating power are available. Modulators at even millimeter-wave frequencies are reported. Although microwave or millimeter-wave modulators have become realistic, the bandwidth is still narrow compared with the potential optical bandwidths.

The integration of several high-speed devices is an interesting technology to obtain high-speed functional devices. A single sideband modulator or optical frequency shifter, a signal processor, an analog-to-digital converter, and a time demultiplexer have been reported.

While one direction toward the faster control of lightwaves is to increase the modulating frequency, another way is to increase the modulation depth. By deep modulation the bandwidth of the modulated light becomes large, thereby obtaining wide-band optical signals. Deeply modulated lightwaves can be processed by suitable means, for example by letting them pass through a dispersive medium to obtain fast optical signals like ultrashort pulses. Recently terahertz optical sidebands have been produced by deep light modulation at microwave frequency.

The electrooptic deflection of a light beam is also an interesting technique for fast electronic control. Generally speaking, however, high-speed and high-resolution deflection is technically rather more difficult than modulation.

Direct modulation of semiconductor lasers is important as high-speed light control. Semiconductor lasers are compact and highly efficient, and have high relaxation frequencies. Accordingly, direct modulation at microwave frequencies is successfully done. However, sometimes the accompanying phase modulation causes trouble.

5 Optical Control of Electric Signals (OE Control)

While light modulation is the electronic control of lightwaves, the optical control of electric signals, the so-called OE control, is another interesting problem. The optical control of electronic circuits goes back to 1873, when the photoconductive effect was discovered. Since that time various optoelectronic devices like photodiodes, phototransistors, and photocouplers have been developed.

One of the important features of OE control is to obtain high-speed electric signals, which cannot be realized by electronic means, by making use of the ultrafast properties of lightwaves. Semiconductor materials like silicon and gallium arsenide possess a fast photoconductive effect. Illumination of electric circuits including these materials by ultrashort light pulses provides some useful high-speed devices, for example an Auston switch. Fast OE switches can be applied to the generation of high-speed electric signals, the cutting out of a small part of electric signals, and so forth.

Recently the optical control of microwave and millimeter-wave circuits is attracting interest. In an example of a light-controlled phase shifter, an optical phase shifter and an optical frequency shifter are integrated on a lithium niobate substrate. The mixing of frequency-shifted lightwaves and phase-shifted lightwaves produce a phase-shifted microwave signal.

6 Lightwave–lightwave Interaction

The speed of interaction between a lightwave and an electric signal is finally limited by the response time of the electric circuits. For faster control, therefore, lightwave–lightwave interaction becomes important. The classic optical Kerr shutter is an example.

Recently optical bistable devices are attracting attention concerning high-speed operation. Bistable devices are obtained by combining nonlinearity and a feed-back mechanism. A Fabry–Perot interferometer containing a nonlinear medium provides such a device. A subpicosecond response will be obtained by an MQW structure. When an electric feed-back circuit is used, the response time is limited by the electric circuit.

An optical waveguide structure is favored for optical nonlinear interaction, since high power density is obtained over a relatively long length. Optical fibers and waveguides on the substrate are being used widely as a nonlinear medium.

7 Conclusion

Various ultrafast optical devices have been briefly reviewed putting emphasis on short pulse generation and high-speed modulation. At the present time, when femtosecond pulses are generated the ultrafast optical technology will reach a turning point. While efforts to obtain shorter pulses, a more reliable method of generation, and the observation of faster phenomena should be continued, it is important to establish the technology to control the fast optical phenomena at will.

Also it is time to apply the ultrafast optical technology to communications, measurements, information processing, and other fields of interest. In particular, the combination of the ultrafast and ultra-parallel properties of lightwaves will bring about novel prosperous applications.

References

[1] R.L. Fork et al., *Opt. Lett.* **12** (2), 483 (1987).
[2] C.H. Lee, ed., *Picosecond Optoelectronics*, Academic Press, New York, 1984.
[3] W. Kaiser, ed., *Ultrafast Laser Pulses and Applications, Topics in Applied Physics*, vol. 60, Springer-Verlag, 1989.
[4] T. Sueta and T. Kamiya, eds., *Ultrafast Optical Electronics*, Bifukan, 1991 [in Japanese].
[5] T. Kobayashi, *J. IEICE* **72** (2), 171 (1989).
[6] Courtesy of Dr A. Morimoto, Osaka University.

1.3

New Concepts in Ultrafast Optoelectronics

Takeshi Kamiya

Abstract

The use of ultrashort optical pulses for transmission of large capacity information is a promising technology for the next decade. New concepts in this regime of optoelectronics were created by the joint research efforts of Japanese university groups. Some selected topics such as soliton pulse propagation, generation of very high order harmonics, and enhancement of nonlinear material response are reviewed.

1 Introduction

For the purpose of extending information handling capacity, the use of optical technology is one of the most attractive tracks because of the intrinsic virtues of optical phenomena in terms of the high speed and wide bandwidth of signal transmission. It is much easier to transmit signals with optical pulses over a long distance in comparison with conventional electrical cable transmission schemes. Presently fiber-optic telecommunication systems only utilize a small part of the intrinsic capability because of insufficient electronics and optoelectronics technology. The need to add new concepts and technologies is very high.

This section is devoted to a review of recent progress in ultrafast optoelectronics devices and materials, potentially influencing the engineering development of ultrafast systems, with special emphasis on the selected activities of Japanese universities.

2 Soliton Effects in Optical Fibers and Applications

2.1 Soliton equation and ultralong distance communication

Since Hasegawa proposed in 1973 to use soliton effects in the transmission of pulse-coded signals through fiber [1], both theoretical and experimental efforts have been added, enabling virtually endless transmission without reshaping as far as the loss in the fiber is compensated by the erbium-doped fiber amplifier (EDFA) booster array. The intrinsic problem of the accumulation of jitter noise originating from the spontaneous emission process (Gordon–Haus noise limitation) can now be removed by the active filtering of noise, as proposed by Nakazawa et al. [2], or by Mollenauer et al. [3]. An extensive compilation of the theoretical aspects of soliton propagation was provided by Agrawal [4]. Efforts at generalization are being made by Hasegawa [5]–[7] to accommodate the dissipative nature of the fiber-EDFA chain, and by Ohsawa and Fujii [8] to formulate some type of distributed soliton coupling.

2.2 Compression of pulses by soliton effects and its application

Another attractive aspect of soliton effects is the evolution of a higher order soliton pulse in the fiber. As pointed out earlier by Hasegawa and Tomita, and later experimentally demonstrated by Nakazawa et al. [9], a sizable compression of the pulse width is realizable. Kamiya and his colleagues [10] extended this approach to construct a stable subpicosecond pulse source with an arbitrary repetition rate by combining a gain switched distributed feedback laser, a piece of fiber for linear compression, an EDFA, and a piece of fiber for higher order soliton compression (0.6 ps). We have successfully applied this scheme to the implementation of a compact electrooptic sampling system for the evaluation of high-speed electron devices and integrated circuits with picosecond time resolution [11].

3 Generation of Very High Harmonics by Strong Phase Modulation

3.1 Principle of time-frequency domain interplay

The Fourier transform connects the dynamic behaviors of devices in the time and spectral domains. The generation of an ultrashort pulse therefore corresponds to an expansion of the spectral bandwidth. As an alternative choice against the established methods of short pulse generation such as mode-locking and pulse compression, Kobayashi et al. [12] proposed a new technique where a cw laser beam is modulated by an electrooptic phase modulator with a very large swing. Then higher order harmonic components are excited without employing an ultra-large bandwidth modulator. Using a 16 GHz driven lithium niobate EO modulator, a harmonic bandwidth as wide as

1.85 THz was achieved. This corresponds to a time domain pulse width of a few picoseconds. The increased modulation efficiency by introducing periodic structures, or resonant cavity effects, are also discussed.

3.2 Frequency comb generation for optical frequency synthesizer

In comparison with the mature technology of microwave instruments, the precision measurement capability of ultrafast optoelectronics is still relatively poor. Putting the realization of an optical frequency synthesizer as the ultimate goal, Kourogi *et al.* [13] made efforts to generate an optical frequency comb at the near-infrared region by using the large swing of an electrooptic modulator. Both a bulk lithium niobate (LN) modulator embedded in a microwave cavity and a monolithically integrated LN waveguide modulator with a coplanar stripline feeder were fabricated, achieving a 6 and 7 THz harmonic bandwidth.

4 Enhanced Nonlinear Interactions in Materials

4.1 Quasi-phase matching by E-beam written periodic structures

Although ultrafast optoelectronic systems depend significantly on the performance of nonlinear optical devices, enabling optical control of optical signals, their major problem is the relative phase mismatch between the incident wave and the outgoing wave due to refractive index dispersion. Only specific crystals with suitable dispersion characteristics could offer high conversion efficiency by satisfying the phase-matching condition.

Recently it was recognized that the incorporation of periodic structures into the nonlinear crystal induces an efficient coupling between the incident and mixed waves, at a certain condition called 'quasi-phase matching'. Then the challenge is how to realize the incorporation of fine, uniform, and well-defined periodic structures without deteriorating the nonlinear optical properties of the material. Ito *et al.* [14] discovered an efficient method for generating periodic domain reversal, using focused electron beam irradiation on a lithium tantalate single crystal platelet. The space charge on the surface, which was generated by an electron beam, induces poling of the inner part of the crystal, resulting in the formation of a vertical domain with a sharp boundary. The quasi-phase-matching technique is useful not only in second harmonic generation but also in improving optical parametric oscillator performance.

4.2 Spin flip transitions in quantum wells for optical switches

All-optical switches are particularly attractive in ultrafast digital optoelectronic systems because subpicosecond electrical switches are still not yet available. To increase the switching sensitivity the nonlinear response of excitons confined in quantum wells of compound semiconductors was pointed out to be promising, although the slow fall time related to the carrier relaxation in the quantum wells put a limitation on increasing the clock frequency. Therefore an ultrafast nonlinear response with a fast recovery time was needed. One way to accelerate the relaxation process is to introduce a type II quantum well structure where the holes are confined in the quantum well, but conduction electrons can escape to the barrier quickly. Kawazoe et al. [15] successfully demonstrated the fast and sensitive switching characteristics of a AlGaAs/AlAs quantum well with 80 GHz pulse trains. A switching sensitivity of 300 nJ/cm^2 was reported.

4.3 Nonlinear organic fibers

The addition of new materials to the menu of nonlinear optical crystals is attractive in designing efficient ultrafast devices. The recently developed organic crystal DAN for efficient second harmonic generation was also found to be effective as a material of third-order nonlinearity by Yamashita [16]. Yamashita and his colleagues demonstrated this by incorporating the DAN fiber into a CPM dye laser, inducing self-phase modulation in the fiber. This configuration turned out to be effective in generating an ultrashort pulse of less than 100 fs. A fiber-optic laser amplifier is also promising in some applications of ultrashort pulses with the capability of manipulating the pulse energy. Sasaki and his colleagues recently fabricated a plastic fiber amplifier using an organic dye Rhodamin 6G as the active molecule of light amplification [17].

5 Conclusion

The use of ultrashort optical pulses for the transmission of large capacity information is particularly attractive because with the time division multiplexing scheme the necessary number of components can be minimized. Then to avoid the technological barriers imposed by the ultrafast electron device and circuit assembly, all optical switching, transformation, and other manipulation of optical signals will be the core technology for ultrafast systems. The physical and technical basis for this resides in the development of nonlinear optical concepts and methodologies. This section reviewed some of the important contributions by the groups of Japanese universities under the support of Monbusho, Ministry of Education, Science and Culture, Japan.

References

[1] A. Hasegawa and F.D. Tappert, *Appl. Phys. Lett.* **23**, 142 (1973).
[2] M. Nakazawa et al., OAA'93, 1993, PD7.
[3] L.F. Mollenauer et al., OFC/IOOC'93, 1993, PD8.
[4] G.P. Agrawal, *Nonlinear Fiber Optics*, Academic Press, New York, 1988.
[5] A. Hasegawa, *Optical Solitons in Fibers*, Springer Verlag, New York, 1989.
[6] M. Matsumoto and A. Hasegawa, *Opt. Lett.* **18**, 897–899 (1993).
[7] M. Matsumoto, H. Ikeda and A. Hasegawa, *Opt. Lett.* **19**, 183–185 (1994).
[8] Y. Ohsawa and Y. Fujii, *J. Phys. Soc. Jpn.* **61**, 3977 (1992).
[9] M. Nakazawa, K. Suzuki and E. Yamada, *Electron. Lett.* **26**, 2038 (1990).
[10] J.T. Ong, R. Takahashi, M. Tshuchiya, S.H. Wong, R.T. Sahara, Y. Ogawa and T. Kamiya, *IEEE J. Quantum Electron.* **29**, 1701 (1993).

[11] R. Sahara, R. Takahashi, J.T. Ong, M. Tsuchiya, Y. Ogawa and T. Kamiya, *CLEO '93*, 1993, CThS87.
[12] A. Morimoto, A. Shibagaki and T. Kobayashi, *CLEO '93*, 1993, 558.
[13] M. Kourogi, K. Nakagawa and M. Ohtsu, *IEEE J. Quantum Electron.* **QE-29** 2693–2701 (1993).
[14] H. Ito, C. Takyu, H. Inaba, *Electron. Lett.* **27** 1211 (1991).
[15] K. Kawazoe, T. Mishina and Y. Masumoto, *Jap. J. Appl. Phys.* **32** L1756 (1993).
[16] M. Yamashita, *Ultrafast Phenomena VIII*, Springer, Heidelberg, 1993, p. 313.
[17] A. Tagaya, Y. Koike, E. Nihei, S. Teramoto, T. Yamamoto, K. Fujii, and K. Sasaki, *CLEO '93*, 1993, CTuF3.

1.4

Photonic Integration

Hiroshi Nishihara

Abstract

Integration is very important for optical devices/systems. Among several types of integration, thin-film waveguide integration is described here. This integration results in so-called optical integrated circuits (ICs). The advantages and disadvantages of optical ICs are discussed. Recent optical IC devices are reviewed, and their features, such as large-scale substrate, high speed, and stable alignment, are discussed. Finally, important subjects and perspectives for future development are summarized.

1 The Need for Integration

Electronic devices have been developed from electron vacuum tubes for semiconductor transistors, integrated circuits (ICs), and large-scale integrated circuits (LSIs) in the last half century. This integration resulted in great progress in electronics technology. This fact has taught optoelectronic engineers the importance of device integration. Optical devices have been developed from He–Ne lasers for semiconductor laser diodes and optical integrated circuits (OICs). Photonic integration will become a more important subject in optical devices and systems.

Usual optical systems are assembled from several discrete optical components such as lenses, prisms, and laser diodes, and require stable and precise alignment in assembly. Only the properly aligned components can perform a designed function. Such alignment can be achieved by integration.

There are several types of photonic integration. The first type is *assembling* the discrete bulk components. One example is an optical disk pickup head for a CD player, as shown in Figure 1.4.1. The alignment of the commercial products is realized by mechanical assembly. Another example is optical-fiber communication equipment

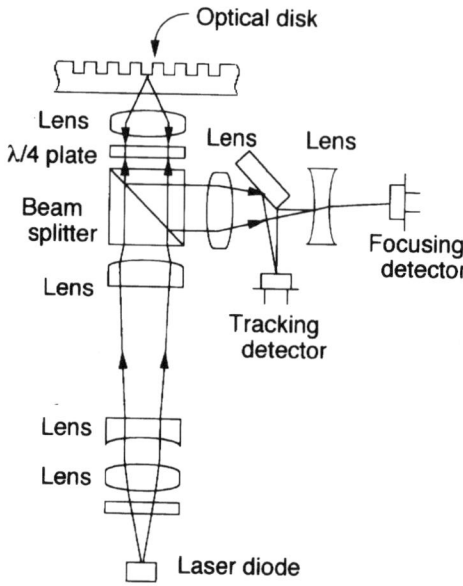

Figure 1.4.1
Assembly-type integration: optical disk pickup assembled with several discrete optical components

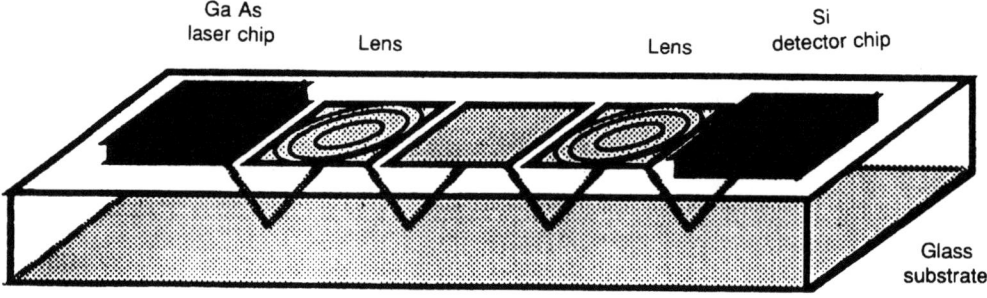

Figure 1.4.2
Planer optics-type integration [1]

which are also commercially available. (*Vertical integration*, as described in section 4.4 of Chapter 4, may be included in this type.) The second type is integration based on so-called *planar optics* [1], in which several thin-film components are integrated onto one surface of a transparent plate, and the plate itself is a medium through which an optical beam connects those components, as shown in Figure 1.4.2. The third type is *thin-film waveguide integration*, in which components are all thin-film waveguide type constructed on a substrate, and are connected through waveguides, as shown in Figure 1.4.3. Various waveguide devices, which are described in this book, have been investigated aimed at this third type of integration. This integration results in so-called *optical integrated circuits*. In this section, therefore, waveguide-type integration is mainly discussed.

Photonic Integration 23

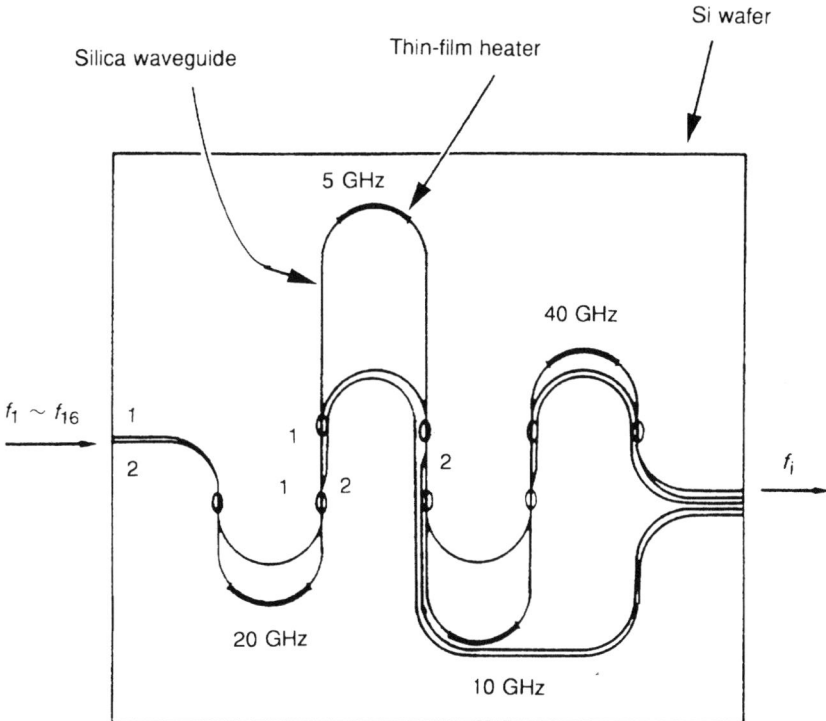

Figure 1.4.3
A 16-channel frequency selection switch [5]

2 Optical Integrated Circuits

The concept of an *optical IC* was proposed by Miller [2] of AT&T Bell Laboratories in 1969. The basic idea is to use thin-film waveguide-type components for optical signal processing. By transmitting optical waves through waveguide structures instead of the free-space or the bulk media, optical ICs have the following advantages [3]:

(1) Features based on single-mode waveguides the widths of which are of the order of micrometers.
(2) Stable alignment (vibration free): the device can withstand vibration and temperature change; this is the greatest advantage of OICs.
(3) Easy control of the guided wave.
(4) Low operating voltage and short interaction length.
(5) High-frequency operation due to shorter electrodes and less capacitance.
(6) Compactness and light weight.

On the other hand, they have the following disadvantages:

(1) Comparatively large transmission loss.
(2) Micron-order fabrication techniques required.

3 Recent Optical IC Devices and their Features

Optical IC devices have been and will be investigated for application to the fields of optical fiber communication, optical signal processing, metrology and sensors, optical sources, and others. The important features of the devices that have been reported recently are classified as follows:

3.1 Large-scale substrate and low-loss waveguide

To upgrade the function of OICs, the number of elements per chip is required to increase, resulting in larger-scale substrate devices. For example, the research group at Plessy reported a 16 × 16 multi-switch [4] for optical communication systems. The switch has 256 switching elements, and all are integrated on a $LiNbO_3$ crystal substrate as large as 70 mm × 2.5 mm. Such a large area device requires advanced photolithography. Another example is a wavelength division multiplexing device which has been reported by a NTT group [5], as shown in Figure 1.4.3. This is a four-stage cascaded asymmetric Mach–Zehnder type interferometer, which performs like a wavelength (frequency)-selective switch for 16 channels. The components are integrated on a 5.2 cm × 4 cm Si wafer. They developed a promising fabrication technique by flame hydrolysis deposition for low-loss 3D silica waveguides using a SiO_2/Si substrate. They found that the propagation loss is as low as 0.01 dB/cm at 1.3 μm wavelength and that the insertion loss is as low as 5.2 dB at the same wavelength.

3.2 High-speed operation

For future coherent communication systems, high-speed modulators with a wider than 10 GHz bandwidth will be required. Such modulators are implemented by OICs on

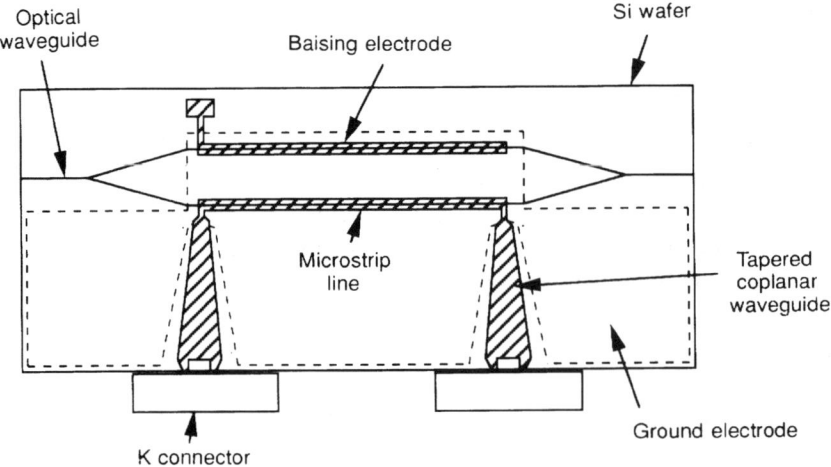

Figure 1.4.4
A 40 GHz response polymeric travelling-wave modulator [6]

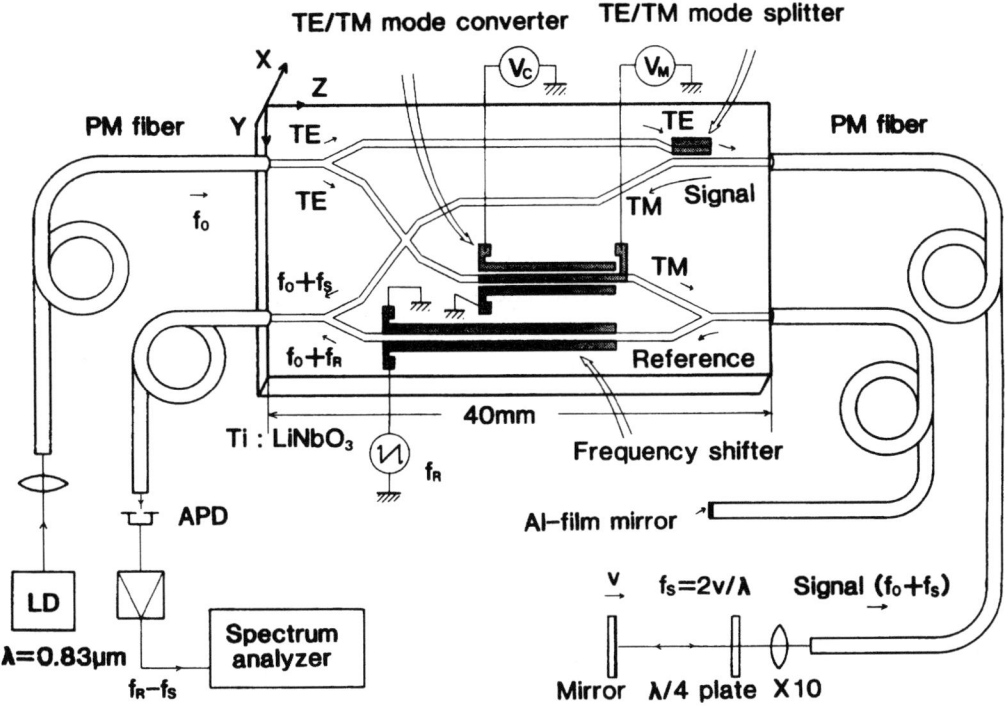

Figure 1.4.5
An integrated-optic chip for a laser Doppler velocimeter [7]

LiNbO$_3$ and nonlinear electrooptic polymers. The research group at Hoechst Celenese Corp. [6] has demonstrated a travelling-wave Mach–Zehnder structure intensity modulator, using EO polymer, with a bandwidth of more than 40 GHz, as shown in Figure 1.4.4.

3.3 Stable alignment

One of greatest advantages of OICs is stable alignment. The integrated-optic version of the optical disk pickup shown in Figure 1.4.1 is illustrated in Figure 4.7.1. The research group at Osaka University investigated an integrated-optic laser Doppler velocimeter [7], as shown in Figure 1.4.5. In these examples several components have been aligned and fixed on chips, resulting in stable alignment.

4 Conclusions

Several important subjects and perspectives for the future development of OICs are summarized in the following.

(1) The establishment of design techniques such as waveguide analysis methods and CAD of waveguide devices.

(2) The establishment of characterization apparatus and techniques.
(3) A reduction in waveguide loss.
(4) The establishment of micro-fabrication techniques which can handle wafers of 50 mms and over.
(5) Multilayer vertically coupled circuits.
(6) The exploration of new materials, new crystals, and new organic materials, especially the development of nonlinear optic materials.

Integrated-optic devices will be required and installed in more systems in various fields, such as coherent optical communications, photonic switching, optical computing, optical interconnection, metrology, and optical functional sources. More photonic ICs will be used in practice in the twenty-first century. For the development of ultrafast and ultra-parallel optoelectronic devices and systems, integration is very important.

References

[1] J. Jahns, *Micro Optics Conference*, Yokohama, K1, 1991, pp. 132–135.
[2] S.E. Miller, *Bell Syst. Tech. J.* **48** (7), 2059–2068 (1969).
[3] H. Nishihara, M. Haruna and T. Suhara, *Optical Integrated Circuits*, McGraw-Hill, New York, 1985.
[4] P.J. Duthie, M.J. Wale and I. Bennion, *Top. Meeting on Photonic Switching*, Kobe, 1990, 13A-3.
[5] K. Oda, N. Takato, T. Kominato and H. Toba, *IEEE J. Selected Areas Commun.* **8**, 1132 (1990).
[6] C.C. Teng, *Appl. Phys. Lett.* **60** (13) 1538–1540 (1992).
[7] M. Haruna and H. Nishihara, *Technical Digest of the 11th Sensor Symposium*, 1992, pp. 111–118.

1.5

Photonic Devices for Ultra-parallel Lightwave Systems

Kenichi Iga

Abstract

In this section we introduce the idea and technology of photonic devices which will be used in ultra-parallel lightwave systems such as parallel optical fiber networks, optical interconnects, and parallel image processing systems. After presenting the concept of parallel optoelectronic devices, we look at the progress in vertical cavity surface emitting lasers as one of the key components in parallel optics. Then we introduce some technologies for integration related to surface emitting lasers. Finally, we discuss optical interconnects as a good example of an applied system.

1 Introduction

We are now facing strong motivation to introduce optical technology into large-scale markets, i.e. new technical areas such as optical computing, optical interconnects, and parallel lightwave systems are being opened up [1]. In particular, recent progress in surface emitting (SE) lasers and parallel functional devices [2] is toward the realization of parallel optical interconnects. We review some possible photonic devices for application to parallel optical systems.

2 Parallel Device Schemes

First, we consider some possible schemes including optical interconnects, parallel fiber-optic subsystems, and so on. By taking full advantage of the parallelism of lightwaves, we will have a choice of parallel optical systems:

(a) multi-fiber transmission;
(b) bit interconnect;
(c) parallel disk access;
(d) displays;
(e) lighting;
(f) multi-spot sensing;
(g) information processing;
(h) pattern recognition.

In optical fiber communication systems, longer than 10 000 km of fibers are considered to be directly connected without any electrical repeaters. Another important advance is a parallel lightwave systems including more than 4000 fibers, for example broad-band ISDN systems.

By taking advantage of the wide band and small volume transmission capability, the optical interconnect is considered to be inevitable in computer technology. A parallel interconnect scheme is needed, and new concepts are being researched. The vertical optical interconnect of LSI chips and circuit boards may be another interesting issue.

Several schemes for optical computing have been considered, but one of the bottle-necks may be a lack of suitable optical devices, in particular two-dimensional surface emitting lasers and surface operating switches. Fortunately, very low threshold surface emitting lasers have been developed, and stack integration together with two-dimensional photonic devices are now actually considered.

In any case, the two-dimensional arrayed configuration of surface emitting lasers and related planar optics will open up a new era of ultra-parallel optoelectronics.

3 Surface Emitting Lasers

The importance of 1.3 μm or 1.55 μm devices is currently increasing, since parallel lightwave systems are really needed to meet the rapid increase in information transmission. However, the GaInAsP/InP system [3] has some substantial difficulties for making SE lasers, due to such reasons that the Auger recombination and inter-valence band absorption (IVBA) are noticeable, the index difference between GaInAsP and InP is relatively small, the valence band offset is large, and so on. As already introduced in Reference [1], we see consistent progress in the reduction of the threshold. Pulsed operation has been obtained at near room temperature, at room temperature, and at 66 °C.

Currently, hybrid mirror technologies are being developed. One uses a semiconductor/dielectric reflector, which is demonstrated by chemical beam epitaxy (CBE). The other is epitaxial bonding of a quaternary/GaAs–AlAs mirror, where 144 °C pulsed operation is achieved by optical pumping. Quite recently an epitaxially bonded mirror made of GaAs/AlAs was introduced into surface emitting lasers operating at 1.3 μm, providing 9 mA of room temperature pulsed threshold.

Thermal problems for CW operation are now extensively studied. A MgO/Si mirror with good thermal conductivity has been demonstrated. To realize a reliable device, the buried heterostructure (BH) is crucial. We have fabricated a BH SE laser exhibiting a relatively low threshold at room temperature pulsed operation.

By improving the heat sinking using a diamond submount and highly reflective mirror, we have achieved room temperature CW operation. The minimum threshold obtained was 22 mA at 14 °C for a 12 μm diameter device size.

In the GaAlAs/GaAs system, an SE laser of 5 μm long and 6 μm in diameter [4] under room temperature CW operation [5] was first realized among other systems. At present, devices exhibiting $I_{th} \cong 2-15$ mA and 10 mW of output power are available in the laboratory level. A very high coupling efficiency to a single mode fiber ($\cong 90\%$) was reported. A spectral linewidth of 50 MHz is obtained with an output power of 1.4 mW.

The GaInAs/GaAs strained pseudo-morphic system grown on a GaAs substrate emitting 0.98 μm exhibits a high laser gain and has been introduced into surface emitting lasers together with using GaAlAs/AlAs multi-layer reflectors. A low threshold (= 1 mA at CW) has been demonstrated [6]. The minimum threshold reported so far is 0.7 mA [7] and 0.65 mA [8]. The minimum J_{th} is 8 mA/μm^2, which is approaching the similar level of stripe lasers. Also, relatively high power — as high as 50 mW — is becoming possible.

Visible surface emitting lasers are extremely important for disk and display applications, in particular, red, green, and blue surface emitters may provide much wider technical areas, if realized. GaInAlP/GaAs SE lasers have been developed and room temperature operation has been obtained [9]. Blue and green SE lasers are much more difficult to realize than any other materials. Some design consideration and fundamental process technology are now being attempted in the authors' laboratory.

By overcoming the technical problems in making tiny structures, and by improving the thermal resistance, we believe that we can obtain a 1 μA threshold device. A lot of improvements in the characteristics of surface emitting lasers have been made, including surface passivation in the regrowth process of buried heterostructures, micro-fabrication, and fine epitaxies. The ohmic resistance of semiconductor DBRs is reduced to the order of 10^{-5} Ωcm^2.

Spontaneous emission control is considered by taking advantage of microcavity structures. The spontaneous emission factor has been estimated on the basis of three-dimensional mode density analysis [10]. The possibility of no distinct threshold devices is suggested.

One another interesting topic for microcavity SE lasers is photon recycling. By covering the side-bounding surfaces of the cavity with a highly reflective material, an amount of spontaneously wasted photons can be recycled. It has been demonstrated that the SE laser device appears to have no distinct threshold. The efficiency of photon recycling has been estimated. Quantum noise characteristics are being studied, such as relative intensity noise (RIN) and linewidths.

4 Photonic Devices and Integration with Surface Emitting Lasers

A wide variety of functions, such as frequency tuning, amplification, and filtering, can be integrated along with surface emitting lasers by stacking. Moreover, a two-dimensional parallel optical logic system can deal with a large amount of image information with high speed. To this end, a surface emitting (SE) laser will be a key device. Optical neural chips have been investigated for the purpose of making optical neuro-computers and a VSTEP integrated device.

High power capabilities from SE lasers feature largely extending two-dimensional arrays. For the purpose of achieving coherent arrays, coherent coupling of these arrayed lasers has been tried by using a Talbot cavity and phase compensation is considered. It is pointed out that two-dimensional arrays are more suitable for making a coherent array than a linear configuration, since we can take advantage of two-dimensional symmetry. Research activity is now progressing to monolithic integration of SE lasers, taking advantage of small cavity dimensions. A densely packed array has also been demonstrated for the purpose of making high power lasers and coherent arrays.

Attempts are now being made to integrate surface operating photonic elements using quantum wells such as an optical switch, a frequency tuner, an optical filter, and superlattices into SE lasers. We demonstrated a 40 Å continuous tuning by the use of an external reflector. A wide variety of functions, such as polarization control, amplification, detecting, and so on can be integrated along with surface emitting lasers by stacking. Polarization control will become very important for SE lasers. One method is presented where a stress effect is incorporated.

5 Schemes of Optical Interconnects

Microoptics is a concept which provides pragmatic schemes for a wide variety of optoelectronic systems. It can be said to be the technology where various demands in optoelectronics for devices are settled by actual methods. A problem is to find a method with no alignment constitution, and one solution may be to employ a two-dimensional microoptic component, stacked planar optics [11], which consists of planar microlenses as shown in Figure 1.5.1. The stacked planar optics is growing by adding new concepts and by integrating active devices such as surface emitting lasers and functional surface operating devices, together with planar microlens arrays. Trial fabrications have been made to realize parallel subsystems. A well-balanced development flow of chip technology to mount/module technology and subsystem technology is desirable. The author would like to advise a necessity of total fine optical technology covering active devices to passive staffs.

In the future, the realization of three-dimensional, large-scale integrated microoptics is desirable. Optical surface mount technology (OSMT) may be one solution. Another method is to use an optical bench using a transparent substrate with a zigzag ray path in it.

An important feature of interconnect in optical spectrum is the coupling of the light from lasers to optical fibers. Butt-joint fiber couplers and the utilization of lensed fibers is one of the most common ways, where a computer-controlled formation of an aspheric surface at the fiber tip is realized resulting in almost no loss coupling. But in the final process of moduling, we need precise alignment.

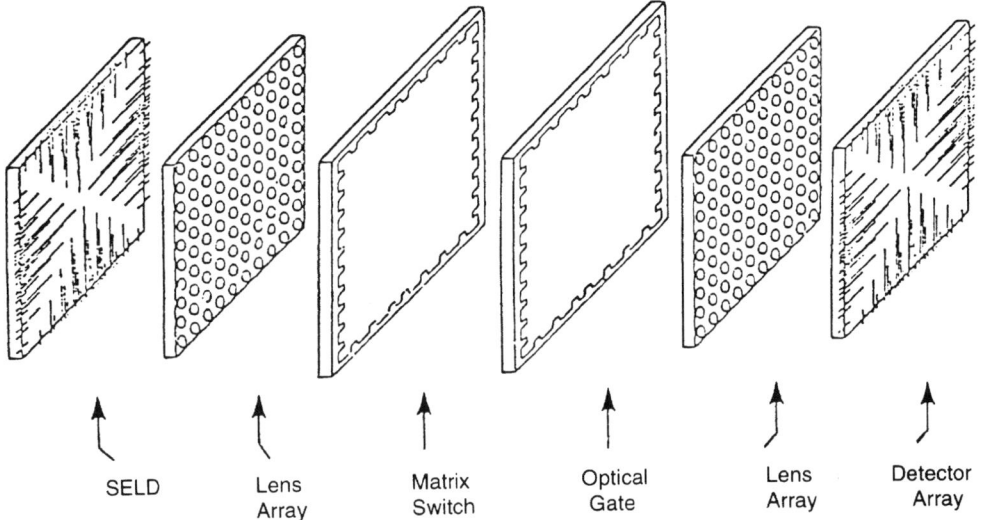

Figure 1.5.1
The concept of ultra-parallel stacked planar optics using a planar microlens array

A put-in microconnector using a planar microlens array provides a solution to this problem, and a put-in microconnector has been proposed [12]. In this scheme we use an array of planar microlenses to focus the light on the fiber array. At the location where a fiber core should be, a small plug is prepared by the self-aligning procedure. A collimated beam was coupled into a 1×4 fiber array by hand without using a precision micro-stage.

Moreover, a much more interesting concept would be self-organizing optics using photo-refractive materials, by which we may build up a lot of interesting devices including an automatic laser to the fiber coupler and so on. We suggested the concept of automatic or self-organizing coupling (SELFOG) of optical beams to optical fibers. Owing to the photo-refractive effect, a kind of optical grating is generated by the pump light beam and reflected beam. Then the light beam can automatically find a way to go toward the location where the light is reflected. Of course we have to have some amount of reflection, and this causes an insertion loss sacrifice.

6 Summary

Vertical optical interconnects of LSI chips and circuit boards, and multiple fiber systems may be the most interesting fields related to surface emitting lasers. From this point of view, the device should be small — as small as possible. The future process technology for it, including epitaxy and etching, will drastically change the situation of SE lasers. Some optical technologies have already been introduced into various subsystems, but the arrayed microoptic technology would be very helpful for advanced systems.

References

[1] K. Iga and F. Koyama, *Surface Emitting Lasers*, Ohm-sha, Tokyo, 1988; *J. Appl. Phys.* **60** (1) 2-13, (1991); K. Iga, *4th Topical Meeting on Integrated Photonics Research*, ITUG1-1, 1993.
[2] K. Iga, F. Koyama and S. Kinoshita, *IEEE J. Quantum Electron.* **QE-24** (9), 1845 (1988).
[3] H. Soda, K. Iga, C. Kitahara and Y. Suematsu, *Jpn J. Appl. Phys.* **18**, 2329-2330 (1979).
[4] K. Iga, S. Kinoshita and F. Koyama, *Electron. Lett.* **23** (3), 134-136 (1987).
[5] F. Koyama, S. Kinoshita and K. Iga, *Appl. Phys. Lett.* **55** (3), 221-222 (1989).
[6] J.L. Jewell, A. Scherer, S.L. McCall, Y.H. Lee, S. Walker, J.J.P. Harbison and L.T. Florez, *Electron. Lett.* **25** (17), 1123-1124 (1989).
[7] R.S. Geels, and L.A. Coldren, *48th Device Research Conference*, June 1990, VIIIA-1; R.S. Geels, S.W. Corzine, J.W. Scott, D.B. Young and L.A. Coldren, *Photonics Lett.* **2** (4), 2345-2346 (1990).
[8] T. Wipiejewski, K. Panzlaff, E. Zeeb and K.J. Ebeling, *18th European Conf. Opt. Comm.* 1992, ECOC'92, PDII-4.
[9] K.F. Huang, K. Tai, C.C. Wu and J.D. Wynn, *Device Res. Conf.*, June 1993, B-7.
[10] T. Baba, T. Hamano, F. Koyama and K. Iga: *IEEE J. of Quantum Electron.* **27** (6), 1347-1358 (1991).
[11] K. Iga, Y. Kokubun and M. Oikawa, *Fundamentals of Microoptics*, Academic Press/Ohm, New York, 1984.
[12] A. Sasaki, T. Baba and K. Iga, *Photon. Tech. Lett.* **4** (8), 908-911 (1992).

1.6

Parallel Optical Computing Systems

Yoshiki Ichioka

Abstract

This section reviews studies on parallel optical computing systems. The fundamental architectures of optical analog computing systems, optical analog/electronic hybrid computing systems, digital optical computing systems, and optical neural computing systems are briefly summarized. Achievable versions of parallel optical digital computing systems are presented.

1 Introduction

The motivation for studies of optical computing systems is the increasing interest in developing a new parallel computing system capable of processing large amounts of data at high speed. In recent years, studies relating to optical computing have been rapidly advanced in various areas, *e.g.* researches on parallel analog and/or digital computing systems, optical parallel hybrid and neural parallel computing systems, the development of logic gate devices and parallel optoelectronic devices, and so on [1].

The potential advantage of optics is the capability of high-speed parallel transmission and processing of structured data with ultra-high resolution. It appears that investigation aimed at the development of digital optical computers is one of the most promising and ambitious fields in optical computing [2,3]. In this section we briefly review fundamental optical computing systems and recent studies of optical digital computers.

2 Optical Computing Systems

Optical computing systems are categorized into four groups: optical analog computing systems; optical analog/electronic hybrid computing systems; optical parallel digital computing systems; and optical neural computing systems.

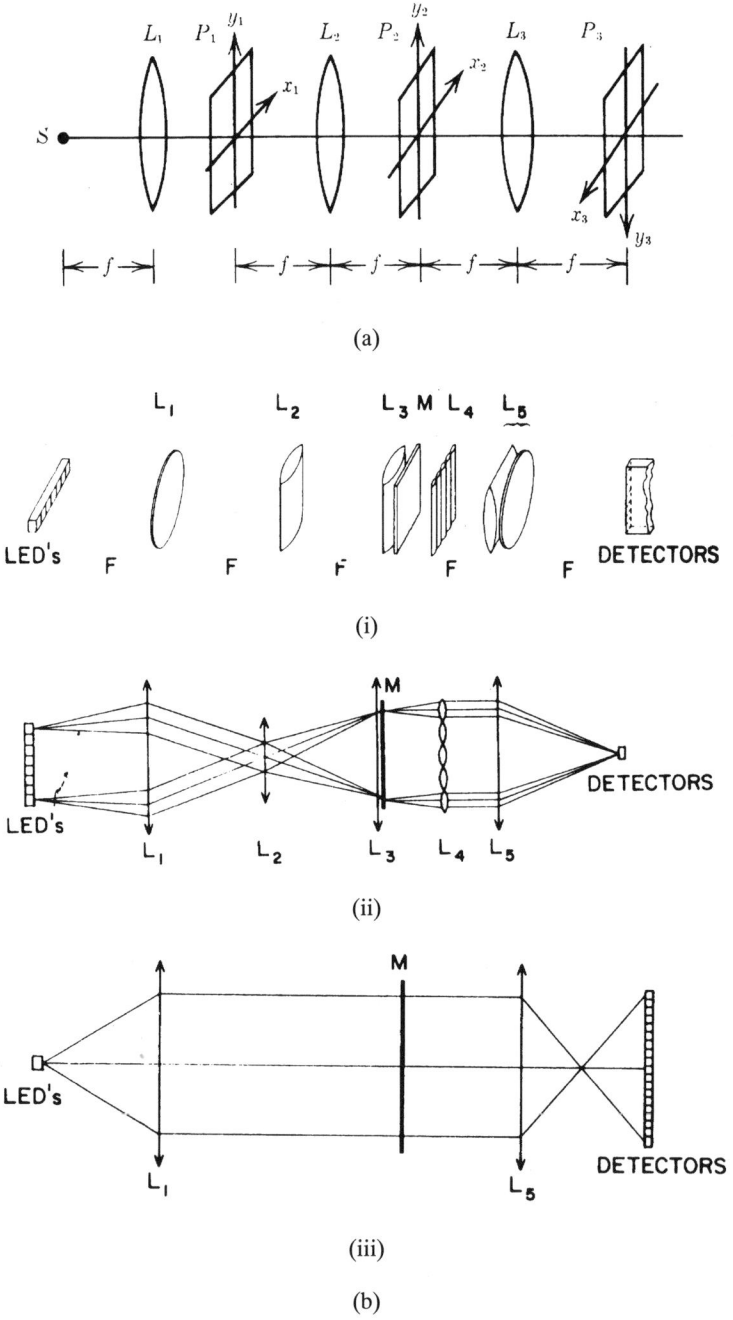

Figure 1.6.1
Analog optical computing systems: (a) coherent optical processing system (4f-system, optical correlator); (b) incoherent optical matrix-vector processor: (i) pictorial view; (ii) top view; (iii) side view [4]

2.1 Optical analog computing systems

Optical analog computing systems are generally special purpose computers using the inherent natures of light, such as parallel analog data transmission and processing at the speed of light. Typical examples are a two-dimensional (2D) Fourier transform processor, a 2D optical correlator [4], and optical matrix–vector processors [5–8]. The fundamental processing systems of these processors are shown in Figure 1.6.1. Recently a number of filtering techniques for 2D optical correlation, which are capable of matching and detecting the stated 2D targets with high sensitivity, have been developed. Phase-only matched filtering is one example [9].

2.2 Optical analog/electronic hybrid computing systems

Optical analog/electronic hybrid computing systems are composed of an analog optical computing system and an electronic processor. Figure 1.6.2 indicates one example of an

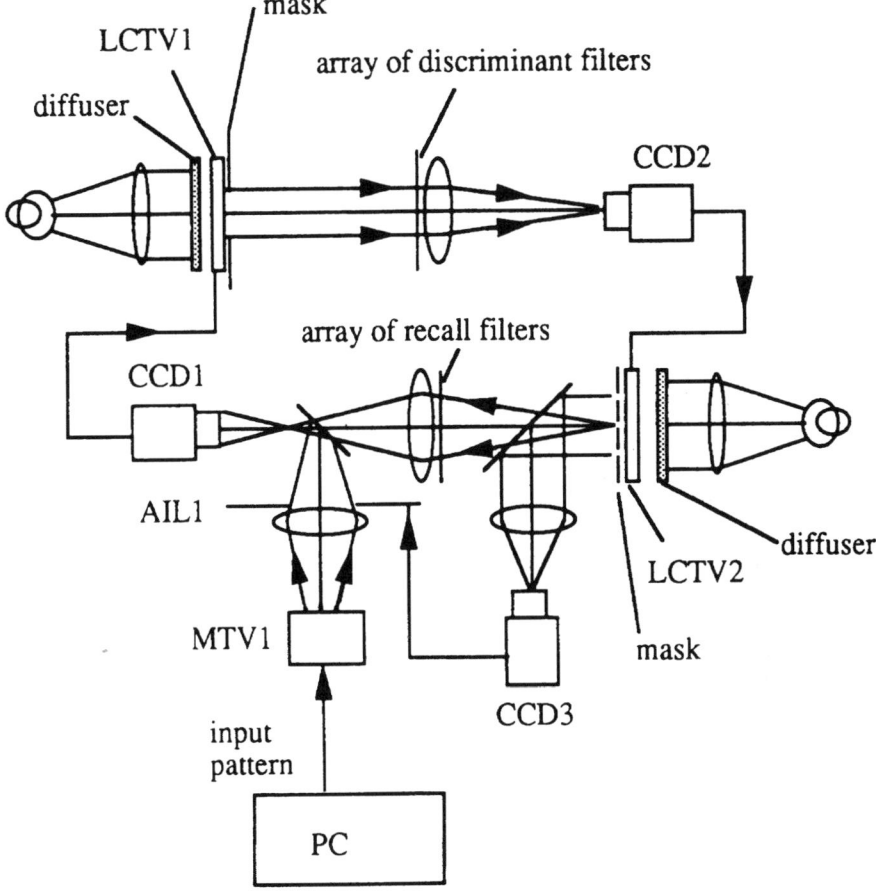

Figure 1.6.2
Incoherent optical/electronic hybrid associative memory system [10]

Figure 1.6.3
Experimental result of association [10]

optical analog/electronic hybrid processor, which is an incoherent associative memory system [10]. This hybrid system forms an incoherent feedback loop processor with two incoherent optical correlators, nonlinear electronic circuits, and two TV cameras. Taking iteratively 2D correlation of the input imperfect hand writing characters or intermediate processed results and the phonto characters in the feedback loop processor, this system can provide clear characters as a final output. Figure 1.6.3 shows an experimental result of association [10].

2.3 Optical parallel digital computing systems

An optical parallel digital computing system is a parallel computer making use of the capabilities of high-speed, ultra-parallel data transmission and processing in optics and flexibility in digital processing [1,11,12]. This category of optical computers will be a powerful tool in the paradigm of parallel information processing in the future.

Optical digital computers would be designed and constructed using new optoelectronic devices such as optical logic gate devices, spatial light modulators, optoelectronic integrated circuits, and LED and/or LD array devices and 2D photo-diode array. Although the concrete architectures of these optical computers have not yet appeared, two directions of studies have been carried out.

One is on an optical interconnection system, making use of the features of optical signal transmission such as high-speed, cross-talkless parallel signal transmission, earthless signal interconnection, and optoelectronic devices [13]. One of the most serious problems restricting speeding up the processing speed of the present digital electronic computer is the delay of signal transmission in electronic communication lines. Delay is produced among IC chips as well as within logic elements. To shorten delay time, signals to be processed with electronic circuits are transformed into optical signals, and they are optically interconnected to other electronic circuits or chips. The use of optical interconnection would be expected to greatly improve the performance of a computing system from the point of view of processing speed and reduction of energy dissipation [13].

The other is on the development of parallel optical digital computing systems. The main thrust of research of optical computers is on optical digital computers [14]. It is natural that the architecture of the first generation of optical digital computers seems to be similar to that of the electronic digital computer, except for parallelism in data transmission. We can use a wide variety of resources relating to digital processing accumulated during the development of digital computers over the past 30 years.

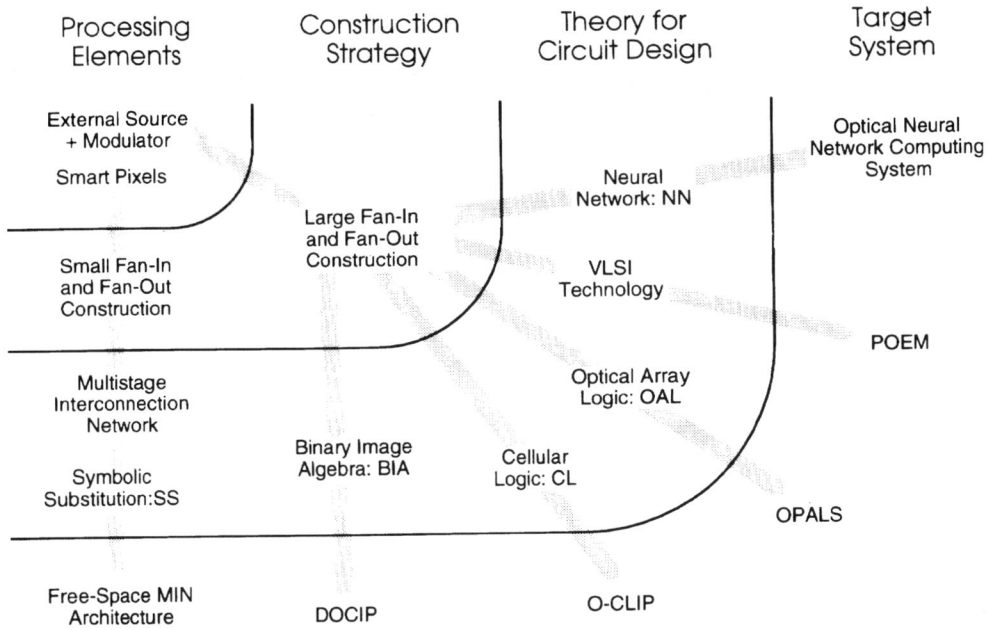

Figure 1.6.4
Flow of development of optical computing systems [14]

In recent years a number of optical computing systems have been proposed. At the present stage, a scenario of the development of optical digital computers which are considered as promising systems are depicted in Figure 1.6.4. As a matter of fact, the directions of research and development of parallel optical computers are not necessarily simple, as shown in Figure 1.6.4.

To develop a new optical computing system as a compound system, a coherent philosophy on the research and development is required. The most important thing for developing the optical computer is to reflect and to make good use of features of optics. The target systems shown in Figure 1.6.4 have been considered under the clarified concepts for the fundamental computing principles. For developing these systems, a circuit architecture, a design method of optical circuit, system architecture and fabrication technology of the system have been clearly shown.

In what follows we briefly summarize the system architectures of the optical parallel digital computers under development.

Multi-stage interconnection network processor

Figure 1.6.5 shows the architecture of an optical computing system called a multi-stage interconnection network with multi-stage logic circuit arrays [15]. This network is a 2D parallel version of the optical processor constructed by piling up regular interconnection networks shown, as in Figure 1.6.6. In this system, input data are fed to the optical logic circuits on the top row of the leftmost array and transferred with free space

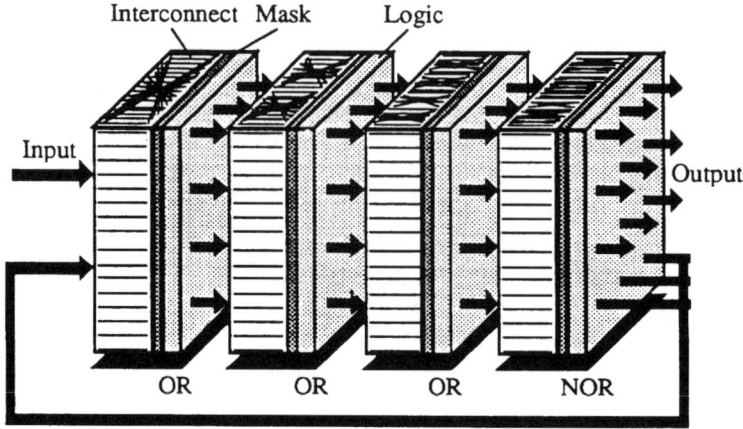

Figure 1.6.5
Multistage interconnection network processor [15]

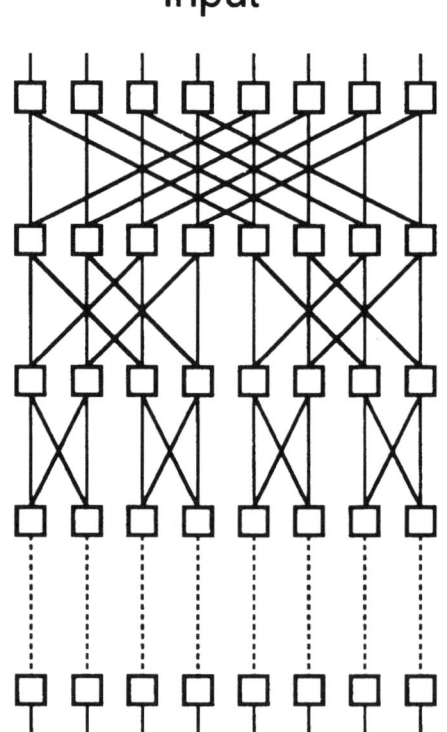

Figure 1.6.6
1D regular network [15]

interconnection from left to right to the logic circuits on the top rows of the sequential stages of the logic circuit arrays. The output signal appeared from the top row of the rightmost logic circuit array are fed back to the second row of the first logic circuit array. In the individual rows, optical logic circuits connected to regular interconnections operate concurrently. This system is regarded as a pipeline system with multi-stages of small task. Optical functional circuits using S-SEED devices together with reflective optical systems have been developed [16].

OPALS

The optical parallel array logic system (OPALS) [17] is an optical parallel digital computing system based on the optical parallel computing principle called optical array logic. Optical array logic is a technique useful for defining and executing parallel neighborhood operations for 2D data. In optical array logic, any neighborhood operation can be uniquely defined by a set of operation kernels. It is simply implemented in parallel using an optical correlation technique together with an image coding method.

Figure 1.6.7 is a conceptual diagram of the OPALS which consists of a loop processor and input/output ports. The OPALS can execute iterative processing for binary data. The main functions implementable on the OPALS are logical neighborhood operations. The OPALS can implement cellular logic in parallel. Cellular logic is an operation which converts structured data into new structures, pixel by pixel, according to the values of the stated pixel and its nearest neighbors in the original data. The functions of the operations can be changed during iteration by programming. The programmability of OPALS is the most important feature.

Several achievable versions of OPALS have been constructed. An experimental system of the optical version of OPALS called P-OPALS is presented in this book.

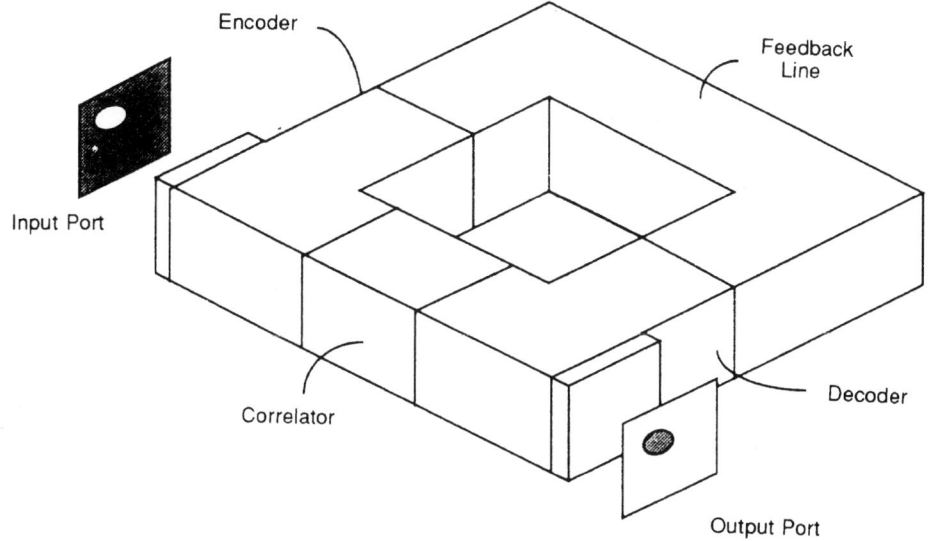

Figure 1.6.7
Block diagram of OPALS

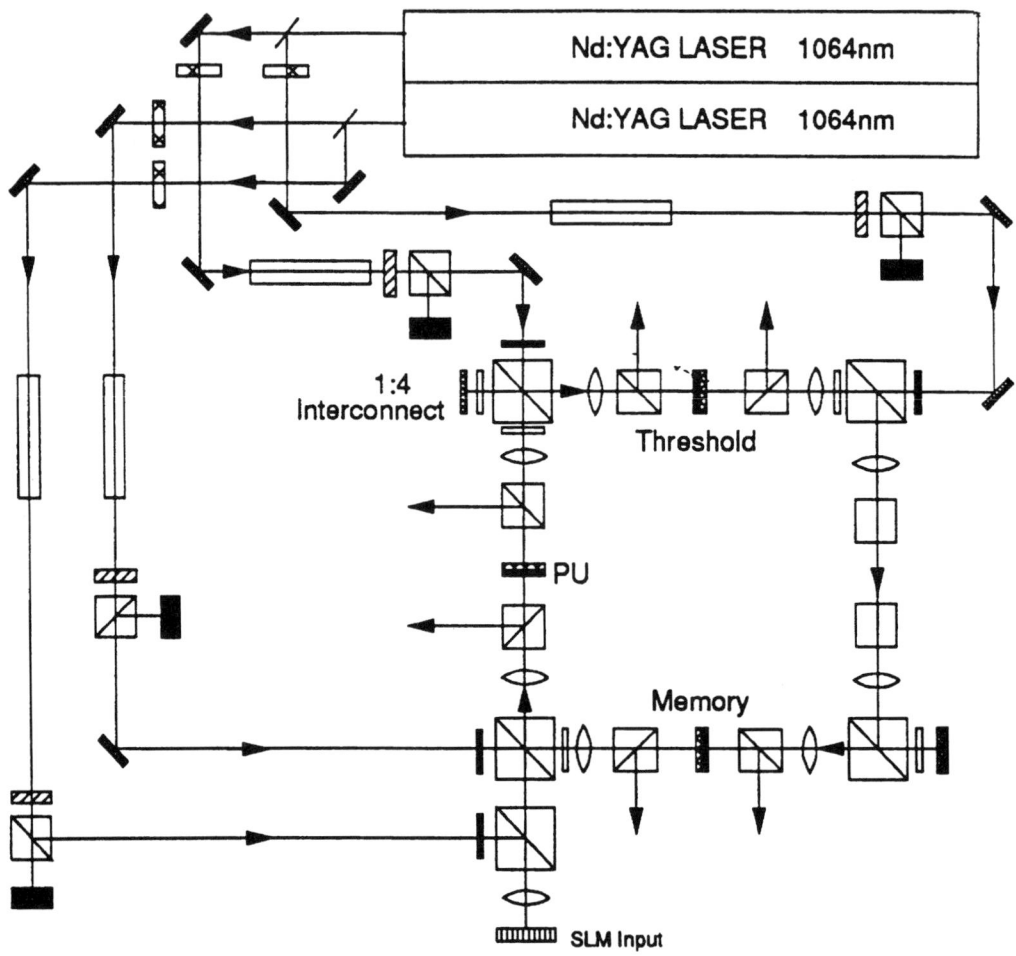

Figure 1.6.8
Block diagram of 16 × 16 O-CLIP [18]

O-CLIP

O-CLIP (optical cellular logic image processor) is the image processor to realize cellular logic using the network with large fan-in and fan-out architecture [18]. Figure 1.6.8 shows the block diagram of O-CLIP. The system is composed of serially connected multi-stages of combination sets of optoelectronic parallel devices and holographic optical elements. The experimental system is constructed by using an optoelectronic ZeSn BEAT device fabricated on an interference filter.

DOCIP

DOCIP (digital optical cellular image processor) [19] is an image processor executing cellular logic in parallel. DOCIP adopts the circuit architecture capable of mapping the logic circuit directly on to the logic gate array and free space optical interconnections. This uses the technique of constructing the desired logic circuits by means of connecting signal processing elements on the logic gate array by the holographic optical element array.

POEM, D-STOP

POEM is a system with the architecture based on optical/electronic hybrid interconnection [20]. In POEM, VLSI processor arrays implement parallel logical processing, and optical interconnection is used for data transmission between processing elements. Interconnection between neighboring processing elements can also be made electronically. Figure 1.6.9 shows the block diagram of POEM.

D-STOP (dual-scale topology optoelectronic processor) is an optical digital computing system based on optical/electronic hybrid interconnection [21]. D-STOP has an H-type electrical network to connect individual processing elements in a VLSI processor. Optical interconnection is utilized for parallel signal connection between individual VLSI processor arrays. D-STOP has been proposed to realize a parallel neural network and to solve the problem of artificial intelligence.

2.4 Optical neural network computing system

An optical neural network computing system is a new generation of optical computers [22]. To realize a neural network computing system, nonlinear processing and

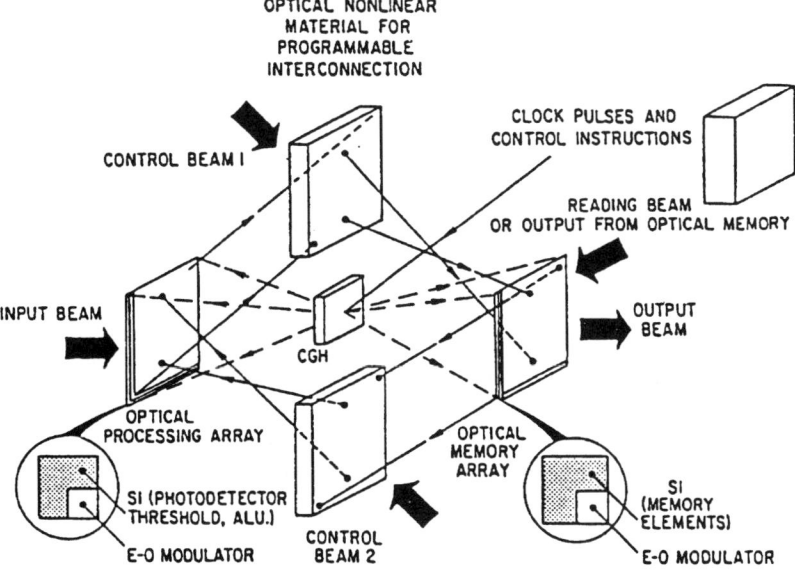

Figure 1.6.9
Block diagram of POEM [20]

interconnection among a number of neurons are required. Parallel data transmission by light matches well this requirement. Fundamentally this category of computing system needs analog optical processing, optical interconnection, and nonlinear thresholding processing. To systematize the optical neural computer, space-variant parallel processing is necessary, which is difficult to implement by traditional optical techniques. A new approach using a holographic technique based on phase conjugation has been attempted. Fast optical associative memory systems applying a neural network have been developed [23].

3 Conclusion

We have reviewed the recent research on parallel optical computing systems. We presented architectures of achievable versions of optical computing systems.

Owing to clear architectures, ease of data handling, and capabilities of parallel programming and system implementation, optical parallel digital computing systems that make good use of the parallel nature of light together with flexibility in digital processing would be candidates for optical parallel computers in the near future.

To construct an actual optical computing system, the development of new functional devices and the high performance of spatial light modulators are indispensable. Recent advances in fabrication technology of VLSI and optoelectronic devices would make it possible to develop these functional devices.

Finally I would emphasize the importance of parallel information processing in various fields as well as in human life in the twenty-first century, because most information to human beings will be provided through human vision systems in parallel. Therefore we must develop a new paradigm capable of freely implementing parallel information processing. In such a paradigm, optical parallel computing systems would play an important role.

References

[1] Special issue on Optical Computing, *Proc. IEEE* **72** (1984); *Optical Computing, 1993 Technol. Digest Series*, **7**, The Optical Society of America, Washington, D.C., 1993.
[2] D.H. Schaefer and J.R. Fischer, *Spectrum* **19**, 32 (1982).
[3] A.A. Sawchuk and T.C. Strand, *Proc. IEEE* **72**, 758–779 (1984).
[4] J.W. Goodman, *Introduction to Fourier Optics*, McGraw-Hill, San Francisco, 1968.
[5] J.W. Goodman, A.D. Dias and L.M. Woody, *Opt. Lett.* **2**, 1 (1978).
[6] W.T. Rhodes and P.S. Guilfoyle, *Proc. IEEE* **72** (7), 820–830 (1984).
[7] H.C. Caulfield, J.W. Rhodes, M.J. Foster and S. Horvitz, *Opt. Commun.* **40**, 86–90 (1981).
[8] D. Casasent, J. Jackson and C. Neuman, *Appl. Opt.* **22**, 115–124 (1983).
[9] J.L. Horner and P.D. Gianino, *Appl. Opt.* **23**, 812–816 (1984).
[10] M. Taniguchi, K. Matsuoka and Y. Ichioka, *Optik 94* **4**, 150–154 (1993).
[11] A. Huang, *Proc. SPIE* **232** (IEEE Catalog 80CH1548-7), Washington, D.C., 1982.
[12] Y. Ichioka and J. Tanida, *Proc. IEEE* **72**, 787–801 (1984).
[13] J.W. Goodman, F.I. Leonberger, S-Y Kung and R.A. Athale, *Proc. IEEE* **72**, 850–866 (1984).

[14] J. Tanida, *Kogaku* **20** (10), 632–641 (1991) [in Japanese].
[15] M. Murdocca, *A Digital Design Methodology for Optical Computing*, MIT Press, Cambridge, MA, 1990.
[16] M.E. Prise, N.C. Craft, M.M. Downs, R.E. LaMarche, L.A. D'Asano, L.M.F. Chirovsky and M.J. Murdocca, *Appl. Opt.* **30**, 2926–2930 (1988).
[17] J. Tanida and Y. Ichioka, *J. Opt. Soc. Am. A* **2**, 1245–1253 (1985).
[18] A.C. Walker, R.G.A. Craig, D.J. McKnight, I.R. Redmond, J.F. Snowdon, G.S. Buller, E.J. Restall, R.A. Wilson, G. MacKinnon, M.R. Taghizadith, J.M. Miller and B.S. Werrett, *Optical Computing, 1991 Technol. Digest Series* **6**, The optical Society of America, Washington, D.C., 1991, pp. 199–202.
[19] K.S. Huang, A.A. Sawchuk, B.K. Jenkins, P. Chavel, J.M. Wang, A.G. Weber, C.H. Wang and I. Glaser, *Proc. SPIE* **963**, 687–694 (1988).
[20] F. Kiamilev, S.C. Esener, R. Raturi, Y. Fainman, P. Mercier, C.C. Guest and S.H. Lee, *Opt. Eng.* **28**, 396–409 (1989).
[21] A.V. Krishnamoorthy, J.E. Ford, G.C. Marsden, G. Yayla, S.C. Esener and S.H. Lee, *Optical Computing, 1991 Technol. Digest Series* **6**, The Optical Society of America, Washington, D.C., 1991, pp. 244–247.
[22] N. Farhat and D. Psaltis, *Technol. Digest of Optical Computing*, Incline Village N.V., 1985, p. WB3-1.
[23] H-Y. S. Li, Y. Qiao and D. Psaltis, *Appl. Opt.* **32**, 5026 (1993).

1.7

Multi-dimensional Optical Information Processing

Yoshihiro Ohtsuka

Abstract

Optical interferometry is a key technology to extract a variety of physical, chemical, and biological parameters carried by an optical wave. Optical multi-dimensional information processing of these parameters is outlined in association with some high-quality optoelectronic devices. Multiplexing methods are also described as another key technology.

1 Introduction

A propagating optical wave is capable of carrying a wide range of multi-dimensional information about physical, chemical, and biological parameters as it interacts with these parameters. We have not possessed any optical detector capable of responding to the very high frequency of light in the past, which has made it almost impossible to detect the optical wave itself. The only way to look into the propagating optical wave is to use a conventional type of photodetector which can offer optical intensity-based information. It should be recalled that no phase information of the optical wave can be extracted from the direct detection of the optical intensity, since the phase does not remain in the direct detection. If there is a need to extract the phase information of the optical wave, it is necessary in general to adopt an optical interferometric or heterodyne technique.

Optical interferometry [1] is a key technique to obtain the optical parameters such as amplitude, phase, and a change in frequency that allow substantially for processing optical multi-dimensional information. Nonetheless, the simultaneous detection of these parameters and their parallel processing are difficult and likely to be almost impossible at present, as long as only the currently available single photodetector is used for the detection of one-dimensional space or time-varying interference fringes. Fortunately, however, much progress has recently been made in optoelectronics technology, which has brought about a variety of high-quality optoelectronic devices. The following devices would be valuable, for example, for use in optical multi-dimensional information processing. Many kinds of semiconductor laser diodes (LD), exploited particularly in the visible wavelength region in the past decade, have become promising light sources for optical information processing. Emphasis should be given to the fact that such a laser diode is capable of emitting a high power beam of light up to the range of 30–50 mW, and its frequency can be scanned as well by controlling an injection current. Also, the super-luminescent diode (SLD) that emits strong power around the 0.8 μm region has been available extensively as a broadband spectral source; its typical center wavelength and spectral width are 840 nm and $\sim$18 nm, respectively. This SLD can thereby be conveniently introduced to promote some interferometric specific applications that substantially need a short-coherence length source. In addition, a surface-emitting laser diode [2] might be available for multi-dimensional information processing in the very near future.

An arrayed photo-detector such as a CCD TV camera works as a position-sensitive area photodetector, which is a greatly promising device to process spatio-temporal optical information, although its frequency bandwidth is not so wide and limited to $\sim$1 kHz at the best. The frequency bandwidth of such an arrayed photodetector is likely to be insufficient for the optical dynamic studies in a higher frequency region. If a much wider frequency bandwidth is required, it is suggested to incorporate a detector of arrayed photodiodes that responds basically as fast as a single photodiode. If one wants to use such a detector, however, parallel outputs from many terminals, arrayed spatially and regularly in two dimensions, have to be processed as conveniently as possible for practical use.

Spatial light modulators have been invented so far for two-dimensional, optical signal processing. For example, there are two kinds of liquid crystal cells working as spatial intensity and phase modulators, respectively. These liquid crystal cells would play an important role as a spatio-temporal modulator to the optical wave in multi-dimensional information processing. Also, a phase-controlling device using a photo-refractive crystal is very fascinating. Since its refractive index changes, depending on the intensity of an illuminating beam of light, an optical-intensity-dependent interference pattern, for example, can be written spatially into such a crystal. Another useful optical device applicable for use in multi-dimensional information processing is the one that makes use of nonlinear optical effects. A methyl-orange-doped polyvinyl alcohol (PVA) film [3] works as such a device to generate a phase-conjugate wave, for example, which functions to cancel the unwanted fluctuating effect of atmospheric turbulence in optical interferometry. Of course, there is another kind of crystal such as $BaTiO_3$ for generating the phase-conjugate wave, but its size seems to be too small to use in practice. In contrast, it is feasible to have a large PVA film fabricated for practical use. The potential use of such a large PVA film would be to make optical computational processing of some moving images.

These optical components and devices will work together with many kinds of currently available electronic devices, and play an important role in multi-dimensional optical information processing.

Multiplexing is another key technology in optical multi-dimensional information processing. A wide range of optical parameters belonging to the optical wave of interest can be multiplexed according to what information processing is needed. In recent years, much progress has been made in the optical fiber sensing technology, particularly in multiplexing techniques [4]. Since such multiplexing techniques would be promising in any other optical technology, it is worth briefly reviewing some typical multiplexing methods, such as spatial multiplexing, time-division multiplexing, wavelength-division multiplexing, frequency-division-multiplexing, and coherence multiplexing.

2 Spatial Multiplexing

A beam of light is spatially divided into the same number of beams as the target objects of interest. Also, the same number of detectors as the target objects must be prepared to receive the split beams of light that are modulated in parameters by the objects. As shown in Figure 1.7.1, information about three objects are extracted by the respective detectors. The major advantage is that all the beams of light are spatially separate to avoid unwanted cross-talk.

3 Time-division Multiplexing

Time-division multiplexing (TDM) is implemented by introducing an optical pulse. As shown in Figure 1.7.2, the pulse is successively divided on propagation in free space into three pulses on reflection at a series of three mirrors, and each new pulse is modulated in height by any one of the objects of interest. After interacting with all the objects, the three pulses impinge together upon one photodetector from which a train of three successive pulses emerge due to the optical path-length differences of the three pulses. Information about any one of the objects can be reflected in the variation of the corresponding electric pulse from the photodetector.

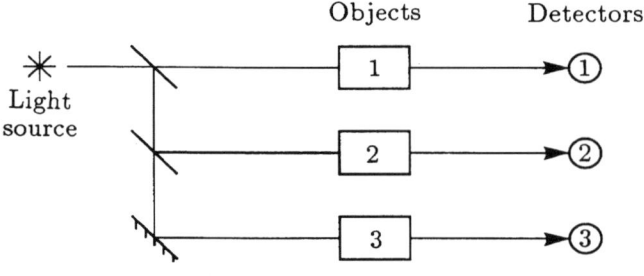

Figure 1.7.1
Spatial multiplexing scheme

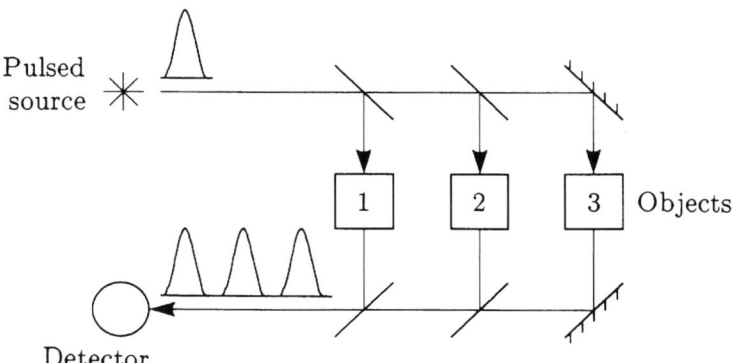

Figure 1.7.2
Time-division multiplexing scheme

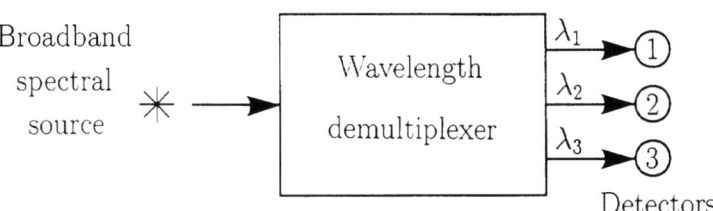

Figure 1.7.3
Wavelength-division multiplexing scheme

4 Wavelength-division Multiplexing

Wavelength-division multiplexing (WDM) requires a broad-band spectral source which is split into spectral components by passing through a wavelength-dependent dispersive device such as an optical grating. Each split spectral component interacts with a corresponding object and is received by a photodetector, as shown in Figure 1.7.3. For the case when remote sensing is preferable, all the beams of the spectral components emerging from the objects of interest are recombined into one beam that propagates in a long distance, which is again split into the spectral components to be detected.

5 Frequency-division Multiplexing

A frequency-modulated carrier-wave (FMCW) multiplexing scheme is explained as a typical frequency-division multiplexing (FDM) technique. Let us consider a semiconductor laser diode, the frequency of which is linearly increasing for a period T by controlling an injection current. As a result, the frequency takes on a form of saw-tooth mode shown as the solid line in Figure 1.7.4(a). A beam of light from the laser diode is launched into an optical interferometer that has three target objects located at separate distances on the one arm, as shown in Figure 1.7.4(b). Any one

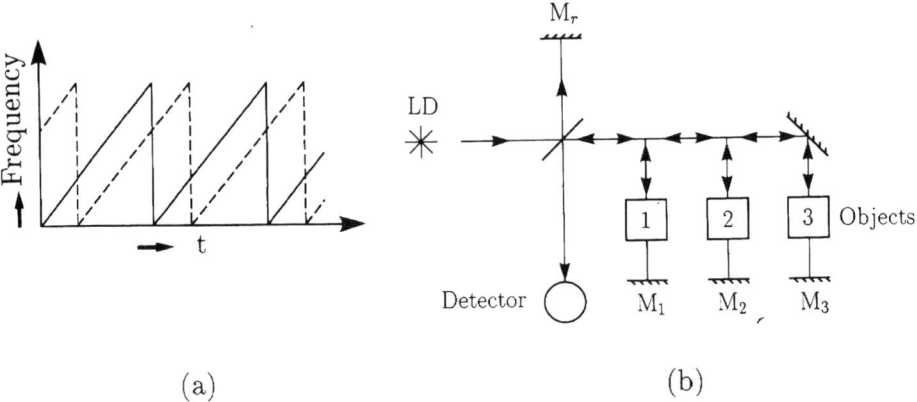

Figure 1.7.4
Frequency-division multiplexing scheme. The dotted sawtooth wave in (a) shows a linear change in frequency for one of the three beams returning from the reflection mirrors M_1, M_2, and M_3. A FMCW multiplexing scheme is shown in (b)

of the three returning beams, modulated in a complex amplitude by the objects, is adjusted to be delayed from the reference returning beam from the mirror M_r, and then a definite frequency difference results between the reference beam and any one of the returning three beams. Since the optical path-length differences between the reference and signal beams returning from the three mirrors M_1, M_2, and M_3 are fixed, the photodetector receives together all the beams with different frequencies to be photomixed, and generates a beat photocurrent consisting of three intermediate beat frequencies. Accordingly, the information about the three objects is involved in the three spectra discriminated in the frequency domain of the photocurrent.

6 Coherence Multiplexing

A coherence multiplexing scheme requires a broadband spectral source the coherence length of which is very short. We now consider the interferometric scheme represented in Figure 1.7.5 which has two target objects on the sensing arm. The optical path distances from the beam splitter BS_1 to the three reflection mirrors M_{r0}, M_1 and M_2 are, respectively, d_0, d_1, and d_2, and the distances from BS_3 to M_{r1} and M_{11} and from BS_4 to M_{r2} and M_{22} are, respectively, d_0, d_1, and d_2. In this scheme, the optical path-length differences $d_1 - d_0$, $d_2 - d_0$, and $d_1 - d_2$ are all much longer than the coherence length of the source. That is to say, no interference takes place at such path-length differences. The measurement is based on observing the fringe shifts, which occur as the path-length differences $l_1 - l_0$ and $l_2 - l_0$ in the sensing part are changed by Δ_1 and Δ_2, respectively, compared with the corresponding fixed lengths $l_1 - l_0$ and $l_2 - l_0$ in the receiving interferometers that have the photodetectors D_1 and D_2.

The introduction of an arrayed type of photodetector into all the multiplexed schemes mentioned above will make it possible to obtain some significant spatio-temporal parameters [5] carried by the propagating optical wave.

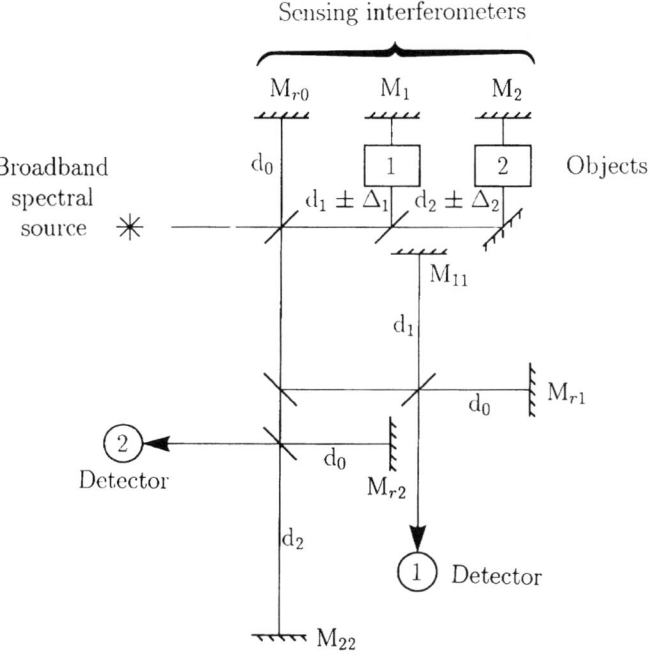

Figure 1.7.5
Coherence multiplexing scheme

References

[1] Y. Ohtsuka, *Trans. Inst. MC* **4**, 115 (1982).
[2] K. Iga, F. Kayama and S. Kinoshita, *IEEE J. Quantum Electron.* **QE-23**, 1845 (1988).
[3] C. Egami, K. Nakagawa and H. Fujiwara, *Jpn J. Appl. Phys.* **29**, L1544 (1990).
[4] J.P. Dakin, *J. Phys. E: Sci. Instrum.* **20**, 954 (1987).
[5] K. Oka and Y. Ohtsuka, *Exp. Mech.* **33**, 44 (1993).

2
BASICS OF ULTRAFAST OPTOELECTRONICS

2.1

Ultrashort Semiconductor Laser Pulse Source

Takeshi Kamiya and **Masahiro Tsuchiya**

Abstract

As an alternative approach to the mode locking scheme, the ultrashort pulse generation by the gain switching of a semiconductor laser is investigated. Taking advantage of the flexibility to choose the repetition frequency arbitrarily, the use of an Erbium doped fiber amplifier together with a soliton compressor enabled the functional optical pulse generator with high power (> 10 pJ), short pulse width (< 1 ps) based on the DFB semiconductor laser.

1 Introduction

One of the potential advantages of optoelectronics technology is the large information-handling capacity in the time domain. Among various laser sources capable of generating pulses with a pulse width less than 1 ps, semiconductor lasers are especially attractive due to their compactness, reliability, and efficiency. Since the earlier efforts of picosecond optoelectronics [1], the expectation to implement systems exploiting the advantage of optical methods is ever growing.

This work is devoted to one of the promising approaches to generate sub-picosecond pulses with a high repetition rate, namely the combination of a gain-switched semiconductor laser and a dispersive fiber for compression [2-4]. This approach has an advantageous feature in comparison with the alternative technique, namely the mode-locking operation of semiconductor lasers. While the repetition frequency is almost

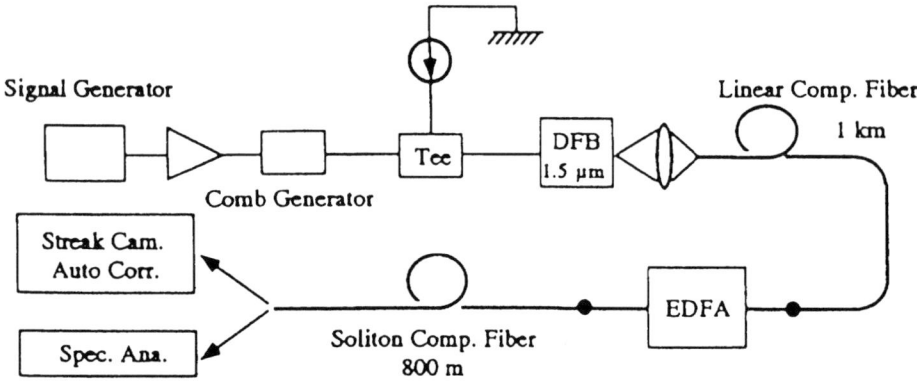

Figure 2.1.1
The experimental setup used for two-stage compression

uniquely determined by the round trip time in the case of a mode-locked laser, the gain-switched laser can operate at an arbitrary repetition rate, suitable for synchronization with an external clock source. As for the progress in mode-locked semiconductor lasers, the recent paper [5] and the references therein are recommended.

The scheme of our experimental trial to generate an ultrashort pulse is depicted in Figure 2.1.1 [6]. The operation frequency of 1.55 μm in wavelength is chosen so that power amplification can be made by an erbium-doped fiber amplifier (EDFA).

2 Gain Switching of Distributed Feedback Lasers

2.1 Modeling gain switch operation

When a semiconductor laser is driven by short current pulses, the temporal change in the optical gain becomes so rapid that evolution of the photon density cannot follow, causing an instantaneous gain in excess of the stationary value. This excess gain induces a steep rise in the photon density to a point far exceeding the stationary output. Correspondingly the induced emission rate becomes larger, yielding a rapid dissipation of stored population inversion. Through all these procedures the optical pulse can be shorter than the injected current pulse. This method of short pulse generation is called gain switching, because of its similarity to Q-switching in other kinds of lasers. The dynamics of a semiconductor laser can be represented by a set of rate equations [7] for the space-averaged carrier density $N(t)$ and photon density $S(t)$ as

$$\frac{dN}{dt} = \frac{I(t)}{edWL_c} - \frac{N}{\tau_e} - v_g G(N) \frac{S}{1 + \varepsilon S} \qquad (1)$$

where $I(t)$ is the injection current, d is the thickness of the active layer, W is the waveguide width, L_c is the cavity length, τ_e is the recombination lifetime, v_g is the

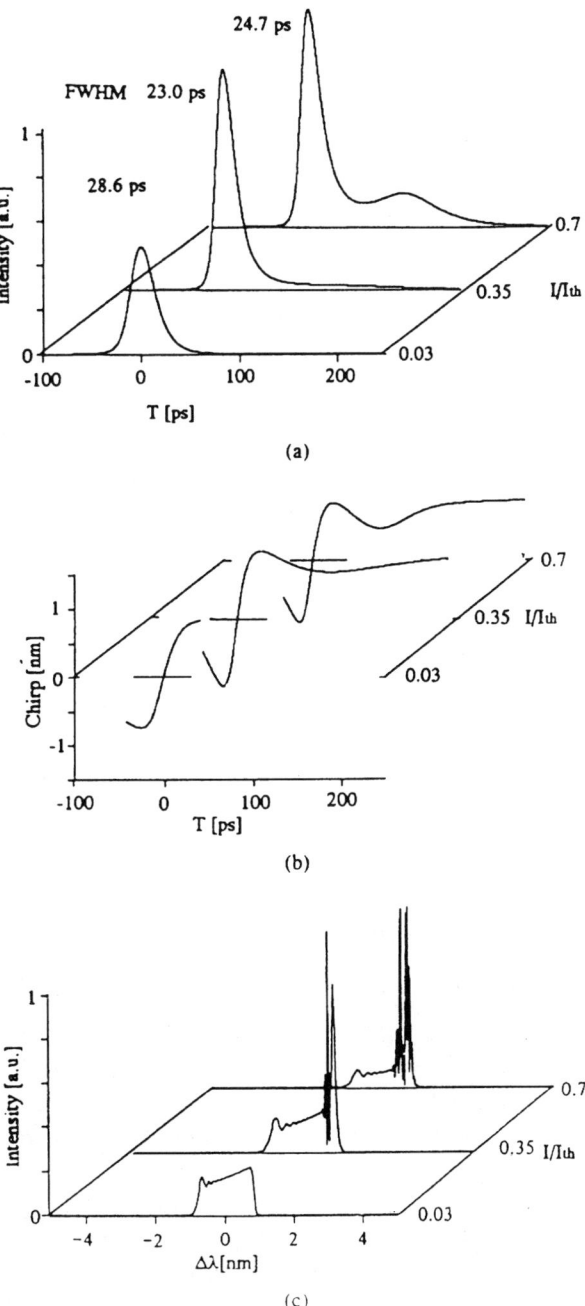

Figure 2.1.2
Theoretical changes in (a) pulse waveform, (b) chirp, and (c) the time-averaged spectrum of the gain-switched pulse with increase in dc bias. The pulse wave form and spectrum are normalized with respect to those of the maximum carrier inversion dc bias. The set of parameters gives a lasing threshold current of $I_{th} = 18$ mA. The maximum carrier inversion occurs around a dc bias of 0.35 I_{th}

group velocity, $G(N)$ is the gain coefficient, and ε is the gain compression factor.

$$\frac{dS}{dt} = \frac{\Gamma_g G(N) S}{1 + \varepsilon S} - \frac{S}{\tau_p} + \beta C_1 N^2 \qquad (2)$$

where Γ_g is the confinement factor, τ_p is the photon lifetime, β is the spontaneous emission factor and C_1 is the radiative recombination coefficient.

The simulated optical pulse waveforms become as shown in Figure 2.1.2, which indicates the possibility of compressing the pulse from 150 ps to 25 ps.

A conspicuous effect of the semiconductor laser transient is the dynamic change in optical frequency due to the negative correlation of the refractive index and the carrier density. The frequency shift, called the chirp characteristic, is expressed as [8]

$$\Delta \omega(t) = -\frac{\alpha v_g a (N - N_0)}{2} \qquad (3)$$

where α is the linewidth enhancement factor, connecting the differential gain and differential dispersion. The simulated frequency shift is also plotted in Figure 2.1.2, in which a decreasing shift in the frequency is shown, often called the 'red shift chirp'.

2.2 Experimental results

A direct evaluation of the pulse waveform together with the chirp characteristics was performed by employing a streak camera combined with a grating monochromator according to Tsuchiya's method. Figure 2.1.3 (a) depicts the plot of an instantaneous

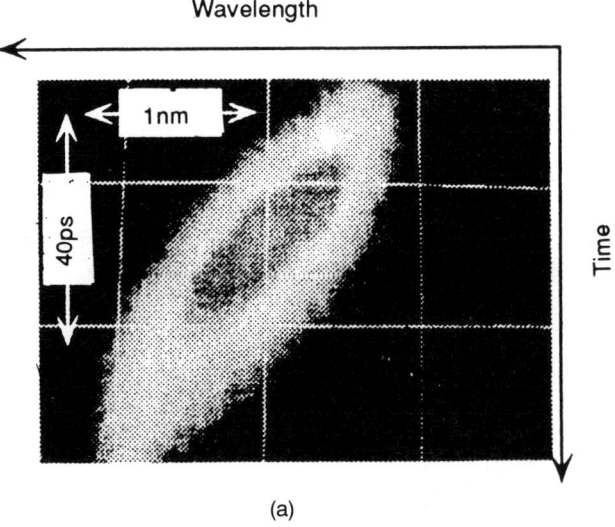

(a)

Figure 2.1.3
Streak camera trace of temporal-spectral transient of a gain switched DFB laser output. (a) spectrally resolved temporal evolution; (b) averaged spectra; (c) averaged waveform

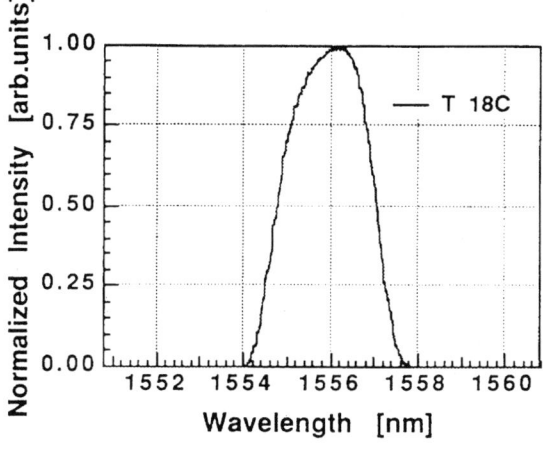

(b)

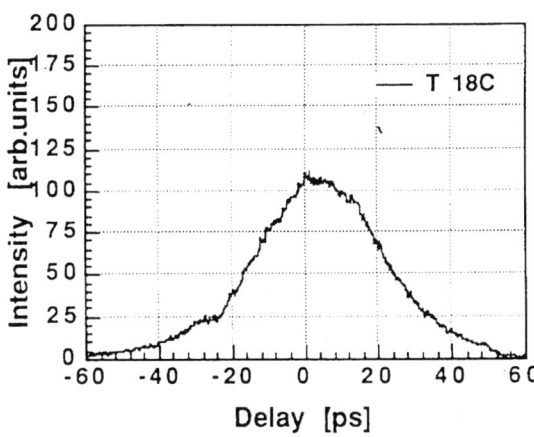

(c)

Figure 2.1.3 (*continued*)

frequency change as a function of time. Summing up the spectral components we obtain the temporal waveform shown in Figure 2.1.3 (b), meanwhile the summation over the time gives us the time-averaged spectra shown in Figure 2.1.3 (c). These latter two coincide with the well-known gain-switching characteristics of a DFB laser. Streak camera measurement identified the fact that the linear downward shift in frequency is dominating, with a small correction for the nonlinear chirp component, depending on the drive conditions. The presence of the nonlinear chirp component is harmful in the linear compensation of chirp by the dispersion of the fiber.

3 Linear Compression of Pulses by Chirp Compensation

3.1 Principle of chirp compensation

As first demonstrated by Takada *et al.* [4], the chirp characteristics can be partially compensated by the addition of the normal dispersion fiber. Intuitively compression of the pulse width can be understood in the following manner. For the red shift chirp pulse, the leading edge has the higher frequency component, which propagates the normal dispersion fiber relatively more slowly than the low-frequency component which dominates the trailing edge of the pulse. After certain length of propagation it is expected that the leading edge is caught up by the trailing edge of the pulse, resulting in a delta function-like output. More precisely, the theoretical formulation should not violate Heisenberg's uncertainty principle of conjugate variables, namely photon energy and time. The treatment is straightforward when we employ the Fourier transformation of the incident pulse. The dispersion characteristics of the fiber being expressed as the transfer function in frequency domain, the inverse Fourier transform of the product of the incident pulse spectra and the fiber transfer function gives us the output signal in the time domain [9].

3.2 Experiments of linear compression

For semiconductor lasers with Fabry–Perot resonators the gain-switched pulse contains many axial mode components, which propagate through the fiber with different group velocities. Therefore the exit signal is in the form of a temporally separated train of pulses, each corresponding to the individual axial mode [9]. To avoid the formation of pulse trains, we used the DFB laser to maintain single axial mode operation even under transient condition [6]. The pulses from the gain-switched semiconductor laser

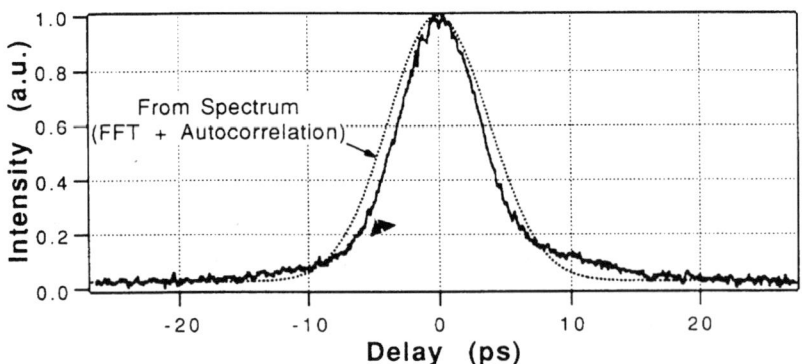

Figure 2.1.4
Autocorrelation trace of a linearly compressed fiber from a filtered fiber. The calculated autocorrelation trace based on the experimental spectra manipulated by FET and autocorrelation procedures is also shown by the dotted curve. The time bandwidth product $\Delta t \Delta v = 0.37$ is obtained

were fed to a normal dispersion fiber of length 1 km with a dispersion parameter of -16.0 ps km^{-1} nm^{-1}. The resultant pulse width was in the range of 5–7 ps depending on the biasing conditions of the laser. From measurement of time-averaged spectra, a time bandwidth product was evaluated to be around 0.6, exceeding the Fourier transform limit value of 0.35.

The existence of nonlinear chirp deteriorates the pulse quality.

It was found that our experimental system contains birefringent characteristics. The associated polarization dispersion characteristics could be controlled by the addition of a quarter-wave plate and a half-wave plate. Combined with a polarizer, this polarization control module acted as a tunable spectral filter, convenient for suppressing the spectral portion responsible for the nonlinear chirp characteristics. After cleaning the spectra the compressed pulse width was 6 ps and the time bandwidth product reduced to 0.37, very close to the transform limit value, as shown in Figure 2.1.4 [10].

4 Nonlinear Compression of Pulses by Soliton Effect

4.1 Soliton propagation in the fiber

Third-order nonlinear susceptibility of the quartz fiber induces nonlinear transmission characteristics of the optical pulses, described by the nonlinear Schrödinger equation for its envelope [11]. The equation has a soliton solution, maintaining the pulse throughout

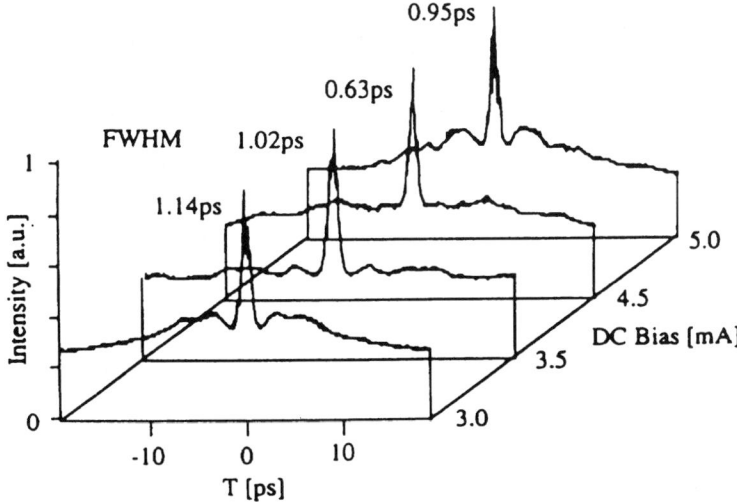

Figure 2.1.5
The experimental autocorrelation waveform of the pulse after the anomalous dispersion fiber. The results show that it is possible to tune the waveform and also the pulsewidth of the soliton pulses simply by varying the dc bias condition of the gain-switched laser. It is also observed that the optimum dc bias point for the given fiber length was around 4.5 mA, yielding an autocorrelation width of 630 fs. If we assume an FWHM ratio of 1.07, the time domain FWHM is 590 fs

the propagation axis, as far as the attenuation loss is compensated. While the fundamental mode soliton solution keeps the original waveform exactly along propagation, the higher order soliton pulse deforms periodically. It is possible to excite a higher order soliton pulse with a relatively broader pulse and extract it with narrowed shape at a fractional length of soliton period. Since the excited soliton order is proportional to the pulse energy, and the compression ratio becomes larger, the stronger excitation is favorable for compression purposes.

4.2 Experimental result

The linearly compressed pulses were amplified by an erbium-doped fiber amplifier with a small signal gain of 30 dB, and the resultant pulses with pulse energy amounting to 30 pJ were obtained. They were fed to an anomalous dispersion fiber of length 200 m with the dispersion parameter of 4.14 ps km^{-1} nm^{-1}. The estimated soliton order was about 6. The shortest pulse width evaluated by the autocorrelation technique was 0.6 ps, as shown in Figure 2.1.5 [6].

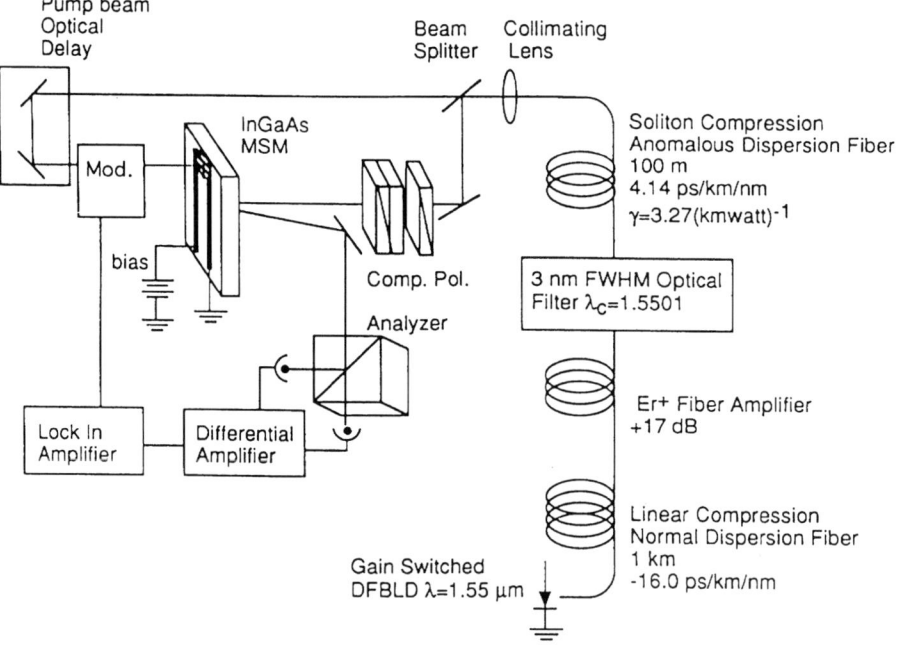

Figure 2.1.6
Experimental layout showing the pulse generation scheme and the electrooptic sampling system

5 Ultrashort Pulses Applied to the Electrooptic Sampling Measurement of the InGaAs Photodiode Rise Time

The optical pulse source thus developed has the advantages of compact size, rapid and arbitrary repetition rate, and an operating wavelength of 1.55 μm, suitable for measurements associated with fiber-optic telecommunication systems.

The experimental setup shown in Figure 2.1.6 depicts a trial electrooptic sampling measurement of a fast MSM photodiode composed of InGaAs, with a pulse response time of 30 ps [12]. The photodiode was fabricated monolithically together with a coplanar strip line on top of a semi-insulating InP substrate, which also acted as an electrooptic probing crystal upon backside illumination of the optical probe pulse. The timing between the excitation pulse at the MSM photodiode and the probe pulse was scanned with a mechanical delay unit. The instantaneous voltage across the substrate at the probe beam position was plotted as a result of the sensitive polarimetry of the reflected probe light, with a voltage sensitivity of 0.1 mV/$\sqrt{\text{Hz}}$. As shown in

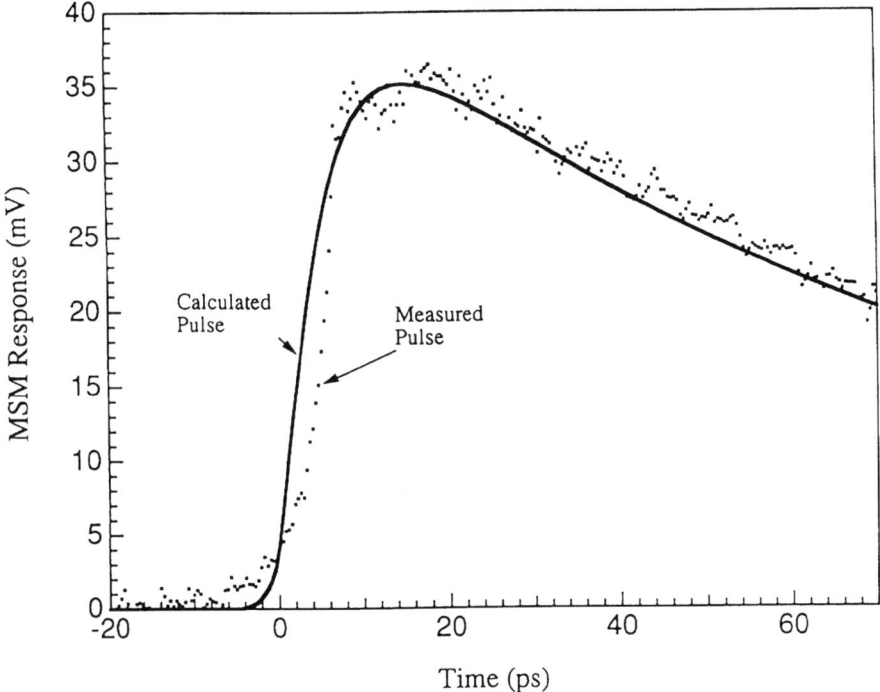

Figure 2.1.7
Sampled response of the MSM photodetector with 1 μm metal fingers and 2 μm semiconductor spaces with a 10–90% rise time of 6.5 ps and the calculated response using the optical pulse with a FWHM of 0.6 ps, a rise time constant of 2.5 ps, and fall time constant of 110 ps

Figure 2.1.7, the rise time of the MSM PD under test was 6 ps, within which the instrumental time constant was evaluated to be about 2 ps. The parasitic capacitance associated with the lumped photodiode is the most probable source responsible for the remaining part of the time constant.

6 Summary

In summary, the scheme of gain switching of DFB laser combined with erbium-doped fiber amplification and soliton effect compression proved to be effective in generating stable pulses of less than 1 ps in pulse width. The waveform quality depends on the linear and nonlinear chirp characteristics, which can be controlled by suitable filtering techniques. Nearly transform limited pulse with time bandwidth product of 0.35 was obtained. The pulse source was applied to a trial electrooptic sampling measurement of an integrated InGaAs MSM photodiode, with a time resolution of 2 ps.

References

[1] Chi-H. Lee, ed., *Picosecond Optoelectronics Devices*, Academic Press, New York, 1984.
[2] M. Nakagawa, K. Suzuki and E. Yamada, *Electron. Lett.* **26**, 2038 (1990).
[3] H.F. Liu, Y. Ogawa and T. Nonaka, *IEEE J. Quantum Electron.* **27**, 1655 (1991).
[4] A. Takada, T. Sugie and M. Saruwatari, *J. Lightwave Technol.* **5**, 1525 (1987).
[5] P.B. Hansen, G. Raybon, U. Koren, B.I. Miller, M.G. Young, M.A. Newkirk, M.D. Chien, B. Tell and C.A. Burrus, *IEEE J. Quantum Electron.* **29** (1993).
[6] J.T. Ong, R. Takahashi, M. Tsuchiya, S.H. Wong, R.T. Sahara, Y. Ogawa and T. Kamiya, *IEEE J. Quantum Electron.* **29**, 1701 (1993).
[7] A. Yariv, *Optical Electronics*, 4th edn, HBJ College Publishers, Fort Worth, 1991, ch. 15.
[8] T. Koch and J. E. Bowers, *Electron. Lett.* **20**, 1038 (1984).
[9] Y.T. Lee, R. Takahashi and T. Kamiya, *Jpn. J. Appl. Phys.* **29**, L89 (1990).
[10] B. Thedrez, H. Takeshita and T. Kamiya, *CLEO' 94*, Anaheim, May 1994, CTuH4.
[11] G.P. Agrawal, *Nonlinear Fiber Optics*, Academic Press, New York, 1989, Chs. 2, 3, 5.
[12] R. Sahara, R. Takahashi, J. T. Ong, M. Tsuchiya, Y. Ogawa and T. Kamiya, *CLEO' 93*, Baltimore, May 1992, CThS87.

2.2

Ultrashort Pulse Generation Method by Electrooptic Modulation

Tetsuro Kobayashi and **Akihiro Morimoto**

Abstract

The materialization of an ultrafast electrooptic modulator or deflector that operates in the picosecond or terahertz range is very difficult. However, the generation of ultrashort optical pulses shorter than picosecond using a modulator or a deflector is possible when large modulation is effectively utilized. Here several kinds of electrooptical generation methods of ultrashort optical pulses are described. After fundamental considerations, most the promising one, EO pulse generators — a Fabry-Perot EO modulator and an improved one, both utilizing multi-interference modulation — are described with the experimental results of subpicosecond pulse generation and high power efficiency.

In addition, velocity matched EO modulators and deflectors using higher mode E-wave and periodical domain inversion are described with the experimental result of extremely wide (~2 THz) sideband. The wide sideband is capable of being converted to subpicosecond pulses. Finally, an ultrafast electrooptical pulse synthesizer is discussed.

1 Introduction

The generation of ultrashort optical pulses, or in a broad sense, the generation of ultrafast optical signals with an ultrashort temporal structure, is one of the useful applications of the ultrafast feature of optical waves. Mode-locking of a laser has been ordinarily used to generate short pulses. Up to now ultimately a short pulse, including only a few optical cycles, has been obtained by mode-locking. Since the pulse width is, however, limited by the laser gain bandwidth in the case of mode-locking, ultrashort pulses (ps) are not always obtained from any kind of laser. Additionally, mode-locking, particularly passive mode-locking, has a weak point in controllability due to the inclusion of some kinds of nonlinearities. On the other hand, direct pulse generation by electrooptic (EO) modulation has the advantage of good controllability and fits the application to electronics, although it does not have satisfactory speed. If the electrooptic method [1] has the ability to generate ultrashort pulses as with mode-locking, the speed of optoelectronics is greatly accelerated. This study has been done with such a purpose.

2 Fundamental Considerations

Ultrashort optical pulses have wide optical frequency spectra as seen from the principle of Fourier transformation. Roughly speaking, for a pulse width of τ, the spectral width $\Delta \nu$ is given by

$$\Delta \nu \geq \frac{a}{\tau}, \quad a(0.1-1) : \text{constant depending on pulse shape}$$

Accordingly, to generate short optical pulses from ordinary laser output by EO modulation, a wide sideband must be produced. For the case of picosecond to femtosecond pulses, the spectral width $\Delta \nu$ is in the several hundreds of gigahertz to several tens of terahertz range. This width is much higher (wider) than the usual driving frequency of an EO modulator or deflector. At present, it is technologically difficult to obtain an ultrafast ($\leq$ ps) electric driving signal and also to realize ultrafast EO modulator operating in the subpicosecond range. So, what can we do to generate ultrashort pulses using a low speed EO modulator? One realistic and useful solution is to utilize deep modulation with a large modulation index.

If the absorption coefficient or refractive index is modulated, the total modulation function of the optical wave is given by

$$e^{-\alpha + \alpha \cos \omega_m t} = \sum e^{-\alpha} I_n(\alpha) e^{jn\omega_m t}, \text{ for modulation of absorption coefficient} \quad (1)$$

or

$$e^{j\Delta\theta \sin \omega_m t} = \sum J_n(\Delta\theta) e^{jn\omega_m t}, \text{ for modulation of the refractive index} \quad (2)$$

respectively. The frequency widths of the sidebands induced by the above modulations are

$$\Delta \nu \approx 2\alpha f_m \text{ for AM} \quad \text{and} \quad \Delta \nu \approx 2\Delta\theta f_m \text{for PM} \quad (3)$$

approximately. For both cases, the sideband widths $\Delta \nu$ can be much larger than the modulating frequency when the modulation indices are very large, namely $\alpha \gg 1$ or $\Delta\theta \gg 1$. Hence if the above wide sidebands are effectively utilized for pulse-forming, ultrashort optical pulses can be generated. In fact, by using the electro-absorption (EA) modulation as shown in equation (1), ultrashort pulses have been directly generated without any other treatment [2], although the energy efficiency is low. On the other hand, for the case of an EO deflector considered as a spatially distributed phase modulator, short pulses can be picked out through a narrow slit for large amplitude deflection, as shown in Figure 2.2.1. The pulse width is given approximately by [3]

$$\tau_p \approx \frac{1}{\Delta v} \quad (4)$$

where Δv is the frequency width of the sideband produced by the deflector.

In addition, we show two examples of simple pulse generation methods using high index phase modulation. The first one is a high repetition pulse generation using selection of the sidebands through a Fabry–Perot interference filter, as shown in Figure 2.2.2. A high bit pulse train (100 Gb/s) has been generated by this method [4]. The second one is direct pulse compression using a cw phase modulated (PM) light with a group delay dispersion element [5], as shown in Figure 2.2.3. Using equation (2), the instantaneous frequency of PM light is obtained as

$$\omega(t) = \omega_0 - \Delta\theta\omega_m \cos(\omega_m t)$$

$$\sim \omega_0 + (-1)^q \Delta\theta\omega_m^2 (t - t_q) \quad \text{for } t \sim t_q \left(= \frac{(q + \frac{1}{2})\pi}{\omega_m} \right) \quad (5)$$

It is seen from equation (5) that linear frequency chirping occurs close by t_q. Hence the amount of group delay dispersion required for optimum pulse compression is

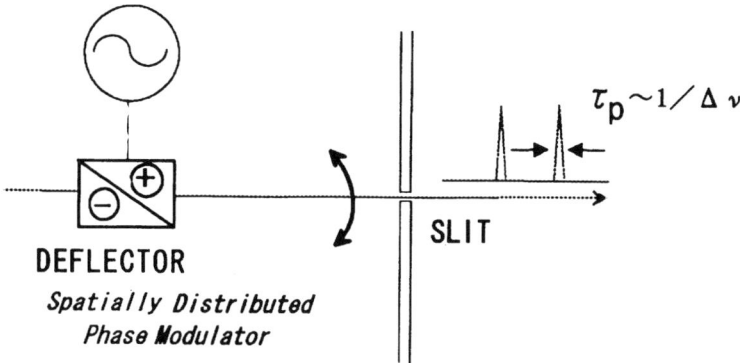

Figure 2.2.1
Short pulse generation by large amplitude deflection

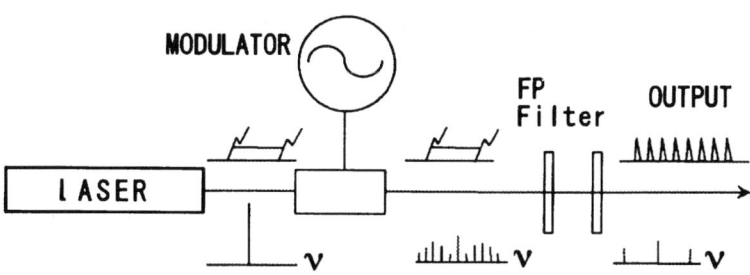

Figure 2.2.2
Generation of high bit pulses by FP filter with PM

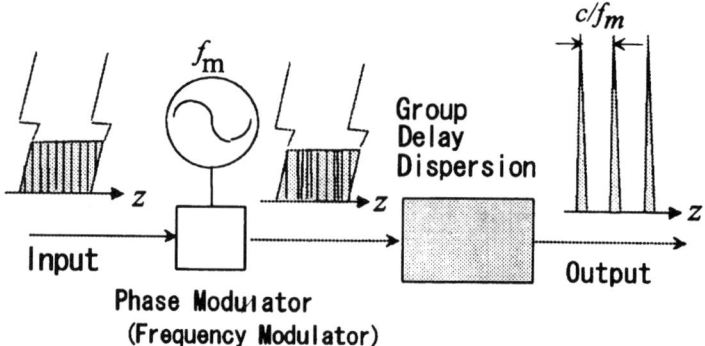

Figure 2.2.3
Pulse generation by direct compression of cw PM light

$$\frac{\partial \tau}{\partial \nu} \sim -\frac{2\pi}{\partial \omega/\partial t} \sim -(-1)^q \frac{2\pi}{\Delta\theta\omega_m^2} \quad (6)$$

and the width of the compressed pulse is approximately $0.7/2\Delta\theta f_m$ [5]. To establish a high modulation index long electrooptic interaction length is required with velocity matching between the E and O waves. For this purpose synchronous multi-stage modulation is useful as well as long traveling-wave modulation with velocity matching.

3 Synchronous Multi-stage Modulation and a FP Modulator and a Modified One

A synchronously driven serial multi-stage modulator chain [1] is very useful to generate short optical pulses. For it several modifications have been proposed as using harmonic modulations or multiple-index modulations. In most cases, large insertion loss is a problem. Mode-locking is also considered as a kind of synchronously driven endless multi-stage modulation with loss compensation by laser amplification. A Fabry–Perot EO modulator, as shown in Figure 2.2.4, is also a type of multi-stage modulator but rather a multiple interference modulator. It is also a kind of interferometer scanned synchronously with a cavity round trip and short pulses are obtained as an output [6,7]. The pulse width depends on the modulation index, the modulating frequency, and the finesse $\mathcal{F}$ of an FP resonator and given by [6]

$$\tau_p \sim \frac{1}{2\Delta\theta f_m \mathcal{F}} \quad (7)$$

and is not limited by the laser gain bandwidth as mode-locking. For large $\mathcal{F}$ the pulse width is very short and subpicosecond pulses have been obtained by this method [7].

In this method the output pulses are picketed out from the input only in short duration. Other parts of the input are reflected back and, as a result, the output power efficiency is low. To improve efficiency a modified Fabry–Perot (MFP) modulator, as shown in

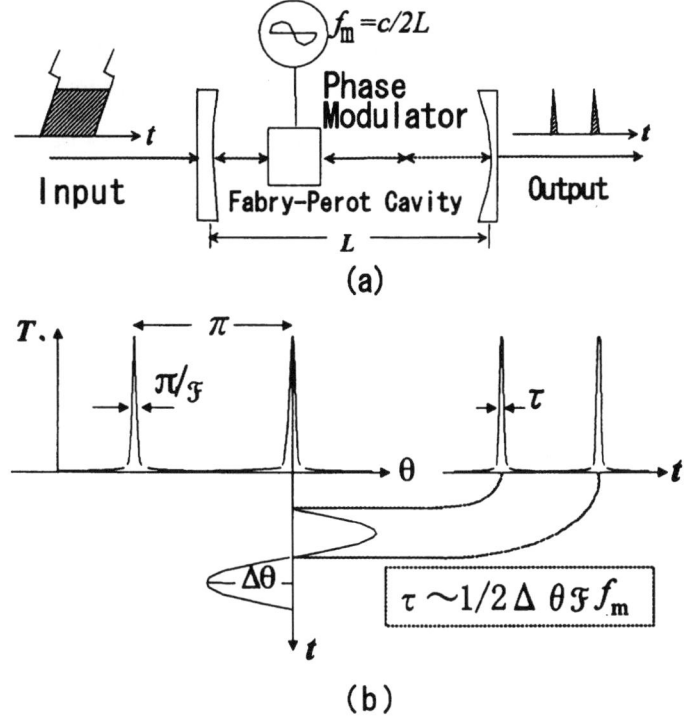

Figure 2.2.4
Fabry-Perot electrooptic modulator

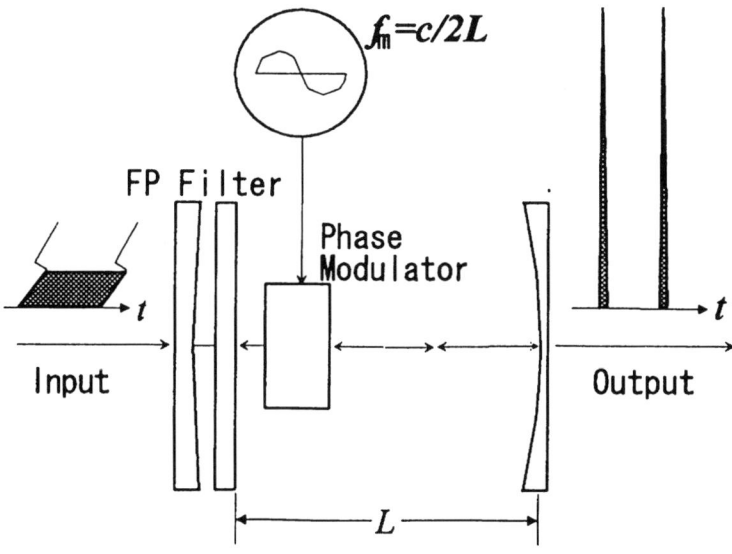

Figure 2.2.5
Modified Fabry-Perot Electrooptic modulator

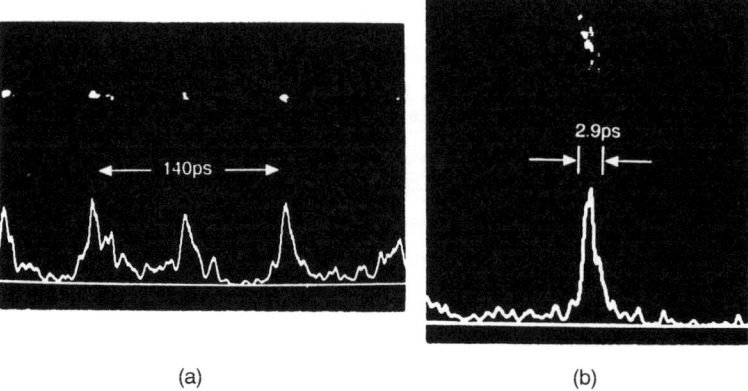

Figure 2.2.6
Typical example of an output pulse from a MFP modulator

Figure 2.2.5, was proposed by us [8]. The MFP modulator is a sort of FP modulator with an input Fabry–Perot filter which lets only the input light (carrier) pass and reflects all other sidebands. Then the input light energy is efficiently converted to the sideband energy by internal phase modulation, so that ultrashort pulses are generated as in the FP modulator. The experiment was carried out with a microwave resonant type LiTaO$_3$ phase modulator ($\Delta\theta = 0.79$ rad) at a 7.1 GHz for a 514.5 nm single-mode argon laser input [9]. The reflectances of two input mirrors and an output mirror are 99% and 98%, respectively. A typical streak trace of the output pulse is shown in Figure 2.2.6. The pulse width is limited by a temporal resolution of 2.5 ps, and is estimated to be 1.7 ps from the spectral measurement. The pulse energy was several ten times larger than that of an ordinary FP modulator. This modulator makes it possible to generate a few tens of terahertz sidebands (as in the ordinary FP modulator) and femtosecond pulses (limited by group delay dispersion in the cavity) in the near future.

4 Velocity Matched Traveling-wave Modulator and Application to Pulse Generation

To obtain a high modulation index using a long-size traveling-wave type EO modulator, velocity matching is essential in the high frequency range above several gigahertz. For the case of broad band modulation, velocity matching must be done between group velocities of light and modulating electric waves. This, however, is very difficult due to the frequency dispersion of the electrooptic materials. On the other hand, for an electrically narrow band modulation, velocity matching may be done between the phase velocity of the E wave and group velocity of light. It is relatively easy. The following are two typical examples suitable for obtaining high modulation indices with velocity matching.

4.1 Velocity matched EO modulator and deflector using a higher mode of strip line

The phase velocity of a higher mode E wave traveling along a strip line is controlled by adjusting the width of the electrode. The electric field in the cross-section of a higher mode waveguide is not uniform. Lens, prism, or grating effects are electrooptically induced. Phase modulation is carried out by passing a light beam along the anti-node of the E-field, and deflection does so along the node [10]. Figure 2.2.7, shows the case using the deflector, where ultrashort pulses are picked out through the narrow slit placed at the farfield.

4.2 Quasi-velocity matching using the periodic domain inversion technique

Figure 2.2.8 shows modulation indices as functions of the EO interaction length. While the modulation index of a velocity matched modulator is proportional to the interaction length (a), that of an ordinary modulator oscillates with a period of $2L(L =$

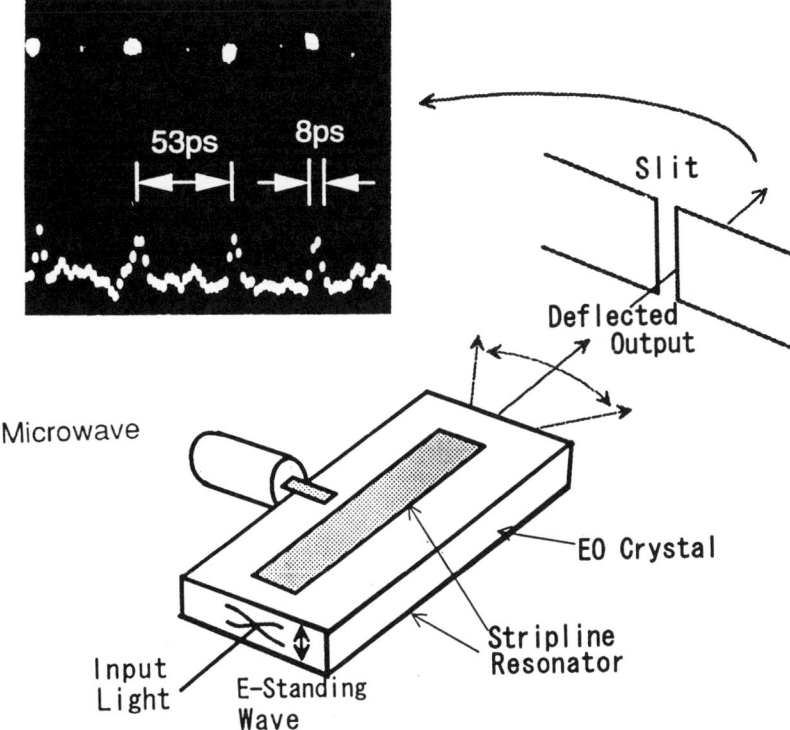

Figure 2.2.7
Velocity matched EO modulator/deflector using a higher mode of microwave strip line guides

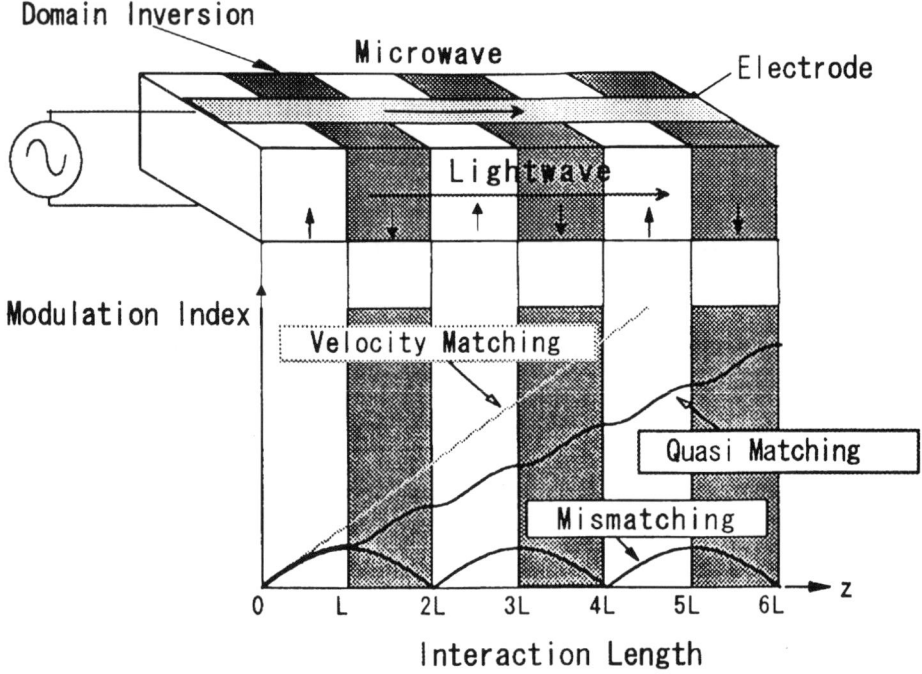

Figure 2.2.8
Quasi-velocity matching using domain inversion of an EO crystal

$1/\{2f_m(v_{mp}^{-1} - v_{og}^{-1})\}$ as the length increases because of the velocity mismatching (b), where f_m the modulating frequency, v_{mp} the phase velocity of the microwave, and v_{og} the group velocity of light. In other words, making the modulator longer than L is fruitless. On the other hand, if the signs of the modulation are inverted with period L, as shown in the upper part of Figure 2.2.8, quasi-velocity matching occurs and the modulation index is almost proportional to the length (c) (although lower than (a) by a factor of $2/\pi$).

To obtain such quasi-velocity matching, employing the domain inversion technique is useful as well as in second harmonic generation. In our case, domains of an $LiTaO_3$ crystal (0.5 mm thick, 30 mm long) were inverted by applying high voltage (> 10 kV/mm on $+z$ surface) to electrodes in an insulating fluid at room temperature. A microwave resonator (16.25 GHz) is formed by a silver strip line on the crystal. The optimum inversion period L was determined experimentally. Modulation by the microwave traveling in the opposite direction to light is negligible in this configuration. An experiment of this modulator was done with a 514.5 nm Ar laser and a multi-kilowatts Ku band magnetron. A typical sideband spectrum is shown in Figure 2.2.9, which is the result of double pass modulation. A theoretical calculation is also shown in the figure. The sideband spectral width is 1.85 THz (FWHM) and the modulating index is 57 rad at 16.25 GHz [11].

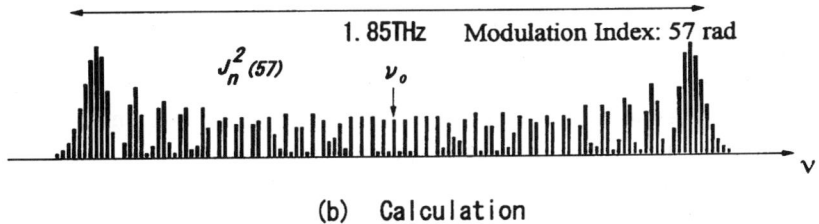

(b) Calculation

Figure 2.2.9
Typical example of an optical sideband spectrum

This sideband width is to our knowledge, the widest ever reported by a simple electrooptic phase modulator. The modulation depth was limited by a discharge on the electrode. By making the crystal longer, we can further improve the modulation index because our magnetron has great reserves of power. A thinner modulator or a waveguide modulator with domain inversion will also improve the modulation efficiency.

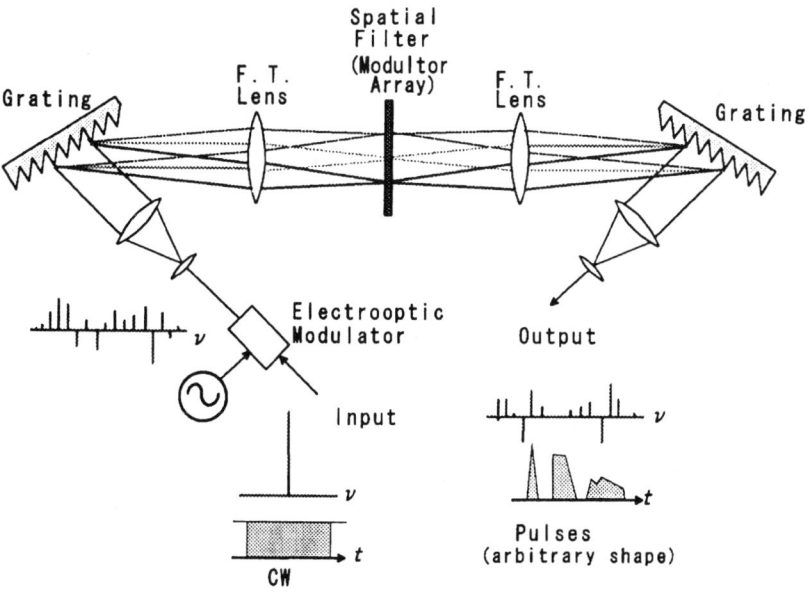

Figure 2.2.10
Optical pulse synthesizer

5 EO Pulse Synthesizer

A more general method to generate short pulses using a wide sideband produced by EO modulation/deflection is an electrooptical pulse synthesizer. It utilizes Fourier analysis and synthesis of the optical sideband in a frequency domain [12,13], which was proposed and developed by us at first [3]. A typical setup of the EO pulse synthesizer is shown in Figure 2.2.10. A similar synthesizer using an ultrashort optical pulse as a source instead of modulated light has also been studied by many authors [14,15]. An EO synthesizer is composed of a modulation section (generation of sidebands), a demultiplexing section (spatial separation of sidebands), an f-domain control section, and a multiplexing section (Fourier composing). For the case of using a deflector, the deflector plays the roles of generator and separator of the sideband at the same time. As a spatial modulator for f-domain control, a liquid crystal spatial filter, a lens, a prism, a slit, and an EO modulator array are usable.

6 Conclusion

Various kinds of electrooptic generation methods of ultrashort optical pulses have been described. As the modulation technology progresses, the applicable area for these methods will be extended to the few tens of terahertz or shorter than 100 fs range. In practical use, however, there remain many problems such as a decrease in the driving power from subkilowatt to subwatt range and of insertion loss, and miniaturizing and integration of devices and systems. Furthermore, to extend the technology to more general ultrafast signal control, such as signal modulation, random access switching without limiting only in pulse generation is also an important problem. Employing optical amplification and enhanced electrooptic effects introduced by the semiconductor quantum well microstructure will hopefully bring us the solution.

References

[1] T. Kobayashi, *Trans. Jpn IEICE*, **J74-C-1**, 387–397 (1991).
[2] M. Suzuki, H. Tanaka, K, Utaka, N. Edagawa and Y. Matsushima, *Electron. Lett.* **28**, 1007–1008 (1992).
[3] T. Kobayashi, H. Ideno and T. Sueta, *IEEE J. Quantum Electron.* **QE-16**, 132–136 (1980).
[4] A. Morimoto, H. Yao, T. Kobayashi and T. Sueta, *XVI Int'l Conf. Quantum Electron.* Tokyo, 1988, ThH-6.
[5] T. Kobayashi, H. Yao, K. Amano, Y. Fukushima, A. Morimoto and T. Sueta, *IEEE J. Quantum Electron.* **24**, 382–387 (1988).
[6] T. Kobayashi, T. Sueta, Y. Cho and Y. Matsuo, *Appl. Phys. Lett.* **21**, 341–343 (1972).
[7] T. Kobayashi, A. Morimoto, T. Fujita, K. Amano, T. Uemura and T. Sueta, *Ultrafast Phenomena, V*, Springer-Verlag, 1986, 134–137.
[8] T. Kobayashi, A. Morimoto, B. Y. Lee and T. Sueta, *Ultrafast Phenomena, VII*, Springer-Verlag, 1990, 41–44.
[9] A. Morimoto, A. Shibagaki and T. Kobayashi, *Conf. on Lasers and Electrooptics, 1993 OSA Technical Digest Series, Vol. 11*, Opt. Soc. America, 1993, 558–559.
[10] B.Y. Lee, T. Kobayashi, A. Morimoto and T. Sueta, *Electron. Lett.* **28**, 330–332 (1992).

[11] A. Morimoto, E. Saruwatari and T. Kobayashi, *Conf. on Lasers and Electrooptics, 1994 OSA Technical Digest Series, Vol. 8*, Opt. Soc. America, 1994, p. 21.
[12] T. Kobayashi, A. Morimoto, M. Doi, B.Y. Lee and T. Sueta, *Ultrafast Phenomena, VI*, Springer-Verlag, 1988, 135–138.
[13] T. Kobayashi and A. Morimoto, *Picosecond Electronics and Optoelectronics, Vol. 4*, Opt. Soc. America, 1989, 81–86.
[14] A.M. Weiner, J.P. Heritage and E.M. Krishner, *J. Opt. Soc. Am.* **B5**, 1563–1572 (1988).
[15] K. Ema, M. Kuwata-Gonokami and F. Shimizu, *Appl. Phys. Lett.* **59**, 2799–2801, (1991).

2.3

Ultrafast Optical Switching by Quantum Well Structures

Yasuaki Masumoto

Abstract

We present a highly repetitive picosecond polarization optical switching operation of type II AlGaAs/AlAs quantum well structures at room temperature. The switching mechanism is due to the fast Γ-X interlayer scattering of electrons and the fast spin relaxation of holes. The type II system realizes both large optical nonlinearity and rapid recovery of absorption saturation. According to our experimental results, the switching time is less than 1 ps, the switching power is as low as 3 nJ/cm^2, and the switching repetition rate is more than 80 GHz. Moreover, a continuous optical switching operation is possible for future applications

1 Introduction

In order to realize ultrafast optical switching devices, optical nonlinear materials that have a sensitive and ultrafast response to optical pulses must be sought. Until now, various materials ranging from semiconductors to organics have been widely investigated. The ultrafast optical switch must satisfy three requirements: ultrafast response; high repetition rate; and low switching energy. We propose three practical conditions. The switching speed must be faster than several tens of picoseconds, since this value has already been achieved by electric-current switching devices. We adopt the values 1 ps and 10 ps as the application criteria for the switching speed and the repetition interval of the ultrafast optical switches, respectively. If the semiconductor lasers are utilized as an optical switching light source, the energy density for the optical switching

Ultrafast Optical Switching by Quantum Well Structures

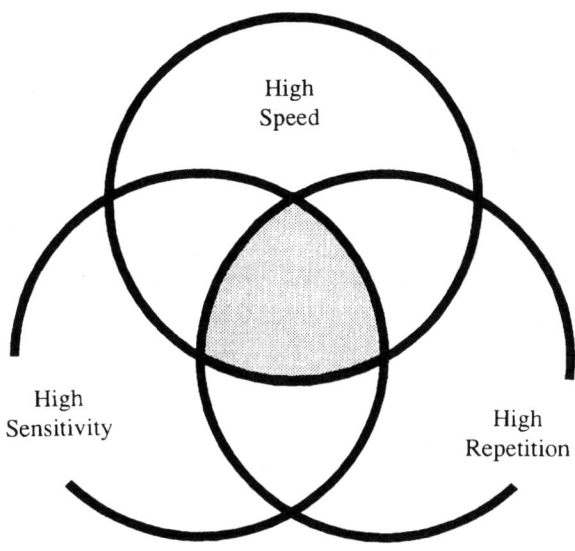

High Speed : < 1 ps(> 1 THz)

High Repetition : > 100 GHz

High Sensitivity : > 1 μJ/cm²

Figure 2.3.1
Three requirements for the ultrafast optical switch

should be less than ~1 μJ/cm², which is equal to the damage threshold of the semiconductor lasers of ~1 MW/cm² multiplied by the pulse width of 1 ps. Therefore, three practical conditions are: (i) an ultrafast response less than 1 ps; (ii) a repetition rate higher than 100 GHz; and (iii) switching energy density lower than ~1 μJ/cm². They should be satisfied simultaneously for the practical application as illustrated in Figure 2.3.1.

Quantum well structures composed of GaAs and related compounds are one of the most promising materials and have been extensively studied so far. In GaAs quantum wells, the enhancements of the oscillator strength and the binding energy of the lowest exciton take place due to the quantum confinement effect. As a result, excitons exist and show large optical nonlinearity even at room temperature [1]. In type I quantum well structures, the generation of carriers and resultant exciton bleaching are faster than 1 ps. However, once the carriers are generated, the recovery time of the system is determined by the recombination lifetime of the carriers, which is of the order of 1 ns, and thus a highly repetitive operation is impossible. Some methods of solving the problem have been proposed. The doping of impurities reduces the recombination lifetime, but also reduces the optical nonlinearity significantly.

The optical Stark effect of the excitons was proposed as an ultrafast optical switching mechanism, since nonresonant excitation does not generate real carriers. However, the optical Stark effect requires a laser energy density of as high as 100 μJ/cm^2 [2]. At the excitation density where sufficient optical Stark shift is achieved, a two-photon absorption process generates considerable numbers of carriers. This prevents highly repetitive optical switching.

We discovered a new type of optical nonlinearity in type II quantum well structures for the first time and found it to be applicable to highly repetitive optical switching [3,4]. In the following, we present the mechanism of the optical nonlinearity and show the experimental demonstration of highly repetitive ultrafast optical switching. We show that the type II system satisfies three practical conditions for ultrafast optical switching at room temperature.

2 The Optical Nonlinearity in Type II Quantum Well Structures

In the type I quantum well system, both electrons and holes generated by photo-excitation are confined in well layers. However, the dynamics of photo-excited carriers in the type II system is completely different. The energy of the X-electron state in a barrier layer is located below that of the Γ-electron state in a well layer. Thus, only

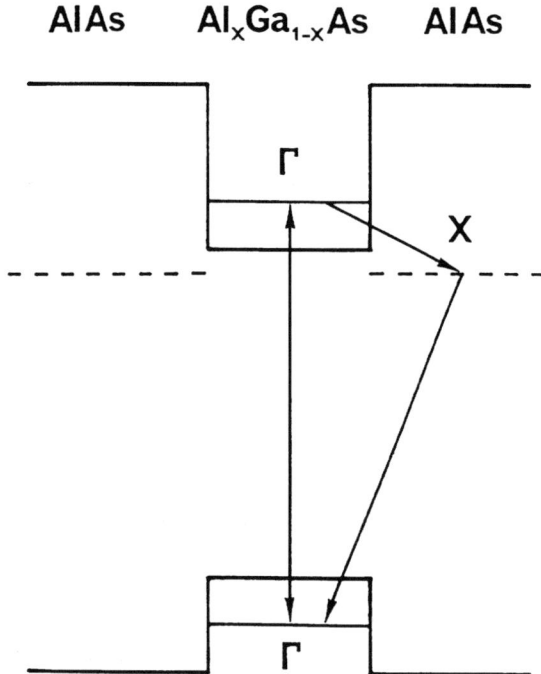

Figure 2.3.2
Schematic energy diagram of type II Al$_x$Ga$_{1-x}$As/AlAs multiple quantum well structures

holes are confined in the well layer, and electrons are quickly scattered into a barrier layer, as schematically illustrated in Figure 2.3.2 [5]. The rate of this Γ–X electron transfer process is controllable by the artificial design of the quantum well, and the typical value ranges from several tens of femtoseconds to several picoseconds [6–8]. Once the electrons are scattered into the barrier layer, the recombination lifetime is very long and reaches microsecond order at 2 K, since electrons and holes are separated in both momentum and real spaces [3]. At room temperature, the recombination lifetime is expected to be shortened to the nanosecond region with the assistance of thermal phonons through a nonradiative path.

The Γ–X scattering process slightly complicates the optical nonlinearity of the type II system. It is useful to consider the optical nonlinearities separately, before and after the Γ–X scattering. Before the scattering, the situation for excitons is similar to that in type I quantum wells. Sensitive and ultrafast bleaching is achieved by the phase space filling effect of excitons. After the scattering, the condition is quite different from that in the type I system. Then the holes are confined in the well layer and contribute to the bleaching through phase space filling of the hole state. As a result, bleaching partially recovers and remains for a long time at low temperature. The long time component of bleaching reflects the slow recombination of carriers. This long time component may disturb the highly repetitive switching.

The long time component is reduced to an order of 1 ns at room temperature. However, the accumulated carriers reduce the optical nonlinearity. To solve these difficulties, we reduced the switching power density by using the spin-polarization detection system of ultra-high sensitivity. If we can reduce the switching energy density, the accumulated carrier density is reduced to less than the practical application level. On the basis of the above-mentioned consideration, we present a new type of optical switch, made of type II AlGaAs/AlAs quantum wells. The optical switch is based on two ultrafast carrier processes: the Γ–X interlayer scattering of electrons, and the spin relaxation of holes. We could realize both a fast switching time and a low light-energy density for switching using the fast Γ–X interlayer scattering of electrons, the fast spin-relaxation of holes, and a highly sensitive detection system. We could also reduce the accumulation effect of holes by decreasing the light-energy density to a value as low as possible.

3 *A Demonstration of Ultrafast Switching*

In what follows, we demonstrate a picosecond repetitive spin polarization switching operation in type II AlGaAs/AlAs multiple quantum well structures at room temperature. This shows a fast switching time of 1 ps, a 80 GHz repetition rate, and low switching energy density of 3 nJ/cm^2. We also refer to the possibility of a continuous switching operation.

The sample used in this demonstration consists of 100 periods of 9.2 nm $Al_{0.34}Ga_{0.66}As$ and 2.7 nm AlAs layers, which form type II ternary alloy multiple-quantum-well structures. Optical characterizations of the sample, including the ultrafast response, have been reported previously [3,5,7,9–11].

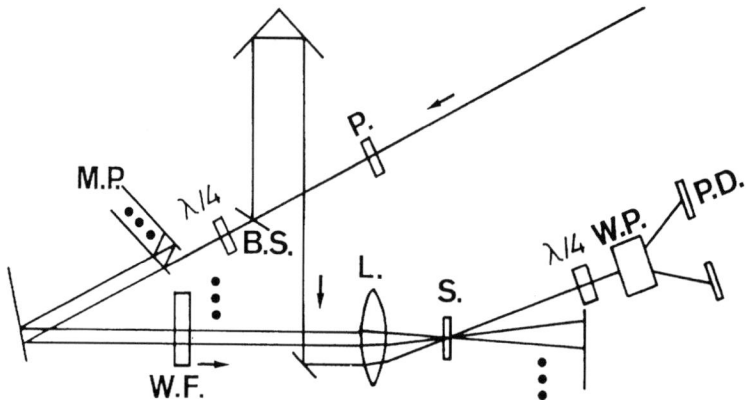

Figure 2.3.3
Experimental setup for the demonstration of ultrafast switching: B.S., beam splitter; P., polarizer; λ/4, quarter-wave plate; W.P., Wollaston prism; P.D., photodiode; M.P., mirror pair; L., lens; S., sample; W.F., wedge filter

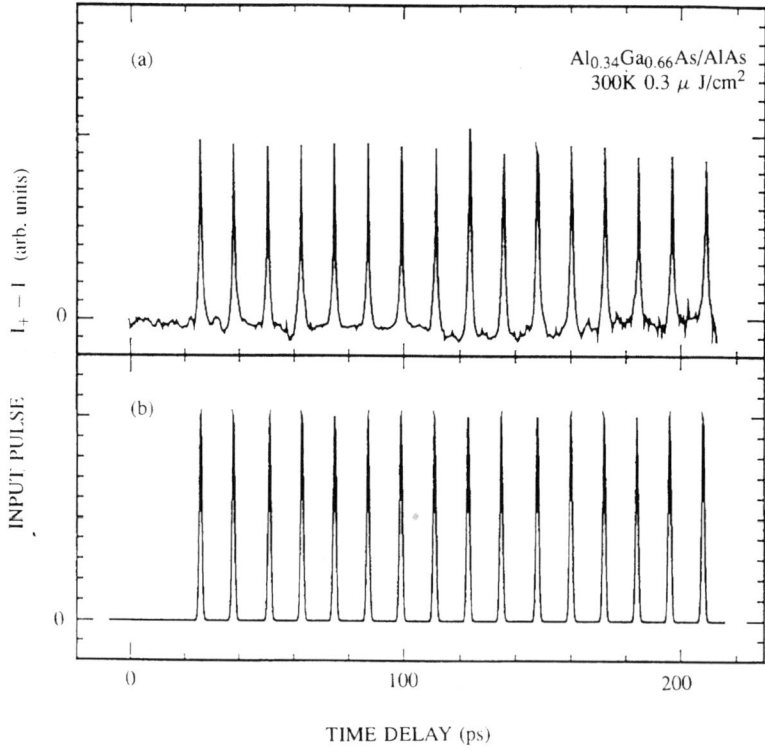

Figure 2.3.4
Experimental demonstration of ultrafast optical switching. (a) The $I_+ - I_-$ signal corresponding to the degree of polarization; (b) The profile of the pump pulse train

The experimental setup shown in Figure 2.3.3 was used for the observation of the ultrafast and highly repetitive optical switching. The laser system used in the experiment was a synchronously pumped cavity-dumped dye laser pumped by a mode-locked Nd^{3+}:YAG laser. The excitation photon energy is tuned to the exciton absorption peak. The temporal pulse width is about 1 ps, and the wavelength is 642 nm. A quarter-wave plate transforms the linearly polarized light into a circularly polarized one. The pump beam passes through an etalon composed of two dielectric mirrors whose reflectivities are 100% and 95%, respectively. The etalon generates a picosecond pulse train. A linear variable attenuator inserted in the pulse train makes the intensity of each pulse constant. The intensity of one pulse and the interval of the pulse train are 0.3 $\mu J/cm^2$ and 11 ps, respectively. The probe beam is a linearly polarized light that is the sum of equal amounts of the right circularly polarized light and the left circularly polarized light. Pump beams and a probe beam are focused onto the sample by using a lens. The probe beam transmitted through the sample passes through another quarter-wave plate. The right circularly polarized light component is turned toward the vertically linearly polarized light, while the left circularly polarized light is turned toward the horizontally polarized light. Two linearly polarized lights are separated by a Wollaston prism and are detected by two balanced photodiodes that have the same sensitivity. We obtained the signal $I_+ - I_-$ by using a simple differential electronic circuit; the signal corresponds to the spin polarization.

Figure 2.3.4 shows the experimental results. The time evolution of the $I_+ - I_-$ signal is shown in Figure 2.3.4(a). Figure 2.3.4(b) shows the intensity profile of the excitation pulse train measured by the second harmonic generation correlation method. The signal shown in Figure 2.3.4(a) copies the excitation pulse train shown in Figure 2.3.4(b). This rapid recovery of the signal is based on the fast Γ–X interlayer scattering of electrons

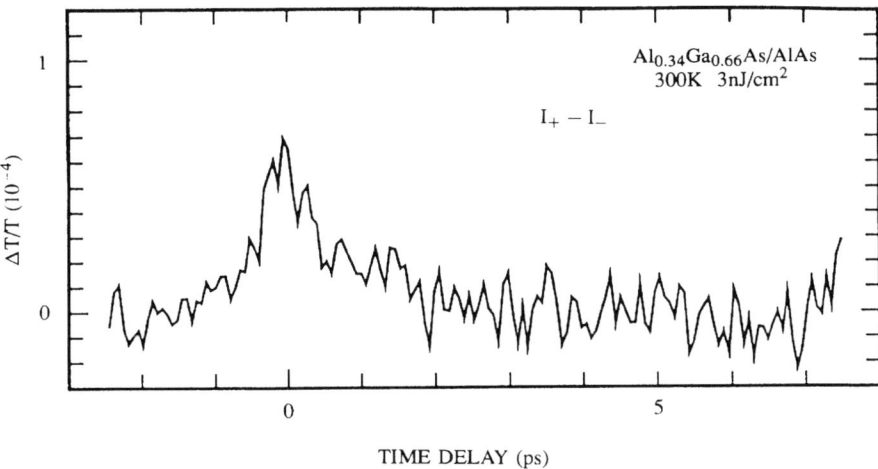

Figure 2.3.5
The differential transmission signal, $I_+ - I_-$, observed under the pump of the right circularly polarized light. The pump energy density is 3 nJ/cm^2

and the fast spin relaxation of the holes at room temperature. The spin relaxation time is shorter than the laser pulse width and is less than 1 ps. Tackeuchi *et al.* [12] measured the spin relaxation time of 30 ps in the type I GaAs/AlGaAs quantum well structures at 300 K by means of pump-and-probe absorption measurement. The spin relaxation time in type II AlGaAs/AlAs quantum well structures is faster than that in type I by more than 30 times. In type I quantum wells, both the electrons and holes memorize the polarization, and the polarization of the electrons is randomized more slowly than that of the holes. On the other hand, in type II quantum wells, the situation is slightly different. Immediately, the electrons and holes memorize the polarization. However, the polarized electrons quickly escape from the well layers because of the fast Γ-X interlayer scattering, and the polarization of holes decays within 1 ps, which is probably due to the frequent hole-phonon scattering process. We used this instant phenomenon for the ultrafast optical switch.

The sample has a large nonlinearity. In addition, the zero-method measurement system reduces the background noise significantly. As a result, we could observe the detectable change in the signal $I_+ - I_-$, the signal-to-noise ratio of which is 4, even at the low light excitation density of 3 nJ/cm^2. The experimental result is shown in Figure 2.3.5.

Next, we demonstrate the ternary-pulse-code switching operation. By placing a quarter-wave plate between two dielectric mirrors composing an etalon, we could obtain

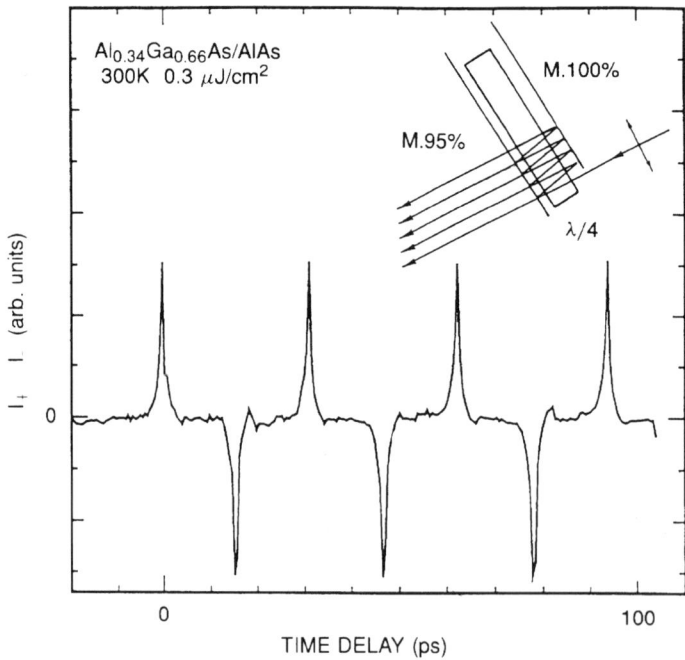

Figure 2.3.6
The $I_+ - I_-$ signal of the sample. Circularly polarized pump pulses change their polarization signs alternately. The inset shows the setup of a pair of mirrors to demonstrate the two-value optical switching

the pump pulses with alternating circular polarization signs. Namely, the first pulse is right circularly polarized light, the second is left, the third is right, and so on. Figure 2.3.6 shows the corresponding experimental result. The left and right circularly polarized lights change the $I_+ - I_-$ signals to negative and positive, respectively. Thus, we could realize the ternary-pulse-code switching operation.

4 Two Problems for Continuously Repetitive Ultrafast Optical Switching

Highly repetitive pump pulses cause the accumulation of holes in wells and therefore may reduce the switching signal intensity. We have carefully checked this effect. The lifetime of holes, τ, was measured by the pump-and-probe method and was found to be 1.5 ns at room temperature. The lifetime is of the order of microseconds at 2 K [3]. The reduced lifetime at room temperature comes from the phonon-assisted nonradiative process. Next, we measured the reduction of the signal intensity as a function of the accumulated hole density, as is shown in Figure 2.3.7. The signal intensity is reduced almost linearly in proportion to the accumulated hole sheet density. The reduction of the

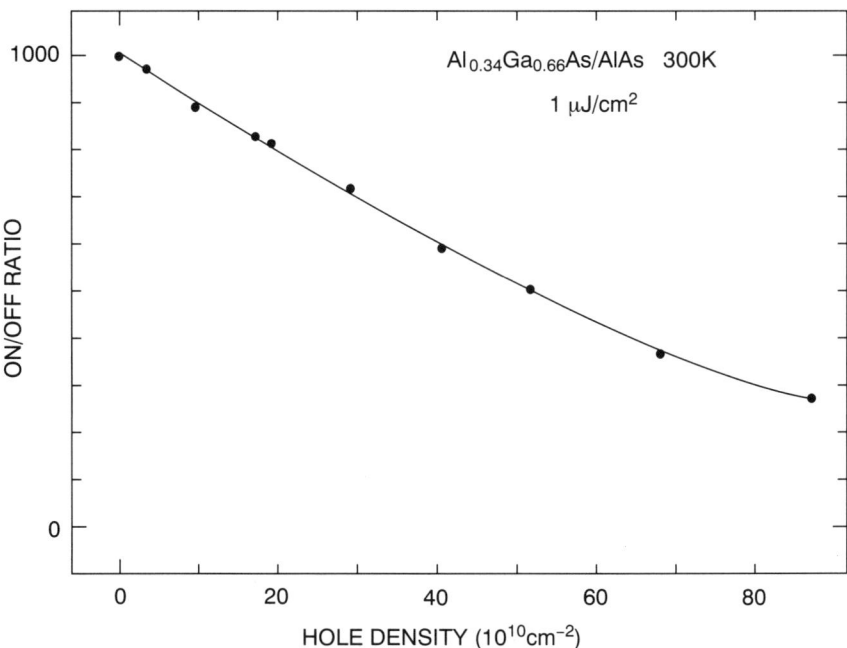

Figure 2.3.7
The on/off signal ratio as a function of accumulated hole density in the $Al_{0.34}Ga_{0.66}As$ well. The experiment was done in the configuration illustrated in Figure 2.3.3 by using three pulses, a hole creation pulse, a pump (switching) pulse, and a probe pulse. The accumulated hole density was estimated from the base line of the pump-probe signal at the negative time delay. The estimation method was established by our previous work [9–11]. The pump pulse energy density was 1 $\mu J/cm^2$

on/off signal ratio by 50% corresponds to the sheet hole density of 5×10^{11} cm^{-2}. Under the excitation by the highly repetitive pump pulses, hole accumulation is balanced with the decay. The balanced hole sheet density of n_s is given by $n_s = gr\tau$, where g is the hole density created by a pulse and r is the pulse repetition rate. The excitation density of 3 nJ/cm^2 corresponds to the sheet hole density of 7.8×10^7 cm^{-2}. If we imagine the 1 THz continuous optical switching operation at the excitation density of 3 nJ/cm^2, n_s is equal to 1.2×10^{11} cm^{-2}. The value of n_s corresponds to the on/off ratio reduction of 13 %. Therefore, we can expect a continuous optical switching operation with a good signal-to-noise ratio. We need not worry about the effect of the accumulation of holes.

Another problem associated with the continuous optical switching operation is the possible rise in temperature of the device. In this experiment, we did not observe any difference in the switching signal intensities for two laser repetition rates, 812 kHz and 4.06 MHz. Nevertheless, for practical applications, temperature rise may be a problem. Assuming that the repetition rate is 1 THz, the quantum-well layers 1 μm thick, the excitation spot area 100 μm^2, and excitation density 3 nJ/cm^2, the estimated calorific value is 3 mW per device. This value is not large compared with several mW per gate, which is currently performed in a TTL integrated circuit. The temperature rise can be reduced by a heat sink.

5 Conclusion

We demonstrated the ultrafast polarization optical switching operation in type II Al$_{0.34}$Ga$_{0.66}$As/AlAs quantum wells. This switching operation is based on the fast Γ–X interlayer scattering of electrons and the fast spin relaxation of holes at room temperature. The switching time is less than 1 ps, the demonstrated repetition rate is 80 GHz, and a good signal-to-noise ratio is observed even at the low excitation density of 3 nJ/cm^2. The possibility of a continuous switching operation is shown. Three requirements for the ultrafast optical switch are simultaneously satisfied.

References

[1] D.S. Chemla, D.A.B. Miller, P.W. Smith, A.C. Gossard and W. Wiegmann, *IEEE J. Quantum Electron.* **QE-20**, 265–275 (1984).
[2] A. Mysyrowicz, D. Hulin, A. Antonetti, A. Migus, W.T. Masselink and H. Morkoç, *Phys. Rev. Lett.* **56**, 2748–2751 (1986).
[3] T. Mishina, F. Sasaki and Y. Masumoto, *J. Phys. Soc. Jpn* **59**, 2635–2638 (1990).
[4] J. Feldmann, E. Göbel and K. Ploog, *Appl. Phys. Lett.* **57** 1520–1522 (1990).
[5] Y. Masumoto and T. Tsuchiya, *J. Phys. Soc. Jpn* **57**, 4403–4408 (1988).
[6] J. Feldmann, R. Sattmann, E.O. Göbel, J. Kuhl, J. Hebling, K. Ploog, R. Muralidharan, P. Dawson and C.T. Foxon, *Phys. Rev. Lett.* **62**, 1892–1895 (1989).
[7] Y. Masumoto, T. Mishina, F. Sasaki and M. Adachi, *Phys. Rev. B* **40**, 8581–8584 (1989).

[8] J. Feldmann, J. Nunnenkamp, G. Peter, E. Göbel, J. Kuhl, K. Ploog, P. Dawson and C.T. Foxon, *Phys. Rev. B* **42**, 5809-5821 (1990).
[9] T. Kawazoe, Y. Masumoto and T. Mishina, *Phys. Rev. B* **47**, 10452-10455 (1993).
[10] T. Mishina and Y. Masumoto, *Phys. Rev. B* **44**, 5664-5667 (1991).
[11] Y. Masumoto, T. Mishina and F. Sasaki, *Optical Properties of Solids*, K.C. Lee, P.M. Hui and T. Kushida, eds., *World Scientific, 1991*, 16-56.
[12] A. Tackeuchi, S. Muto, T. Inata and T. Fujii, *Appl. Phys. Lett.* **56**, 2213-2215 (1990).

2.4

Generation of Ultra-wide Band Optical Frequency Grids

Motoichi Ohtsu

Abstract

The generation of ultra-wide band optical frequency grids is important in both continuous-wave highly coherent tunable light sources and ultrashort light pulse generation. We show the feasibility of diode-laser-based generation of optical frequency grids in 1 PHz frequency span. By using AlGaAs, InPGaAs lasers and their sum and difference frequency generations in KTP crystals, frequency grids from 170 THz (1.7 μm) to 600 THz (0.5 μm) have been obtained. A frequency stability of 10^{-9}-10^{-10} for the generated frequency grids has been realized by locking the frequency to saturated the potassium line and molecular iodine absorption lines. Highly efficient optical frequency comb generators in the 0.8 μm region with sideband spans wider than terahertz have also been developed for generating fine frequency grids

1 Introduction

Continuous-wave highly coherent frequency-tunable light sources have become more important in ultra-high-speed and ultra-parallel optoelectronics and other different fundamental research fields, such as ultra-high resolution spectroscopy, quantum optics, chemistry, and medicine. Semiconductor lasers were paid attention for their intrinsic tunable, compact, and efficient characteristics. Frequency tuning and frequency/intensity modulation of diode lasers can be easily carried out, and extremely low amplitude noise compared with most other laser sources is an inherent advantage in the above-mentioned applications. Furthermore, in recent years remarkable results in the improvement of coherence, tunability, and diode-laser-based optical phase locking have been achieved, and these developments make it possible to develop a diode-laser-based wideband coherent optical frequency sweep generator (OFSG) which could offer highly coherent light (spectrally and spatially) with tunability capable of covering the 1 PHz frequency span, and its realization is in progress [1,2].

On the other hand, the generation of ultrashort light pulses is also very important in ultra-high-speed and ultra-parallel optoelectronics for investigating ultrafast processes in physics, chemistry, etc. The principle of Fourier synthesis of light pulses is quite simple while leaving a lot of technical challenges to be realized. By generating ultra-wide frequency grids (see Figure 2.4.1) of precisely equidistant optical frequencies with high precision frequency accuracy and coherently locked phases, one can expect to generate light pulses with a duration inversely proportional to the entire span of the participating frequency components. It also provides more options in the synthesized pulse shape because we can control the frequency components independently. In fact, Fourier synthesis of light pulses has been investigated since 1977 using independent CO_2 lasers. In 1990 Hänsch proposed a subfemtosecond pulse synthesizer using separate phase-locked lasers [3]. The scheme is generalized to contain $(n-1)$ mutually phase-locked fundamental lasers with frequencies of $nf, (n+1)f, \ldots, (2n-1)f$, and a set of equidistant frequency grids, $f, 2f, \ldots (4n-2)f$, can be generated using only second-order nonlinear conversion, i.e. second-harmonic generation, sum-frequency, and difference-frequency generations. By these frequency grids, light pulses with a repetition rate of f and a pulse duration of $1/(4n-2)f$ are expected. Similar to the proposed YAG and color-center laser-based six-component synthesizer system [3], i.e. $n=2$, we can also use only diode lasers to realize it so as to make full use of the advantages of diode lasers. In such a scheme, two fundamental lasers are chosen to be at 1.32 μm ($2f$) and 0.88 μm ($3f$). Then components at 2.64 μm (f), 0.66 μm ($4f$), 0.53 μm ($5f$) and 0.44 μm ($6f$) have to be generated using nonlinear frequency conversions.

Although our work on the OFSG does not directly aim at pulse synthesis, experimental purpose will differ from what is needed for the above scheme. However, the experimental results in the generation of frequency reference grids in a wide frequency span show significant technical improvement and maturity. Furthermore, the phase-locking-based frequency grids in the 1 PHz span provide the possibility of synthesizing subfemtosecond light pulses using only diode lasers.

2 Principle of a Wideband Optical Frequency Sweep Generator

2.1 Construction of an optical frequency sweep generator

Figure 2.4.1 is a block diagram showing a systematic configuration of the OFSG which is capable of carrying out simultaneously absolute frequency stabilization, continuous frequency tuning, and precision measurements of the frequency difference.

In contrast to the tuning mechanisms of the conventional tunable lasers, this system contains multiple highly coherent diode lasers with their own tunability. Furthermore, absolute frequency stabilization is introduced into the system. Coherent light generation

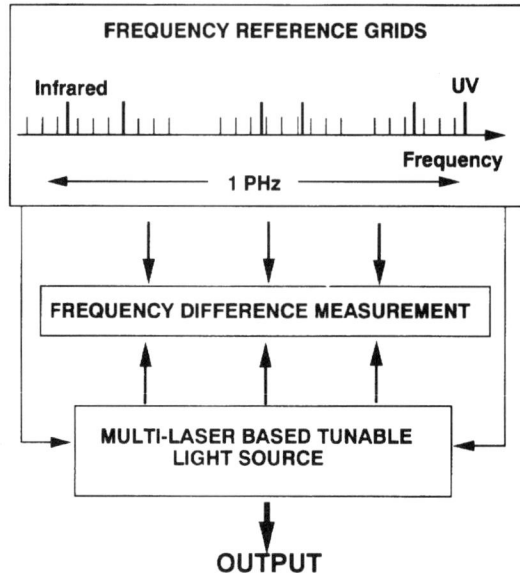

Figure 2.4.1
Schematic explanation of the construction of the OFSG. The frequency reference grids consist of coarse and fine frequency reference grids generated by diode lasers and their frequency conversions and by optical frequency comb generators, respectively. The frequency stabilizations are performed by using atomic/molecular absorption resonances. The multi-laser-based tunable light source contains various diode lasers and their frequency converters. The frequency difference measurement consists of optical heterodyne schemes

to provide frequency reference grids in a wide frequency span can be realized by using diode lasers and their frequency conversions with the help of atomic or molecular transitions. Potassium titanyl phosphate ($KTiOPO_4$; KTP) is found to have a wide phase matchable range the upper limit of which is ~680 THz (0.45 μm) by sum-frequency generation, while difference-frequency generation can provide infrared light from 300 THz (1 μm) to a frequency as low as ~100 THz (3 μm). Furthermore, generation of the light in the above-mentioned entire frequency span can be possible by using only type II angle tuning at room temperature. Because the presently available diode lasers do not satisfy such a continuous lasing spectrum, a multi-reference scheme is necessary for the OFSG. To measure precisely the frequency difference between the tunable output and the frequency reference, the tunable output is heterodyne phase-locked to one of the frequency reference grids in a corresponding region. The frequency sweeping accuracy can be sufficiently high with the help of optical phase locking.

2.2 Frequency noise characteristics in nonlinear frequency conversion

The frequency noise of the generated light in frequency conversion is an important characteristics in the system. Because typical frequency noise spectra of diode lasers have been shown to have an almost flat profile from megahertz to gigahertz, the white noise model is considered as a good approximation. It was found that the fluctuations of phase mismatch, which can be caused by temperature fluctuations and vibrations as well as the frequency fluctuations of the fundamental lasers, make no contribution to the phase fluctuations of the generated wave in a first-order approximation. In the case of sum- or difference-frequency generation, the linewidth of the generated wave is the summation of these fundamental waves, i.e. $\Delta \nu_3 = \Delta \nu_1 + \Delta \nu_2$, while in the case of second-harmonic generation using one laser, i.e. the two fundamental waves come from the same laser, the linewidth of the second-harmonic wave takes the value of $\Delta \nu_{SHG} = 4 \Delta \nu_{laser}$. Therefore the linewidth of the generated field in nonlinear frequency conversion is determined by that of the fundamental lasers.

3 Generation of Frequency-tunable Light in a Wide Frequency Span

Because the available wavelengths of the present diode lasers exist incontinuously in the region from 0.6 to 1.5 μm with some gaps, we should use frequency conversions to extend coverage of the frequency to the region where direct diode laser spectra are not available. Second-harmonic generation [4], sum-frequency generation, and difference-frequency generation or parametric amplification [5,6], have been experimentally performed by using AlGaAs (0.78/0.8 μm) and InGaAsP (1.5 μm) DFB lasers. Since the continuous tuning range of a DFB laser can overlap with that of a conventional Fabry–Perot (FP) type diode laser, even though the FP type laser experiences mode hops, use of the DFB laser in frequency conversions offers a continuous frequency tuning range wider than 1 THz and ensures the continuous frequency coverage of the entire frequency-tunable span determined by the two lasers.

Figure 2.4.2 shows the system configuration for both sum- and difference-frequency generations. Multi-electrodes corrugation-pitch-modulated MQW-DFB lasers [7] at 1.5 μm with a maximum output power of 50 mW and a MHz-linewidth, single-mode AlGaAs lasers at 0.8 μm (SDL-5311) and at 0.78 μm (HL7852) were employed as the fundamental laser sources. The linewidths of AlGaAs lasers in experiment were narrowed to less than 100 kHz by using optical feedback from an external confocal FP cavity [8].

In sum-frequency generation, a maximum green light power of 0.68 μW was obtained and the linear dependence of the output power at sum-frequency on either of the fundamental powers was confirmed. The measured frequency tuning bandwidths tuned by the 1.5 μm laser were $\sim$100 GHz in cases when another fundamental laser was at 0.78 and 0.8 μm. The calculated values are 85 and 90 GHz for a 1-cm KTP, respectively. The calculation also gives a frequency tuning bandwidth of 350 GHz, for a 1-cm KTP when the 0.78/0.8 μm laser is tuned. By using a combination of these diode lasers, highly coherent frequency-tunable green light has been obtained from 0.51 to 0.56 μm (50 THz) with a continuous tuning range of $\sim$1 THz.

It should be noted that in using of diode lasers, the elliptical Gaussian beams will lead to some deviation of the obtained results from the calculation for circular beams. Especially in the presence of walk-off, like the case of KTP angle phase matching, the

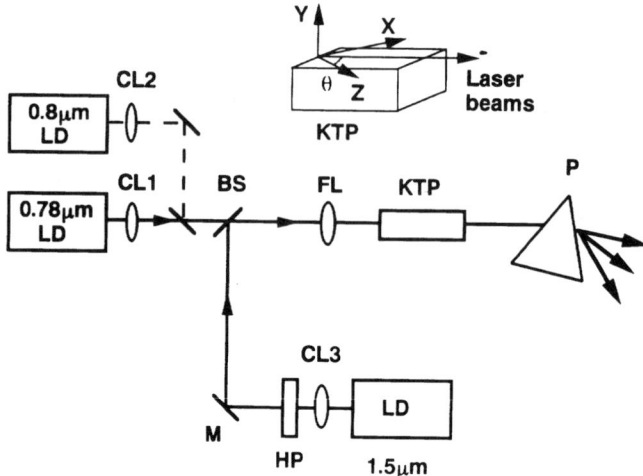

Figure 2.4.2
Experimental configuration for both sum- and difference-frequency generations. CL1-3, collimating lenses; M, reflecting mirror; BS, beam splitter; FL, focusing lens; P, prism; HP, half-wave plate for satisfying the type II phase matching condition. A 10 mm long KTP was put in the θ plane ($\phi = 0°$)

arrangement of the orientation of the elliptical beam shape can result in a 2–3-fold power difference in output [5].

In difference-frequency generation, 0.3 μW output power was obtained when two fundamental lasers were at 1.5 and 0.78 μm. The difference-frequency was around 1.6 μm and can be observed with an optical spectrum analyzer (Anritsu MS9702B). The tunable range was larger than 5 THz (1.58–1.62 μm) which corresponded to the wavelength range of the AlGaAs laser from 0.78 to 0.79 μm by controlling the operation temperatures and currents of the lasers. To confirm the feasibility of difference-frequency generation in a wide frequency span using diode lasers, we performed difference-frequency generation with the help of a Ti:sapphire laser which replaced the AlGaAs laser in the above experiment [6]. Both experiment and calculation show that the frequency tuning bandwidth was 200 GHz for a 1-cm KTP by tuning the 1.5 μm DFB laser. The obtained tunable range was from 1.38 to 1.67 μm (38 THz) which corresponded to the pump wavelength from 0.73 to 0.80 μm. The maximum available tuning range can be as wide as 600 nm (65 THz) around 1.54 μm, which is limited only by the present crystal size when the Ti:sapphire laser was used.

4 Atomic/Molecular Resonance Stabilized Frequency Reference Grids

To provide the reference frequency grids in a wide frequency span in the OFSG, absolute frequency stabilization of these frequencies has also to be performed, in which atomic/molecular absorption resonances are used as frequency references. Two schemes are used to generate the coarse frequency grids, while the fine frequency grids are generated by the frequency comb generators.

4.1 Frequency reference and linking using optical double resonance in potassium

Potassium (^{41}K) was chosen for this scheme by which the coarse frequency reference grids can be obtained at 0.77, 1.54, and 0.51 μm simultaneously [9]. The pump-probe spectroscopy scheme is shown by Figure 2.4.3. The second-harmonic wave of a 1.54 μm DFB laser was used as the probe. The fundamental power was 40 mW and the detected power after the spectroscopic system was measured to be 20 nW. A saturation-spectroscopy scheme was arranged to lock the laser frequency to the saturated absorption resonance. The atomic cell containing ^{41}K (2 cm long) was installed inside an oven to maintain the temperature close to 60 °C.

Frequency stabilization of the pump laser was carried out by locking the laser frequency to the saturated cross-over absorption peak of the ^{41}K–D$_1$ line by using the phase-sensitive technique. The residual frequency fluctuations were estimated to be less than 50 kHz from the error signal, implying a frequency stabilization as high as 10^{-10}.

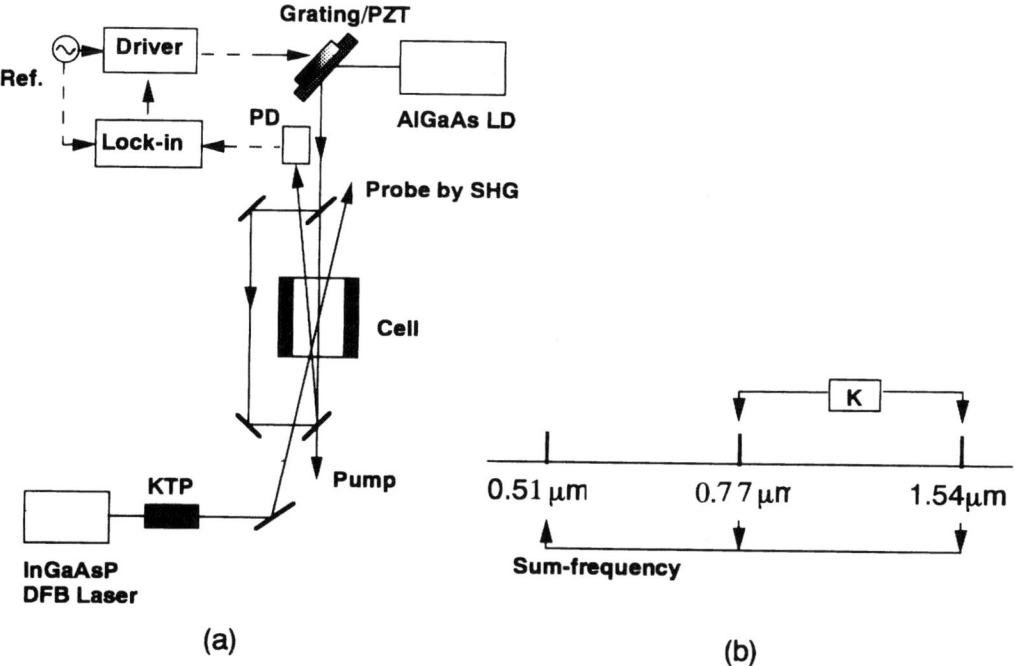

Figure 2.4.3
(a) Experimental configuration for the pump-probe spectroscopy. PD, photodetector; K, atomic potassium. The pump laser was a grating-feedback type diode laser with an output power of 1.4 mW the frequency of which was stabilized using the saturation-spectroscopy scheme. The laser facet was coated with an anti-reflection film to extend the tunability using a grating (1200 lines/mm). All parts were fixed on an invar plate to improve thermal stability. (b) Explanation of frequency grids generation in the corresponding regions

For the pump-probe spectroscopy, the second-harmonic wave was arranged as a probe co- and counter-propagating to the pump beam, as shown in Figure 2.4.3. A high contrast (> 60%) of the nonlinear absorption with a resonance width of ~10 MHz was obtained in the co-propagating scheme, which was preferable for frequency stabilization.

4.2 Frequency references using molecular iodine absorption resonances

In the previous scheme, because the pair of fundamental lasers must be linked to each other through the second-harmonic generation, the frequency range is limited. To extend the tunable frequency span further, different kinds of AlGaAs laser, *e.g.* in the 0.8 μm region, are mixed with the InGaAsP laser. A second scheme, shown by Figure 2.4.4, to provide frequency references is considered, which does not limit the other laser frequency but needs a frequency reference in the generated frequency region. Molecular iodine (I_2) was chosen for the frequency reference for our experiment because of its plethora of absorption lines covering a very large frequency range in the green region [10,11].

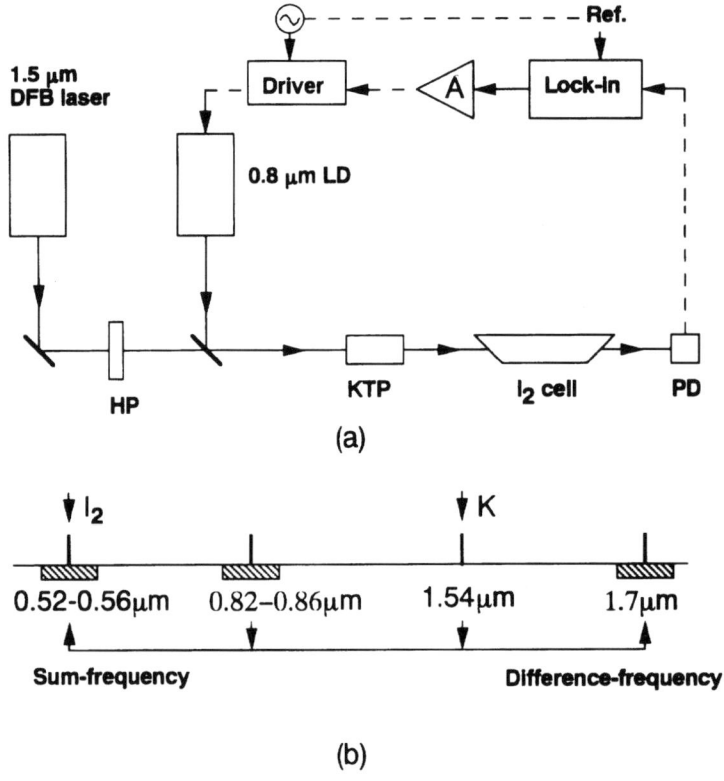

Figure 2.4.4
(a) Experimental scheme for stabilizing sum-frequency by controlling only one fundamental laser. PD, photodetector. The feedback is applied to the 0.8 μm LD using the error signal from the lock-in amplifier. (b) Explanation of frequency grids generation in the corresponding regions

For frequency stabilization, the generated green light passed through a 15 cm long I_2 cell which contained natural iodine. A resonance with an absorption of 55% and a width of 1.2 GHz was used as the reference at 536.32 nm (560 THz). This width was much wider than the Doppler width ($\sim$400 MHz) because of the hyperfine splitting inside the observed resonance.

A phase-sensitive technique was employed to stabilize the 0.82 μm laser while the 1.54 μm laser was free-running. The residual frequency fluctuations were estimated to be within ±0.5 MHz from the error signal, implying a normalized frequency stability better than 10^{-9} at the optical frequency of 560 THz.

Using a combination of the above two schemes, the generation of frequency grids can be realized at 0.51, 0.54, 0.77, 0.83, 1.54, and 1.7 μm through sum- and difference-frequency generations.

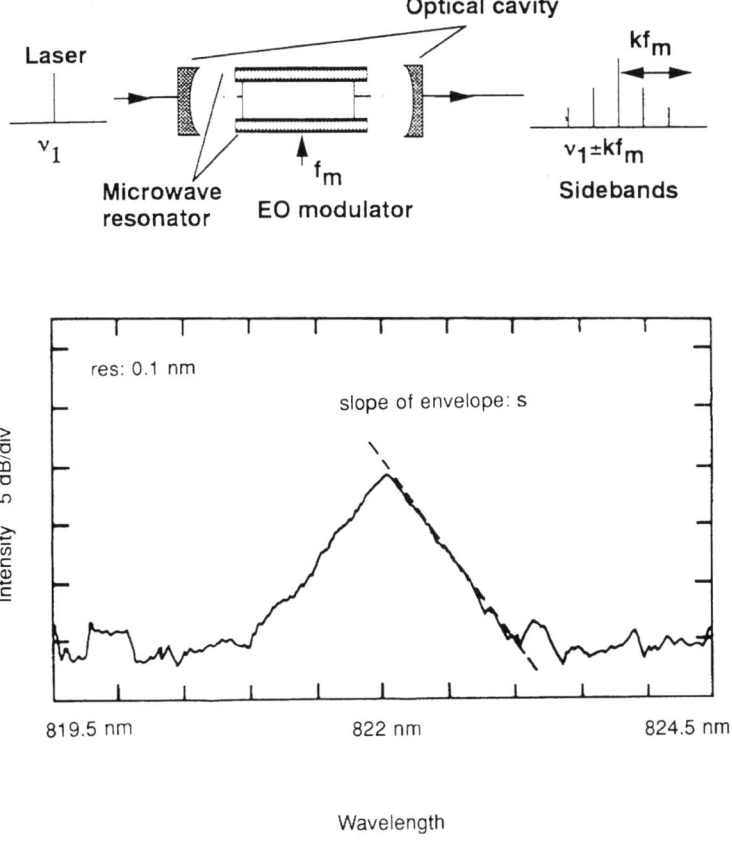

Figure 2.4.5
(a) Optical frequency comb generator. A lithium niobate (1 mm × 1.25 mm × 20 mm) with anti-reflection coating ($\sim$0.1% around 0.8 μm) was installed inside a microwave guide (1 mm × 15 mm × 25 mm), and the driven microwave power was $\sim$8 W. Then they were put in a symmetrical optical cavity (reflectivity of mirrors is $\sim$99.3% from 760 to 860 nm). (b) Envelope of sidebands generated from the optical frequency comb generator at 0.82 μm, which was observed by an optical spectrum analyzer. The slope, s, of the envelope in the log-scale was 32 dB/THz

4.3 Optical frequency comb generation at 0.8 μm wavelength

As described above, the fine frequency grids are indispensable for providing frequency references in the OFSG. Here we present the first result of an optical frequency comb generator at 0.8 μm.

The principle of the OFC generation used here is based on the enhanced generation of modulation sidebands of an EO modulator [12]; see Figure 2.4.5(a). For a phase velocity matched microwave guide in which an EO crystal is installed, the power of each sideband can be expressed as $P_k = P_0 \exp(-\pi(\nu - \nu_0)/mFf_m)$, where P_0 is the power coefficient determined by modulation and cavity efficiency, k is the order of the sideband, m is the modulation index, F is the finesse of the FP cavity, f_m is the modulation frequency, ν_0 is the central frequency (carrier), and ν is the frequency of the kth sideband. Therefore the slope of the envelope in a log-scale is $s = \pm \pi/mFf_m$. It should be noted that the modulation index m is proportional to the optical carrier frequency when other conditions are fixed; thus one can expect a higher modulation efficiency at a shorter wavelength region. The measured transmission efficiency and finesse of the cavity were 2% and 175, respectively. The observed slope of the envelope of the modulation-generated sidebands, as shown in Figure 2.4.5(b), was 32 dB/THz at a modulation frequency of 5.9 GHz when the input laser power was 2 mW. On the basis of the incident laser power used and the slope of the envelope obtained, this frequency comb can provide a usable two-sided span wider than 4 THz since a single-mode diode laser with a power of 100 mW at this wavelength region is available.

5 Concluding Remarks

From the results of the nonlinear frequency conversions using diode lasers and the generation of frequency reference grids, we can conclude that the present frequency reference grids with a frequency stability of 10^{-9}–10^{-10}, as well as a frequency-tunable output, can be extended from the present 600 THz (0.5 μm) to 900 THz (~0.3 μm) by adding InGaAlP visible diode lasers and InGaAsP lasers at the 1.3 μm region to the present system. Further improvements are expected after extension of the continuous frequency tuning range by using a grating-extended cavity scheme with a range of 10 THz, a frequency stability up to 10^{-11} by using FM spectroscopy techniques, and power enhancement by using, for example, the coherent addition technique [2].

References

[1] M. Ohtsu, *Highly Coherent Semiconductor Lasers*, Artech House, Inc., Boston/London, 1992, chs. 2–5.
[2] M. Ohtsu, K. Nakagawa, M. Kourogi and W. Wang, *J. Appl. Phys.* **73**, R1–R17 (1993).
[3] T.W. Hänsch, *Opt. Commun.* **80**, 71–75 (1990).
[4] W. Wang, K. Nakagawa, Y. Toda and M. Ohtsu, *Appl. Phys. Lett.* **61**, 1886–1888 (1992).
[5] W. Wang and M. Ohtsu, *Opt. Commun.* **102**, 304–308 (1993).
[6] W. Wang and M. Ohtsu, *Opt. Lett.* **18**, 876–878 (1993).

[7] M. Okai, T. Tsuchiya, K. Uomi, N. Chinone and T. Harada, *IEEE Photo. Technol. Lett.* **2**, 529-530 (1990).
[8] B. Dahmani, L. Hollberg and R. Drulliger, *Opt. Lett.* **12**, 876-878 (1987).
[9] W. Wang, A. M. Akulshin and M. Ohtsu, *IEEE Photon. Technol. Lett.* **6**, 95-97 (1994).
[10] A. Arie and R. L. Byer, *Appl. Opt.* **32**, 7382-7386 (1993).
[11] W. Wang and M. Ohtsu, *Jpn. J. Appl. Phys.* **33**, 1648-1651 (1994).
[12] M. Kourogi, K. Nakagawa and M. Ohtsu, *IEEE J. Quantum Electron.* **QE-29**, 2693-2701 (1993).

2.5

Control of Femtosecond Optical Pulses Using Nonlinear Organic Materials

Ryuji Morita and **Mikio Yamashita**

Abstract

To obtain large $\chi^{(3)}$ materials, we discuss the relationship between $|\chi^{(3)}|$ and $|\chi^{(2)}|$ in the nonresonant region with optical nonlinearities only due to the electronic polarization. Moreover, using organic crystal DAN cored fibers, we have carried out the experiment of nonamplified femtosecond pulse compression and nonlinear femtosecond pulse propagation. 39 fs laser output pulses have been directly compressed to 22 fs, and the delayed nonlinear response time is evaluated to be $\sim$30 fs. Furthermore, we have numerically investigated the effect of the delayed nonlinear response on pulse compression using an organic crystal cored fiber in the normal dispersion region. Using up to third-order phase compensation both in the presence and in the absence of the delayed nonlinear response time, a 100 W, 100 fs hyperbolic-secant pulse is compressed to $\sim$10 fs

1 Introduction

In recent years, some organic materials have attracted considerable attention because of their extremely large optical nonlinearities [1]. Extensive studies on the nonlinearities of those organic materials have been stimulated by a number of potential applications, such as second-harmonic generation, frequency mixing, parametric amplification, optical bistability, pulse compression, etc. Moreover, they are expected to respond in the femtosecond time-region since their nonlinearities are ascribed to π-electrons.

The purpose of our study is to control femtosecond optical pulses using those highly nonlinear and fast responding organic materials. We describe mainly femtosecond pulse compression using an organic crystal cored fiber as one of the examples. Section 2 presents an experiment of nonamplified pulse compression using an organic crystal cored fiber. In section 3 we derive a relationship between second- and third-order nonlinear susceptibilities. Section 4 describes an experiment of the nonlinear pulse propagation and evaluation of the nonlinear delayed response time. Finally, in section 5 we analyze the effect of the delayed nonlinear response on femtosecond pulse compression.

2 Nonamplified Pulse Compression Using an Organic Crystal Cored Fiber

First, we explain the principle of pulse compression. The process of pulse compression consists of two stages. When an optical pulse propagates in the normal-dispersion regime of a third-order nonlinear optical fiber, nearly linear, positive chirp (frequency increasing toward the trailing side) is imposed on the pulse, associated with a broadening of the spectral width, through the combined effect of group velocity dispersion (GVD) and self-phase modulation (SPM). SPM is the change in the phase of an optical pulse due to the intensity-dependent nonlinearity of the refractive index of the medium. This chirp leads to dispersion-induced pulse broadening since different frequency components of the pulse travel at different speeds in the presence of positive GVD. Subsequently, the grating pair following the fiber provide anomalous GVD or negative chirp, resulting in cancellation between the positive and negative chirp. Thus an output pulse can be nearly transform-limited and compressed to the extent of the inverse of the spectral width.

Conventionally, a single-mode fused silica fiber has been used for femtosecond optical pulse compression. However, because of its low nonlinearity, compression using a fused silica fiber needs high amplification of laser output pulses by means of a large and complex system. This leads to some problems concerning the temporal and spatial pulse instability, a reduction in the pulse repetition rate, optical damage to the fused silica fiber, and difficulty in the precise adjustment of a large amplifier system. As one approach to solve those problems, we apply a highly cubic nonlinear organic fiber.

Femtosecond output pulses from a colliding-pulse mode-locked (CPM) dye laser have been directly compressed using a 5 mm long 4-(N,N-dimethylamino)-3-acetamidonitrobenzene (DAN) fiber followed by a grating pair [2–5]. DAN is one of the organic materials that have the largest optical nonlinearities. The output pulse width and average power of the CPM laser have been 39 fs and 28 mW at a central wavelength of 623 nm with a spectral width of 11.0 nm, respectively, as shown in Figure 2.5.1. The total transmission efficiency including the coupling and propagation losses of the DAN fiber has been typically 18%. While monitoring the laser output and compressed pulses by two autocorrelators and the fiber output spectrum by an optical spectrum analyzer, the grating separation has been carefully adjusted to obtain the optimum amount of GVD. Consequently, a compressed pulse width of 22 fs has been generated with a spectral broadening of 17.9 nm, as shown in Figure 2.5.2. This is the shortest pulse, to our knowledge, among the pulses compressed without any amplification of laser output pulses, and the first application of an organic fiber to pulse compression.

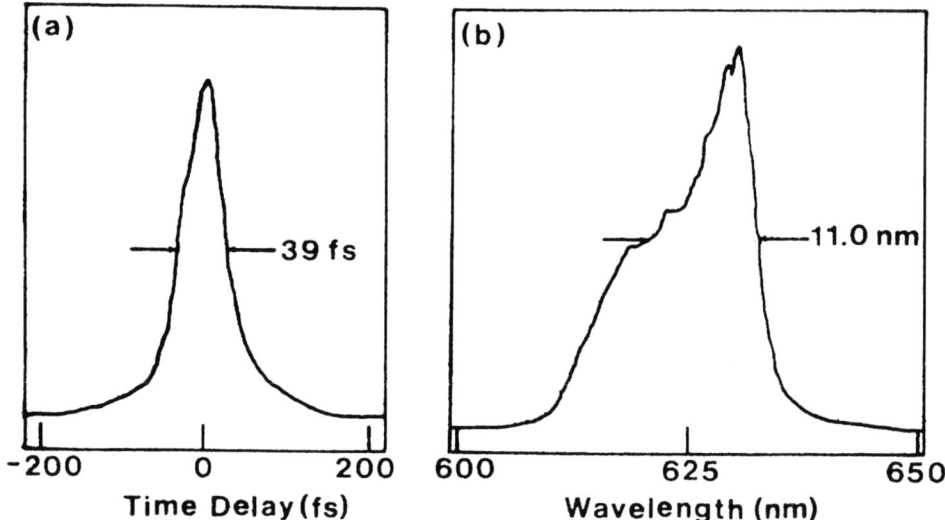

Figure 2.5.1
(a) and (b) are autocorrelation trace and corresponding spectrum of the input pulse to a DAN fiber

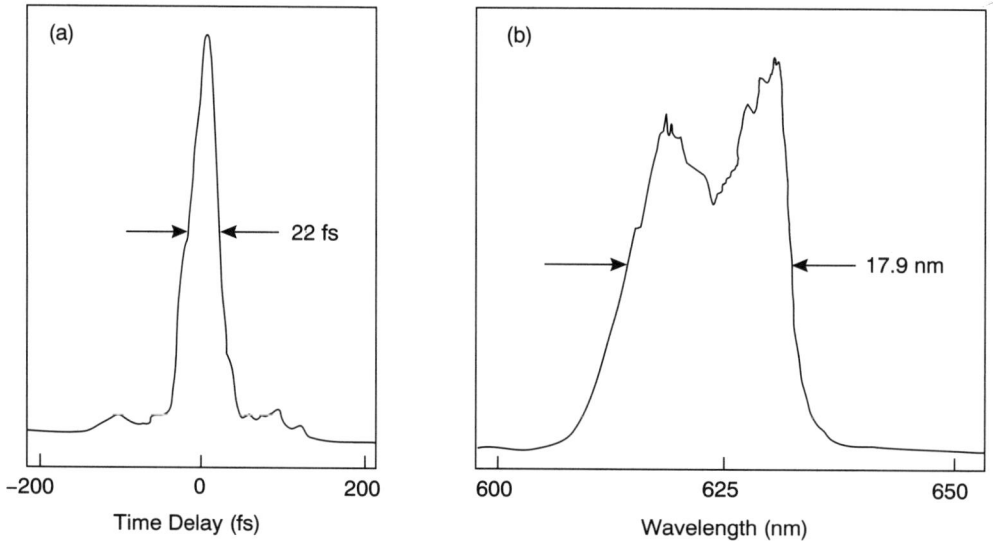

Figure 2.5.2
(a) and (b) are autocorrelation trace and corresponding spectrum of the pulse compressed using a DAN fiber

3 Relationship between Second- and Third-order Nonlinear Susceptibilities

Although the way to design materials having large quadratic nonlinearities $\chi^{(2)}$ has been well established, the dogma of design materials having large cubic nonlinearities $\chi^{(3)}$ has not yet been made clear. For example, it has not been revealed how large the $\chi^{(3)}$ values of large $\chi^{(2)}$ materials are. In this section we discuss a semi-theoretical relation between second- and third-order nonlinear optical susceptibilities in the nonresonant region for noncentrosymmetric inorganic and organic crystals the nonlinearities of which arise from electronic polarization. That is, we consider only the optical nonlinearities responding to femtosecond pulses, which are essential for ultrafast nonlinear optical devices such as femtosecond pulse compressors [2–6]. Thereby, the relatively slow response nonlinearities due to the charge-separated polarization in LiNbO$_3$, the molecular orientational polarization in CS$_2$, the semiconductor electron–hole excitonic polarization in GaAs, etc. are excluded.

From the theory of quantum mechanics, expressions for linear and nonlinear susceptibilities can be derived. Here, we apply the two-level model introducing the effective level parameters $g^{(i)}$ ($i = 1, 2, 3$) instead of the many excited-level effect. Thus, linear and nonlinear optical susceptibilities are expressed by [7]

$$\chi^{(1)} = g^{(1)} F_1 \frac{2N}{\varepsilon_0 \hbar} \frac{|\mu|^2}{\Omega} \tag{1}$$

$$\chi^{(2)} = g^{(2)} F_2 \frac{3N}{\varepsilon_0 \hbar^2} \frac{|\mu|^2 \Delta\mu}{\Omega^2} \tag{2}$$

$$\chi^{(3)} = g^{(3)} F_3 \frac{4N}{\varepsilon_0 \hbar^3} \frac{|\mu|^2 [(\Delta\mu)^2 - |\mu|^2]}{\Omega^3} \tag{3}$$

where the electronic transition frequency $\Omega = \omega_{ng}$ from the ground state g to the excited state n, the electronic transition dipole moment $\mu = \mu_{ng}$, and the dipole moment difference $\Delta\mu = \Delta\mu_n$.

We now derive a quantitative relation between nonlinearities of different orders from equations (1) and (2). That is, $|\chi^{(3)}|$ can be given by

$$|\chi^{(3)}| = \left| \frac{8}{9} \frac{g^{(3)} g^{(1)}}{(g^{(2)})^2} \frac{F_3 F_1}{F_2^2} \frac{[\chi^{(2)}]^2}{\chi^{(1)}} - \frac{g^{(3)}}{(g^{(1)})^2} \frac{F_3}{F_1^2} \frac{\varepsilon_0}{N\hbar\Omega} [\chi^{(1)}]^2 \right|. \tag{4}$$

Estimation of the parameters $g^{(i)}$ ($i = 1, 2, 3$) from the already measured values $\chi^{(i)}$ ($i = 1, 2, 3$) gives

$$g^{(1)} F_1 \sim 10, \quad g^{(2)} F_2 \sim 1, \quad g^{(3)} F_3 \sim 10$$

From equation (4), $|\chi^{(3)}|$ thus becomes

$$|\chi^{(3)}| \sim \left| 10^2 \frac{[\chi^{(2)}]^2}{\chi^{(1)}} - \frac{1}{10} \frac{\varepsilon_0}{N\hbar\Omega} [\chi^{(1)}]^2 \right| \tag{5}$$

Figure 2.5.3 shows the already measured $|\chi^{(3)}|$ versus $|\chi^{(2)}|$ maximum values in the nonresonant region for noncentrosymmetric crystals including organics with optical nonlinearities due only to electronic polarization. The theoretical result from equation (5) is also represented as the solid line in this figure. The experimental and theoretical results are quantitatively in good agreement. This suggests that relation (5) is also valid for organics. Since the N value is typically $\sim 10^{28}$ m^{-3}, $\chi^{(3)}$ is proportional to $[\chi^{(2)}]^2$ for the case that $\chi^{(2)} \gtrsim 10^{-12}$ m/V. This predicts that some organic materials such as 2-methyl-4-nitroaniline (MNA), 3,5-dimethyl-1-(4-nitrophenyl)pyrazole (DMNP), and N-(4-nitrophenyl)-N-methylaminoacetonitrile (NPAN) with larger $\chi^{(2)}$ values than that of 4-(N,N-dimethylamino)-3-acetamidonitrobenzene (DAN) have larger $\chi^{(3)}$ values.

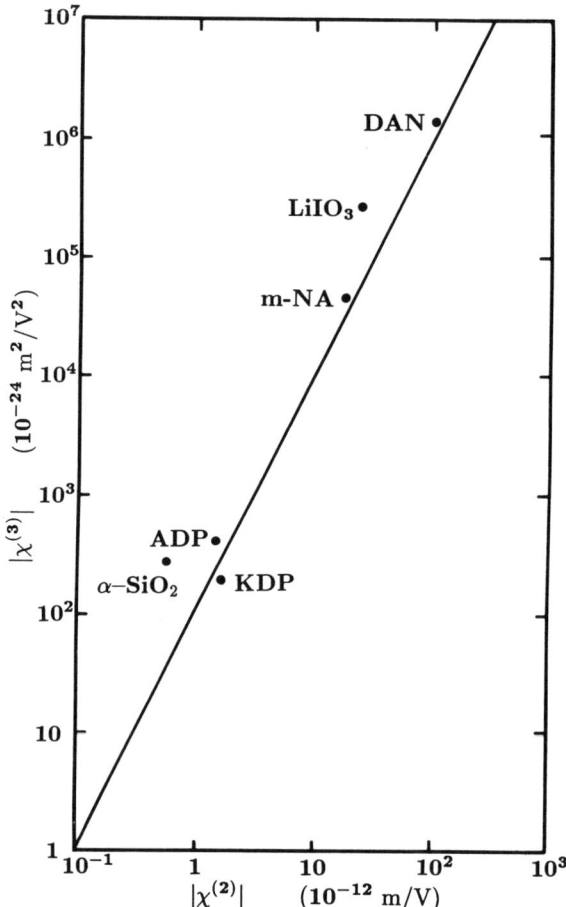

Figure 2.5.3
$|\chi^{(3)}|$-$|\chi^{(2)}|$ relationship in the nonresonant wavelength region. The solid line represents the theoretical result ($\chi^{(1)} \sim 1$ is assumed). Maximum $\chi^{(2)}$ and $\chi^{(3)}$ values measured are plotted (α-SiO$_2$: $\chi^{(2)}_{111}$, $\chi^{(3)}_{1111}$; KDP(KH$_2$PO$_4$): $\chi^{(2)}_{312}$, $\chi^{(3)}_{1111}$; ADP(NH$_4$H$_2$PO$_4$): $\chi^{(2)}_{312}$, $\chi^{(3)}_{1111}$; m-NA(m-nitroaniline): $\chi^{(2)}_{311}$, $\chi^{(3)}_{1111}$; LiIO$_3$: $\chi^{(2)}_{311}$, $\chi^{(3)}_{3223} = \chi^{(3)}_{3232} = \chi^{(3)}_{3322}$; DAN(4-(N,N-dimethylamino)-3-acetamidonitrobenzene)): $\chi^{(2)}_{233}$, $\chi^{(3)}$ is effective value)

4 Evaluation of the Delayed Nonlinear Response Time

In this section we describe the input pulse power dependence of the fiber output spectra in the femtosecond time region in order to understand the ultrafast nonlinear propagation mechanism in the DAN fiber [8].

Figure 2.5.4 shows output spectra from a 4 mm long single-mode DAN fiber. The temporal and spectral widths of slightly down-chirped pulses from the CPM laser, which are propagated in the fiber, are 80 fs (FWHM) and 5.4 nm (FWHM), respectively, as shown in (a). As the input power increases, the output spectrum shifts to the red side, and then its shorter wavelength component grows gradually as in (b)–(d). Finally, the spectrum broadens to 11.3 nm at the input peak power of 75 W, as shown in (e). The red-shift and the spectral broadening are due to the delayed nonlinear response and dispersive SPM, respectively.

To make these behaviors of the spectra clear, we apply a modified nonlinear Schrödinger equation as follows:

$$\frac{\partial u}{\partial \xi} + \Gamma u + \frac{i}{2}\frac{\partial^2 u}{\partial \tau^2} - \delta\frac{\partial^3 u}{\partial \tau^3} = \frac{i}{\tau_R}\left[uJ(\tau;\tau_R) + is\frac{\partial}{\partial \tau}\{uJ(\tau;\tau_R)\}\right] \quad (6a)$$

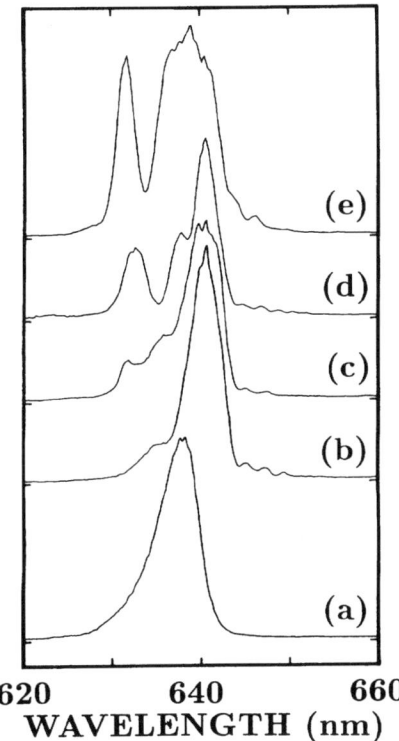

Figure 2.5.4
(a) Input spectrum (pulse width 80 fs, repetition rate 85.5 MHz); (b)–(e) output spectra from a DAN fiber (input peak power 2.3, 13, 41, 75 W, respectively)

$$J(\tau; \tau_R) \equiv \int_0^\infty d\tau' \exp(-\tau'/\tau_R)|u(\tau - \tau')|^2 \tag{6b}$$

where the complex variable u is the normalized amplitude of the pulse envelope, and the parameters are defined by

$$\xi = \frac{|\beta_2|z}{T_0^2}, \quad \tau = \frac{t - z/v_g}{T_0}, \quad \delta = \frac{\beta_3}{6|\beta_2|T_0}$$

$$s = \frac{2}{\omega_0 T_0}, \quad \tau_R = \frac{T_R}{T_0}, \quad \Gamma = \frac{\alpha L_D}{2}$$

$$L_D = \frac{T_0^2}{|\beta_2|} \quad (\beta_2 > 0)$$

z is the distance along the fiber. The time variable τ is a normalized retarded time measured in the frame of reference moving along the fiber at the group velocity v_g. ω_0 is the central optical frequency, T_0 an initial pulse width parameter (FWHM $T_p = 2\ln(1 + \sqrt{2})T_0 \simeq 1.763 T_0$), α the propagation loss, and β_2 and β_3 the second- and third-order dispersion, respectively. β_2 is assumed to be positive. The nonlinear refractive index n_2 is included in the definition of the normalized amplitude u. The response function is assumed to be in exponential form specified by the response time T_R in equation (6b). Moreover, the instantaneous response part of the nonlinear refractive index is assumed to be negligible.

The fiber-input pulses are assumed to have an amplitude shape

$$u(\xi = 0, \tau) = \text{sech}(\tau) \exp \frac{-iC\tau^2}{2} \tag{7}$$

where C is the parameter representing initial linear frequency chirp.

The parameters we used for a DAN crystal cored fiber are the central wavelength $\lambda_0 = 630$ nm, the core diameter 2.5 μm, $n_2 = 2.1 \times 10^{-18}$ m^2/V^2 [6], $\beta_2 = 1.85 \times 10^{-24}$ s^2/m, $\beta_3 = 3.79 \times 10^{-39}$ s^3/m [6,9], $\alpha = 14$ dB/cm ($\delta = 6.0 \times 10^{-3}$, $\Gamma = 2.8 \times 10^{-1}$), and the input pulse width $T_{p,\text{in}} = 80$ fs ($T_0 = 45.4$ fs).

On the basis of the equation, we calculate the output spectra. We vary the T_R values so that a calculated red-shift may agree with that in the experiment. Thus, the nonlinear response time of DAN is evaluated to be ~30 fs. To the best of our knowledge, this is the first experimental evaluation of the response time in the nonresonant region for nonlinear organic crystals.

5 Compensation for the Effect of the Delayed Nonlinear Response on Pulse Compression

Conventionally, most analyses of pulse compression in the normal dispersion regime have been performed without taking account of the delayed nonlinear response, while taking in the soliton-effect compression regime. This is because fused-silica fibers are usually utilized for pulse compression, and their response time, which is evaluated to be 2–4 fs [10], is much shorter than the input pulse widths in many cases. However, when the pulse widths are comparable with the response time, taking account of the nonlinear response time is significant. In section 4, the delayed nonlinear response time of DAN is found to be relatively large. This fact stimulates us to consider the effect of the response time for femtosecond optical pulse compression using a DAN fiber. Hence, to investigate the effect of delayed nonlinear response on pulse compression, we carry out a calculation of up to third-order phase compensation for output pulses from fibers to obtain compressed pulses, after solving equation (6) using the split-step Fourier method [11].

In other words, we determine the compensated phase ϕ_c in the frequency domain

$$\phi_c = \phi(\omega) - \{a_2(\omega - \omega_0)^2 + a_3(\omega - \omega_0)^3\} \tag{8}$$

where $\phi(\omega)$ is the pre-compensated phase of the fiber output pulses, so that compensated pulse widths are minimized by adjusting the parameters a_2 and a_3. Adjustment is usually done using prism and grating pairs in a practical experiment.

Results of the numerical calculation are summarized in Tables 2.5.1 and 2.5.2 [12]. Table 2.5.1 shows the dependence of the compressed pulse width $T_{p,c}$ on the response

Table 2.5.1
Dependence of the compressed pulse width $T_{p,c}$ on the response time T_R and the chirp parameter C

	Compressed pulse width $T_{p,c}$ (fs)			
	Response time T_R			
C	0 fs	10 fs	20 fs	30 fs
−2	12.1	10.5	10.3	10.4
−1	11.5	10.9	10.9	9.54
−0.4	11.3	10.5	9.77	10.4
0	11.1	9.85	10.5	11.2
+0.4	10.9	10.8	11.0	8.99
+1	10.6	11.2	9.13	9.72
+2	9.92	9.85	9.17	8.67

Table 2.5.2
Dependence of the optimum propagation length z_{opt} on the response time T_R and the chirp parameter C

	Optimum propagation length z_{opt} (mm)			
	Response time T_R			
C	0 fs	10 fs	20 fs	30 fs
−2	0.552	0.460	0.500	0.401
−1	0.493	0.487	0.474	0.454
−0.4	0.516	0.493	0.431	0.426
0	0.460	0.572	0.401	0.421
+0.4	0.447	0.379	0.416	0.516
+1	0.454	0.368	0.506	0.500
+2	0.546	0.513	0.473	0.651

time T_R for the DAN crystal cored fiber (input peak power $P_0 = 100$ W, input pulse width $T_{p,in} = 100$ fs, and chirp coefficient $C = 0, \pm 0.4, \pm 1, \pm 2$). What is evident from Table 2.5.1 is that even in the presence of the delayed response, a 100 W, 100 fs hyperbolic-secant pulse is compressed to ~ 10 fs as well as in the absence of the delayed response, in the range of $|C| \leq 2$. However, if T_R exceeds a critical value, the effect of the delayed response degrades compression. That is, further calculation shows the critical T_R value is ~ 70 fs ($\tau_R \simeq 1.2$) under the conditions $T_{p,in} = 100$ fs and $P_0 = 100$ W. The dependence of the propagation length z_{opt} on the response time T_R, under the same conditions as in Table 2.5.1, is shown in Table 2.5.2. From Table 2.5.2 the optimum propagation lengths z_{opt} are found to be around 0.5 mm, where the propagation loss is low enough.

In Figure 2.5.5(a) we show representative output spectra from the fiber at $T_R = 0$ and 30 fs with an input spectrum. The blue-side broadening in the spectrum at $T_R = 0$ fs due to self-steepening is clearly seen, whereas the frequency down-shift (corresponding to the red-shift) occurs due to the delayed response effect and suppresses the blue-side broadening at $T_R = 30$ fs. In addition, the effect of third-order dispersion results in

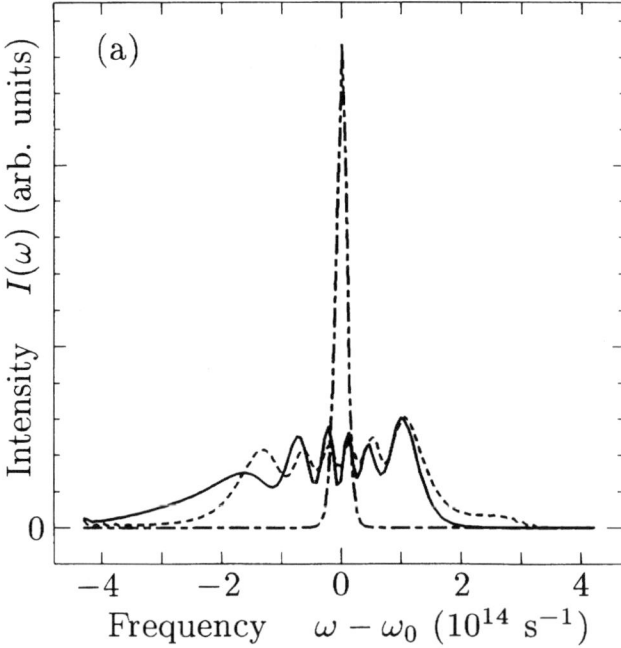

Figure 2.5.5
(a) Typical fiber output spectra for the case of the response time $T_R = 30$ fs (solid line) and $T_R = 0$ fs (dashed line) at the distance $z = 0.421$ mm. The dashed-and-dotted line represents the spectrum for the input pulse (the input power $P_0 = 100$ W, the pulse width $T_{p,in} = 100$ fs, and the chirp parameter $C = 0$). (b) Typical fiber output pulses for the case of $T_R = 30$ fs (solid lines; (A) the pre-compensated pulse, (B) the compensated pulse) and $T_R = 0$ fs (dashed line; the pre-compensated pulse) at $z = 0.421$ mm. The dashed-and-dotted line represents the input pulse. T is the time in the coordinate system moving at the average group velocity of the pulse

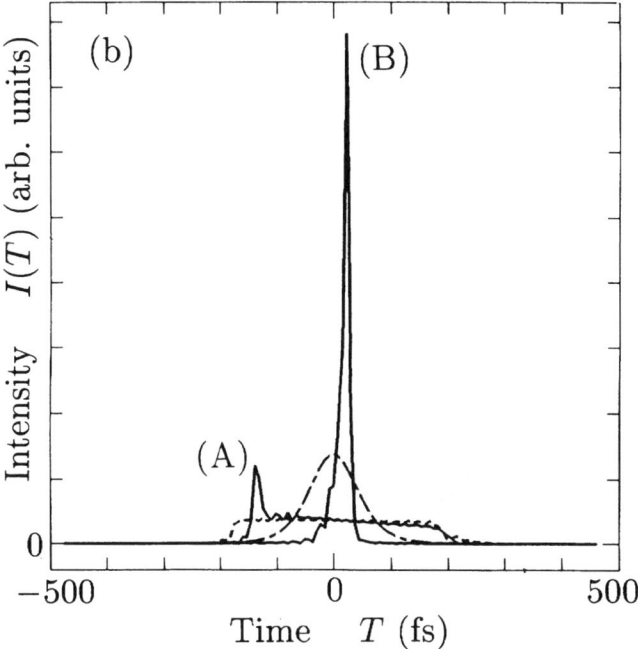

Figure 2.5.5 (*continued*)

spectral asymmetry for both cases. Figure 2.5.5(b) shows one of the typical compressed pulse profiles ($T_R = 30$ fs) with the fiber input and output pulses ($T_R = 0$ and 30 fs) as a function of time. It is found that the pre-compensated pulse at $T_R = 30$ fs has a small peak in the leading edge due to the delayed response. This is understood physically by considering the following fact. The red components travel faster than the blue components in the normal-dispersion regime. As a result, the red-shifted broad peak in the frequency domain due to the delayed response corresponds to the forming of a small peak at the leading edge in the time-domain.

The compressed pulse shape is slightly asymmetric and the compressed pulse quality $\bar{Q}_c$ (the ratio of compressed to pre-compensated pulse energy) is about 0.73 at $T_R = 30$ fs.

To understand the mechanisms of phase compensation, we consider the phase profile of the fiber output in the frequency domain. Figure 2.5.6 illustrates the pre-compensated phase $\phi(\omega)$ of the fiber output pulse in the frequency domain at $C = 0$ in the absence or presence of the delayed response ($T_R = 0$, 10, 20, and 30 fs). From Figure 2.5.6 it is clear that the shape of $\phi(\omega)$ becomes more asymmetric with increasing T_R value. For efficient pulse compression in the presence of the delayed response, therefore, third-order phase compensation is needed. By adjusting sufficiently the third-order phase compensation through equation (8), pulses are efficiently compressed even in the presence of the delayed response, as shown in Table 2.5.1. This proves that third-order phase compensation is essential for efficient pulse compression in the presence of the delayed response. It has been believed that the effect of the delayed response

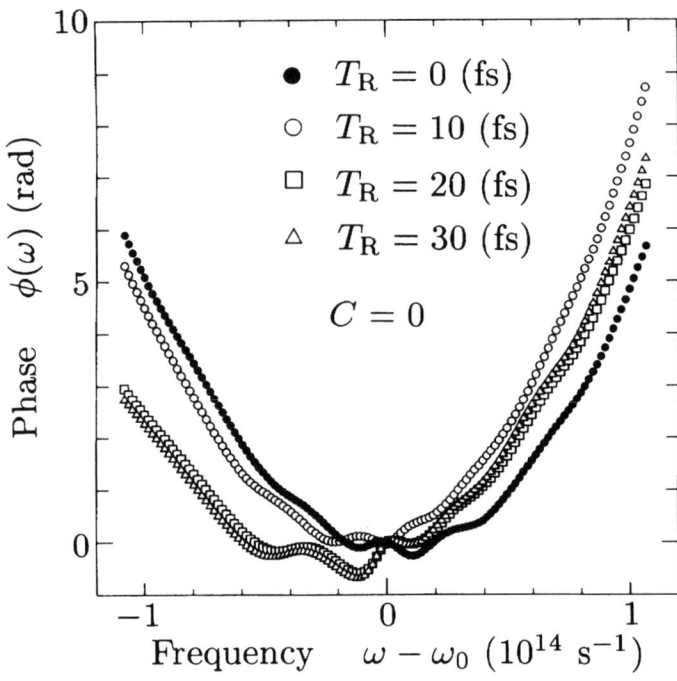

Figure 2.5.6
Fiber output phase $\phi(\omega)$ in the frequency domain at the optimum propagation lengths z_{opt}. The parameters are $P_0 = 100$ W, $T_{p,\text{in}} = 100$ fs, and $C = 0$

degrades pulse compression. This is because only second-order phase compensation has so far been taken into account. Third-order phase adjustment accompanied by second-order compensation is efficient for pulse compression in the presence of delayed response time.

6 Summary

To obtain large $\chi^{(3)}$ materials, we have discussed the relationship between $|\chi^{(3)}|$ and $|\chi^{(2)}|$ in the nonresonant region with optical nonlinearities due only to electronic polarization. Moreover, using organic crystal DAN cored fibers, we have carried out an experiment for nonamplified femtosecond pulse compression and nonlinear femtosecond pulse propagation. 39 fs laser output pulses have been directly compressed to 22 fs, and the latter has enabled us to evaluate the delayed nonlinear response time to be ~30 fs. Furthermore, we have investigated numerically the effect of the delayed nonlinear response on pulse compression using an organic crystal cored fiber in the normal-dispersion region. Using up to third-order phase compensation even in the presence of a delayed nonlinear response time, a 100 W, 100 fs hyperbolic-secant pulse is compressed to ~10 fs as well as in the absence of it. Third-order phase adjustment can compensate for the phase asymmetry in the frequency domain caused by the effect of the delayed nonlinear response. It must be emphasized that, similar to such an example of femtosecond optical pulse compression, organic materials are promising for other controls of femtosecond optical pulses.

References

[1] See, for example, D.S. Chemla and J. Zyss, *Nonlinear Optical Properties of Organic Molecules and Crystals*, Academic Press, 1987.
[2] M. Yamashita, K. Torizuka, T. Uemiya and J. Shimada, *Appl. Phys. Lett.* **58**, 2727 (1991).
[3] M. Yamashita, K. Torizuka, and T. Uemiya in *Nonlinear Optics, Fundamentals, Materials and Devices*, S. Miyata, ed., North-Holland, Amsterdam, 1992, p. 525.
[4] M. Yamashita, *Ultrafast Phenomena VIII*, Springer, 1993, p. 313.
[5] M. Yamashita, *Optoelectronics* **8**, 379 (1993).
[6] M. Yamashita, K. Torizuka and T. Uemiya, *Appl. Phys. Lett.* **57**, 1301 (1990).
[7] R. Morita and M. Yamashita, *Jpn. J. Appl. Phys.* **32**, L905 (1993).
[8] R. Morita, H. Sone, C. Ohshima and M. Yamashita, in *Conference on Laser and Electrooptics*, Vol. 11 of 1993 OSA Technical Digest Series, Optical Society of America, Washington, DC, 1993, p. 468.
[9] P. Kerkoc, M. Zgonik, K. Sutter, Ch. Bosshard and P. Günter, *Appl. Phys. Lett.* **54**, 2062 (1989).
[10] A.B. Grudinin, E.M. Dianov, D.V. Korobkin, A.M. Prokhorov, V.N. Serkin and D.V. Khaĭdarov, *JETP Lett.* **46**, 221 (1988).
[11] G.P. Agrawal, *Opt. Lett.* **15**, 224 (1990).
[12] R. Morita and M. Yamashita, *Opt. Lett.* **19**, 1459 (1994).

3
NONLINEAR OPTICS: MATERIALS, PROCESSES AND APPLICATIONS

3.1

Image Processing by Phase Conjugation in Organic Dye-doped Polymer Films

Hirofumi Fujiwara, Tomoaki Takeda and **Kazuo Nakagawa**

Abstract

Dye-doped polymer films such as an erythrosin-B-doped polyvinyl alcohol (EB/PVA) film, a methyl-orange-doped polyvinyl alcohol (MO/PVA) film and an EB- and MO-doped PVA ((EB+MO)/PVA) film can work not only as a phase conjugator but also as a hologram at a power level of 1 W/cm^2.

Using these dye-doped films, we demonstrate real-time image processing for detecting the phase and amplitude difference between two objects, and detecting a time-varying portion from an input scene.

1 Introduction

Optical phase conjugation has received much attention in the fields of optical image processing, adaptive optics, optical communication, laser cavity, interferometry, and so on. A phase conjugate (PC) wave is one whose phase is conjugate with respect to that of a wave (= a probe wave). The generally available method to generate the PC wave uses a backward degenerate four-wave mixing (BDFWM) arrangement, in which four waves have the same frequency. When the probe wave onto which object information is encoded and two counterpropagating pump waves impinge upon a phase conjugator, the generated PC wave retraces the path of the probe wave. Optical phase conjugation in photorefractive crystals and dye-doped films has been studied for applications to some all-optical parallel image processing [1–3].

Dye-doped polymer films such as an erythrosin-B-doped polyvinyl alcohol (EB/PVA) film, a methyl-orange-doped polyvinyl alcohol (MO/PVA) film, and an EB- and

MO-doped polyvinyl alcohol ((EB + MO)/PVA) film, can generate PC waves at powers of lower than 1 W/cm^2 emitted from a cw argon-ion laser and are useful also for recording a hologram [2,3]. With a new view on optical phase conjugation in these dye-doped films, we demonstrate some all-optical parallel image processing such as the detection of phase and amplitude differences between two objects and the detection of a newly-appearing or time-varying portion of an object by eliminating the static background.

2 PC Wave Generation in Dye-doped Films

Dye-doped films such as EB/PVA, MO/PVA, and (EB + MO)/PVA exhibit optical nonlinearity by absorbing light. Their absorption spectra are shown in Figure 3.1.1. In the (EB + MO)/PVA film the ratio of EB to MO in number density was ~1/2. The three dye-doped films absorb a 515 nm light from an argon-ion laser and the absorption spectrum of the (EB + MO)/PVA film is given by a linear combination with the absorption spectra of the EB/PVA and the MO/PVA. This holds if the dye concentration is not so high (as is in our case). The three dye-doped films had a thickness of ~50 μm. All-optical PC image processing was performed at 515 nm.

Dye-doped films such as EB/PVA, MO/PVA, and (EB + MO)/PVA work not only as a phase conjugator but also as a hologram. We consider two polarization configurations of probe and pump waves for the generation of PC waves: the polarization direction of the probe wave is orthogonal to that of the pump waves (orthogonal polarization configuration) and parallel to that of the pump waves (parallel polarization configuration).

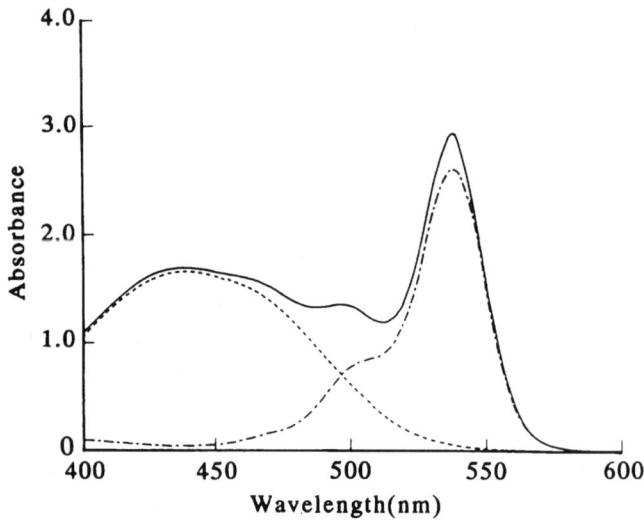

Figure 3.1.1
Absorption Spectra of EB/PVA film (— · — · — · -), MO/PVA film (------------), and (EB+MO)/PVA film (————)

In the parallel polarization configuration, two sets of interference fringe patterns formed by the probe wave and each pump wave produce population gratings owing to the mechanism of saturable absorption in an EB/PVA film [4]. On the other hand, in the orthogonal polarization configuration, the probe wave and each pump wave periodically modulate the polarization state in the EB/PVA film upon which the interaction strength of light with the EB dyes depends. When the dye molecules are randomly oriented and fixed in a transparent solid matrix, the generated PC wave is linear-polarized parallel to the polarization direction of the probe wave in both polarization configurations [5]. We call such a PC wave a saturable absorption component. The response time is determined primarily by the life-time of the first excited triplet state of the EB dyes. Figure 3.1.2(a) shows the transient PC signals for the EB/PVA film in two polarization configurations. At a pump power of ~ 2.2 W/cm^2 (almost equal to the saturable intensity) and at 515 nm, the response time was ~ 1 ms and the PC reflectivity was $\sim 0.5\%$ for the parallel polarization configuration. For pump intensities lower than the saturable intensity of the EB dye, the PC reflectivity is higher in the parallel polarization configuration than in the orthogonal one [5].

MO dyes in the *trans* state orienting along the polarization direction of an incident argon-ion laser light change their structure from *trans* form to *cis* form by the interaction of light with the dyes. Since the two forms have different absorption coefficients and refractive indices, excitation by the laser light induces optical anisotropy in a MO/PVA film [6]. Figure 3.1.2(b) shows the transient PC signals in the MO/PVA film for both polarization configurations. The more efficient PC wave is generated for the orthogonal polarization configuration rather than the parallel one. The PC reflectivity reached $\sim 10\%$ and the response time was ~ 1 s for the orthogonal polarization configuration. We call this PC wave a *trans–cis* isomerization component for want of a better name.

The experimental result for transient PC signals for an (EB + MO)/PVA film is shown in Figure 3.1.2(c), where the probe wave was linear-polarized at an angle of 45° with respect to the polarization direction of the pump waves. The PC waves were observed through an analyzer set to transmit the PC waves polarized parallel to and orthogonal to the polarization direction of the pump waves. Their response times were ~ 1 ms and ~ 1 s, respectively. The fast and slow responses result from mainly the time responses of the EB and MO dyes, respectively (refer to Figure 3.1.2(a) and (b)). The PVA film in which both EB and MO dyes are doped produces a PC wave that maintains the optical nonlinearity of each kind of dye.

So far we have discussed the generation of PC waves in the BDFWM configuration. Table 3.1.1 summarizes the PC efficiencies and the response times of PC waves for three kinds of dye-doped films.

These dye-doped polymer films have another optical function. The EB dyes exhibit a chemical change and their color fades under exposure to intense light. The semipermanent intensity-dependent photochemical change in the EB/PVA film is available for hologram recording without the chemical developing process [2,3]. A PC wave is generated by this hologram. We call this PC wave a holographic component. A holographic record is possible also in a MO/PVA film through the mechanism of *trans–cis–trans* isomerization [6].

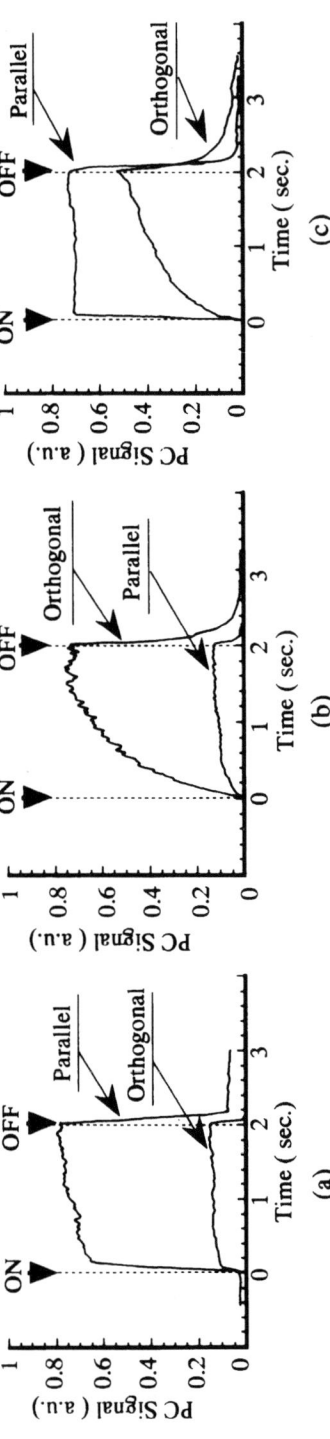

Figure 3.1.2
Transient PC signals in parallel and orthogonal-polarization configurations for (a) EB/PVA film, (b) MO/PVA film, and (c) (EB+MO)/PVA film. Pump intensity was 1 W/cm^2 at 515 nm. A probe wave was turned on at ON and off at OFF

Table 3.1.1
PC efficiency and response time for three kinds of dye-doped films

Kind of dye-doped film	Response time of PC wave	PC Efficiency Parallel polarization	Orthogonal polarization
EB/PVA	~1 ms	intense	weak
MO/PVA	~1 s	weak	intense
(EB+MO)/PVA	~1 ms	intense	
	~1 s		intense

3 PC Parallel Image Processing

Dye-doped polymer films such as EB/PVA, MO/PVA and (EB+MO)/PVA work not only as a phase conjugator but also as a hologram, so that they can simultaneously generate two kinds of PC waves. Table 3.1.2 reviews some of the applications for all-optical parallel image processing in combination with two kinds of PC waves: the detection of the phase difference between two objects with the MO/PVA or EB/PVA film, the detection of the amplitude difference between two objects with the EB/PVA film, and also the detection of the time-varying portion of an object with the (EB + MO)/PVA film. The first two methods require two processing steps, similar to real-time holographic interferometry [3].

3.1 Detection of the phase difference between two objects

Detection of the phase change in a time-varying object or the phase difference between two objects in real time is explained by referring to Figure 3.1.3.

Let two counterpropagating pump waves E_1 and E_2 and the probe wave carrying object information of amplitude T_1 or T_2 be linearly polarized normal to the paper. The amplitudes are defined as $T_j = a_j \exp(ib_j)$ ($j = 1, 2$), where a_j and b_j are real.

Table 3.1.2
All-optical image processing such as detections of phase and amplitude difference and detection of a moving object in combination with two kinds of PC waves

PC component	Polarization of probe and pump waves	Applications
Saturable absorption	orthogonal	detection of phase difference EB/PVA film
	parallel	MO/PVA film
Holography	orthogonal	detection of amplitude difference EB/PVA film
	parallel	
Trans-cis isomerism	orthogonal	novelty detection (EB+MO)/PVA film
	parallel	

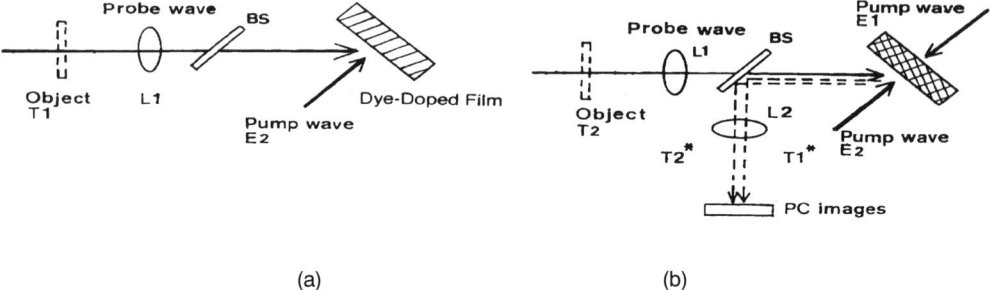

Figure 3.1.3
Schematic diagram of the optical configuration used for all-optical PC image processing. (a) Hologram-recording and (b) interference between two PC waves. L_1, L_2:lenses; BS:beam splitter. In (b) the polarization direction of the probe wave was parallel to, orthogonal to, and at an angle of 45° with respect to that of the pump waves for the detection of the phase difference, the detection of the amplitude difference, and the detection of a time-varying object, respectively

In the first step, the interference pattern between one of the pump waves E_2 and the probe wave of amplitude T_1 was holographically recorded in an EB/PVA film. This film works not only as a phase conjugator but also as a hologram in the second step. When two pump waves and the probe wave of amplitude T_2 impinge upon the same film in the second step, it simultaneously generates two coherent PC waves of amplitudes T_1^* and T_2^*, where an asterisk denotes the complex conjugation. The resulting intensity is proportional to $|T_1^* + T_2^*|^2 = a_1^2 + a_2^2 + 2a_1a_2 \cos(b_1 - b_2)$, so that the phase difference between T_1 and T_2 can be determined from the interfering term. A MO/PVA film also is useful for the detection of the phase difference because it serves as a hologram-recording medium (refer to Table 3.1.2).

In the optical arrangement where the probe wave of amplitude T_1 itself interferes with its PC wave T_1^*, the resulting intensity of the interference pattern is proportional to $|T_1 + T_1^*|^2 = 2a_1^2(1 + \cos 2b_1)$. Thus the fringe-detecting sensitivity is doubled with respect to the present method [7]. Recently detection of the phase difference has been carried out using a composite material which was made by coating an azo dye-doped PMMA film on a Fe-doped LiNbO$_3$ crystal plate [8]. The dye-doped film worked as a phase conjugator and the LiNbO$_3$ plate as a hologram.

3.2 Detection of the amplitude difference between two objects

The first-step processing for the detection of the amplitude difference is similar to the processing in the first step used for the detection of the phase difference. The hologram of an object with amplitude T_1 is recorded on an EB/PVA film. In the second step, the probe wave of amplitude T_2 was polarized orthogonal to the polarization direction of the pump waves (orthogonal polarization configuration). Then the EB/PVA film simultaneously produces two orthogonally polarized PC waves proportional to both T_1^* and T_2^*. Their polarization directions are parallel to those of the pump waves and the probe wave, respectively. The amplitude difference between two orthogonally polarized

PC images was taken by means of an analyzer [3]. The resulting intensity at the image plane is proportional to $|T_1^* - T_2^*|^2$.

We now consider a special case where the object used in the first step or in the second step has a spatial constant T_1 or T_2. Since the intensity distribution is proportional to $|T_1^* - T_2^*|^2$, the contrast in the resulting image can be reversed at the light intensity level corresponding to $T_1 = T_2$.

Exposure to intense light in the second step may thermally expand the dye-doped films and thus change temporally the relative phase between two coherent PC waves. To reduce this thermal effect, overall exposure in the second step should be as low as possible. A detailed discussion of this problem is found in Reference [9].

3.3 Novelty detection

The elimination of an undesirable portion from an object will be frequently required for high-speed image processing. We demonstrate the real-time detection of a time-varying or moving portion by eliminating the static background, called novelty filtering [10], with an (EB+MO)/PVA film. By referring to the second step in Figure 3.1.3, the polarization direction of the probe wave was inclined at an angle of 45° with respect to that of the pump waves. The EB and MO dyes in the (EB + MO)/PVA film generate two orthogonally polarized PC waves having different response times, as shown in Figure 3.1.2(c). An analyzer was set to cancel out only two orthogonally polarized PC waves corresponding to the static background. Figure 3.1.4 shows the transient PC signal passing through the analyzer when the probe wave is turned on at ON and off at OFF. The novelty detection was verified from the fact that the output PC signal becomes more intense just after the probe wave is turned on and off. Furthermore, Figure 3.1.4 shows that newly-varying portions of the probe wave appear bright whether they become bright or dark. For a fast-moving object, its entire image is reconstructed because of the contribution of only the EB dyes in the (EB + MO)/PVA film to the generation of the corresponding PC wave.

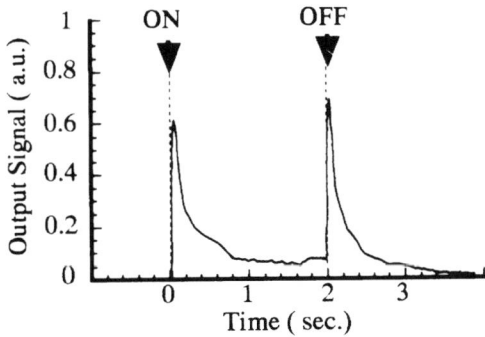

Figure 3.1.4
Transient PC signal in an (EB+MO)/PVA film. The analyzer was set to cancel out two orthogonally polarized PC waves for a static probe wave

A few modified methods for novelty detection may be considered. One is to take the amplitude difference between a probe wave and its PC wave with a finite response time. The other is to use a composite of both an EB/PVA film and an MO/PVA film. In the first case, only the PC wave has a phase change because the phase conjugator expands with absorption of the pump waves. In the second case the two orthogonally polarized PC waves suffer different phase changes when the two dye-doped films absorb the pump waves, thus the phase difference between two PC waves changes temporally [9,11]. A technique to overcome this is to use an optical arrangement where the optical paths are common for two orthogonally polarized PC waves or to generate two orthogonally polarized PC waves from the same dye-doped film. The (EB + MO)/PVA film is suitable for this purpose.

4 Conclusion

We have demonstrated the real-time detection of phase and amplitude differences between two objects and a moving object using dye-doped polymer films in combination with two kinds of PC waves. The advantages of the dye-doped films are as follows: (1) they act not only as a phase conjugator but also as a hologram; (2) they can produce simultaneously or separately two orthogonally polarized PC waves; (3) mixing different dyes in the same film may provide the possibility of a new nonlinear optical function; and (4) large and thin films can be easily prepared. The PC reflectivity, however, is low compared with that of typical photorefractive crystals.

References

[1] P. Günter and J.P. Huignard, *Photorefractive Materials and Applications*, II, P. Günter and J.P. Huignard, eds., Springer-Verlag, 1989, p. 205.
[2] K.K. Sharma, K.D. Rao and G.R. Kumar, *Opt. Quantum Electron.* **26** (1) 1 (1994).
[3] K. Nakagawa and H. Fujiwara, *Opt. Commun.* **70** (2) 73 (1989); H. Fujiwara, K. Nakagawa and T. Suzuki, *Opt. Commun.*, **79** (1) 6 (1990).
[4] Y. Silberberg and I. Bar-Joseph, *Opt. Commun.* **39** (4) 265 (1981).
[5] W.R. Tompkin, M.S. Malcuit, R.W. Boyd and J.E. Sipe, *J. Opt. Soc. Am. B* **6** (4) 757 (1989).
[6] T. Todorov, L. Nikolova, N. Tomova and V. Dragostinova, *IEEE J. Quantum Electron.* **QE-22** (8) 1262 (1986).
[7] I. Bar-Joseph, A. Hardy, Y. Katzir and Y. Silberberg, *Opt. Lett.* **6** (9) 414 (1981).
[8] R.K. Mohan, S. Balan, P.S. Narayanan and C.K. Subramanian, *Opt. Commun.* **106** (1, 2, 3) 84 (1994).
[9] K. Kawano, K. Nakagawa, T. Takeda and H. Fujiwara, *Opt. Commun.* **102** (5, 6) 421 (1993).
[10] D.Z. Anderson and J. Feinberg, *IEEE. J. Quantum Electron.* **25** (3) 635 (1989).
[11] T. Takeda, M. Yamada, K. Nakagawa and H. Fujiwara, *Frontiers in Information Optics*, **92**, April 1994, Kyoto, 5A-15.

3.2

Nonlinear Interactions in Periodic Domain Reversals

Hiromasa Ito

Abstract

The quasi-phase matching (QPM) achieved by a periodic domain structure has evoked much interest recently because of its efficient nonlinear interactions. Two different fabrication methods for these structures were proposed and studied using $LiNbO_3$ and $LiTaO_3$, and 500 μm-thick domain volume gratings were successfully fabricated by both methods of electron beam writing and static electric field application. For e-beam writing using $LiNbO_3$ and $LiTaO_3$, the behaviors of surface ions compensating spontaneous polarization charges closely relate with the domain reversal mechanism as well as the feature of domain reversal. By the static electric field application, 500 μm-thick volume domain gratings with 7.8 μm period were fabricated using $LiTaO_3$. The SHG experiments were also demonstrated to verify their potentialities. This technique opens new possibilities for fabricating miniature nonlinear optical devices as well as miniature electrooptic devices with many functions.

1 Introduction

An efficient nonlinear process at relatively low power level has attracted much attention recently not only for the coherent light source in unexploited wavelength regions but also for new applications such as antibunched light generation for optical communications and optical information processing. For these purposes, the quasi-phase matching (QPM) achieved by a periodic domain structure evokes much interest because of its efficient nonlinear interactions. Controlling the domain of ferroelectric materials is one of the ideal ways to realize QPM.

In general, the domain change in ferroelectric materials has been known to occur by many reasons, such as local stress, local electric field, heat [1], impurity doping [2], and so on. The periodic domain grating of LiNbO$_3$ has been fabricated mostly by titanium (Ti) indiffusion through the $+z$ surface. In this case, a high temperature process is necessary, and the depth of the inverted area is restricted by the impurity diffusion properties. In addition, a refractive index change occurs due to the diffusion process.

Recently, the periodic reversed domain structure in LiNbO$_3$ and LiTaO$_3$ by means of e-beam writing at room temperature has been reported by us. The domain reversed area is deep enough to penetrate through the substrate from the top to the bottom [3,4]. By using the e-beam writing process to produce the domain grating, the influence of a refractive index change is minimized.

However, there is a problem with this method in that the shape of the domain reversals tends to be a dot-by-dot pattern in nature. To avoid this problem, we developed a method using an external electric field [5].

In this section two electrical domain inversion methods are reported: an e-beam writing method and a static electric field application method. The experimental results of a highly efficient second harmonic generation (SHG) are also presented.

2 Quasi-phase Matching by Domain Reversal

Many nonlinear optical interactions have conventionally been demonstrated using a variety of phase matching techniques including birefringence and so on. To obtain efficient and wide ranging interactions, QPM is really a novel method. Since QPM can use the largest nonlinear optical coefficient of materials and is completely free from the dispersion limitation which is usually a big problem, it is possible to realize an efficient nonlinear interaction and the operation wavelength range covers the entire material transparent range.

QPM is achieved by periodic modulation of the nonlinear polarization. The condition for QPM is given by

$$T = 2m \times \frac{\lambda_{\text{SH}}}{(n_{\text{SH}} - n_{\text{F}})} \tag{1}$$

where T is the domain period, λ_{SH} is the harmonic wavelength in a vacuum, n_{F} and n_{SH} are the refractive indices at the fundamental and harmonic wavelengths, respectively, and m is an integer for the order of the QPM interaction.

Nonlinear conversion efficiency depends directly on the effective nonlinear coefficient of the periodic domain structure. In general, the ratio of the domain reversed region to the period could be any arbitrary value. The parameter ξ is defined for this ratio. The effective nonlinear coefficient for the mth-order QPM is calculated by the Fourier expansion of the periodic structure, and is given by

$$d_{\text{eff}}^{(m)} = \left(\frac{2d}{m\pi}\right) \times \sin(m\pi(1-\xi)) \tag{2}$$

Figure 3.2.1 summarizes the calculations. Apart from the conditions $\xi = 0.5$ and $m = 1$, the efficiency decreases monotonically. It is seen that the efficient operation for a higher order interaction is realized when $m = 2$ and $\xi = 1/4$ followed by $m = 3$ and $\xi = 1/6$, 6/5 or 1/2.

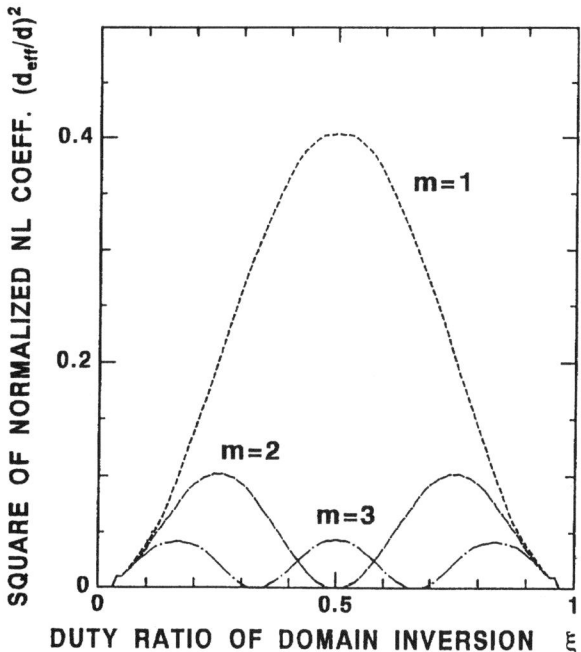

Figure 3.2.1
Square of effective nonlinear coefficient for a periodic domain structure as a function of the duty ratio of domain reversal ξ. The parameter indicates the order of the QPM interaction

It should be noted that control of the domain width as well as that of the period is essential to obtain a high conversion efficiency for QPM operations.

3 Periodic Domain Reversal Methods

3.1 Electron beam writing

Domain reversal with a period of few microns has been made by means of e-beam writing, and the reversed area is deep enough to penetrate through a substrate of 500 μm from the top to the bottom. Figures 3.2.2(a) and (b) show typical examples of etched patterns to reveal domain structures of $+z$ surfaces of LiNbO$_3$ for the 38 μm period and 4 μm period, respectively. It is clearly seen that a sample with a long period tends to be almost continuous; however, it is not for a short period sample, and it often shows a dot-by-dot feature. The e-beam writing method has the advantage of simplicity in fabricating the periodic domain reversals, but this dot-by-dot domain reversal pattern decrease the nonlinear interaction efficiency.

The mechanism of domain reversal has also been studied [6], especially for charge balance. The spontaneously polarized negative charges ($-P_s$) on the $-z$ surface are normally compensated by positive ions. In the case when the domain reversal occurs from the $+z$ to the $-z$ surface due to an electric field caused by injected electrons, the total electrons must be twice the total amount of the spontaneously polarized charge

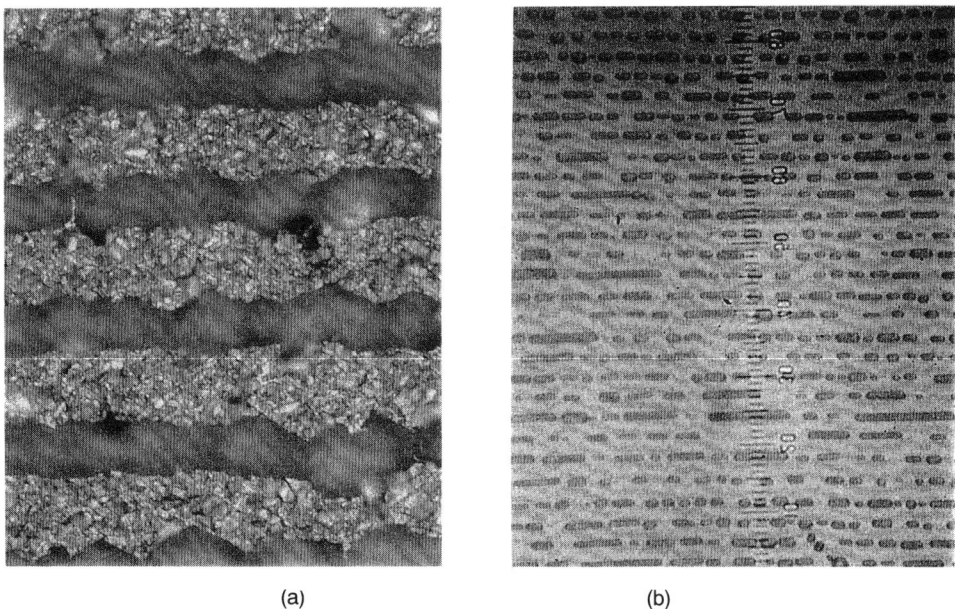

Figure 3.2.2
Periodic domain reversal of LiNbO$_3$ after chemical etching by electron beam writing. Period is (a) 38 μm, and (b) 4 μm

($-2P_s$), since injected electrons compensate for both spontaneously polarized charges and adsorbed positive ions.

Therefore, dot-by-dot reversed domain features could be explained as follows. Supposing that an e-beam is scanned continuously, domain flipping occurring at the position of e-beam injection. A qualitative measurement of the relation between the injected electrons and the inverted volume showed us that an e-beam domain control tends to occur at a slightly lower level of electron charges than the neutrization condition [6]. So uncompensated positive ions remain and move toward the surroundings due to the electric repellents. Due to these displaced ion fields the injected electrons in this area cannot induce enough electric field to flip the domain. The beam then moves continuously to a position beyond the displaced positive ions and domain reversals again occur. This happens in a cascade, and eventually the feature of domain reversals tends to be dot-by-dot patterns.

In the case of e-beam writing, therefore, it is necessary to control the surface ions of a substrate in order to realize the optimal domain reversed patterns.

3.2 Static electric field application

To avoid the dot-by-dot domain patterns produced by e-beam writing, the application of a static electric field through a patterned metal mask is investigated. The breakdown voltages of $LiNbO_3$ and $LiTaO_3$ had been believed to be lower than the domain reversed voltage at room temperature [7], however a successful domain reversal of $LiNbO_3$ was demonstrated by means of the pulsed electric field method close to room temperature [8]. To realize volume domain reversals for the purpose of efficient nonlinear interactions, a domain reversal by means of an applied static electric field was studied.

The method of an applied static electric field is as follows. A 500 μm-thick $LiTaO_3$ wafer was first processed to fabricate a periodic electrode of Cr-Au on the $+z$ surface by photolithography with a desired period. The $-z$ surface was covered by a uniform metal electrode. A sample was placed in a vacuum chamber or an insulating oil bath to prevent a discharge around the electrodes. An injected current was monitored and recorded by a computer, and simultaneously the total amount of the injected charge was calculated.

Since the domain reversal occurs quite close to the breakdown voltage at room temperature, control of the applied voltage is usually crucial. Figure 3.2.3 shows a typical example of recorded traces of the current and applied voltage change. It is seen that the domain reversed current began to flow at about 10.5 kV, and the voltage was kept constant at this level. The current ceased about a minute later automatically, and then the voltage was decreased to zero. The total injected charge in this case was 27.6 μC from the time of integration of the current, while the calculated value of

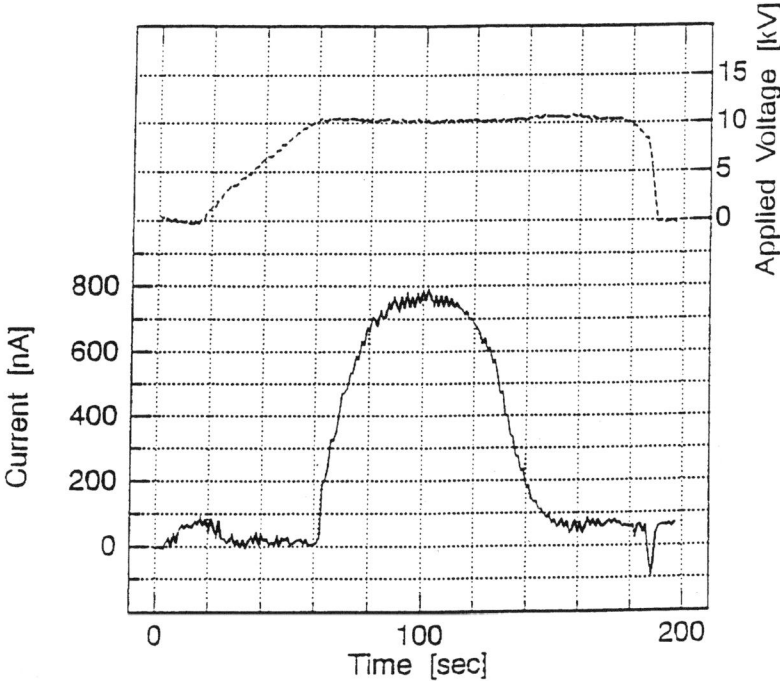

Figure 3.2.3
Temporal behavior of the absorbed current and the applied voltage on the volume domain grating fabrication by an applied static electric field

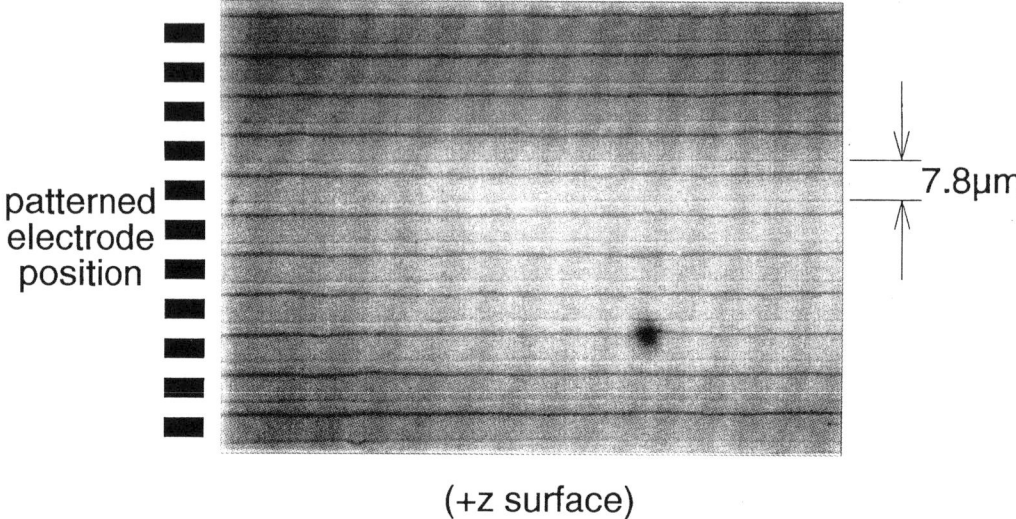

Figure 3.2.4
Periodic domain reversal of $LiTaO_3$ after chemical etching by an applied static electric field. Thickness and period are 500 μm and 7.8 μm, respectively

2 × (spontaneous polarization) × (electrode area) gave 28.0 μC, which was in good agreement with the measured one. This implies that the domain reversed mechanism mentioned above is experimentally verified. Figure 3.2.4 shows an optical microscope picture of the $+z$ surface after etching, the period of which is 7.8 μm. A domain grating is formed inside a $LiTaO_3$ substrate from the top to the bottom, so that a volume domain grating is fabricated by this method.

For many nonlinear optical applications, a volume domain grating is more attractive than a surface domain grating, since the former has larger possibilities to realize nonlinear optical interactions using narrower or wider periods as well as a thicker substrate.

4 Second-harmonic Generation

QPM-SHG was investigated to check the basic characteristics of the devices [9]. A set of domain gratings fabricated by a static electric field on a substrate had three different periods of 7.5 μm, 7.8 μm, and 8.1 μm, and each size was typically 2(W) × 4(L) mm². Then the sample was cut and polished at both ends. The fundamental beam was focused by an $f = 7.5$ mm lens and entered normal to the surface.

The observed second harmonic intensity dependence on the fundamental wavelength is shown in Figure 3.2.5. It is seen that a single tuning peak for each of the three domain periods was observed. The width was about two times wider than that of the theoretical one. This would be partly due to the diffraction effect of a fundamental beam in the domain reversal area.

Figure 3.2.6 summarizes the results between the fundamental wavelength and the QPM domain period for $m = 1$ and $m = 2$. The curves and circles represent the

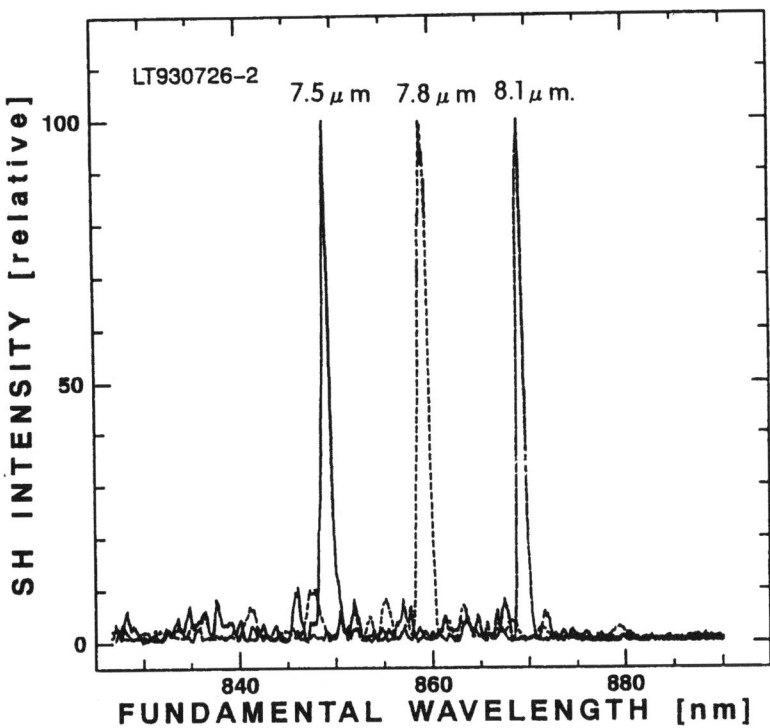

Figure 3.2.5
QPM-SH intensities vs. fundamental wavelength for 7.5 μm, 7.8 μm, and 8.1 μm period LiNbO$_3$ domain grating

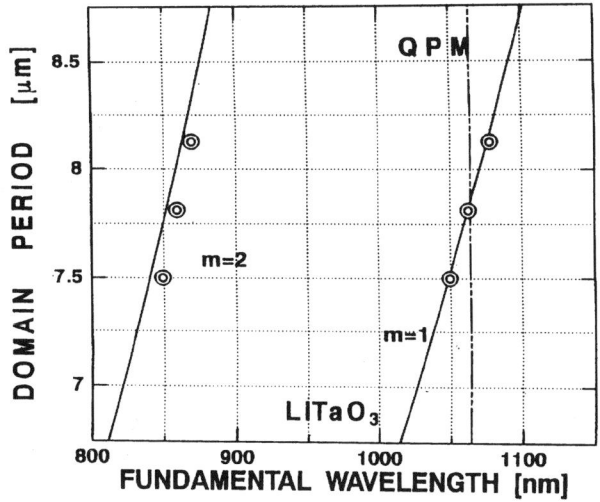

Figure 3.2.6
Domain period vs. fundamental wavelength. The parameter indicates the order of the QPM interaction

theoretical values obtained using the Sellmeier equation for $LiTaO_3$ and the experimental ones, respectively. It is clearly seen that experimental data correspond well to the theoretical ones.

Using the e-beam writing method, the measured normalized conversion efficiency is two orders less than the theoretical one. This degradation of efficiency would be mainly due to the irregularity of the domain reversed features. Using the static electric field application method, a normalized conversion efficiency of 1.3%/W-cm is experimentally obtained, while the theoretical value is 1.6%/W-cm.

5 Conclusion

Two domain reversal methods, e-beam writing and static electric field application, have been developed by investigating the domain reversal mechanism, and QPM-SHG devices using these two methods have been prepared and examined.

The e-beam method revealed the possibility of a volume domain grating with a thickness of 500 μm. The behaviors of the surface ions for this method closely relate with the domain flipping mechanism as well as the feature of dot-by-dot domain reversal. Using the applied static electric field method, a volume domain grating of 500 μm thickness and 7.8 μm period for $LiTaO_3$ were fabricated with a smooth domain grating structure. Additionally, a high conversion efficiency for QPM-SHG was demonstrated.

This implies that efficient nonlinear interactions could be possible by utilizing QPM through the periodic volume domain grating, and this technique opens new possibilities for fabricating miniature nonlinear optical devices as well as miniature electrooptic devices using domain reversed structures.

References

[1] J. Webjorn, F. Laurell and G. Arvidsson, *IEEE Photon. Tech. Lett.* **1**, 316–318 (1989).
[2] E.J. Lim, M.M. Fejer, R.L. Byer and W.J. Kozlovsky, *Electron. Lett.* **25**, 731–732 (1989).
[3] H. Ito, C. Takyu and H. Inaba, *Electron. Lett.* **27**, 1221–1222 (1991).
[4] R.W. Keys, A. Loni, R.M. De La Rue, C.N. Ironside, J.H. Marsh, B.J. Luff and P.D. Townsend, *Electron. Lett.* **26**, 188–190 (1990).
[5] C. Takyu, M. Ohashi, M. Sato and H. Ito, *The International Conference on Optically Nonlinear Organic Materials and Applications*, Hawaii, 1994, 18D11.
[6] M. Ohashi, C. Takyu, H. Ito and K. Taniguchi, *IEICE Technical Report*, OQE93-40, 1993, pp. 8–13.
[7] M. Yamada, N. Nada and K. Watanabe, *Integrated Photonics Research Technical Digest*, **10**, New Orleans, 1992, TuC2-1.
[8] M. Yamada, N. Nada, M. Saitoh and K. Watanabe, *Appl. Phys. Lett.* **62**, 435–436 (1993).
[9] H. Ito, M. Sato, C. Takyu and M. Ohashi, International Symposium Ultrafast and Ultraparallel Optoelectronics, Chiba, 1994.

3.3

Three-dimensional Optical Memory Using Photorefractive Materials

Satoshi Kawata and **Yoshimasa Kawata**

Abstract

Research into three-dimensional optical memory with photorefractive materials is described for ultra-high density/capacity memory exceeding the classical limit of a conventional optical recording system. Bit data are recorded as highly localized refractive index variations in three-dimensional volume using a focused laser beam. A confocal laser-scan microscope is employed to record the sequence of dot data. In particular, we show an optical memory recording/writing system and discuss the various methods of writing and reading data in photorefractive materials and polymers. The devices needed to realize three-dimensional memory recording and reading of optical bit data are also presented. Some results obtained from preliminary experiments using polymers and $LiNbO_3$ as recording media are also shown.

1 Introduction

Optical memories such as compact disks and magneto-optical disks are becoming essential in high technology products such as audio and visual disks, and the external computer memory disk. In these memory devices a laser beam is used to record and read information. Since the laser spot can be focused to within 1 μm scale, optical memory can attain density and capacity values that are higher than magnetic memory.

The optical memory is ultimately limited by the diffraction of electromagnetic waves. Present techniques have almost reached this limit in optical memories that are commercially sold as compact disks or magneto-optic disks. Even with an infinitely large objective lens, the best achievable bit data resolution distance for recording and reading is never smaller than half the beam wavelength.

Consequently, current efforts in optical memory devices are geared towards the development of durable short wavelength compact lasers that emit blue or green light. Doubling the frequency (or halving the wavelength) of the laser output reduces the beam spot radius by two, thereby increasing the density by four when recording in two dimensions. If one aims to increase the memory density by one hundred of the current benchmark, laser diodes with output wavelengths that are ten times shorter than those currently available have to be employed. This requirement is obviously impossible because no safe compact laser diodes and optical components, particularly lenses, can be manufactured in the 70–80 nm wavelength range.

Alternative methods exist that can overcome the above-mentioned technical problem. One is to record using evanescent waves that are nonisotropic and localized, and whose wavelength is shorter than the wavelength of the normal propagating beam [1–3]. The use of evanescent waves has been thoroughly explored in superresolving microscopy. It has also been applied to optical memory recording. The nonlinear response of the material to the laser beam power in both writing and reading can be utilized to attain superresolution in optical memory recording and reading.

In this section we describe another method to overcome the density limit, namely by introducing an additional axial dimension in the recording process [4–7]. The z or longitudinal axis is utilized in addition to the surface dimensions (xy space) of conventional optical memory. The data are thus written not on the material surface but within the three-dimensional (3D) thick volume. The materials that can be used are: (1) photopolymers originally developed for 3D (Lippmann) holographic applications such as write-once memory [8], and (2) photorefractive crystals commonly used in erasable real-time holography, phase-conjugation, and wave-coupling [9].

2 Optical Systems for Writing and Reading

Figure 3.3.1 shows an optical system for recording 3D optical memory. The laser beam is focused through a microscope objective lens onto a point in the photorefractive thick medium. To record a bit data series, either the laser beam or the medium is scanned in three dimensions. The recording material changes its refractive index locally at the illuminated point. An acoustic–optic modulator controls the focused beam intensity and implements the two (bistable) logic states necessary in bit data recording.

To read recorded data the same optical arrangement is used except that a detection part is added to the system and a phase-contrast objective lens is utilized together with an annular pupil for imaging the refractive-index variations. Figure 3.3.2 shows an optical setup for reading bit data in a 3D optical memory. This system is simply a laser-scanning confocal microscope appropriate for phase-contrast imaging [10,11]. Point light source illumination reduces unnecessary scattered light because the point detector of a confocal microscope only detects the light intensity from a specific point of interest in the thick sample and rejects the scattered light produced from other nonfocused points. High contrast images are therefore observed, and unwanted crosstalk between planes is low. Better performance is obtained in comparison with the images produced using a conventional optical microscope. Spatial resolution is also improved because of the nonlinear frequency-response of the photorefractive materials.

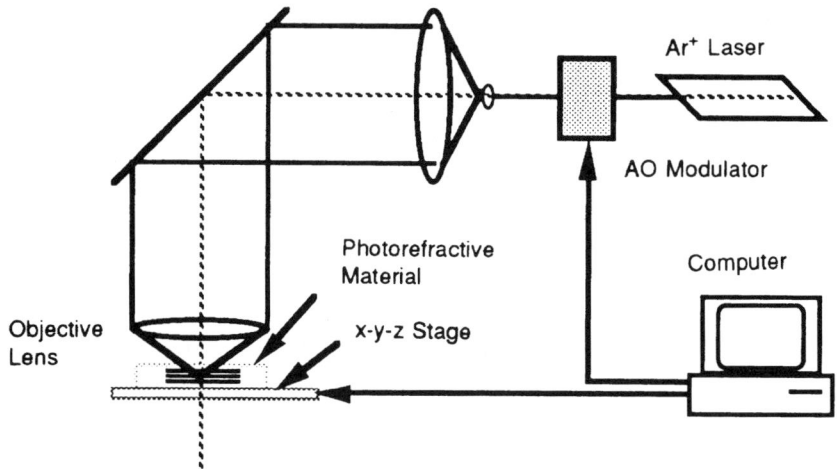

Figure 3.3.1
Optical recording system for three-dimensional optical memory

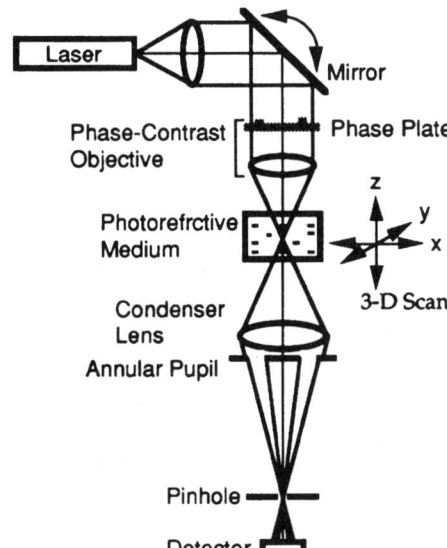

Figure 3.3.2
Optical reading system for three-dimensional optical memory

3 Photopolymer Materials for Read-only Memory

For recording and reading the data within a thick volume, the recording medium must be almost transparent to the laser beam. Photosensitive materials suitable for such an application are photochromic materials with two-photon excitation/fluorescence features, photorefractive ferroelectric crystals, and photopolymerizable materials.

Photopolymers have been successfully used in holographic recordings. They are superior to silver halides and dichromated gelatin because holographic development by wet processing is not required. Photopolymers also exhibit high diffraction efficiency. Because of their high transparency, photopolymers also do not scatter the recording beam thereby producing high contrast holograms. We therefore use this material for writing and reading for 3D optical memory in bit data format. Strickler and Webb have also used polymers in 3D optical memory recording and reading based on a two-photon process [12].

4 Experimental Results of 3D Recording and Reading

As a recording medium we used a monomer mixture of Methacryl and Allyl compounds with Benzil as an initiator and Michler's ketone as a dye sensitizer [8]. The Methacryl compound polymerizes faster than the Allyl compound, and hence the refractive index of the polymerized Methacryl compound alone is higher ($n \sim 1.60$) than the polymerized Methacryl compound ($n \sim 1.50$).

The photopolymerizable solution is prepared in a bounded space formed by microscope cover glasses of 170–500 μm thicknesses that are placed on top of a slide glass. The upper surface of the solution-filled spacer is then covered by a thin glass (170 μm thickness). The recording area is weakly photopolymerized (weak solidification) by pre-exposure to uniform illumination by an ordinary fluorescent lamp.

The medium is placed on a microscope stage that is driven by a computer-controlled x–y–z actuator with 0.02 μm precision. The data are written by a 5 mW Argon ion laser emitting a 488 nm line at exposure times of 60 ms per point. The microscope used was a Carl Zeiss Axiophoto with an objective lens of NA = 1.0 (oil imulsion, 40×).

Figure 3.3.3 shows an example of 3D optical memory recording. Bit data were written every 2 μm × 2 μm in a plane, and the longitudinal separation between the data planes was 15 μm. The total number of layers recorded for the optical memory shown in Figure 3.3.3 was 15, which cover a total longitudinal length of 225 μm. The images were obtained using an ordinary phase-contrast microscope.

Figure 3.3.4 shows a longitudinal cross-section of another 3D memory recording that was read using a laser-beam scanning microscope (Olympus LSM-GB200) with a phase-contrast objective (100×). Results indicate that the separation between layers can be reduced to as small as 3 μm. Good separation was achieved because the photopolymerization process exhibits high-pass spatial-filtering characteristics. This is due to the restricted diffusion length of the monomer during photopolymerization. A typical value for a maximum spatial response is ~1 μm. The low spatial frequency components that produce defocused bit images in out-of-focus layers are therefore not recorded.

Figure 3.3.5(a) presents an example of bit data read using a confocal microscope which we built, together with the Carl-Zeiss Axiophoto which was also used for data writing. A He–Ne laser (632.8 nm) was used together with a phase-contrast objective

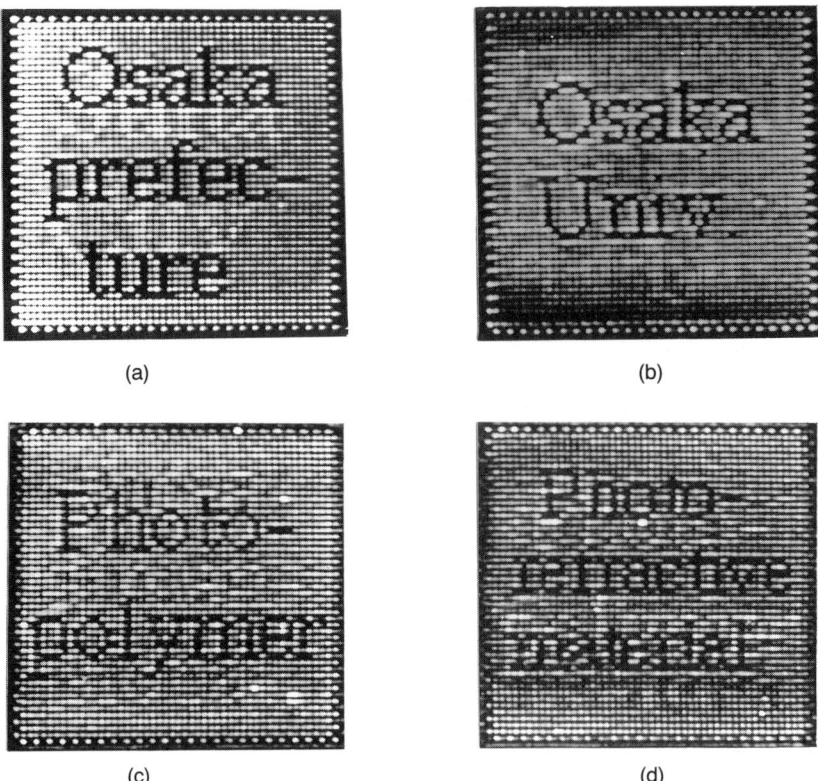

Figure 3.3.3
Experimental result of recorded bit data. Bit data were written every 2 μm × 2 μm in a plane, and the longitudinal separation between the data planes was every 15 μm. A total of 50 × 50 × 15 data points were recorded

and an annular pupil for phase-contrast (dark field) imaging. For comparison the same segment of the data was read using a conventional microscope with the same objective lens, and is shown in Figure 3.3.5(b). The results demonstrate the advantages of confocal microscopy for high-contrast and high-resolution imaging of 3D structures. Since only the light intensity in the conjugate pair of the point of interest in the thick sample volume (or the focused point of the laser beam in the volume) is detected in a confocal microscope, scattered light produced by other nonfocused points does not contribute to the detected signal. Hence, the signal contrast of the images is excellent and the cross-talk between planes is negligible compared with images obtained using a nonconfocal microscope. Spatial resolution is also better because of the nonlinear spatial response (product of illumination point-spread functions and the detection amplitude point-spread function) [10].

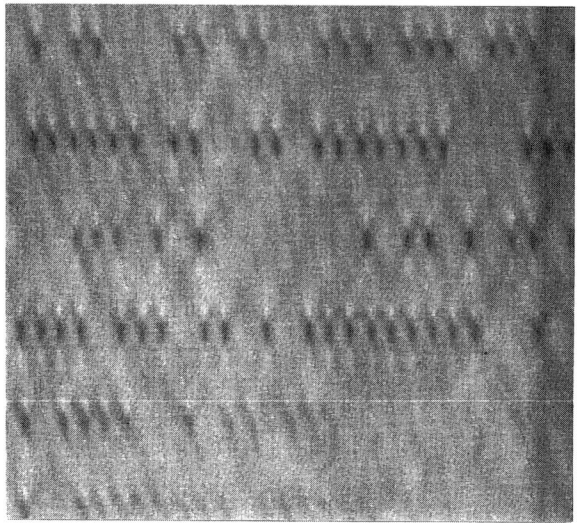

Figure 3.3.4
A longitudinal cross-section of another memory recording where a bit data point was written every $2 \times 2 \times 15$ μm

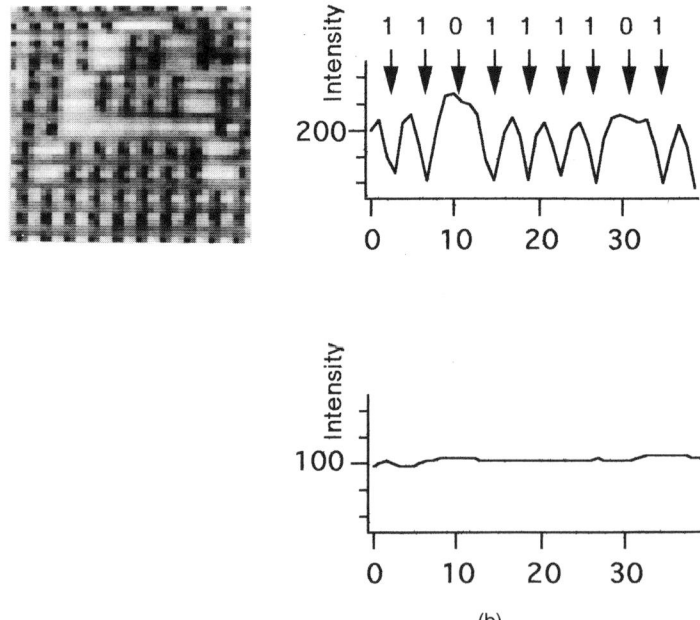

Figure 3.3.5
Reading of recorded data (a) by confocal microscope, and (b) by a conventional microscope with a phase-contrast objective and annular pupil for phase contrast (dark field)

5 Photorefractive Crystals for Erasable Memory

We also investigated the use of photorefractive crystals for application in 3D erasable memory recording. Figure 3.3.6 shows bit data images written in a Fe-doped Lithium Niobate crystal ($LiNbO_3$:Fe). The distance between bits in a plane is 4 μm × 4 μm, and adjacent planes are separated by 22 μm. The letters L, A, and B in the square frames are easily seen from top to bottom, respectively. Cross-talk from defocused images in other layers are not significant in the imaged layer. Data were recorded using the same optical arrangement as the one used for photopolymer memory except that the objective lens utilized has an NA = 0.75.

Figure 3.3.7 illustrates the mechanism of photorefractive effect [9]. When the laser beam is focused inside the crystal, the electrons near the focused point are excited from the donor level to the conduction band of the crystal. The excited electrons diffuse and drift until they are recombined with vacant donor sites. The number of electrons locally excited to the conduction band depends on the intensity distribution of the focused

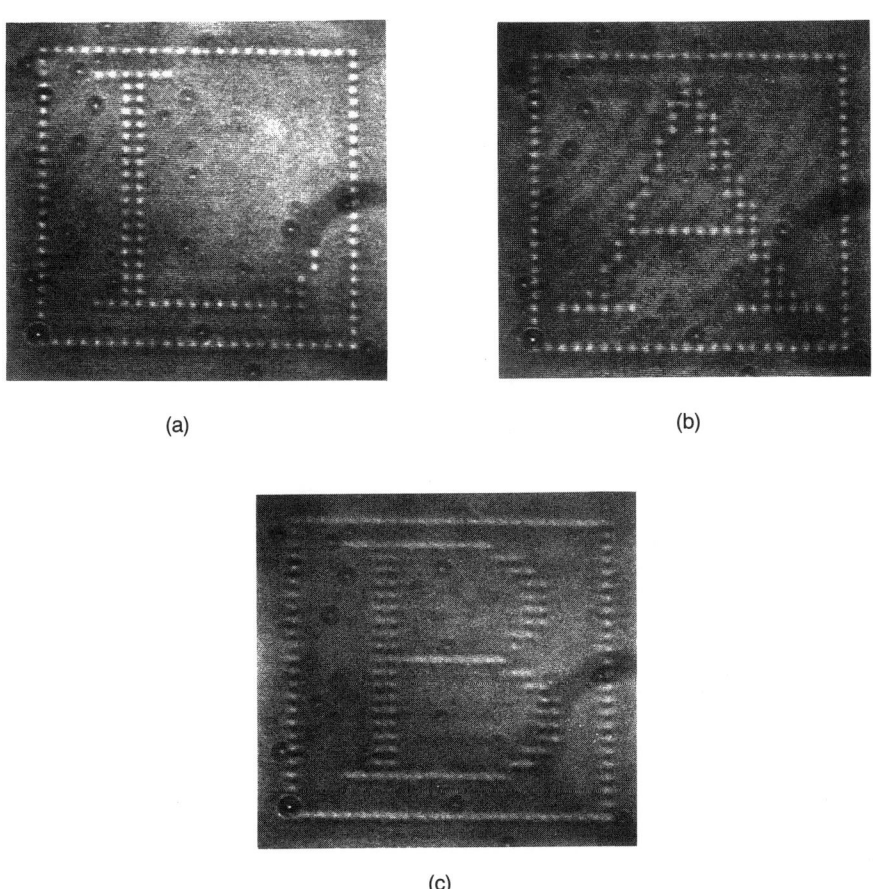

Figure 3.3.6
Bit data image written in a Fe-doped Lithium Niobate crystal

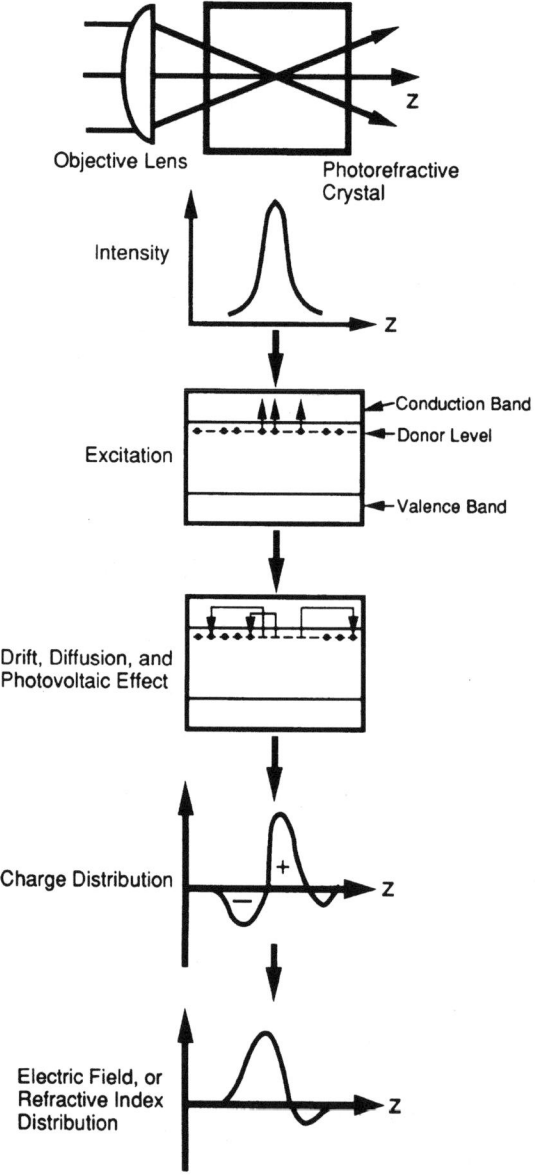

Figure 3.3.7
Mechanism for recording bit data in a photorefractive crystal

beam when diffusion and drift occur homogeneously. As a result, the distribution of the recombined electrons in the sample is not uniform, resulting in the generation of a locally distributed electric field.

Since the photorefractive crystals exhibit an electrooptical effect (Pockels effect), the local electric field produces a refractive-index distribution in the crystal and a small index modulated dot is induced by the focused beam.

Recorded bit data can be erased locally by uniform beam illumination, because a uniform beam homogenizes the local charge density of the donor level and hence removes the refractive index differences. As a result, memory can be written and read at will. The typical life-time of the $LiNbO_3$:Fe crystal as a memory device is several months.

6 Discussion

Holographic 3D memory recording is a common application of photopolymer materials and photorefractive crystals. However, bit data recording possesses clear advantages over holographic memory recording. In terms of the signal-to-noise ratio, holographic memory is not robust to speckle noise unlike 3D memory recording by the focused-beam scanning which is essentially imaging using spatially-incoherent illumination. Moreover, signal contrast in images using confocal imaging is better than that obtained by the hologram read-out because pinhole detection eliminates the scattered light that causes background noise. Moreover, random, access bit recording and reading is possible only when using scanning optics.

We used oil emulsion objective lenses to reduce spherical aberration in the experiments except when using photorefractive materials as recording media. However, the use of oil is not recommended, particularly during data read out. Spherical aberration can also be corrected reasonably without the use of oil-emulsion objectives, by servo-controlling the relative position of the point-of-interest to the lens with respect to the positions of the source and the detector.

The future use of 3D optical memory in consumer products also needs the development of 3D servo tracking mechanisms and data coding schemes. Even without two-photon excitation for bit-data recording, we have shown good depth separation between recorded planes in 3D optical memory because of the nonlinear effect of the photochemical process and the spatially high-pass filtering response of photopolymerization process. However, the use of multi-photon processes is also desired.

References

[1] S. Kawata, *Kogaku (J. Opt. Soc. Jpn)* **21**, 766–779 (1992) [in Japanese].
[2] Y. Inouye and S. Kawata, *Opt. Lett.* **19**, 159–161 (1994).
[3] E. Betzig, J.K. Trautman, R. Wolfe, E.M. Gyorgy and P.L. Finn, Appl. Phys. Lett. **61**, 142–144 (1992).
[4] S. Kawata, T. Tanaka, Y. Hashimoto and Y. Kawata, *SPIE Proc.* **2042**, 314–325, Quebec, 1993.
[5] Y. Kawata, H. Ueki, Y. Hashimoto and S. Kawata, *Appl. Opt.* (in press).
[6] Y. Hashimoto, Y. Kawata and S. Kawata, *Proceedings of Japan Optics '92*, Kyoto, 1992, pp. 39–40.
[7] Y. Kawata, Y. Hashimoto and S. Kawata, *Proceedings of Semi-Ann. Mtg. of Jpn. Soc. Appl. Phys.*, Tokyo, 1993, p. 900.
[8] H. Tanigawa, T. Ichihashi and A. Nagata, *Kogaku (J. Opt. Soc. Jpn)* **20**, 227–231 (1991) [in Japanese].
[9] Y. Kawata, S. Kawata and S. Minami, *J. Opt. Soc. Am. B* **7**, 2362–2368 (1990).

[10] T. Wilson and C.J.R. Sheppard, *Theory and Practice of Scanning Optical Microscopy*, Academic Press, London, 1984.
[11] C.J.R. Sheppard, Min Gu, Y. Kawata and S. Kawata, *J. Opt. Soc. Am. A.* **11**, 593–598 (1994).
[12] J.H. Strickler and W.W. Webb, *Opt. Lett.* **16**, 1780–1782 (1991).

3.4

Second-harmonic Generation in Multilayered Semiconductors

Nagaatsu Ogasawara

Abstract

A highly efficient semiconductor multilayered second-harmonic generator is proposed based on a new phase-matching scheme for intracavity second-harmonic generation (SHG). The device consists of half-the-harmonic-wavelength thick alternating layers of two different semiconductors inserted in a high-Q cavity. The device is efficient because of the coexistence of two phase-matched SHG processes, the standing-wave quasi-phase-matched SHG and the diffraction-assisted phase-matched SHG, occurring through the modulation of the second-order nonlinear optical coefficient and the refractive index, respectively, along the longitudinal axis of the cavity.

1 Introduction

There is currently much interest in second-harmonic generation (SHG) using laser diodes as fundamental light sources because of the growing demand for compact short-wavelength light sources in modern photonics technologies. III–V compound semiconductors, the diode laser material, are attractive also as a nonlinear optical medium for SHG because of their high second-order nonlinearity comparable with those of $LiNbO_3$ and $LiTaO_3$, the most representative nonlinear optical crystals. In addition, a number of epitaxial growth and processing technologies established regarding III–V semiconductors can be utilized to fabricate compact and integrable second-harmonic generators [1,2].

To realize efficient SHG, phase matching, i.e. equality of wavenumber between the harmonic radiation and the nonlinear polarization, is essential [3]. To date, a variety of phase-matching techniques have been proposed [4–7]. Among them, the quasi-phase matching (QPM) technique [4], where phase-matching is achieved, in effect, by making use of the periodic modulation of the nonlinearity to compensate for the wavenumber difference, is attractive in practice since it is applicable to normally nonphase-matchable but highly nonlinear materials and since the phase-matching is obtained at an arbitrary wavelength and temperature by choosing the modulation period appropriately. In most studies of QPM reported so far, attention has been focused on the configuration where the fundamental wave travels through the nonlinear medium from one end to the other, generating nonlinear polarization with a wavenumber twice as large as that of the fundamental wave. Efficient SHG also requires the use of intense fundamental waves since the conversion efficiency is proportional to the fundamental optical power density in the nonlinear medium [3]. The nonlinear medium is often set inside a high-Q cavity so that the intense fundamental wave resonating in the cavity can be utilized. There, the forward- and backward-propagating fundamental waves, forming a standing wave, induce a nonpropagating nonlinear polarization in addition to forward- and backward-propagating nonlinear polarizations.

In this section we discuss a new phase-matching scheme that uses both nonpropagating and propagating components of the nonlinear polarization in a cavity as a source of harmonic generation. A multilayered semiconductor SHG device is proposed where alternating layers of two different semiconductors are inserted in a high-Q cavity to implement spatial modulation of optical constants along the longitudinal axis of the cavity. It is shown that the conversion efficiency of an SHG device based on the new phase-matching scheme incorporating half-the-harmonic-wavelength thick GaP/AlP multiple layers in a cavity is considerably higher than that of a device with the conventional QPM structure [8].

2 Phase Matching in Periodic Structures

We discuss a phase-matching scheme that takes advantage of the artificial modulation of optical constants, i.e. the second-order nonlinear optical constant d and the refractive index n of the nonlinear medium, for a configuration where a fundamental wave passes through the nonlinear medium in one direction (2.1) and for a configuration where counterpropagating fundamental waves form a standing wave in the nonlinear medium (2.2).

2.1 Conventional QPM for the fundamental travelling wave

In the conventional QPM [4], the fundamental wave relevant is a travelling wave with a wavenumber $k^{(\omega)}$ and generates a nonlinear polarization with a wavenumber $2k^{(\omega)}$, twice as large as that of the fundamental wave, as shown in the upper part of Figure 3.4.1. The difference between the wavenumber of the second-harmonic wave $k^{(2\omega)}$ and that of the propagating nonlinear polarization $2k^{(\omega)}$, $\Delta k = k^{(2\omega)} - 2k^{(\omega)}$, is compensated for by modulating d at a period of the coherence length $l_c = 2\pi/\Delta k = \lambda^{(2\omega)}/(n^{(2\omega)} - n^{(\omega)})$, where $n^{(2\omega)}$ and $n^{(\omega)}$ are the refractive indices for the harmonic and fundamental

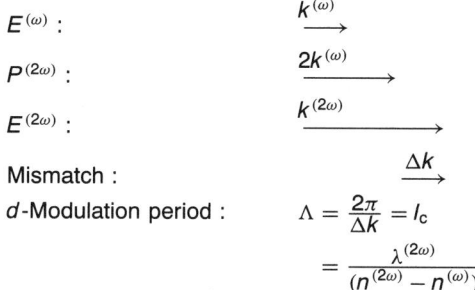
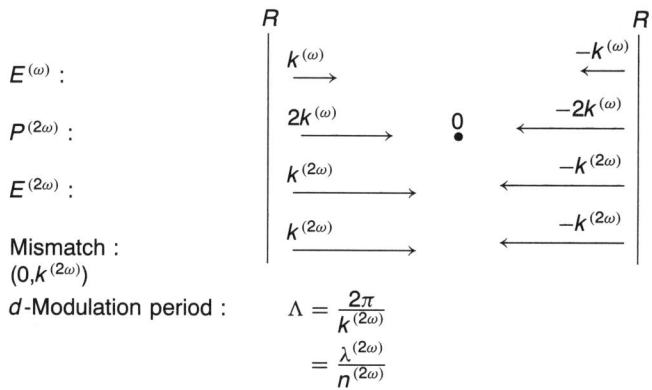

Figure 3.4.1
Configuration of a conventional QPM for a fundamental travelling wave and that of a new QPM for a fundamental standing wave. $E^{(\omega)}$ the fundamental field; $P^{(2\omega)}$ the nonlinear polarization; $E^{(2\omega)}$ the harmonic field

frequencies, respectively, and $\lambda^{(2\omega)}$ is the harmonic wavelength. An index grating with a period of l_c can also be utilized to compensate for the wavenumber difference [5].

2.2 New phase-matching scheme for the fundamental standing wave [8]

We describe here two types of new phase-matching, the standing wave QPM and the diffraction-assisted phase matching, achieved in a configuration where two counter-propagating fundamental waves are interacting with linear and nonlinear gratings with a period of one wavelength of the harmonic wave $\lambda^{(2\omega)}/n^{(2\omega)}$.

Standing-wave QPM

The nonlinear polarization induced by the coupling between forward- and backward-propagating fundamental waves includes, in addition to forward- and backward-propagating components, a nonpropagating (zero-wavenumber) component, as shown

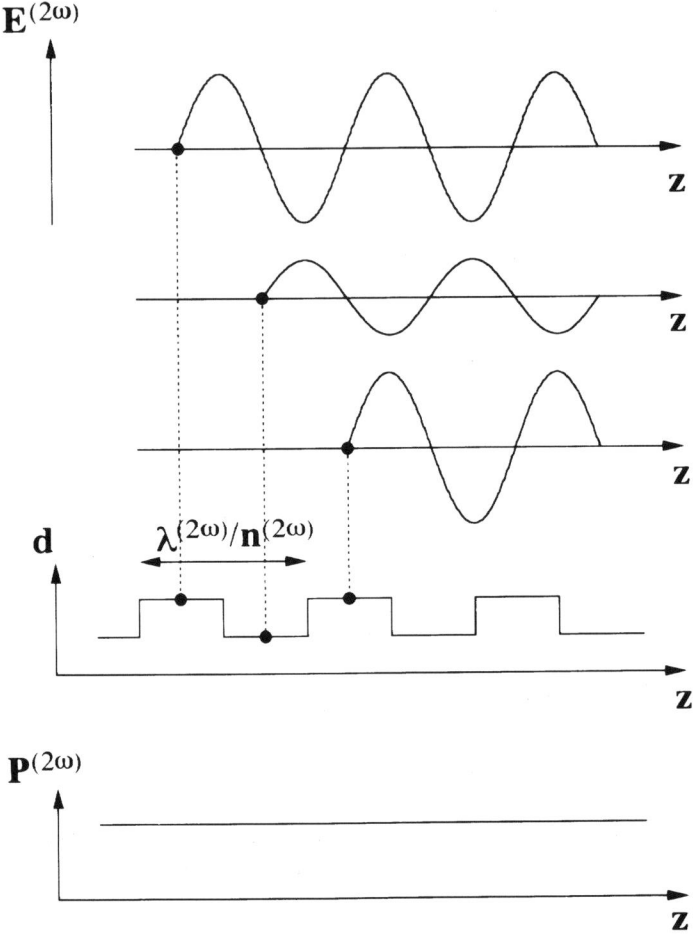

Figure 3.4.2
Schematic illustration of a standing-wave QPM SHG

in the lower part of Figure 3.4.1. The zero-wavenumber component can be utilized for phase-matched SHG by modulating d at a period of one wavelength of the harmonic wave; the difference between the wavenumber of the second-harmonic wave $k^{(2\omega)}$ and that of the nonpropagating nonlinear polarization 0 is $k^{(2\omega)}$ and this is just compensated for by the modulation of d with a period of $\lambda^{(2\omega)}/n^{(2\omega)} = 2\pi/k^{(2\omega)}$. This QPM, the standing-wave QPM, is illustrated in Figure 3.4.2. The phase of the nonlinear polarization utilized here is homogeneous and harmonic waves of an identical phase are generated simultaneously at each point in the nonlinear medium. Then, harmonic waves generated at two points separated from each other by one wavelength of the harmonic wave are superimposed spatially with an identical phase, while those generated at two points separated from each other by half the wavelength of the harmonic wave are superimposed spatially with reversed phases. Therefore, a modulation of d with a period of one wavelength of the harmonic wave introduces an imbalance of amplitudes between in-phase and anti-phased harmonic waves, giving rise to a net growth in the harmonic wave. The standing-wave QPM is the first phase-matched SHG process achieved in the new structure.

Diffraction-assisted phase matching

When the modulation of d is implemented by stacking different materials, the index n is also modulated to form an index grating which diffracts the fundamental wave, as shown in Figure 3.4.3. Then, the forward-propagating fundamental wave, for example, is a superimposition of the components with wavenumber $k^{(\omega)}$ and wavenumber $-k^{(\omega)} + k^{(2\omega)}$; the latter arises owing to the diffraction of the backward-propagating fundamental wave with wavenumber $-k^{(\omega)}$ by the index grating with a period of $2\pi/k^{(2\omega)}$. The coupling between the undiffracted ($k^{(\omega)}$) and diffracted ($-k^{(\omega)} + k^{(2\omega)}$) fundamental waves combined with the DC component of d produces a nonlinear polarization with a wavenumber $k^{(2\omega)}$ that is identical to that of the harmonic radiation, leading to a phase-matched SHG. This diffraction-assisted SHG is the second phase-matched SHG process achieved in the new structure.

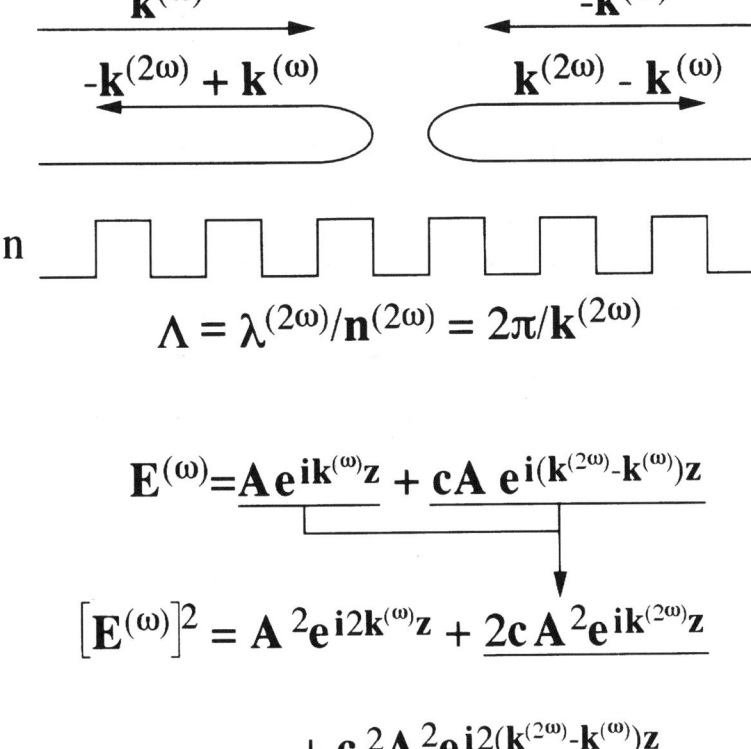

Figure 3.4.3
Schematic illustration of a diffraction-assisted phase-matched SHG

3 Multilayered Semiconductor SHG Device [8]

3.1 Structure

Alternating layers of two semiconductors with different d and n, formed by an epitaxial growth, can be utilized to achieve the phase-matching conditions described in the preceding section. Figure 3.4.4 shows schematically a second-harmonic generator composed of semiconductor multiple layers sandwiched between high-reflectivity multilayered dielectric mirrors forming a Fabry–Perot cavity. A combination of GaP and AlP, which have relatively wide band gaps among III–V compounds, is a good choice of nonlinear medium for short-wavelength light generation. The difference in the magnitudes of d and n between GaP and AlP provides linear and nonlinear gratings along the longitudinal axis of the cavity. The GaP/AlP layers are (111) oriented so that the nonlinear polarization parallel to the layers, from which a harmonic wave emanates along the longitudinal axis of the cavity, is induced through the d tensor of the $\overline{4}3m$ symmetry crystal by a fundamental wave incident normally on the layers.

The lower part of Figure 3.4.5 shows the profiles of d in the cavity for the conventional QPM structure (c) and for the new structure (n). The segments with the higher

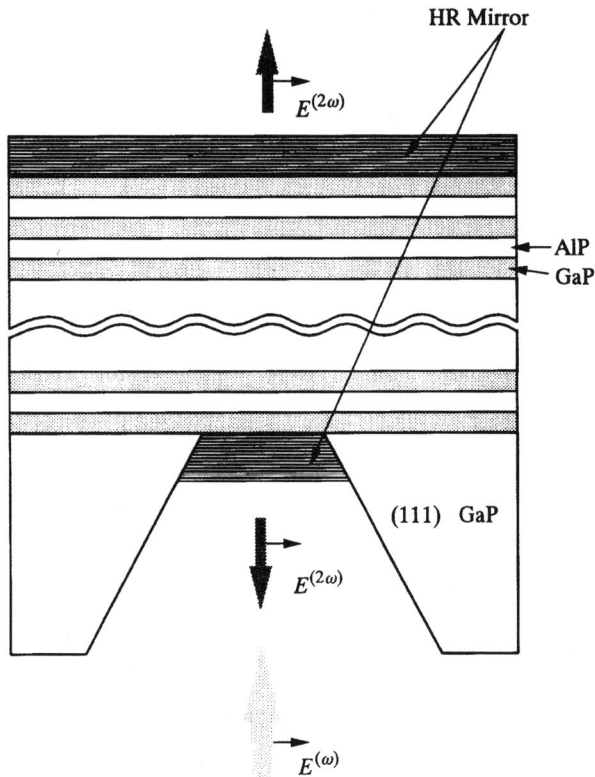

Figure 3.4.4
SHG device incorporating (111) GaP/AlP layers in a cavity [8]

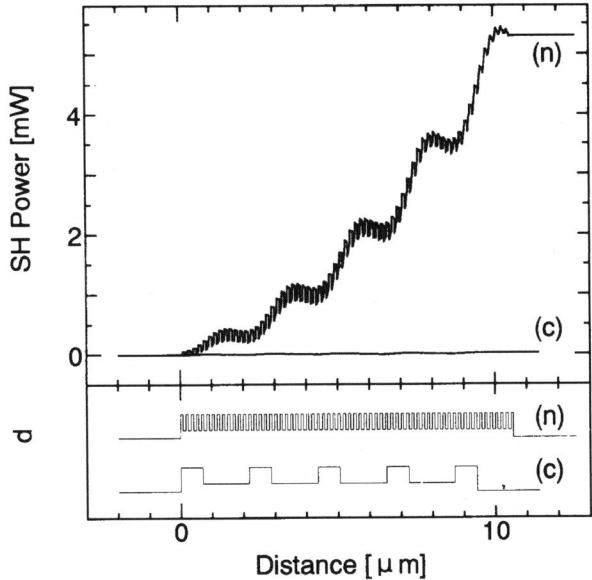

Figure 3.4.5
Profiles of d (lower part) and rightward-propagating second-harmonic power (upper part) in the new device (n) and the conventional QPM device (c) [8]

Table 3.4.1
Material parameters ($\lambda^{(\omega)} = 1.064$ μm) [8]

Materials	$n^{(\omega)}$	$n^{(2\omega)}$	$\alpha^{(\omega)}$ (cm^{-1})	$\alpha^{(2\omega)}$ (cm^{-1})	d (pm/V)
GaP	3.11	3.49	0	125	45
AlP	2.76	2.94	0	0	15

nonlinearity are GaP layers and those with the lower nonlinearity are AlP layers. In the conventional QPM structure for SHG with a fundamental wavelength of $\lambda^{(\omega)} = 1.064$ μm, the thickness of each segment $l_c/2$ is calculated to be 0.7 μm for GaP and 1.48 μm for AlP using the refractive index values in Table 3.4.1. In the new structure, the thickness of the segment $\lambda^{(2\omega)}/(2n^{(2\omega)})$ is 0.0762 μm for GaP and 0.0905 μm for AlP.

3.2 Analysis of SHG efficiency

The efficiency of the conventional and new QPM devices can be analyzed utilizing the procedure described in Reference [9]; the reflection of both the fundamental and harmonic waves at heteroboundaries is taken into account by means of the transfer matrix technique and the harmonic generation from the given fundamental field is evaluated, neglecting pump depletion, based on Green's function formalism. The material parameters are tabulated in Table 3.4.1. Here, the magnitude of d of GaP was determined from an absolute measurement based on the Maker fringe method for $\lambda^{(\omega)} = 1.3$ μm [10]; the dispersion of d was taken into account assuming Miller's rule, i.e. the constancy of the quantity $d/(\chi^{(2\omega)}(\chi^{(\omega)})^2)$ for different ω, where $\chi^{(\omega)}$ and $\chi^{(2\omega)}$ denote the linear susceptibility for the fundamental and harmonic frequencies, respectively. The d value of AlP was obtained making use of the ratio $d(\text{AlP})/d(\text{GaP}) = 0.33$ for $\lambda^\omega = 1.064$ μm [9], which was determined by the method of reflected second harmonics [11]. The origin of the other parameters in Table 3.4.1 are found in Reference [8]. Let us take the reflectivity of the end mirrors to be 0.99 and 0 for the fundamental and harmonic waves, respectively, and choose the phase shift upon reflection of the fundamental wave so that the cavity resonance is achieved. These conditions can be simultaneously satisfied by properly designing the multilayered dielectric mirrors.

The upper part of Figure 3.4.5 shows the profile of the rightward-propagating harmonic power for a fundamental power of 200 mW with a cross-section of 10 μm^2 incident on the left end of the device. It can be seen that the efficiency of the new device (n) is markedly higher than that of the conventional device (c); a harmonic output of ~5 mW is obtained from the new device only 10 μm thick, while the output of the conventional QPM device is ~26 μW. An equal amount of harmonic power is also obtained from the left end of the device.

3.3 Discussion

Figure 3.4.6 helps us discuss the efficiency of these devices, where the ordinate is the harmonic field amplitude instead of the harmonic power in Figure 3.4.5. The curve (n1) depicts the contribution to the harmonic field in the new device arising from the standing-wave QPM SHG and (n2) is the contribution from the diffraction-assisted, phase-matched SHG. It can be seen that the magnitudes of both contributions grow with rightward propagation. (n) is the sum of (n1) and (n2). Therefore it can be said that the high efficiency of the new device stems from the fact that both the nonpropagating and propagating components of the nonlinear polarization are effectively utilized as a source of harmonic generation. The undulation in the curve (n2) is a Maker fringe arising from the nonlinear polarization component with a wavenumber $2k^{(\omega)}$.

The low efficiency of the conventional structure (c) is understood as follows. The layer thickness of the conventional device, $l_c/2 = \lambda^{(2\omega)}/2(n^{(2\omega)} - n^{(\omega)})$, depends on the magnitude of index dispersion. Since the dispersion of AlP ($n^{(2\omega)} - n^{(\omega)} = 0.18$) is half of that of GaP (0.38), AlP layers are twice as thick as GaP layers. Consequently, although the d of GaP is three times as large as that of AlP, the harmonic wave generated in GaP layers is severely cancelled out by the anti-phased harmonic wave generated in the thicker AlP layers. In the new structure, on the other hand, the thickness is determined by $n^{(2\omega)}$ and the difference in the thickness between GaP and AlP is less than 20%. Therefore the cancellation of the harmonic wave in the less-nonlinear but thicker AlP layer is greatly alleviated, as clearly shown in the curve (n1).

It is worth mentioning that the new SHG device discussed here should be distinguished from the recently proposed semiconductor surface-emitting SHG device [1,2], where the harmonic wave radiating normally to the surface is generated

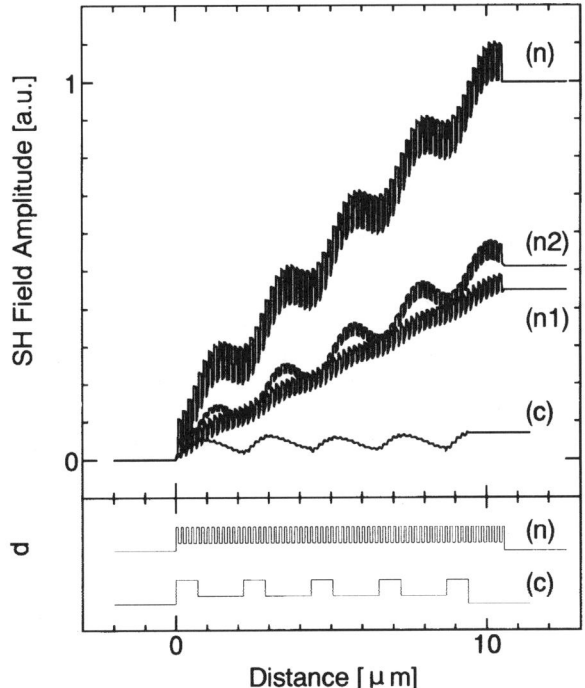

Figure 3.4.6
Profiles of rightward-propagating second-harmonic field in the new device (n) and the conventional QPM device (c). (n) is the sum of the second-harmonic fields arising from the standing-wave QPM SHG (n1) and the diffraction-assisted phase-matched SHG (n2) [8]

from nonlinear polarization induced by counterpropagating fundamental waves running along a waveguide formed parallel to the surface. In the latter device the interaction length, which is determined by the waveguide thickness, can be enlarged only at the expense of lowering the fundamental power density; the conversion efficiency is subject to a trade-off between the interaction length and the fundamental power density. On the other hand, in the colinear configuration employed in the new SHG device, where both the fundamental and harmonic waves propagate normally to the surface, the conversion efficiency grows in proportion to the square of the device thickness.

4 Conclusion

We have discussed a phase-matching scheme utilizing periodic gratings. A new scheme has been presented where two phase-matched SHG processes, the standing-wave QPM SHG and the diffraction-assisted phase-matched SHG, are induced through the nonlinear and linear gratings with a period of one wavelength of the harmonic wave inserted in a cavity. An SHG device incorporating half-the-harmonic-wavelength thick GaP/AlP layers in a cavity has been shown to be much more efficient than a conventional device composed of half-the-coherence-length thick layers. The new SHG device can be fabricated by taking advantage of such epitaxial growth techniques featuring high controllability of layer thickness as molecular beam epitaxy and metalorganic vapor phase deposition.

References

[1] R. Normandin, S. Létourneau, F. Chatenoud and R.L. Williams, *IEEE J. Quantum Electron.* **27**, 1520-1530 (1991).
[2] D. Vakhshoori and S. Wang, *IEEE J. Lightwave Technol.* **9**, 906-917 (1991).
[3] A. Yariv, *Quantum Electronics*, John Wiley & Sons, 1988, Ch. 16.
[4] J.A. Armstrong, N. Bloembergen, J. Ducuing and P.S. Pershan, *Phys. Rev.* **127**, 1918-1939 (1962).
[5] T. Suhara and H. Nishihara, *IEEE J. Quantum Electron.* **26**, 1265-1276 (1990).
[6] S. Umegaki, T. Kondo, R. Morita, N. Ogasawara and R. Ito, *Nonlinear Optics* **1**, 253-272 (1991).
[7] T. Kondo and R. Ito, in *Molecular Nonlinear Optics*, Academic Press, 1994 Ch. 5.
[8] H. Takahashi, M. Ohashi, T. Kondo, N. Ogasawara, Y. Shiraki and R. Ito, *Jpn. J. Appl. Phys.*, **33**, L1456-1458 (1994).
[9] M. Ohashi, T. Kondo, S. Fukatsu, S.S. Kano, Y. Shiraki and R. Ito, *Proc. SPIE*, **1983**, 836-837 (1993).
[10] T. Kondo, I. Shoji, K. Yashiki and R. Ito, *International Conference on Quantum Electronics Technical Digest Series 1992*, **9**, 242-244 (1992).
[11] M. Ohashi, T. Kondo, R. Ito, S. Fukatsu, Y. Shiraki, K. Kumata and S.S. Kano, *J. Appl. Phys.* **74**, 596-601 (1993).

3.5

Second-harmonic Generation and Light Amplification in Organic Waveguides

Keisuke Sasaki

Abstract

We present two research items: (1) poled-polymer waveguides of two copolymer systems with long relaxations of quadratic nonlinear coefficients for blue second-harmonic generations down to 405 nm by Cerenkov-type phase matching, and (2) high-level amplification of a nanosecond order pulse signal in the visible region up to about 1 kW power for few watts input into organic dye-doped graded index polymer optical fibers. The former is promising as a compact blue-green light source for multi-purpose applications, and the latter also can be utilized directly as a tunable coherent light amplifier and for various extended applications.

1 Introduction

Organic materials will be widely utilized in optoelectronics as liquid crystals in display devices. They have many virtues, for example mass production, low-cost performance, and easy processessing. In particular, the freedom of material design and synthesis are the most remarkable potentials in advanced applications. We present two main functional devices in optoelectronics: second-harmonic generation (SHG) and light amplification by waveguides (fibers) of organic materials. In the SHG study two copolymer systems, namely dispersed red 1 (DR 1)/polymethyl-methacrylate (PMMA) and para-nitroaniline (p-NA)/polyvinylalcohol (PVA), for thin film waveguides, are utilized by corona-poling [1–3]. The quadratic SHG tensor components of both waveguides, d_{33}, measured by the Maker-fringe method, were almost the same level, between 10 and 20 pm/V, with long relaxation characteristics. Frequency doubling in a Ti: Sapphire laser by Cerenkov-type phase-matching to pursue a compact blue-green laser was confirmed.

A polymer optical fiber amplifier (POFA) was realized by an organic laser dye-doped grade index fiber structure [4] for the first time [5-7].

A graded index preformed rod of the fiber was prepared using the interfacial sol-gel polymerization process with a dye-dissolved methacrylate (MMA) monomer inside the PMMA tube. The preformed rod was thermally pulled to form a fibre with 0.5 mm outer diameter.

The dye concentration in the raw monomer was quite low, *e.g.* 10 ppm/cm^3. Such a low concentration led to the very slow degradation of the dye material for strong pumping excitation of pulsed SHG Nd:YAG laser, 532 nm wavelength. As a typical amplifying characteristic we measured a 27 dB gain at 1 W signal input.

2 Poled-polymer Waveguides for Compact Blue-green Light Sources

2.1 A poled-polymer waveguide for a SHG device of DR 1/PMMA

Skillful structures such as cross-linking and the anchoring of low-order nonlinear molecules to main-chain polymers are very effective in preparing physically and chemically stable poled-polymer systems [8]. We present a one-pendant-type poled-copolymer system which is the famous second order nonlinear molecule, disperse red 1 (DR1) anchored to polymethyl-methacrylate (PMMA) as the main-chain polymer (Figure 3.5.1). In this figure the parameter x is the molar fraction of DR1 to the PMMA main chain. The prepared polymer was dissolved in an acetone solution with appropriate viscosity for spin-coating onto a Pyrex substrate of suitable thickness. A ridge-type channel waveguide structure, 20 μm wide, (Figures 3.5.2(a) and (b)) was prepared by the usual photoresist process with reactive-ion-etching (RIE) in advance of poling, as shown in Figure 3.5.3. Anisotropic absorption spectra and refractive indices after poling were measured as shown in Figures 3.5.4 and 3.5.5, respectively. In both figures the blue wavelength ranges are located in anomalous dispersion regions.

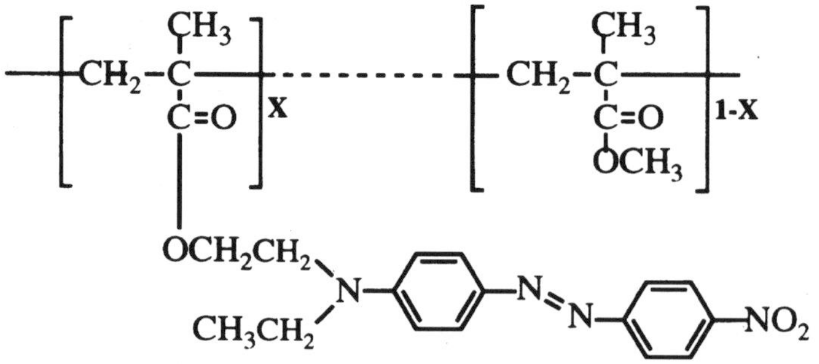

Figure 3.5.1
Chemical structure of the DR1/PMMA pendant-type copolymer system

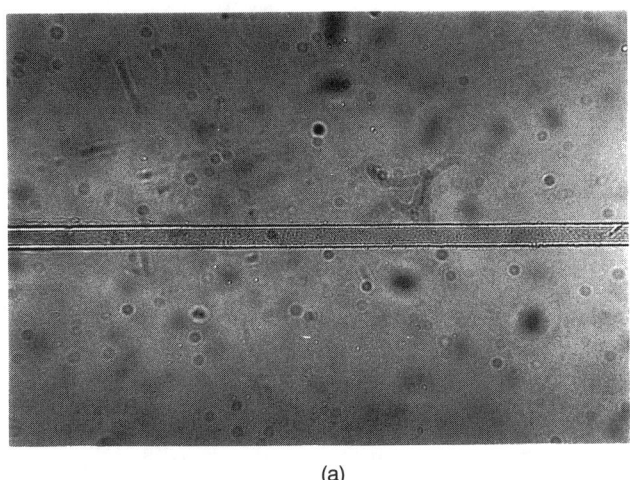

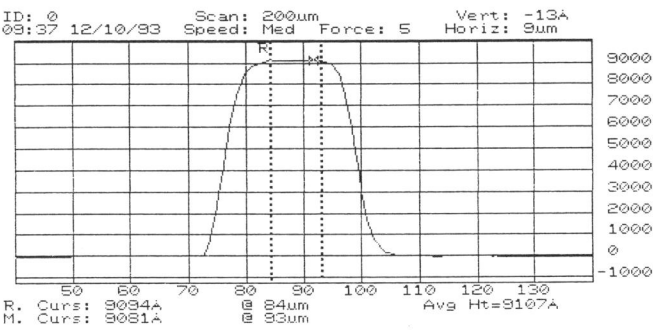

Figure 3.5.2
Ridge-type SHG waveguide. (a) Magnified look-down picture of waveguide. (b) Cross-sectional dimension of the ridge-type waveguide

Table 3.5.1
d_{33} coefficients of the molar fraction of DR1

Sample no.	Molar fraction	d_{33} (pm/V)	n (633 nm)	n (840 nm)	n (1064 nm)
1	0.014	7.0	1.504	1.479	1.489
2	0.023	9.0	1.509	1.501	1.495
3	0.034	16.1	1.531	1.511	1.504
4	0.045	23.2	1.550	1.527	1.518
5	0.073	37.5	1.567	1.539	1.529

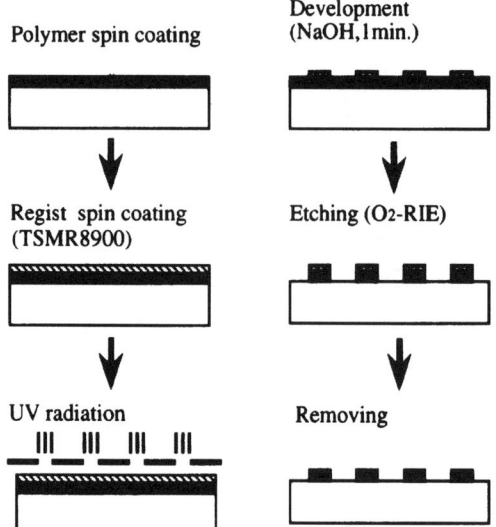

Figure 3.5.3
Schematic illustration of the patterning process of the ridge-type waveguide

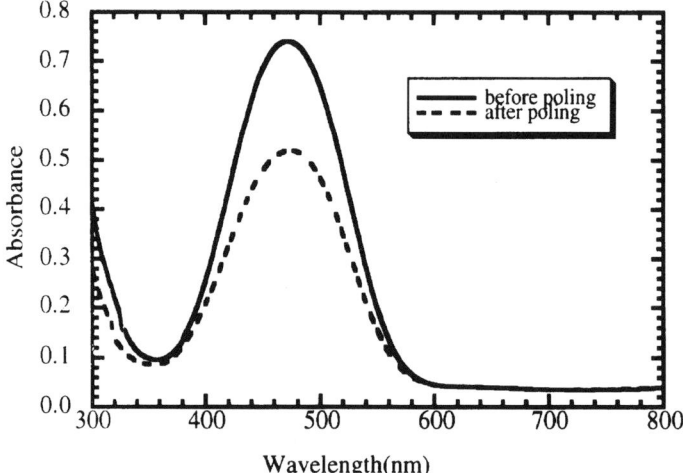

Figure 3.5.4
Anisotropic absorption of the DR1/PMMA poled polymer film

The poling effect on the quadratic nonlinear coefficients d_{33} with molar fractions of DR1 are given in Table 3.5.1. Higher molar fractions gave higher d_{33} values. However, in that case the film lost its smooth surface and was unsuitable as a waveguide. In the frequency doubling experiment sample waveguide no. 2 was mainly used.

Cerenkov-type SHG of Ti:sapphire 840 nm was observed as shown in Figure 3.5.6. In this picture an interference pattern depending on the confinement of the SHG output beam in the channel waveguide was recognized.

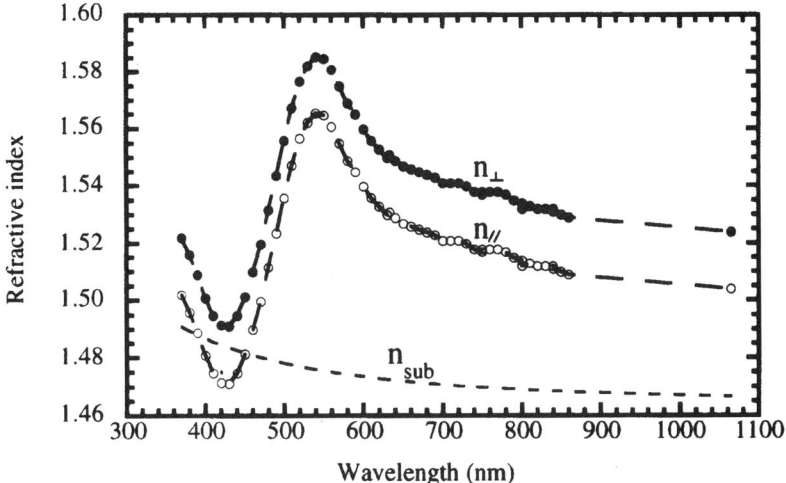

Figure 3.5.5
Anisotropic dispersion of the refractive indices in the DR1/PMMA film

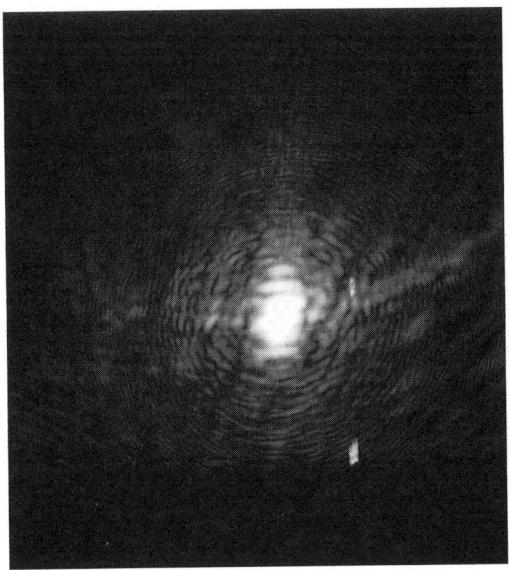

Figure 3.5.6
Far-field pattern of blue (420 nm) Cerenkov-type SHG from the channel waveguide

2.2 A poled-polymer waveguide for a SHG device of p-NA/PVA

We next present another application of a poled-polymer for a SHG waveguide using the para-nitroaniline (p-NA)/polyvinylalchohol (PVA) system. The molecular structure of the system is illustrated in Figure 3.5.7. In this figure x is the molar fraction of p-NA pendants introduced into the PVA chain and is equal to 80%. The glass transition temperature of this material system is about 115 °C with accompanying photoresistive properties, which is advantageous for patterning by electron beam and UV light device preparation. The absorption spectra of the spin-coated film for normal incident light are changed remarkably by bleaching due to molecular alignment during corona-poling, as shown in Figure 3.5.8.

A typical relaxation curve for the d_{33} coefficient is extremely stable, as shown in Figure 3.5.9. After 6 months d_{33} is still 16 pm/V, which is a high enough value for

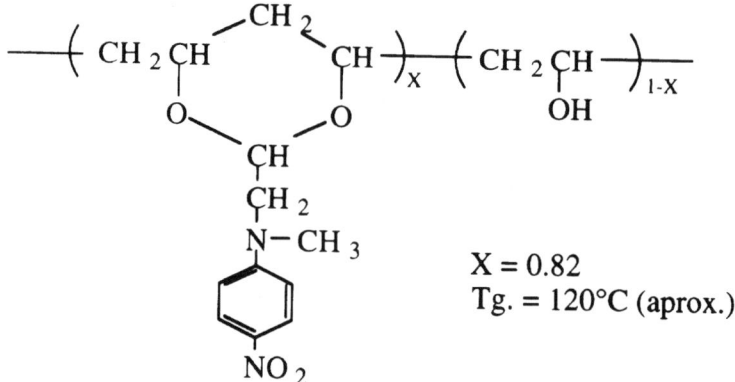

Figure 3.5.7
Molecular structure of the p-NA/PVA pendant-type copolymer

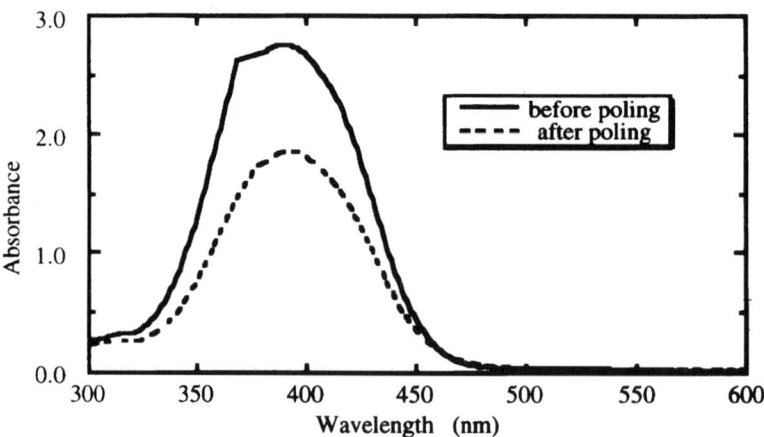

Figure 3.5.8
Bleaching of the absorption spectrum by corona-poling of the p-NA/PVA copolymer film

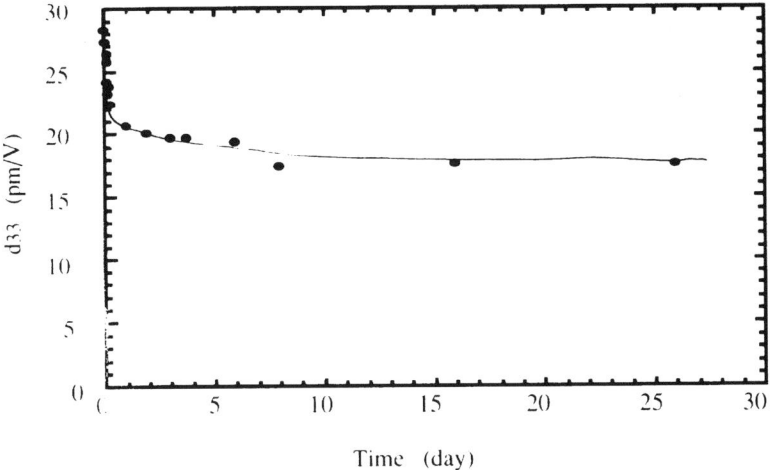

Figure 3.5.9
Relaxation procedure of the d_{33} coefficient

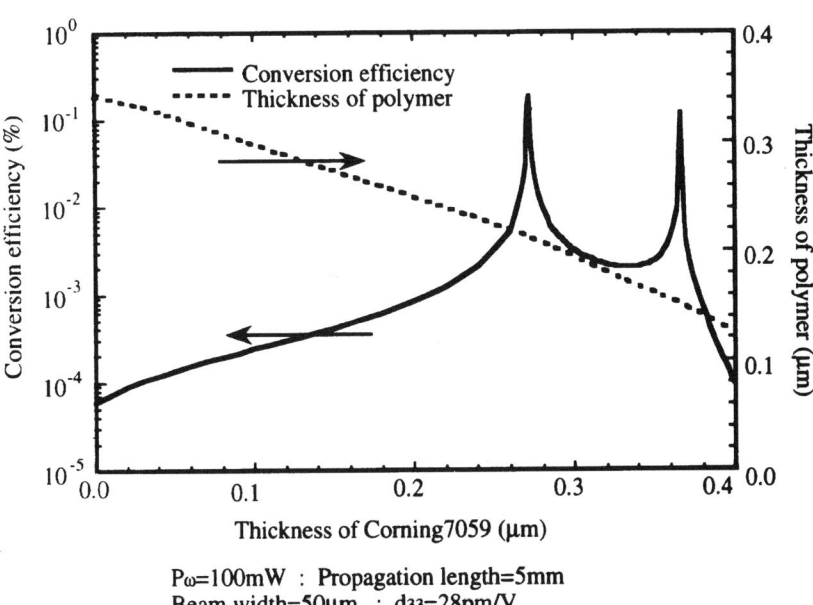

$P\omega$=100mW : Propagation length=5mm
Beam width=50μm : d_{33}=28pm/V

$n_f(\omega)$=1.674:$n_f(2\omega)$=1.819 $n_g(\omega)$=1.570:$n_g(2\omega)$=1.583
$n_{fz}(\omega)$=1.576:$n_{fz}(2\omega)$=1.656 $n_s(\omega)$=1.467:$n_s(2\omega)$=1.476

Figure 3.5.10
Thickness-dependent optimal conversion efficiency for the Corning 7059 buffer layer

practical device application of frequency doubling waveguide. In this study we designed a purposeful hybrid-waveguide structure which consisted of four layers, including a buffer layer with two adjustable parameters, film thickness and refractive index. The buffer layers were deposited on a Pyre substrate by rf sputtering of Corning 7059 glass and mixed $(Ta_2O_5)_x/(SiO_2)_{1-x}$. The refractive index of the rf sputtered buffer layer of the mixture is adjustable in the range from 1.45 ($x = 0$) to 2.2 ($x = 1.0$). It can be seen that the introduction of this freedom contributes dramatically to the improvement of the conversion efficiency in frequency doubling from the theoretical estimation. Figure 3.5.10 shows an example of the calculated optimal conversion efficiency on the Corning 7059 buffer layer thickness. In this figure the efficiency is improved by four orders of magnitude as compared with the no buffer layer case.

Figure 3.5.11 gives the field distributions at the optimal film thicknesses on the buffer layer and the polymer waveguide. Effective overlapping of both fields is realized in the figure. Another buffer layer, $(Ta_2O_5)_x/(SiO_2)_{1-x}$, was prepared to improve conversion efficiency. Insertion of the optimal buffer layer improves the overlapping of both fields at the phase matching between the fundamental first mode and the second harmonic second modes of guided waves, as shown in Figure 3.5.12. In this figure field distributions have the same polarities as in the nonlinear layer. In this case the efficiency is improved by about 30 times.

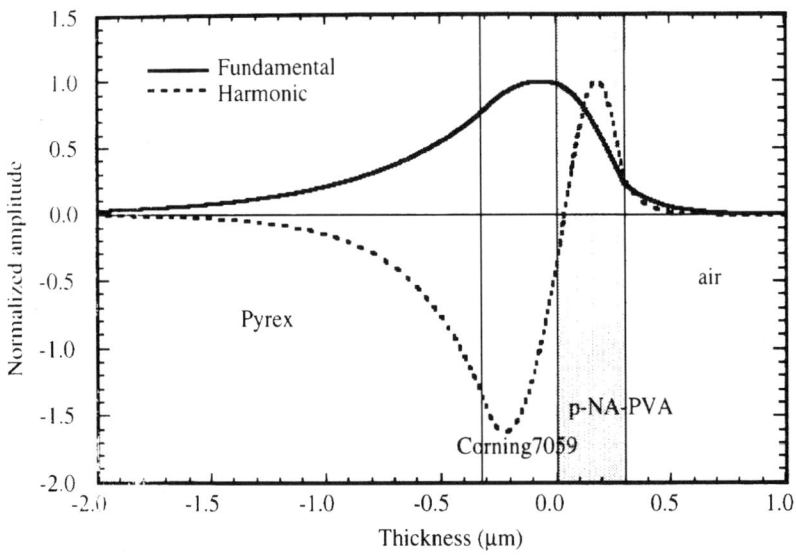

(air/p-NA-PVA/corning7059/Pyrex)

Polymer thickness = 0.30 μm   glass thickness = 0.32 μm   Neff = 1.503

Figure 3.5.11
Field distribution in the hybrid waveguide

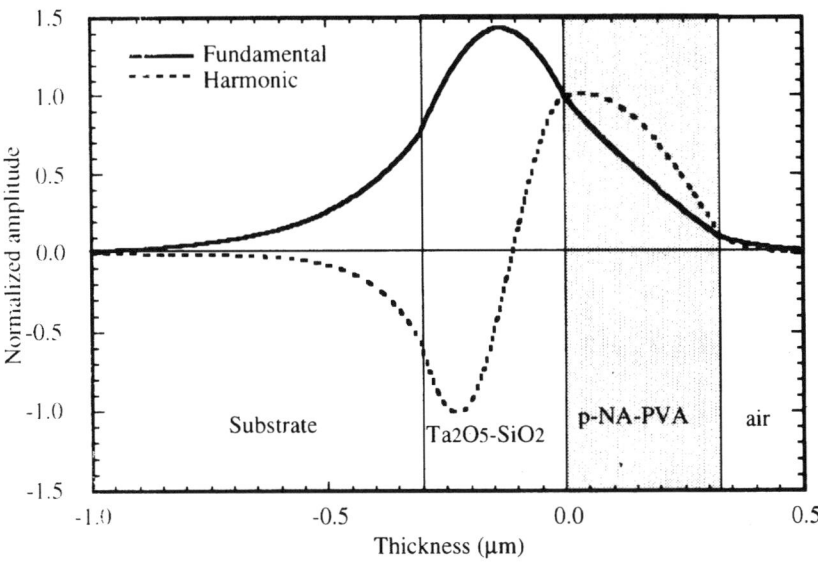

Figure 3.5.12
Insertion effect of the index-controllable $(Ta_2O_5)_x/(SiO_2)_{1-x}$ layer on SHG conversion efficiency

3   Organic Dye-doped Polymer Optical Amplifier

In long-distance light communication Er-doped silica fibers are utilized in practical networks as an essential element device. In short-distance communications, such as local area networks (LANs) and inner-building communications, polymer optical fibers are applied to networking. In this study we present a new polymer optical amplifier [9] which will be utilized not only in optical communication but also in various kinds of tunable coherent light sources for scientific instrumentations or medical applications. The polymer optical amplifier was prepared by the interfacial gel-polymerization technique in a tube of polymethylmethacrylate (PMMA) as a performed rod. A mixture of methylmethacrylate monomer, 1,1-*bis*(t-buthylperoxy)3,5,5-trimethylcyclohexane as initiator, n-buthyl mercaptan, benzyl n-buthyl phthalate (BBP), and organic dye was placed in the tube and heated in a furnace at 95 °C. Polymerization in the tube proceeded toward the center by making graded-index (GI) structure measured by the interference method [10,11] as shown in Figure 3.5.13.

We used three organic dyes, Rhodamine B, Rhodamine 6G, and Perylene red with concentrations from 0.01 to 10 ppm. Maintaining these concentration ranges is extremely important in order to stop dye deterioration due to dimerization aggregation

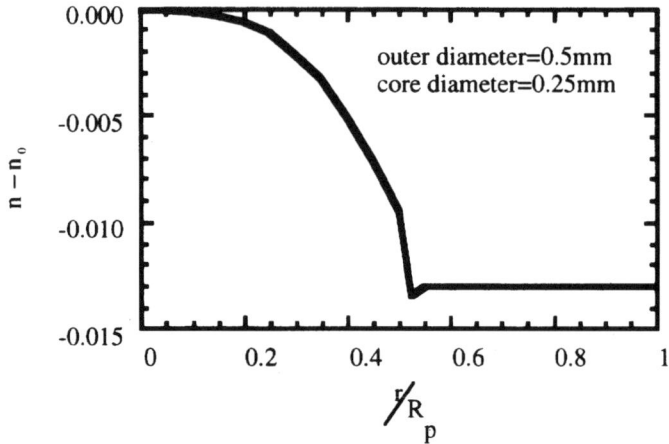

Figure 3.5.13
Radial graded-index structure of the preformed rod

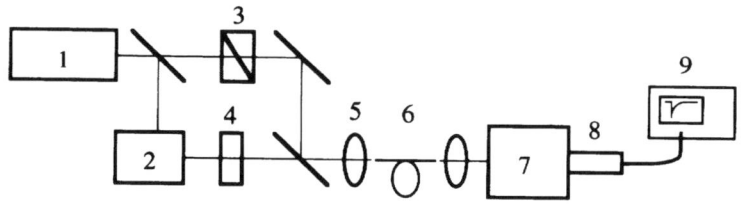

1: YAG SH Laser, 2: Dye Laser, 3: Polarizer, 4: Attenetor
5: Microscope objective, 6: GI POFA, 7: Monocrometer
8: PMT, 9: Oscilloscope

Figure 3.5.14
Experimental setup for light amplification in the dye-doped polymer fiber

of the dye molecules. The performed rod was heat-drawn into a fiber of 250 μm core diameter at 190–250 °C by taking up the reel.

The amplification experiment was carried out using the setup shown in Figure 3.5.14. A double Q-switched Nd:YAG pulsed laser, 532 nm (10 Hz, FWHM: 5 ns) was divided into two paths for fiber pumping and for dye laser excitation to prepare a signal source. Pumping and signal were coaxially coupled into the fiber, as shown in Figure 3.5.14. The amplified output signal was measured by a photomultiplier with a monochrometer. Results are given in Table 3.5.2 for three kinds of dyes. The highest gain, 28 dB (absolute power was 620 W), was realized by the Rhodamine B dye-doped fiber at 591 nm signal wavelength for 10 kW pumping power even though optimal designing condition has not yet been established.

A three-level model of the dye was applied to estimate the amplification process with exponential absorption of pumping power along the direction of propagation in

Table 3.5.2
Measured small signal gains for fiber amplifiers with three dyes

Dye	Rhodamin 6G	Rhodamin B	Perylene red
Concentration (ppm)	5	1	10
Signal wavelength (nm)	571	591	621
Input signal level (mW)	10	40	10
Fiber length (mm)	750	750	300
Gain (dB)	24	18	13

the fiber by using the following equation for small signal operation:

$$G = 10 \log e \left[\frac{\{(S_{es} + S_{as})/S_{es}\} \ln\{S_{ap}I_p(0) + (1/t)\}}{\{S_{ap}I_p(0) \exp(-nS_{as}L) + (1/t)\}} - nS_{es}L \right]$$

where S is the cross-section and the subscripts p, s, e, and a stand for excitation, signal, emission, and absorption, respectively; n, L, t, $I_p(0)$ are the dye concentration, length of fiber, relaxation time, and the coupled pumping power normalized by the core cross-section, respectively.

Gain calculation for 0.01 ppm of Rhodamine B dye concentration in PMMA with 5 ns of relaxation time is simulated as shown in Figure 3.5.15. In the figure the gain has a plateau peak on the fiber length and pumping power. An amplified output power of 620 W is an absolute maximum large value, which can be never realized by Er-dope silica fibers and this original virtue can be promisingly applied to many research fields which require tunable coherent high-power light sources.

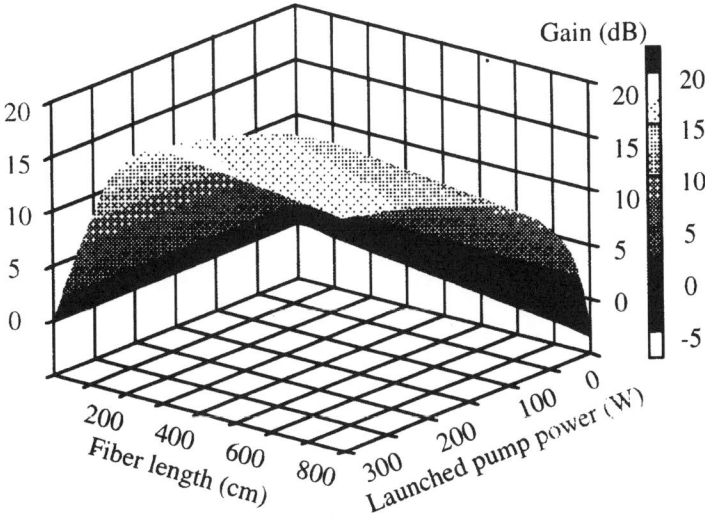

Figure 3.5.15
Estimated amplified output depending on fiber length and pumping power

4 Conclusions

In this study we reported the promising optoelectronic potentials of organic materials. They have many virtues, for example large optical nonlinearities for second-harmonic generation due to the large dipole moments of nonlocalized pi-electron systems in molecular structures and good compatibility between guest active molecules and host polymer materials in our polymer optical fiber amplifier.

In general up to now it has been said that organic materials are unstable from the chemical, physical, and mechanical point of views and are unsuitable for optoelectronic applications except for liquid crystal display devices.

The virtues of optoelectronic organic materials will be effectively utilized by the development of novel processing technologies for those materials.

References

[1] K.D. Singer, J.E. Sohn and S.J. Lalama, *Appl. Phys. Lett.* **49**, 248 (1986).
[2] K.D. Singer, M.G. Kuzyk and J.E. Sohn, *J. Opt. Soc. Am. B* **4**, 968 (1987).
[3] O. Sugihara, T. Kinoshita, M. Okabe, S. Kunioka, Y. Nonaka and K. Sasaki, *Appl. Opt.* **30**, 2957 (1991).
[4] Y. Koike, *Polymer* **32** (10), 1737 (1991).
[5] A. Tagaya, Y. Koike, T. Kinoshita, E. Nihei, T. Yamamoto and K. Sasaki, *Appl. Phys. Lett.* **63** (7), 883 (1993).
[6] A. Tagaya, Y. Koike, T. Kinoshita, E. Nihei, T. Yamamoto and K. Sasaki, *Technical Digest of CLEO'93*, 1993, CTuF3.
[7] T. Yamamoto, K. Fujii, S. Teramoto, A. Tagaya, E. Nihei, T. Kinoshita, Y. Koike and K. Sasaki, *Proc. SPIE* **2024**, 42 (1993).
[8] T. Kinoshita, Y. Nonaka, E. Nihei, Y. Koike and K. Sasaki, *Proc. 5th TOYOTA Conference*, 1991, p. 479.
[9] K. Sasaki, T. Saito, M. Serizawa, S. Furukawa and O. Hamano, Appl. *Phys. Lett.* **39**, 300 (1981).
[10] Y. Koike, Y. Sumi and Y. Ohtsuka, *Appl. Opt.* **25**, 3356 (1986).
[11] Y. Ohtsuka and Y. Koike, *Appl. Opt.* **19**, 2866 (1980).

3.6

Multi-photon Diffraction Switching Using a Resonant Two-level System

Naohiro Tan-no and **Osamu Satoh**

Abstract

We report on the theoretical and experimental study of multi-photon diffraction switching, with both phase-conjugate and nonconjugate properties, using two or three beams incident onto a coherent, resonant two-level system. The generation of this multi-photon scattering is described by theoretical analysis using a multi-wave optical Bloch equation, and the diffraction switching with the phase-conjugated wavefronts is confirmed by experiments using a dye laser and an Na-atomic vapor system.

1 Introduction

Coherent transient effects [1], which we investigated for a laser pulse with a pulse width shorter than the coherence relaxation time of a resonant two-level system, have been widely reported as transient phenomena of four-wave mixing [2,3], and as the generation of a phase-conjugated wavefront [4,5]. We first investigated this coherent interaction by using an optical Bloch equation associated with spatial-phase information of multi-wave mixing for two- and three-wave excitation. Macroscopic nonlinear polarizations were analytically introduced by Rab's solution of the optical Bloch equation. The polarizations produced multi-photon diffraction switching with multiple phase-conjugate and nonconjugate properties: $(n+1)\phi_1 - n\phi_2$ or $(n+1)\phi_2 - n\phi_1$ for two incident phases of ϕ_1 and ϕ_2. The multiple phase-superimposing configurations enabled us to create either multi-beam diffraction switching or spatial-phase logic operations. We achieved experimental confirmation of these phenomena by using an N_2-pumped tunable wavelength dye laser and an Na-vapor system.

2 Theoretical Analysis

We first investigate multi-photon scattering produced by the coherent interaction associated with spatial-phase information, since the optical Bloch equation has already been investigated [6] to describe coherent transient multi-photon scattering.

We consider the intersection of two beams that simultaneously interact with an isolated resonant two-level system. A sufficiently small crossing angle between the two beams is assumed. The configuration of the two beams incident onto a nonlinear medium is shown in Figure 3.6.1.

We write the total incident optical field as

$$E = \sum_{j=1}^{2} e_j E_j(t) \cos(\omega t - \varphi_j) \qquad (1)$$

where e_j is the unit polarization vector, ω is the optical angular frequency, $\varphi_j = k_j r - \phi_j(r)$, and $\phi_j(r)$ is the spatial-phase information.

An interacting Hamiltonian H is written as follows, where we omit the relaxation terms of the two-level system in order to consider the coherent transient effects

$$H = H_0 - \tfrac{1}{2} P_{12} \sum_{j=1}^{2} e_j E_j(t) \cos(\omega t - \varphi_j) \qquad (2)$$

We obtain an effective Hamiltonian H_e by using a unitary transformation and a rotative approximation for the Hamiltonian H. The equation of motion for the density matrix is given by

$$\frac{\partial \rho}{\partial t} = \frac{i2\pi}{h}[\rho H_e - H_e \rho] \qquad (3)$$

The optical Bloch equation that was deduced by the above expression, equation (3), was derived as follows, by defining the Bloch vectors $\rho(u, v, w)$:

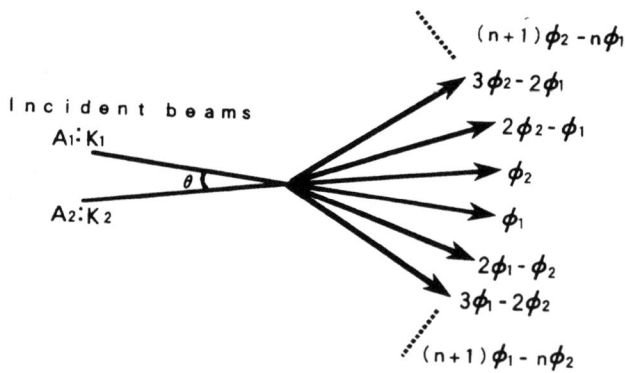

Figure 3.6.1
Schematic diagram of multi-diffraction directions and superimposing multiple phases for two input beams

$$u = \rho_{12} + \rho_{21}, \quad v = i(\rho_{12} - \rho_{21}); \quad w = \rho_{11} - \rho_{22} \tag{4}$$

$$\frac{\partial}{\partial t}\begin{bmatrix} u \\ v \\ w \end{bmatrix} = \begin{bmatrix} 0 & -\Delta\omega & -\chi_s \\ \Delta\omega & 0 & \chi_c \\ \chi_s & -\chi_c & 0 \end{bmatrix}\begin{bmatrix} u \\ v \\ w \end{bmatrix} \tag{5}$$

where

$$\Delta\omega = (\Omega_1 - \Omega_2) - \omega, \quad \chi_s = \sum_{j=1}^{2}\chi_j \sin\varphi_j, \quad \chi_c = \sum_{j=1}^{2}\chi_j \cos\varphi_j$$

and

$$\chi_j = \frac{2\pi p_{12} e_j E_j(t)}{h} \tag{6}$$

The solution to equation (5) is obtained under the assumption of the resonance condition $\Delta\omega = 0$, as follows

$$u = \frac{1}{\sqrt{2}}\sin\sqrt{2}(\theta_1 + \theta_2), \quad v = \frac{1}{\sqrt{2}}\sin\sqrt{2}(\theta_1 + \theta_2), \quad w = -\cos\sqrt{2}(\theta_1 + \theta_2) \tag{7}$$

where

$$\theta_1 = \int_0^t \chi_s dt', \quad \theta_2 = \int_0^t \chi_c dt' \tag{8}$$

The term $\sqrt{2}(\theta_1 + \theta_2)$ corresponds to Rabi's frequency and is spatially modulated by the spatial-phase information, which differs from the usual Bloch solution. The macroscopic polarization of the nonlinear medium is evaluated as follows, by using the unitary inverse transformation and assuming that the two input field areas are equal ($\theta_1 = \theta_2$)

$$P(t, r) = N_0 p_{12} R_e \{(u + iv) \exp(i\omega t)\}$$

$$= 2N_0 p_{12} R_e \left\langle \sum_{n=0}^{\infty}(-1)^{n+1} J_{2n+1}(\theta)[\exp\{-i[(n+1)\varphi_1 - n\varphi_2]\} \right.$$

$$\left. + \exp\{-i[(n+1)\varphi_2 - n\varphi_1]\}]\exp(i\omega t) \right\rangle \tag{9}$$

where $J_{2n+1}(\theta)$ is the Bessel function of order $2n + 1$.

The lower-order expansion of equation (9) is written as follows

$$P(t, r) = 2N_0 p_{12} Re\langle -J_1(\theta)[\exp(-ik_1 r + i\phi_1) + \exp(-ik_2 r + i\phi_2)]$$
$$+ J_3(\theta)\{\exp[-i(2k_1 - k_2)r + i(2\phi_1 - \phi_2)] + \exp[-i(2k_2 - k_1)r$$
$$+ i(2\phi_2 - \phi_1)\} - J_5(\theta)\{\exp[-i(3k_1 - 2k_2)r + i(3\phi_1 - 2\phi_2)]$$
$$+ \exp[-i(3k_2 - 2k_1)r + i(3\phi_2 - 2\phi_1)\} + J_7(\theta)\{\exp[-i(4k_1 - 3k_2)r$$
$$+ i(4\phi_1 - 3\phi_2)] + \exp[-i(4k_2 - 3k_1)r + i(4\phi_2 - 3\phi_1)\}\rangle. \tag{10}$$

Equation (9) shows that higher-order diffractions give rise to multiple phase-superimposition on the direction of many diffraction angles, as shown in Figure 3.6.1. These diffraction phases possess either characteristic multiple phase-conjugation or multiple nonconjugation, and/or multiple mixing of phase-conjugation and nonconjugation for the two input spatial phases. For three incident beams we obtain more complicated multiple phase-superimposition diffractions

$$P(t, r) = \sqrt{2} N_0 p_{12} Re \left\{ \left[J_0(q) \sum_{n=0}^{\infty} J_{2n+1}(\theta) \sin(2n+1)\left(\varphi_1 + \frac{\pi}{4}\right) \right. \right.$$

$$+ 2 \sum_{n=0}^{\infty} \sum_{m=1}^{\infty} J_{2n+1}(\theta) J_{2m}(\theta) \sin(2n+1)\left(\varphi_1 + \frac{\pi}{4}\right) \cos 2m\left(\varphi_2 + \frac{\pi}{4}\right)$$

$$\left. \left. + J_0(\theta) \sum_{n=0}^{\infty} J_{2n+1}(\theta) \sin(2n+1)\left(\varphi_2 + \frac{\pi}{4}\right) \right] \exp(i\omega t) \right\} \qquad (11)$$

The multiple phase terms of the lower-order expansion of equation (11) were obtained as follows

$$\varphi_4 = (k_1 + k_2 - k_3)r - \phi_1 - \phi_2 + \phi_3, \quad \varphi_5 = (k_1 + k_3 - k_2)r - \phi_1 - \phi_3 + \phi_2$$
$$\varphi_6 = (k_2 + k_3 - k_1)r - \phi_2 - \phi_3 + \phi_1, \quad \varphi_7 = (2k_1 - k_3)r - 2\phi_1 + \phi_3$$
$$\varphi_8 = (2k_2 - k_3)r - 2\phi_2 + \phi_3, \quad \varphi_9 = (2k_3 - k_2)r - 2\phi_3 + \phi_2 \qquad (12)$$

3 Experiments with Sodium Atomic Vapor

To verify experimentally the theoretical analysis, we employed sodium vapor for the degenerative multi-wave mixing medium in which the $3S_{1/2}$ and $3P_{3/2}$ levels consisted of an inhomogeneously broadened two-level system. The experiment was performed using the arrangement shown in Figure 3.6.2.

An N_2-laser pumped dye laser using Rhodamin 6G generated the single mode input beam with a pulsewidth of 6 ns, a variable power of up to 500 kW/cm², and

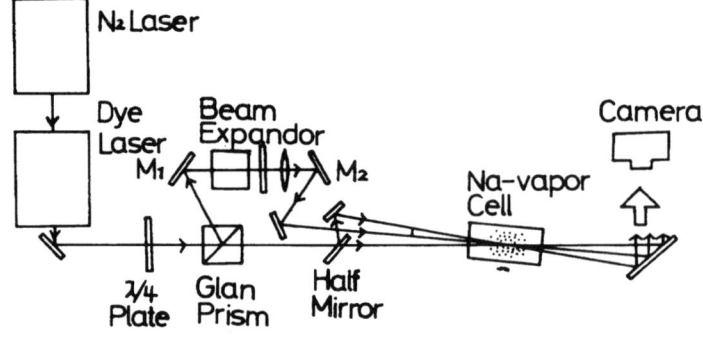

Figure 3.6.2
Schematic arrangement of a dye laser and a Na vapor system

a bandwidth of 9 GHz. The input beam was split into two and one linearly polarized beams orthogonal to each other using a glan-prism and a half-mirror. Two or three beams traversed with small crossing angles (*e.g.* 1 mrad) a heat-pipe-type, 2 cm long, vapor cell containing Na-vapor with a density of 1×10^{13} cm^{-3}.

We first observed significantly intensified higher-order diffraction for the two beams incident shown in Figure 3.6.3. Figure 3.6.3(a) shows a higher-order diffraction switching pattern in which the numbers on the upper side indicate the order of diffracted beams. The first-order pattern is two incident beams. The characteristics of this diffracted pattern agree well with the theoretical prediction. In order to appreciate the phase-conjugation property of these diffracted beams, we observed the third-order diffraction pattern by transporting the character 'C' onto one incident beam, as shown

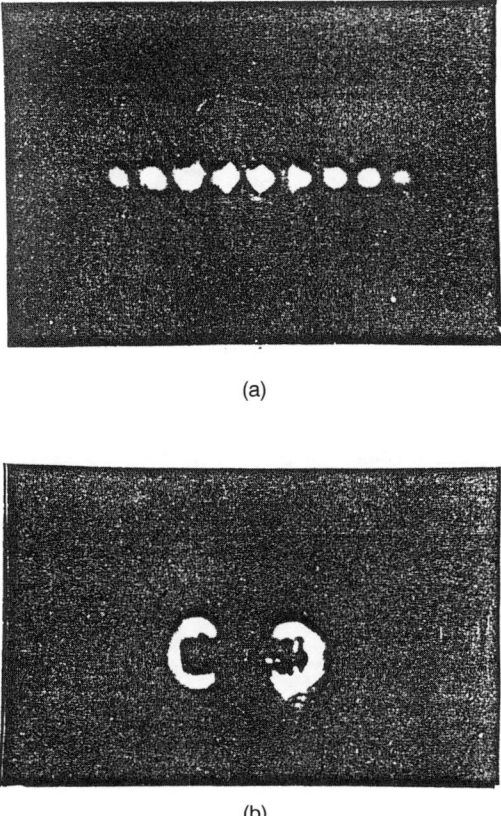

Figure 3.6.3
Observed patterns of multi-photon diffraction switching for two-beam mixing. (a) Higher-order diffraction patterns. The numbers on the upper side indicate the order of diffraction lights. (b) The phase-conjugate property of an incident pattern 'C'

in Figure 3.6.3(b). Thus, we confirmed the forward phase-conjugate property by an inverse pattern of the input 'C'.

We next observed more complicated patterns for three incident beams, shown in Figure 3.6.4. One of the three beams was a linearly polarized light orthogonal to that of the other beams. Figure 3.6.4(a) shows typical cross-sectional spots in which three bright spots near the center indicate incident beams. We also confirmed the conservation of polarization among the diffraction beams. Figures 3.6.4(b) and 3.6.4(c) show the characteristics of polarized multi-diffraction waves with horizontal components and vertical components, respectively. These diffraction switchings were achieved in a time of 6 ns pulsewidth. We can expect a higher speed switching time because coherent polarization is limited by neither the longitudinal relaxation time nor the transverse relaxation time of the two-level system.

Figure 3.6.5 shows the observed intensity patterns of the diffraction beams of Figure 3.6.3(a) by using a CCD linear array sensor. The two arrows in Figure 3.6.3

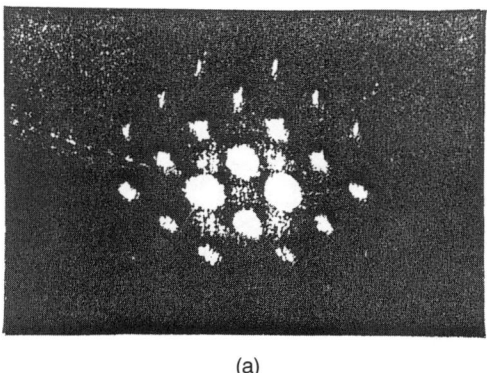

(a)

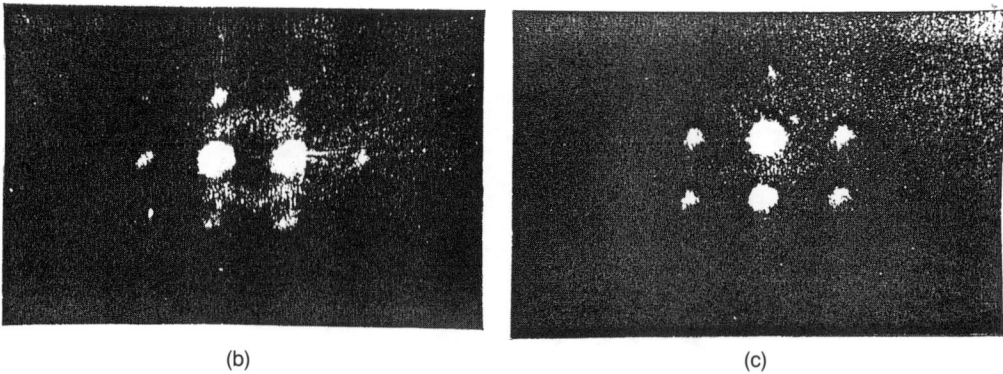

(b) (c)

Figure 3.6.4
Observed patterns of multi-photon diffraction switching for three-beam mixing. (a) Observed multi-diffraction cross-sectional pattern. The three bright spots near the center indicate incident three beams. (b) and (c) show the characteristics of polarized multi-diffraction waves with horizontal and vertical components, respectively

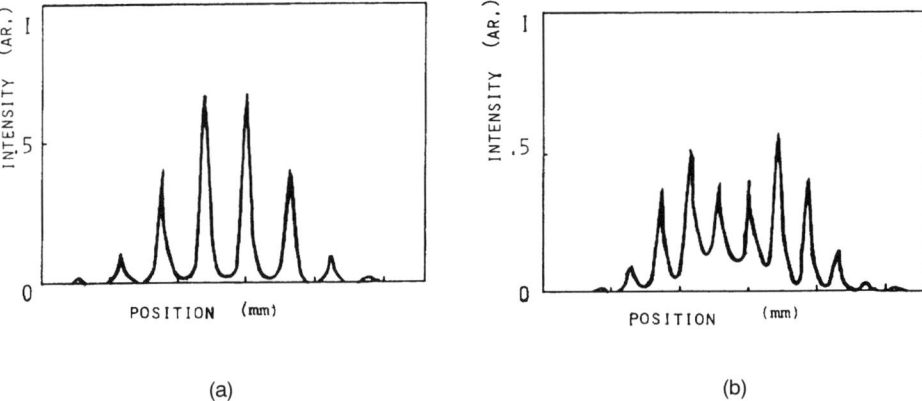

Figure 3.6.5
Multi-diffraction intensity patterns for Figure 3.6.3(a) observed with a linear array sensor. (a) Typical ordering intensity pattern for the higher-order diffraction switching for a certain input intensity. (b) Enhanced third-order diffraction intensities for a different input intensity

indicate the input positions of the two beams. When the input pulse intensities were changed, we observed either normal gradient patterns for ordering of the higher-order diffraction scattering shown in Figure 3.6.5(a) or a significant coherent effect: enhanced higher-order scattering than the first-order transparent input intensity, as shown in Figure 3.6.5(b). The reason why the coherent effect conspicuously appears for a certain input intensity is that diffraction intensities depend on the Bessel functions as a function of the input field area θ as described in equation (9) or equation (10).

4 Conclusion

We have described theoretical and experimental approaches to a novel application for coherent multi-photon scattering which enhances multi-superimposition-phase diffraction switching.

References

[1] Y.R. Shen, *The Principle of Non-Linear Optics*, John Wiley & Sons, 1984, p. 379.
[2] B.R. Suydam and R.A. Fisher, in *Optical Phase Conjugation*, R.A. Fisher, ed., Academic Press. New York, 1983, p. 79.
[3] P. Ye and Y.R. Shen, *Phys. Rev.* **A25**, 2183 (1982).
[4] H. Gwinful and H. Marburger, *Appl. Phys. Lett.* **36**, 613 (1980).
[5] A. Blouin, M.M.D. Roberge and P. Galaneau, *J. Opt. Soc. Am.* **8**, 578 (1991).
[6] N. Tanno, K. Ohkawara and H. Inaba, *Phys. Rev. Lett.* **46**, 1282 (1981).

4
FUNCTIONAL PHOTONICS DEVICES AND INTEGRATION

4.1

Mode-locked Fiber Laser Using a Fiber-optic Phase Modulator

Masaaki Imai and **Shinya Satoh**

Abstract

A Nd^{3+}-doped fiber laser operating at 1.06 μm with a fiber-optic phase modulator is developed for frequency-modulation mode-locking. A 100 MHz bandwidth response of the modulator that consists of a Nd^{3+}-doped optical fiber jacketed with a radially poled piezoelectric copolymer, vinylidene fluoride (73 mol%)/trifluoroethylene (27 mol%) is also discussed. Multiple peaks of phase modulation at 100 MHz region are found to be suitable for highly efficient mode-locking. The principle and approaches to a mode-locked fiber laser using the phase modulator are reported.

1 Introduction

All-fiber technology for modulating light signals in optical fibers has recently received considerable attention, mainly because of the strong potential benefits that fiber-optic phase and polarization modulators may play an important role in the field of optical fiber communication and optical fiber sensing [1,2]. The conventional and most widely used modulator consists of a jacketed optical fiber wound on a hollow cylindrical piezoelectric transducer (PZT). In contrast, piezoelectric plastics such as polyvinylidene fluoride (PVDF) and its copolymer have been used as transducer elements with which the fiber is coated [3]. This type of fiber-optic phase modulator has the advantages of easy assemblage, light weight, flexibility of design shape, and low drive voltage. So far, we have discussed and developed the feasibility of jacketing a single-mode optical fiber with a piezoelectric copolymer, vinylidene fluoride/trifluoroethylene, hereafter abbreviated as

VDF/TrFE [4–7]. The optical performance of the VDF/TrFE copolymer jacketed fiber was found to be suitable for a phase modulator at high frequencies above 100 MHz [8]. Moreover, it was also pointed out that the radial resonance frequency of the jacketed fiber is desirable for efficient phase modulation at a rate equal to the longitudinal mode spacing in a rare-earth doped silica fiber.

In this section the technique of optical phase modulation in a Nd^{3+}-doped single-mode fiber is proposed and extended to the FM mode-locking of short pulses propagating in the active waveguide doped with Nd^{3+} ions. The aim of studying FM mode-locking in a Nd^{3+}-doped fiber laser is to develop an all-fiber technology, such as a fiber-optic mode locker capable of supporting high-power subpicosecond optical pulses in the bandwidth limited regime [9].

2 Fiber-optic Phase Modulator

The schematic view of a fiber-optic phase modulator is shown in Figure 4.1.1. The copolymer jacketed optical fiber has the geometry of a long, multilayered cylinder with the piezoelectric plastic sandwiched between the inner and outer electrodes. When an electric field is applied to the jacket, strains are induced in the jacket and transmitted to the optical fiber. This strain effectively modifies both the length and the refractive index of the fiber [4,5]. The sample fiber used has a 4.4 μm core and 88.4 μm cladding diameter with a 10.9 μm thickness for the copolymer jacket. A 8–9 cm long fiber sample was annealed and radially poled at room temperature using a corona discharge technique. The frequency response of the optical phase shift (rad $V^{-1}m^{-1}$), normalized with an applied voltage and poled length, was measured using a homodyne detection scheme with a drift compensator circuit [6,7]. The measurements are plotted in the wide frequency range of 100 kHz–100 MHz as seen in Figure 4.1.2. A flat response

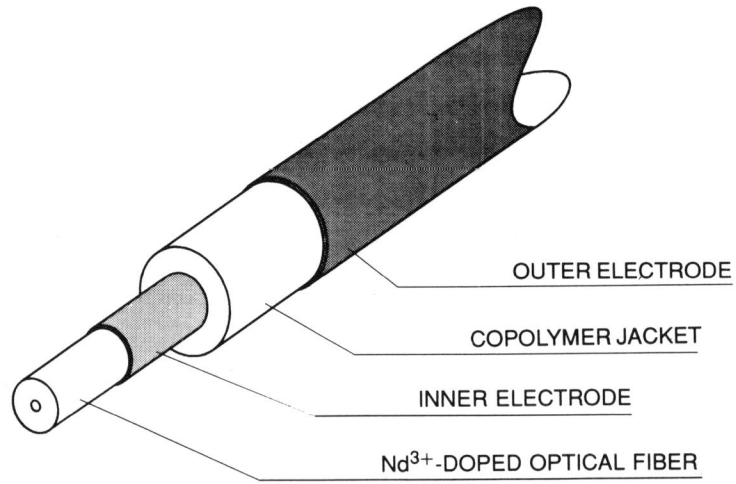

Figure 4.1.1
Schematic view of a fiber-optic phase modulator

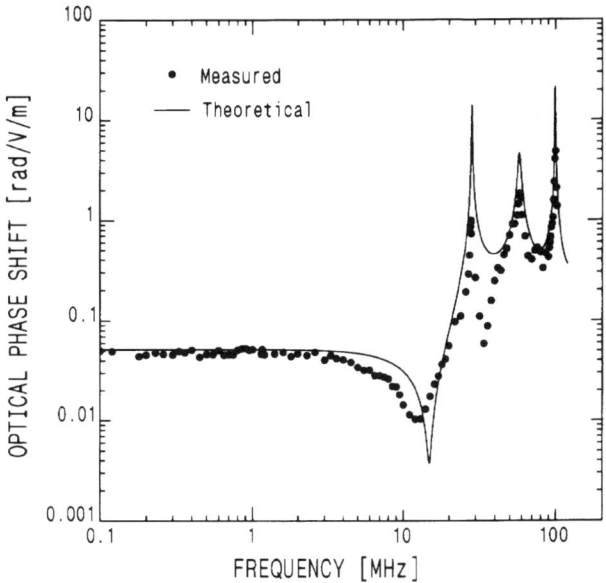

Figure 4.1.2
Frequency response of optical phase shift

Table 4.1.1
Electromechanical properties of glass fiber, metallic electrodes, and VDF (73 mol%)/TrFE (27 mol%) as well as specifications of the fiber-jacket composite

Mechanical loss factor (copolymer)	0.06*
Mechanical loss factor (nickel)	0.0
Mechanical loss factor (aluminum)	0.0
Electromechanical coupling coefficient (copolymer)	0.15*
Density (glass fiber)	2200 kg/m^3 [†]
Density (nickel)	8850 kg/m^3
Density (copolymer)	1880 kg/m^3 [†]
Density (aluminum)	2690 kg/m^3 [†]
Cladding diameter (glass fiber)	88.4 μm [‡]
Inner electrode thickness (nickel)	2.2 μm [‡]
Piezoelectric jacket thickness (copolymer)	10.9 μm [‡]
Outer electrode thickness (aluminum)	0.2 μm [‡]
Elastic constants (glass fiber)	$c_{11} = 7.85 \times 10^{10}$ N/m^2 [†]
	$c_{12} = 1.61 \times 10^{10}$ N/m^2 [†]
Elastic constants (nickel)	$c_{11} = 2.66 \times 10^{11}$ N/m^2
	$c_{12} = 1.14 \times 10^{11}$ N/m^2
Elastic constants (aluminum)	$c_{11} = 1.10 \times 10^{11}$ N/m^2 [†]
	$c_{12} = 5.81 \times 10^{10}$ N/m^2 [†]
Elastic constants (copolymer)	$c_{11} = 9.10 \times 10^9$ N/m^2 [§]
	$c_{12} = 4.0 \times 10^9$ N/m^2 [§]

*K. Kimura and H. Ohigashi, *J. Appl. Phys.*, **61**, 4749–4754 (1987).
[†]Reference [7].
[‡]Measured.
[§]Private communication.

with a phase sensitivity of approximately 4.9×10^{-2} rad $V^{-1}m^{-1}$ was observed from 100 kHz to 5 MHz. At frequencies higher than 10 MHz, the response is dominated by radial resonances of the fiber-jacket composite. The peak frequency of the first-order resonance is found at 28.0 MHz. The second- and third-order resonances can be read as 58.0 and 100 MHz, respectively.

As a result of the changes in both the fiber length and the refractive index, an optical phase shift is induced in the fiber. To calculate the induced phase shifts, data [8] are needed on the elastic constants of the silica fiber, metallic electrodes, and copolymer jacket as well as the piezoelectric constants of the copolymer. These data are listed in Table 4.1.1 with source references. The frequency response of the phase shift was calculated so as to fit the measured data at both the flat level and the peak frequencies of the first- to third-order resonances. The theoretically predicted phase shift is shown by the solid curve in Figure 4.1.2. As can be seen from the figure, qualitative agreement exists between analysis and experimental plots, especially at higher frequencies.

3 Mode-locked Fiber Laser

Recently, FM mode-locking of a Nd^{3+}-doped fiber laser has been demonstrated [9] using a bulk electrooptic phase modulator ($LiNbO_3$) placed close to the output coupling mirror (see Figure 4.1.3(a)). A major limitation of this arrangement was the requirement for intracavity optical elements. This increased intracavity loss and severely restricted the lasing bandwidth through étalon formulation. To overcome such a drawback, an integrated fiber phase modulator [10–12] is used for FM mode-locking. One of the cavity configurations is schematically shown in Figure 4.1.3(b), which is direct modulation of the phase of the fiber mode's electric field. The arrangement simplifies laser cavity design, removing the limitations referred to above. The intrafiber phase modulation yields a low-loss cavity since the single-mode fiber laser and the phase modulator are made from the same sample of a rare-earth-doped fiber. An experimental setup for the mode-locked fiber laser is shown in Figure 4.1.4. The fiber laser used had a $Al_2O_3/GeO_2/SiO_2/F$ core doped with 180 parts in 10^6 of Nd^{3+}, a cutoff wavelength of 0.88 μm, and an index difference of 0.73%. It was single-mode at the lasing wavelength of 1.06 μm. The optical path length of the fiber required to match the optimum mode-locking frequency was 184.37 and 107.14 cm, for 58.0 MHz (second-order resonance) and 100.0 MHz (third-order resonance). The fiber ends were cleaved and butted to a thin substrate mirror (width 0.25 mm) at the input end of the pump power. This permitted > 80% transmission at the pump wavelength of 804 nm emitted from a LD light source, while being highly reflecting ($R = 99.5\%$) at the lasing wavelength. The output end was similarly butted to an output coupling mirror with a reflectivity of 95%. The dichroic mirror was wedged and antireflection coated on the rear surface to suppress multiple reflections. The pump beam was focused and launched into the fiber with a coupling efficiency of 10%. Absorption loss due to Nd^{3+}-ion doped in the optical fiber was 4.4 dB at 804 nm. Therefore, nearly 84% of the pump power was absorbed in the sufficient fiber length of $\sim$ 1.8 m. A typical lasing characteristic is shown in Figure 4.1.5 and a lasing threshold was observed for an absorbed pump power

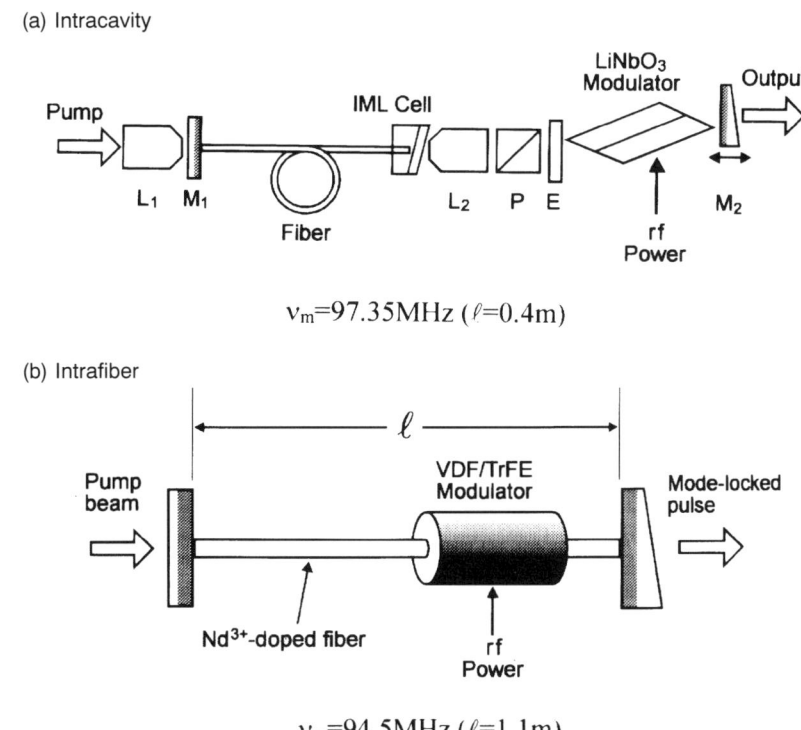

Figure 4.1.3
Schematic diagram of a FM mode-locked fiber laser: (a) intracavity phase modulation and (b) intrafiber phase modulation

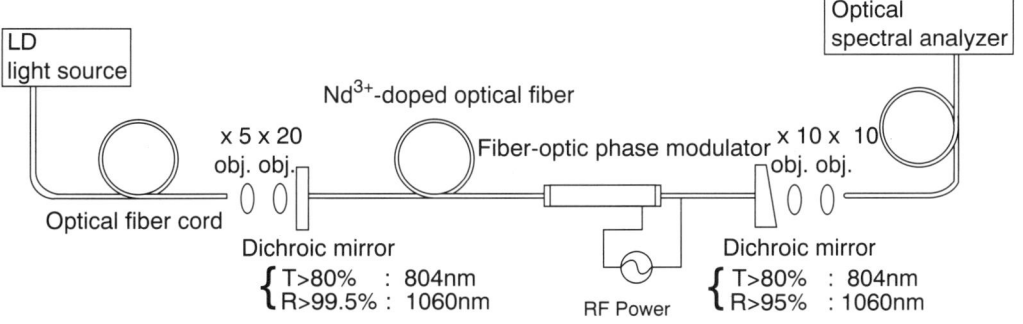

Figure 4.1.4
Experimental setup for the FM mode-locked fiber laser

of 0.75 mW. The whole measured data are seen to be distributed around a linear characteristic well above threshold. The maximum output power was 6.4 μW for 4.4 mW of absorbed pump power. It may be possible to increase the lasing output power by improving the experimental configuration, especially the focusing and launching optical system in order to efficiently couple a pump beam into the fiber laser.

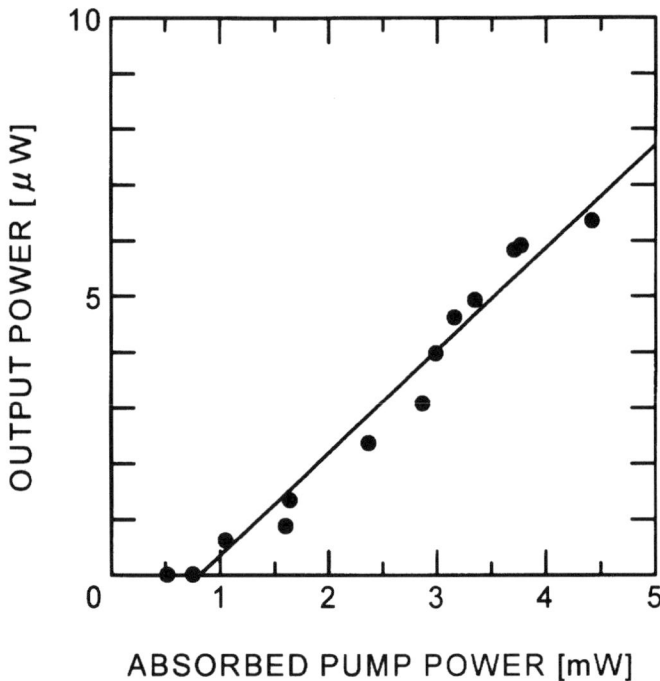

Figure 4.1.5
Absorbed pump power versus output lasing power

4 Conclusions

We have fabricated a Nd^{3+}-doped single-mode fiber jacketed with a piezoelectric VDF/TrFE copolymer to measure the optical performance of the jacketed fiber over a wide frequency range of 100 kHz–100 MHz. Theoretical analysis of the induced phase shifts was performed on the basis of the vibration theory of elasticity with the exact geometry and composition of the multilayered cylindrical structure taken into account. Multiple peaks of phase modulation at 60–100 MHz region can be employed for efficient FM mode-locking of short pulses propagating in the active glass fiber. The proposed technique will result in a low-loss cavity configuration since the optical fiber and the phase modulator are made of the same Nd^{3+}-doped single-mode waveguide. This is under investigation and will be reported elsewhere.

References

[1] R.H. Stolen and R.P. DePaula, *Proc. IEEE*, **75**, 1497–1511 (1987).
[2] J.-P. Goure, I. Verrier and J.-P. Meunier, *J. Phys. D: Appl. Phys.*, **22**, 1791–1805 (1989).
[3] V.S. Sudarshanam and R.O. Claus, *J. Lightwave Technol.*, **11**, 595–602 (1993).
[4] M. Imai, S. Fujiwara, Y. Ohtsuka and A. Odajima, *Opt. Lett.*, **13**, 838–840 (1988).
[5] A. Odajima and M. Imai, *Ferroelectrics*, **92**, 41–46 (1989).

[6] M. Imai, T. Yano, Y. Ohtsuka, K. Motoi and Y. Odajima, *IEEE Photon. Technol. Lett.*, **2**, 727–729 (1990).
[7] M. Imai, T. Yano, K. Motoi and A. Odajima, *IEEE J. Quantum Electron.*, **28**, 1901–1908 (1992).
[8] M. Imai, S. Satoh, T. Sakaguchi, K. Motoi and A. Odajima, *IEEE Photon. Technol. Lett.*, **6**, 956–959 (1994).
[9] M.W. Phillips, A.I. Ferguson and D.C. Hanna, *Opt. Lett.*, **14**, 219–221 (1989).
[10] M.W. Phillips, A.I. Ferguson, G.S. Kino and D.B. Patterson, *Opt. Lett.*, **14**, 680–682 (1989).
[11] G. Geister and R. Ulrich, *Appl. Phys. Lett.*, **56**, 509–511 (1990).
[12] L.A. Zenteno and H. Po, *Opt. Lett.*, **16**, 315–317 (1991).

4.2

Vertical Integration Technology for Fiber-optic Circuits

Shojiro Kawakami

Abstract

This section reviews a novel approach to integrating several functional devices directly into optical fibers. This technology is capable of lens-free, alignment-free integration and is useful for mass production. The integration of optical devices, such as an isolator and a switch, and their components, are demonstrated.

1 Introduction

For advanced optical communication systems several kinds of optical functional devices such as isolators, switches, filters, and amplifiers are required to be integrated into optical fibers [1]. Most of the optical devices are composed of various materials different from optical fibers, such as dielectrics, magnetics, and semiconductors. If the devices have waveguide structure, mode-matching and high precision alignment (0.1–0.5 μm) between the fibers and the devices are strongly desired. The author's group have proposed a novel way of integrating optical devices, as shown in Figure 4.2.1 [2]. The objective of this technology is alignment-free, lens-free integration of optical devices into fibers. The key techniques for this integration are as follows:

(1) to enlarge the field diameter by utilizing thermal diffusion of the dopant in the fibers; thermally-diffused expanded (TEC) fibers;
(2) to make the optical path length of each component as short as possible; for example, a laminated polarization splitter (LPS);
(3) to embed the device into a fiber array.

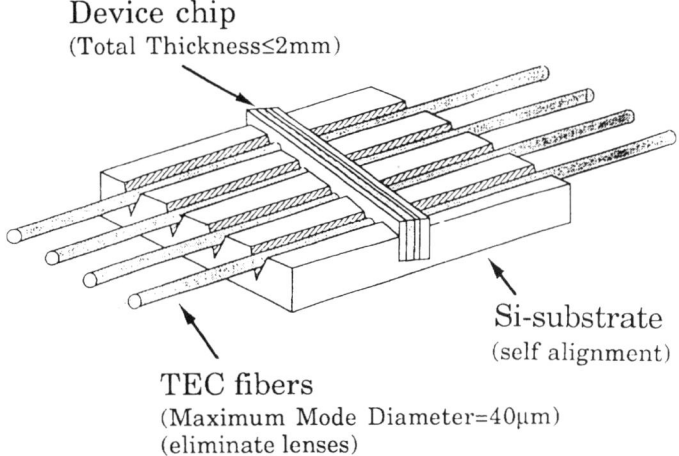

Figure 4.2.1
Vertical integration of arbitrary optical devices into an optical fiber array

This technology is called 'vertical photonics' because the light is propagated vertically to the plane of the planar device. This section reviews our research on vertical photonics and recent progress in components and devices. Two important components, a TEC fiber and an LPS, are represented showing a specific example of vertical integration. Moreover, the vertical integration of an isolator and a switch is demonstrated.

2 Components

In order to discuss the TEC fiber and an LPS as components for vertical integration, a polarization-independent optical isolator is shown in Figure 4.2.2 as a specific example. The principle of operation is represented by lines and arrows which indicate the direction of light propagation. The LPS splits the forward light into an ordinary ray and an extraordinary ray. The two rays are coupled into the output fiber after propagating through the Faraday rotator, the half plate, and the second LPS. On the other hand, the backward light arrives at the points that are offset from the core of the input fiber. To decrease insertion losses, which are caused by diffraction losses between the two fibers, the following two components are essential.

2.1 TEC fiber

Fundamental

A thermally-diffused expanded core (TEC) fiber has a large mode field diameter (MFD), which reduces diffraction loss in the gap between the fibers [3]. To start with, we roughly show the characteristics of TEC fibers assuming that the field distribution has a Gaussian

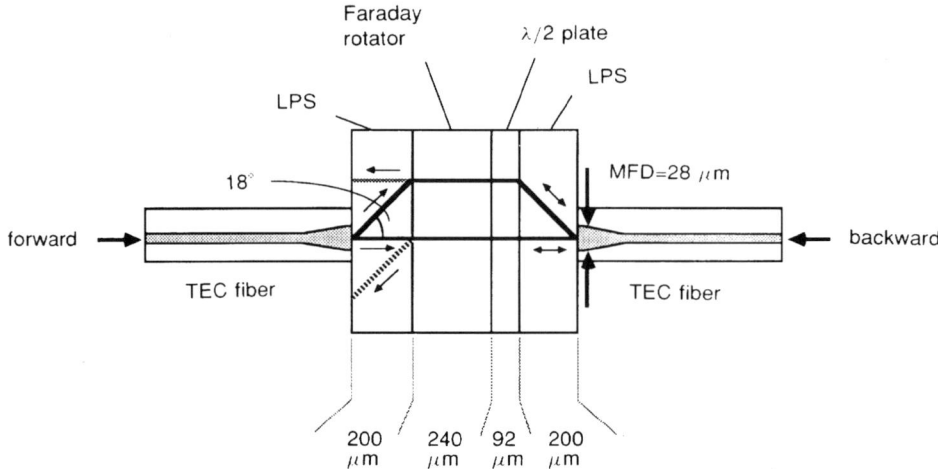

Figure 4.2.2
Configuration of a fiber-integrated polarization-independent isolator

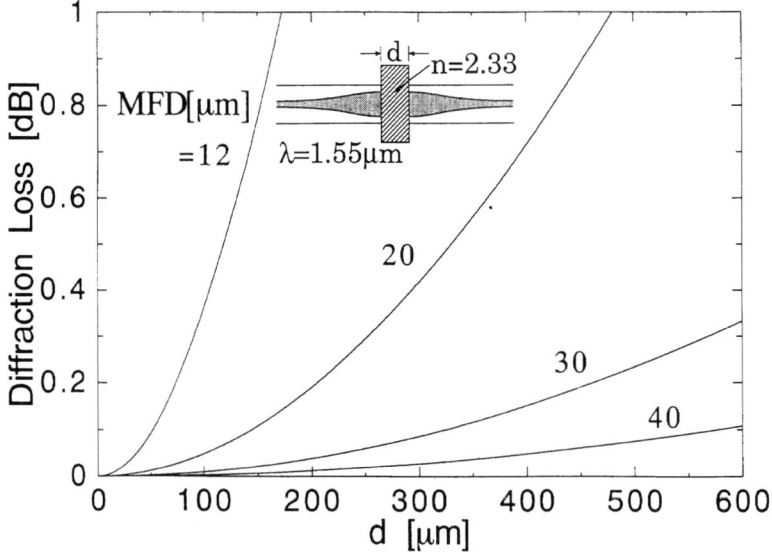

Figure 4.2.3
Calculated diffraction losses as a function of distance between fiber endfaces for various MFDs at $\lambda = 1.55\ \mu$m

profile. The diffraction loss is expressed by

$$\eta = 10\log\left[1 + \frac{d\lambda}{2\pi n w^2}\right] \quad (1)$$

where w denotes the field diameter, d the distance between fiber end faces, n the refractive index of the material in the gap, and λ the wavelength. Figure 4.2.3 shows

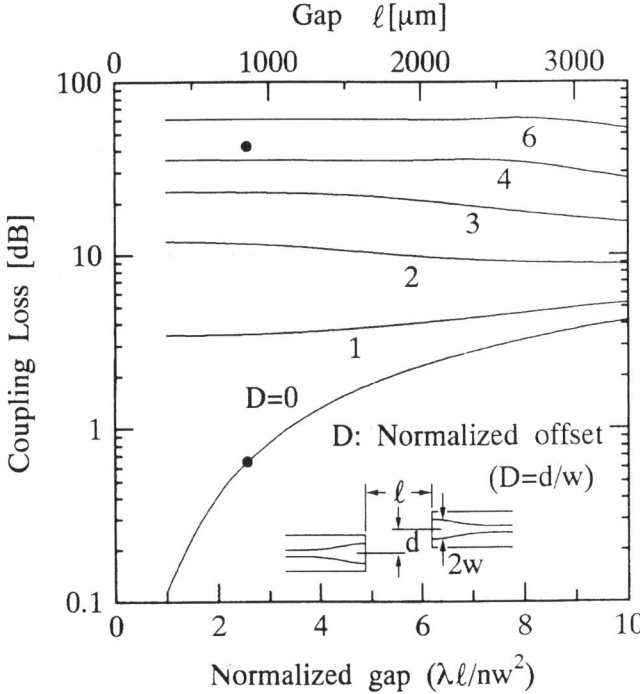

Figure 4.2.4
Coupling losses between TEC fibers as a function of normalized gap and normalized offset, where $V = 2.4$; n denotes the refractive index of the medium filling the gap. Numerical examples are also given for $n = 2.3$, $2w = 28$ μm, and $d = 1.55$ μm

the diffraction loss as a function of d. The loss reduces dramatically by increasing the mode field diameter. Strictly speaking, the modal field distribution of a TEC fiber is broader than the Gaussian profile due to the broadening of the index profile. This decreases the coupling loss caused by offset, such as the backward loss of the isolator. For the precise design of the fiber-integrated devices, the characteristics of TEC fibers should be calculated exactly. The eigenmode in the taper of a TEC fiber was studied by the propagating beam method (PBM) and the coupling efficiency between TEC fibers with a gap or offset was calculated by the overlap integral [4]. Figure 4.2.4 shows the coupling loss between the TEC fibers with a large normalized frequency ($V = 2.40$) as a function of the normalized gap and normalized offset. The dots show the characteristics of the isolator in Figure 4.2.2. An insertion loss of 0.5 dB and a backward loss of 43 dB could be attained.

Experiments

In our experiments use was made of commercially available GeO_2 doped silica single-mode fibers (SMFs) for $\lambda = 1.55$ μm. For thermal diffusion of the dopant, a furnace heater made of SiC was used. (Other heat treatments have been reported [5].) After

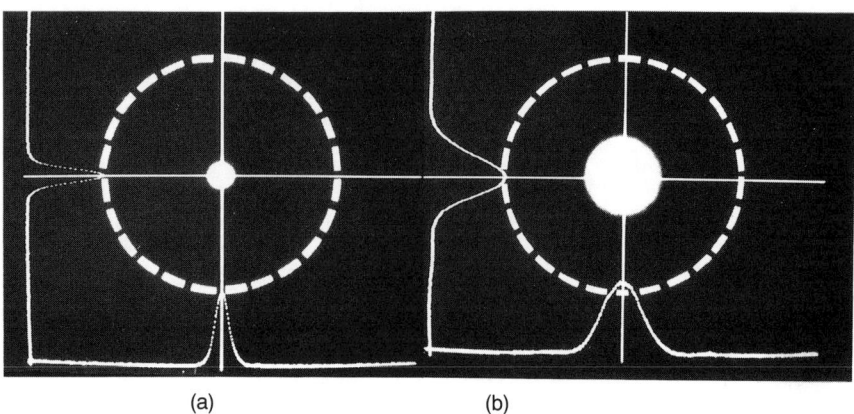

Figure 4.2.5
Photographs of the NFP at λ = 1.55 μm (a) before and (b) after heat treatment; 1280 °C, 45 h. White dotted line shows the contour of the cladding

stripping off the protective coating layer, the fibers were put into a silica tube, which was evacuated in the successive process. By using a furnace heater, 10 fibers or more can be treated at one time. When the temperature at the center of the fibers is below 1280 °C and treatment time is less than 45 h, the outer diameter does not change and high reproducibility is attained. Figure 4.2.5 shows the near-field patterns of the fiber, (a) before and (b) after the heat treatment mentioned above. The MFD is expanded to as large as 40 μm [6].

2.2 LPS

Structure and performance

Figure 4.2.6 schematically shows the structure and performance of a laminated polarization splitter (LPS), consisting of alternately laminated transparent materials with a high refractive index (n_1) and a low index (n_2), the period of alternation being sufficiently small compared with the wavelength λ. Optical properties of an LPS are similar to those of a uniaxial crystal, such as rutile or calcite, because the multilayer has form birefringence [8]. By choosing proper values for n_1/n_2 and θ, the angle between the normal of the layers and the z direction, a large polarization splitting angle, ϕ, can be attained. The maximum value of splitting angle, is given by

$$\tan \phi = \frac{(r^2 - 1)^2}{4r(1 + r^2)} \tag{2}$$

where $r = n_1/n_2$ and $\tan \theta = 2r/(r^2 + 1)$. Figure 4.2.7 shows the dependence of the splitting angle on r. For a large splitting angle of an LPS (at λ = 1.55 μm), hydrogenated amorphous silicon (a-Si:H, $n_1 > 3.2$) and silica (SiO$_2$, $n_2 = 1.45$) are suitable materials. (Silicon carbonate (SiC) and silica are used at λ = 1.3 μm.)

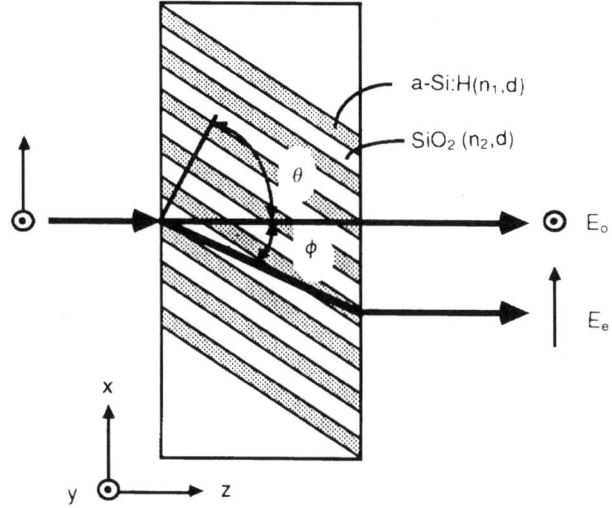

Figure 4.2.6
The structure of a laminated polarization splitter

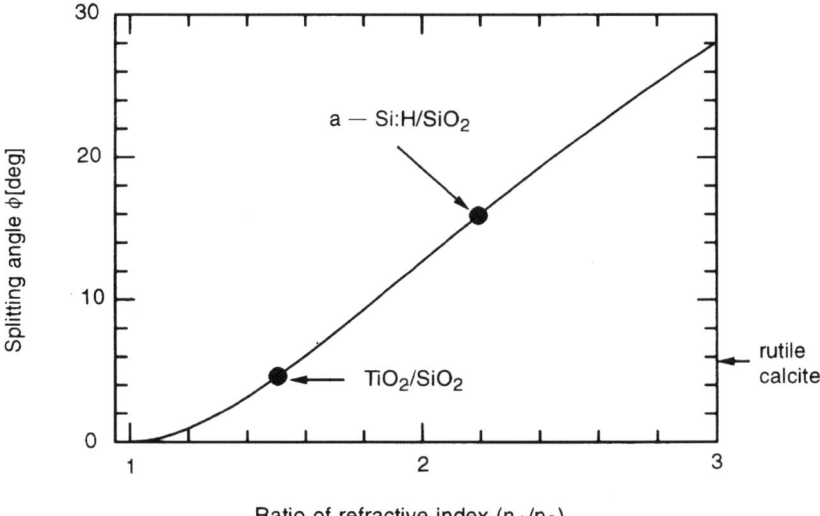

Figure 4.2.7
Maximum splitting angle as a function of the refractive index ratio (n_1/n_2)

Fabrication

To attain high performance of the isolator using LPSs, insertion loss < 1 dB and isolation > 40 dB, the following characteristics of LPSs are desired: a low insertion loss (about 0.1 dB/100 μm) and a large splitting distance (> 50 μm). For the expected characteristics of the LPS, the following optical and mechanical properties of the films are required:

(1) a higher refractive index for a-Si:H films and a lower index for SiO$_2$ films to have a large splitting angle;
(2) low absorption coefficients for a low insertion loss;
(3) flat layer boundaries to restrain scattering losses [9]; and
(4) stresses low enough to obtain a thick multilayer (> 150 μm), which makes it possible to have a large splitting distance.

rf sputtering is an effective method. The optical properties (1) and (2) are obtained by choosing the deposition conditions such as the sputtering gas flow and the substrate temperature. Applying bias rf power to the substrate electrode has a good effect on (3). However, extra bias power increases the film stresses. For condition (4), the bias power was applied at the deposition only near the surfaces of each SiO$_2$ layer. After this investigation, we succeeded in deposition of multilayers having thicknesses of 150 μm and 200 μm with the properties (1)–(4). As a result, an LPS with a sample thickness of 200 μm and a splitting angle of 67 μm was obtained. Figure 4.2.8 represents the

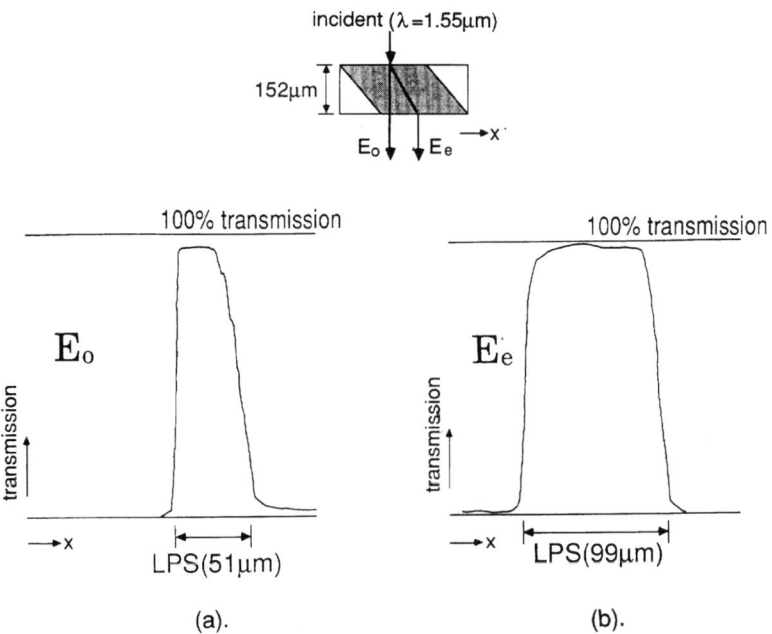

Figure 4.2.8
The near-field patterns for the output of (a) an ordinary wave and (b) an extraordinary wave, respectively. The loss figures, excluding the reflection losses of E_o and E_e, are 0.12 dB/100 μm and 0.09 dB/100 μm, respectively

near-field patterns for the output of the LPS. The insertion losses are 0.12 dB/100 μm for the ordinary ray and 0.04 dB/100 μm for the extraordinary ray excluding the reflection loss, respectively. These values are enough to use the LPSs in the fiber-integrated isolator.

3 Fiber Integrated Devices

3.1 Isolator

Integration of polarization-independent optical isolators into a fiber array was demonstrated [10]. Since LPSs were under development at that time, rutile plates were used as polarization splitters. Using rutile plates degrades the insertion loss because the thickness of the isolator chip becomes as large as 2.3 mm, which is about three times larger than in the case of using LPSs. The constructed isolator, removed from a magnet, is shown in Figure 4.2.9. The fabrication procedure was as follows:

(1) eight V-grooves were formed on a Si wafer;
(2) four pairs of TEC fibers were fixed in the V-grooves;
(3) a square groove was made using a rotating blade;
(4) an isolator chip was fixed in the square groove.

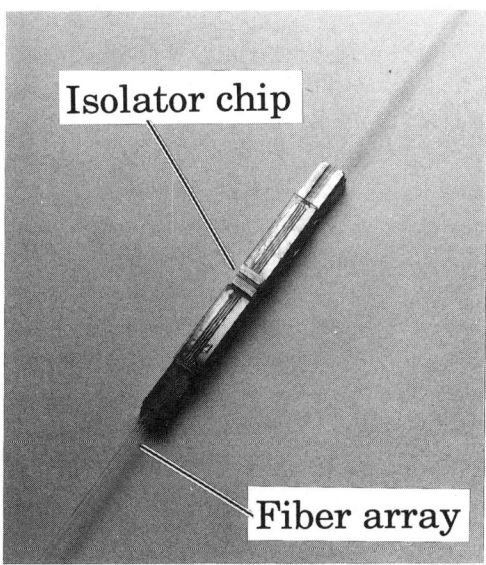

Figure 4.2.9
Constructed fiber-integrated isolator using rutile plates

Table 4.2.1
Characteristics of the isolator

	Design	Measurement
Forward loss	2.2 dB	3.5 dB
Backward loss	48 dB	45 dB
Return loss	—	47 dB

The theoretical and experimental characteristics of the isolator are summarized in Table 4.2.1. An isolator using LPSs integrated into a fiber is currently under development. The performance as an isolator was experimentally verified. The characteristics will be upgraded by an improvement of the fabrication techniques.

3.2 *Liquid crystal switch*

A schematic configuration of a fiber-integrated liquid crystal (LC) optical switch is shown in Figure 4.2.10 [4]. A liquid crystal cell, which is very useful for low switching

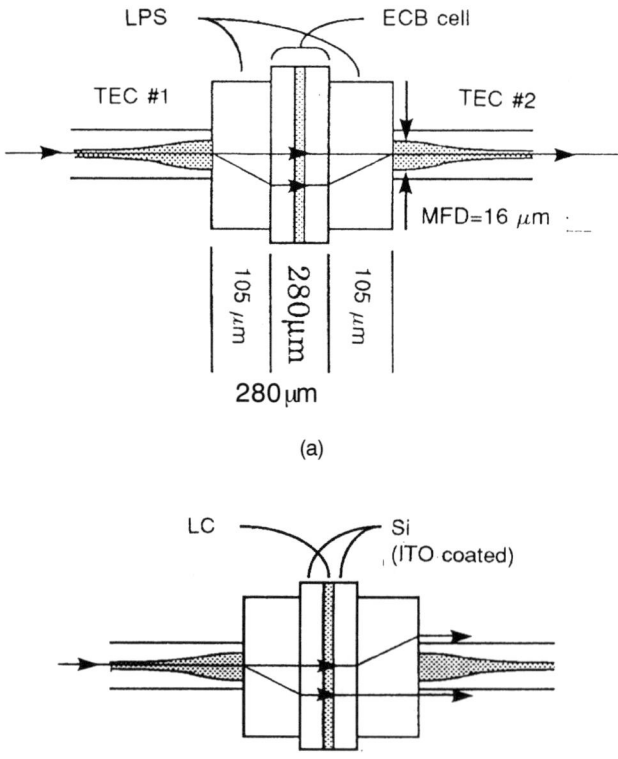

Figure 4.2.10
Design of liquid crystal optical switch; (a) and (b) are the ON and OFF states, respectively

voltage and decreasing an optical pathlength, is utilized as an electrical controlled birefringence cell. The incident light is split into two orthogonally polarized beams by the first LPS. When a voltage is not applied to the LC cell, the polarization state is rotated at an angle of 90°. Then, the beams after the second LPS are not coupled into the output fiber. On the other hand, when the proper voltage is applied, the LC turns into an isotropic material. The polarization states are not rotated and the beams are coupled into the output fiber. According to the calculation with PBM, an insertion loss of 0.1 dB and an extinction ratio of 40 dB are expected assuming that the diameter of the TEC fiber is 28 μm and that the thicknesses of the LPS and the LC are both 100 μm. A prototype switch was fabricated and an extinction ratio of 27 dB was obtained. The characteristics can be improved by better fabrication techniques.

4 Conclusion

Vertical integration technology for fiber-optic circuits is discussed. The objective of this technology is the lens-free, alignment-free integration of optical devices and it is useful for mass production. The performance of the components and the integrated devices are experimentally verified.

Reference

[1] J. Noda, in *Proceedings of Conference on Optical Fiber Communication/International Conference on Integrated Optics and Optical Fiber Communication*, San Jose, California, 1993, paper TuK, p. 39.
[2] S. Kawakami, *Proceedings of 9th Optical Fiber Sensors Conference*, Firenze Italy, 1993, Th2.12, p. 451.
[3] K. Shiraishi, Y. Aizawa and S. Kawakami, *IEEE J. Lightwave Technol.*, **8**, 1151-1161 (1990).
[4] Y. Ohtera, O. Hanaizumi, T. Sato and S. Kawakami, *Technical Report of Institute of Electronics, Information and Communication Engineers, OPE94-11*, 1994, pp.61-66.
[5] H. Hanafusa and M. Horiguchi, *Electron. Lett.*, **27**(21), 1968-1969, (1991).
[6] O. Hanaizumi, Y. Aizawa, H. Minamide and S. Kawakami, *IEEE Photon. Technol. Lett.*, **6**, 842-844 (1994).
[7] K. Shiraishi and S. Kawakami, *Opt. Lett.*, **15**, 516-518 (1990).
[8] M. Born and E. Wolf, *Principles of Optics*, Pergamon, Oxford, 1975, pp. 705-708.
[9] T. Sato, T. Sasaki, K. Tsuchida, K. Shiraishi and S. Kawakami, *Appl. Opt.*, **33**, 6925-6934 (1994).
[10] K. Shiraishi, T. Chuzenji and S. Kawakami, *IEEE J. Lightwave Technol.*, **10**, 1839-1842 (1992).

4.3

Three-dimensional Optical Interconnects Using Stacked ARROW Guided Wave Circuits

Yasuo Kokubun

Abstract

ARROW-type waveguides are functional single mode waveguides and have many features advantageous to integrated photonics. In this study we developed a new stacked configuration of ARROWs to establish three-dimensional optical interconnects. First, a multilevel stacked configuration of slab ARROWs was designed and fabricated. The crosstalk-free propagation and power exchange in the stacked ARROW-B waveguides were experimentally demonstrated. Next, a compact configuration of three-dimensional optical interconnects using channeled ARROW-B was developed. An ultrashort coupling length of 200 μm and large tolerance of core thickness were demonstrated.

1 Introduction

Integrated photonic circuits require a large area owing to the restrictions on the curvature radius of the bends and branches. This will be an obstacle to the dense integration of photonic circuits. The stacked configuration of photonic circuits can enable the dense monolithic integration of photonic devices in a single substrate. In addition, the stacked configuration is important to the flexible layout of photonic circuits and optical interconnects. To establish this, it is necessary to realize the stacked configuration of waveguides, in which waveguides are isolated from each other in the guiding region and are coupled to each other in the crossconnect region. However, because the confinement factor of conventional waveguides is not large enough to avoid crosstalk due to light coupling between neighboring waveguides, these waveguides are not suitable for dense optical interconnects. Besides this, conventional waveguides have no light control function other than as a light-guiding function.

To solve these problems, the authors developed ARROW (Antiresonant Reflecting Optical Waveguide) [2–3] and ARROW-B [4]. We call these waveguides ARROW-type waveguides. The fundamental structures of ARROW-type waveguides are shown in Figures 4.3.1(a) and (b). An interference cladding, which consists of the first cladding layer and the second cladding layer, is sandwiched by the core and the high index semiconductor substrate. We define the symbols of refractive index and thickness of each layer as shown in Figures 4.3.1(a) and (b). The material of the second cladding layer is usually the same as that of the core, i.e. $n_2 = n_c$, and the thickness of this layer, d_2, is usually designed as

$$d_2 \simeq \frac{d_{ce}}{2} \tag{1}$$

where d_{ce} is the equivalent thickness of the core including the effect of field penetration outside the core. Approximate expressions for d_{ce} will be given in the following sections for ARROW and ARROW-B. However, since the dependence of the propagation characteristics on d_2 is not so large, we can design $d_2 \simeq d_c/2$ in most practical cases. The material of the first cladding layer must have a refractive index much higher than those of the core and the second cladding for ARROW, and much lower for ARROW-B.

These waveguides are promising for integrated photonics owing to many features:

(1) performance of useful functions such as polarization and wavelength selective characteristics in ARROW [5,6];
(2) low loss;
(3) large light confinement;

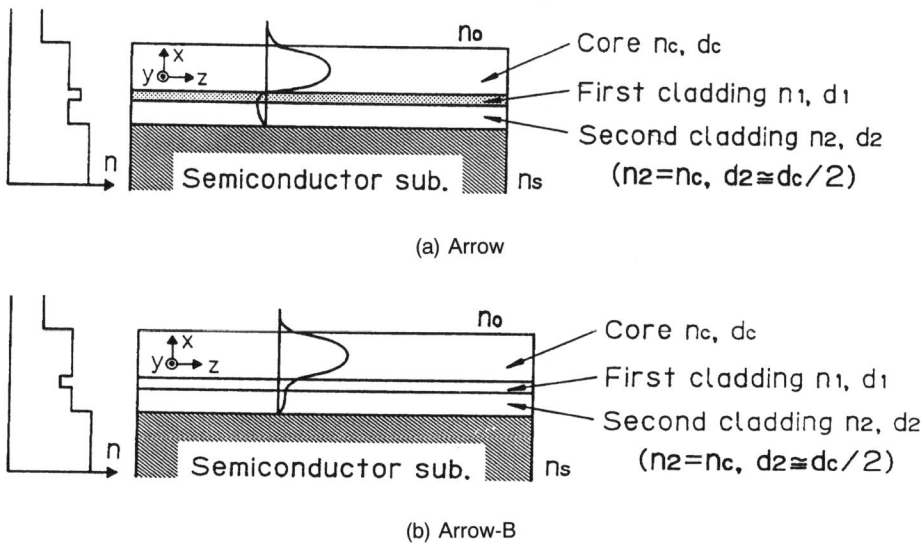

Figure 4.3.1
Fundamental structure of ARROW-type waveguides: (a) ARROW; (b) ARROW-B

(4) effective single-mode propagation even though a large core size;
(5) relatively large core size suitable for efficient connection to single-mode fibers;
(6) ease of fabrication process due to thin cladding between the core and the substrate;
(7) lack of need for precise control of refractive indices;
(8) large tolerance for thicknesses and refractive indexes of layers in the interference cladding due to the above features;
(9) ease of design due to simple expressions for propagation constant and loss;
(10) various choice of waveguides materials, etc.

Owing to (3) and (6) above, ARROW-type waveguides are suitable for stacked interconnects [7,8]. In addition, the degree of coupling can be controlled by regulating the waveguide separation [9]. In this study we propose and demonstrate a new stacked configuration of ARROW-type waveguides for three-dimensional optical interconnects.

2 Fundamental Characteristics of ARROWs

In ARROW-type waveguides, light propagates through the core by repeating the total internal reflection at the upper air–core boundary and ultra-high reflection (> 99.9%) from the interference cladding. The amount of light power passing through the interference cladding corresponds to the radiation loss. Since the reflection from the interference cladding is a kind of multiple-Fresnel reflection, the fundamental mode suffers intrinsic radiation loss. However, this loss can be reduced to a practical level by an appropriate design. In addition, because this radiation loss depends on the propagation angle of each mode, higher order modes are filtered out by loss discrimination due to the low reflectivity of the interference cladding, and thus an effective single-mode propagation can be realized.

In ARROW, because the refractive index of the first cladding is much higher than those of the core and the second cladding, the radiation loss depends strongly on the polarization of guided light. This results in the polarization dependence of radiation loss in ARROW.

In ARROW-B, the refractive index of the first cladding is much smaller than those of the core and the second cladding. If the first cladding is thick enough, total internal reflection occurs at the interface between the core and the first cladding. However, this layer is so thin that the evanescent field reaches the second cladding and interferes with this layer. Thus this reflection is also a kind of interference reflection. Since the first reflection at the interface between the core and the first cladding dominates the overall reflection, the radiation loss in ARROW-B is less dependent on the polarization than in ARROW.

2.1 ARROW

Figure 4.3.2 shows the dispersion characteristics of ARROW versus the thickness of the first cladding. This figure and Figure 4.3.3 were calculated by using the interference matrix method [7]. SiO_2 and TiO_2 are assumed to be the materials of the core and the first cladding, respectively. The propagation constant β in the ordinate is expressed by

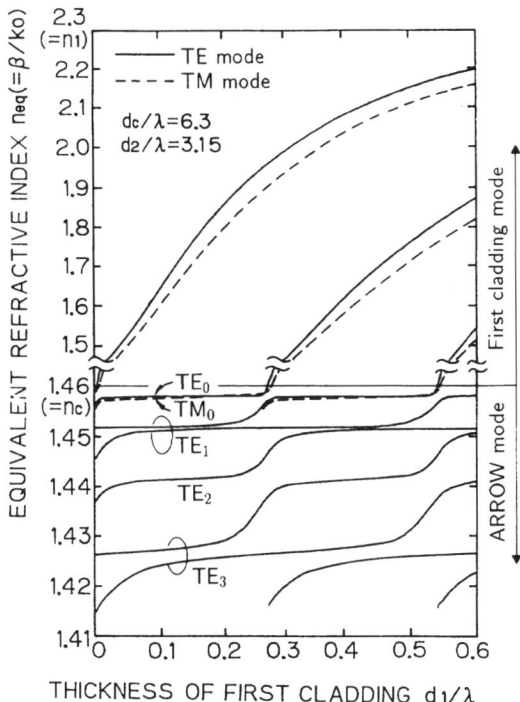

Figure 4.3.2
Dispersion characteristics of ARROW versus normalized thickness of first cladding d_1/λ

the equivalent index $n_{eq}(=\beta/k_0)$. ARROW supports the following three types of modes due to its peculiar leaky structure:

$$\left.\begin{array}{ll} \text{First cladding modes:} & n_c \leq n_{eq} < n_1 \\ \text{ARROW modes:} & n_0 \leq n_{eq} < n_c \\ \text{Radiation modes:} & 0 \leq n_{eq} < n_0 \end{array}\right\} \quad (2)$$

First cladding modes are guided modes in the first cladding layer, guided in the same manner as in the conventional three-layer slab waveguides. ARROW modes are leaky modes into which first cladding modes are transformed when they reach their cutoffs. Considering the practical usage of ARROW, we label the ARROW modes independently of the first cladding modes.

Figure 4.3.3 shows the radiation loss normalized by λ versus d_1/λ. The low loss range of the TE fundamental mode is quite broad and the radiation loss of the TM modes and higher order TE modes is much larger than that of the TE fundamental mode. Thus ARROW is an effective single-mode and single polarization waveguide utilizing the loss discrimination.

The minimum loss of the TE fundamental mode (and also the even modes) is realized when the first cladding satisfies the following antiresonant condition*

---

* The low loss condition corresponds to the antiresonant state of the interference cladding.

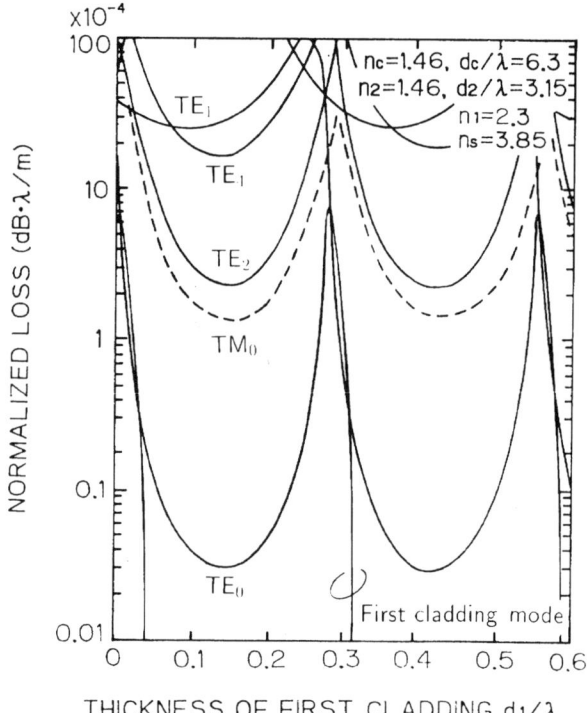

Figure 4.3.3
Radiation loss characteristics of ARROW modes versus normalized thickness of first cladding, d_1/λ

$$\frac{d_1}{\lambda} \simeq \frac{1}{4n_1}\left[1-\left(\frac{n_c}{n_1}\right)^2+\left(\frac{\lambda}{2n_1 d_{ce}}\right)^2\right]^{-1/2} \cdot (2N+1) \quad (N=0,1,2,\ldots) \quad (3)$$

where d_{ce} is the equivalent thickness of core and is approximately given by [2]

$$d_{ce} \simeq d_c + \xi_0 \frac{\lambda}{2\pi\sqrt{n_c^2-n_0^2}} \quad (4)$$

The parameter ξ_i is the polarization factor of the ith layer defined by

$$\xi_i = \begin{cases} 1 & \text{(TE modes)} \\ \left(\dfrac{n_i}{n_c}\right)^2 & \text{(TM modes)} \end{cases} \quad (5)$$

The radiation loss characteristics of the TE fundamental mode also has a broad tolerance against the normalized thickness of the second cladding d_2/λ. The antiresonant condition of the second cladding for even modes is given by equation (1).

When the antiresonant condition is satisfied, the propagation constant β_ν and radiation loss α_ν of the νth even ARROW mode ($\nu = 0, 2, 4, \ldots$) are approximately given by [3]

$$\beta_\nu = k_0 n_c \left[1 - \left\{ \frac{(\nu+1)\lambda}{2n_c d_{ce}} \right\}^2 \right]^{1/2} \tag{6}$$

and

$$\alpha_\nu \lambda \simeq 0.543 \chi \left(\frac{\lambda}{d_c} \right)^5 \frac{(\nu+1)^4}{n_c(n_1^2 - n_c^2)\sqrt{n_s^2 - n_c^2}} \quad (\text{dB} \cdot \lambda/\text{m}) \tag{7}$$

respectively, where χ is the polarization factor defined by

$$\chi = \begin{cases} 1 & (\text{TE modes}) \\ \left(\dfrac{n_1^2 n_s}{n_c^3} \right)^2 & (\text{TM modes}) \end{cases} \tag{8}$$

It is easily seen from equations (7) and (8) that the radiation losses of even order higher modes and TM modes are much larger than that of the fundamental mode.

2.2 ARROW-B

In contrast to ARROW, the dispersion and radiation loss characteristics of ARROW-B have no periodicity against d_1/λ [3], [4]. The n_{eq} of all modes converge to constant values as d_1/λ. This is easily seen from the fact that the structure of ARROW-B is identical to a three-layer symmetric waveguide when $d_1 \to \infty$. Since the loss of each mode decreases monotonically with d_1/λ and there is no singular thickness, ARROW-B can be used in a broad wavelength range. However, the radiation loss characteristics have a periodic dependence on d_2/λ similar to those of ARROW.

The equivalent thickness of the core of ARROW-B is expressed by

$$d_{ce} \simeq d_c + \xi_0 \frac{\lambda}{2\pi\sqrt{n_c^2 - n_0^2}}$$
$$+ \xi_1 \frac{\lambda}{2\pi\sqrt{n_c^2 - n_1^2} \tanh\{(2\pi/\lambda)d_1\sqrt{n_c^2 - n_1^2}\}} \tag{9}$$

Using equation (9), the propagation constant and radiation loss of the fundamental mode are approximately expressed by

$$\beta_0 = k_0 n_c \left[1 - \left(\frac{\lambda}{2n_c d_{ce}} \right)^2 \right]^{1/2} \tag{10}$$

and

$$\alpha_0 \lambda = \chi \frac{2.17\lambda^3 \left\{ 1 - \tanh^2 \left(\frac{2\pi}{\lambda} d_1 \sqrt{n_c^2 - n_1^2} \right) \right\}}{d_{ce}^3 n_c \sqrt{n_s^2 - n_c^2} \left[\frac{1}{\xi_1^2} + 4(n_c^2 - n_1^2) \left\{ \frac{d_{ce}}{\lambda} \tanh \left(\frac{2\pi}{\lambda} d_1 \sqrt{n_c^2 - n_1^2} \right) \right\}^2 \right]} \quad (\text{dB} \cdot \lambda/\text{m})$$

(11)

2.3 Three-dimensional ARROWs by adopting a stripe lateral confinement structure

The channeled structure of ARROW-type waveguides can be realized by using some conventional structures such as a rectangular core and a ridge core. However, in the case of a ridge structure, because the confinement factor of ARROWs is very close to unity, deep etching of the core is required to obtain a lateral index difference large enough for lateral optical confinement. This problem can be solved by adopting a Stripe Lateral Confinement (SLC) structure [10], as shown in Figure 4.3.4.

In Figure 4.3.4 the regions A and B correspond to the core and the cladding regions in the lateral direction, respectively. In region B, a thin intermediate layer having lower refractive index than the core layer is inserted in the center of the vertical core layer. Since the field profile of the fundamental ARROW mode has a maximum near the center of the vertical core layer, the equivalent index of region B is effectively reduced from that of region A. In addition, because the intermediate layer is much thinner than the core layer, the top surface of this channeled structure can be easily planarized by using a bias-sputtering techineque. Therefore, this structure is suitable for the stacked configuration of optical interconnects and waveguide-type devices, as described in the following sections.

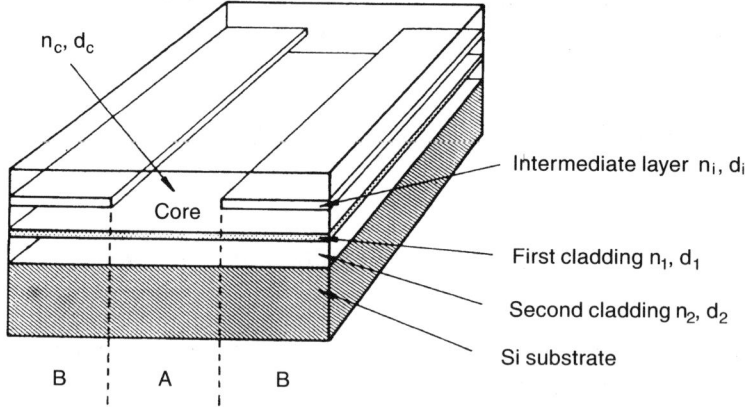

Figure 4.3.4
Channeled structure of ARROW-type waveguides by using the Stripe Lateral Confinement (SLC) structure

3 Multilevel Stacked Optical Interconnects Using Planar ARROWs

Figure 4.3.5 shows the cross-sectional structure of a stacked configuration using ARROW-type waveguides. The stacked waveguides are symmetric ARROWs that have two pairs of interference cladding above and below the core. The ARROW-type waveguides are separated by a thin metal layer less than 0.1 μm thick. Because higher order modes are radiated and absorbed in the metal separation layer, each waveguide can achieve single-mode propagation. In addition, because the separation layer prevents the light from leaking into the adjacent waveguide, crosstalk is completely suppressed. The loss increase due to the use of this metal layer is negligibly small.

Another advantage of this configuration is that a coupling region with a three coupled waveguide structure can be easily constructed by partly eliminating the separation region, because the separation layer is very thin and the thickness of the second cladding is half that of the core.

The propagation and coupling characteristics of this stacked configuration were analyzed by solving the Maxwell's equation directly for the whole waveguide structure and by using the quasi-guided mode approximation [11]. We adopted ARROW-B, which consists of a NA45 glass core ($n_c = 1.54$, $d_c = 3$ μm), a SiO$_2$ first cladding ($n_1 = 1.46$, $d_1 = 0.5$ μm) and a NA45 glass second cladding ($d_2 = 1.5$ μm). The wavelength was assumed to be 0.633 μm for ease of experiment. TE polarization was assumed in this analysis.

We simulated the case in which the fundamental mode in the upper waveguide #2 is transformed into that in the lower waveguide #1 through the coupling region. When the fundamental mode of waveguide #2 is incident on the coupling region, it is expanded into three coupled modes, TE_0^c, TE_1^c, and TE_2^c. Figure 4.3.6 shows the field distributions of these coupled modes just after the incidence on the coupling region ($z = 0$). The field

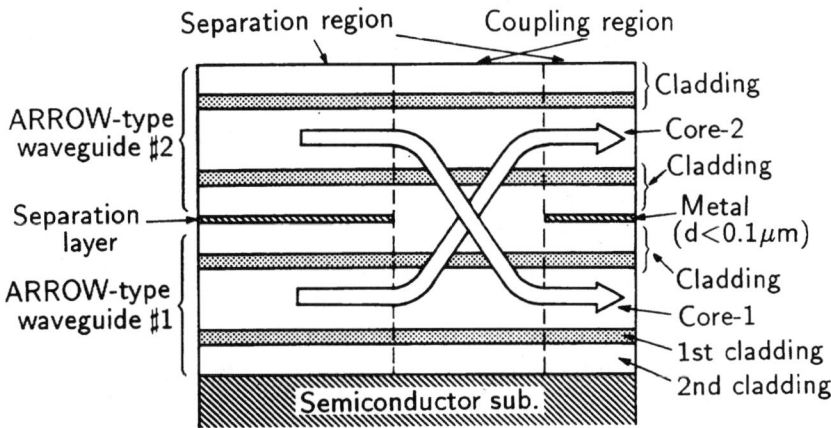

Figure 4.3.5
Cross-sectional structure of multilevel stacked interconnects using planar ARROW-type waveguides

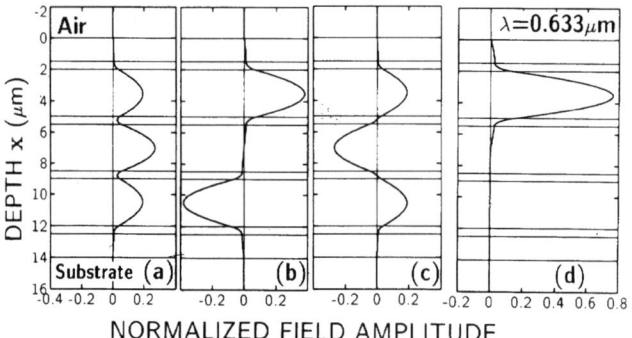

Figure 4.3.6
Field profiles of coupled modes excited by TE_0 mode in ARROW-B #2 at $z = 0$. (a) TE_0^c mode. (b) TE_1^c mode with phase difference $\Delta \psi_{01} = 0$. (c) TE_2^c mode with phase difference $\Delta \psi_{02} = 0$. (d) Total field of (a), (b) and (c). Difference of propagation constants between the TE_0^c and TE_1^c modes is $\Delta \beta = 8.24 \times 10^2$/m.

amplitudes of these coupled modes in Figure 4.3.6 are weighted with the coupling efficiency. Figure 4.3.6(d) shows the actual field, i.e. the superposition of the three coupled modes at $z = 0$. Because this is very close to the fundamental mode of ARROW-B #2 in the separation region, the coupling efficiency from the fundamental mode of waveguide #2 to the field shown in Figure 4.3.6(d) reaches $\eta = 0.997$. The difference between the propagation constants of the TE_0^c and TE_1^c modes $\Delta \beta$ is almost the same as that between the TE_1^c and TE_2^c modes. To achieve power exchange from waveguide #2 to #1, the coupling region should have length $L = \pi/|\Delta \beta|$. L for $d_c = 3$ μm is calculated to be 3.7 mm and that for $d_c = 2$ μm is 2.1 mm.

When the phase difference between the TE_0^c and TE_1^c modes, $\Delta \phi_{01}$, is $-\pi$ at the end of the coupling region, the difference between the TE_0^c and TE_2^c modes $\Delta \phi_{02}$, is -2π and the field is localized in the bottom core, as shown in Figure 4.3.7(d). In addition, this total field is very close to the fundamental mode in waveguide 1 in the separation region, so that the coupling efficiency from the coupling region to waveguide #1 is as high as $\eta = 0.997$. Thus the overall power coupling efficiency between the upper and lower waveguides is theoretically greater than 0.99. The same results are obtained for the case of coupling from waveguide #1 to #2 as well as for the stacked configuration with ARROW.

The stacked configuration with the ARROW-B structure was fabricated by the RF sputtering technique, and the coupling characteristics were measured. The thickness of the NA45 glass core d_c was 2 μm and that of the SiO_2 first cladding d_1 was 0.6 μm. The metal separation layer of 0.06 μm thick was formed by vacuum evaporation using Cr ($n = 2.6 + j3.0$). In addition, the coupling region was formed by eliminating part of the separation layer. Although the optimum length of the coupling region was theoretically $L = 2.1$ mm for $d_c = 2$ μm, various lengths were prepared in consideration of the

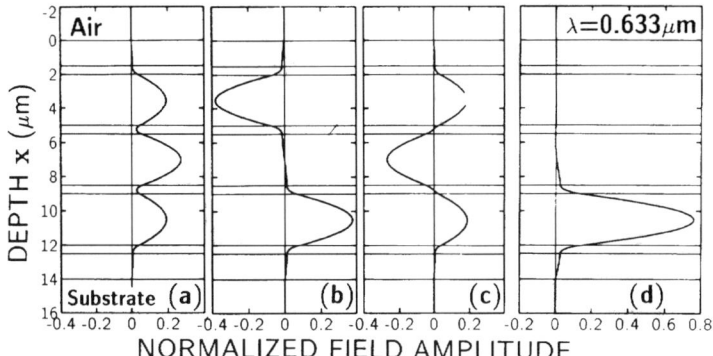

Figure 4.3.7
Field profiles of coupled modes at $z = L = 3.7$ mm. (a) TE_0^c mode. (b) TE_1^c with $\Delta\phi_{01} = -\pi$. (b) TE_2^c with $\Delta\phi_{02} = -2\pi$. (d) Total field. The coupling efficiencies of the fundamental modes in ARROW-B #2 and ARROW-B #1 in the separation region are $\eta = 0.997$ and $\eta = 0.0016$, respectively

fabrication error in the film thickness. In the separation region, the theoretical value of the propagation loss of the TE fundamental mode is as small as 0.2 dB/cm, and those of the other TE higher order modes are larger than 80 dB/cm, which ensures the quasi-single mode.

The coupling characteristics in the stacked waveguide were measured at $\lambda = 0.633$ μm. Almost complete light coupling between the upper and lower waveguides was observed [8]. However, the length of the coupling region was 6 mm. This discrepancy between the theoretical and experimental results seems to be caused by a fabrication error in the film thickness. In fact, it was found from the numerical simulation that the coupling characteristics, such as the efficiency and the coupling length, are quite sensitive to the fabrication error of the core thickness. Therefore highly precise control of the core thickness is required to maintain high coupling efficiency. In addition, the coupling length is rather long, over several millimeters, and this structure was the slab waveguide. Thus next, we aimed at reducing the coupling length and increasing the fabrication tolerance in a three-dimensional ARROW structure.

4 Compact Three-dimensional Optical Interconnects Using Channeled ARROW-B

According to numerical calculation, the tolerance of the core thickness in the structure shown in Figure 4.3.5, giving a half value of the maximum coupling efficiency, is less than 1%. This small tolerance is caused by the loss of symmetry in the field profiles of the coupled modes. If there is no fabrication error, they are symmetric and the coupling efficiency takes its maximum. However, when there is some fabrication error, the profiles of the coupled modes become asymmetric and the coupling efficiency decreases.

To solve this problem, we proposed and demonstrated an improved three-dimensional stacked configuration of ARROWs, as shown in Figure 4.3.8, which has a large tolerance of the coupling characteristics and a short coupling length [12]. This three-dimensional structure consists of two stacked ARROW-Bs and has two coupled cores in the crossconnect region. The upper and lower waveguides are formed to make a three-dimensional structure using the use of Stripe Lateral Confinement (SLC) structure. In the input and output regions, three pairs of ARROW-B-type interference claddings are used above the lower core, and below the upper core. The third cladding, above the lower core and below the upper core, acts as SLC layer. In the input and output regions, the upper and lower waveguides are separated laterally by 200 μm and approach the crossconnect region through curved waveguides. After propagating through the crossconnect region, they are separated again by curved waveguides. Employing a thin intermediate cladding, the field profiles of the coupled modes maintain symmetry compared with the previous structure even when there is some fabrication error, and a large tolerance in the coupling characteristics can be obtained. The thickness of the NA45 glass core d_c was designed to be 2 μm, and those of the SiO$_2$ first cladding layer d_1, intermediate layer d_{mid}, and SLC layer d_{SLC} were designed to be 1 μm, 0.2 μm, and 0.05 μm, respectively.

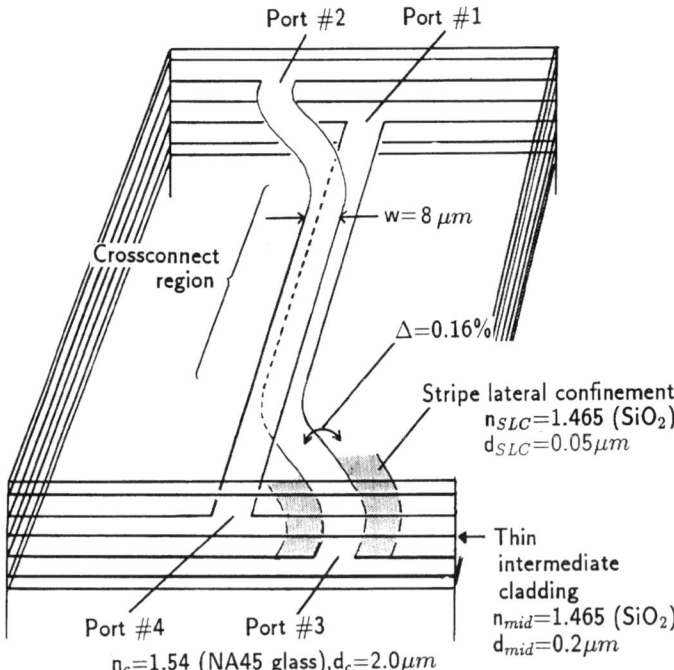

Figure 4.3.8
Three-dimensional optical interconnects using SLC channeled ARROW-B waveguides

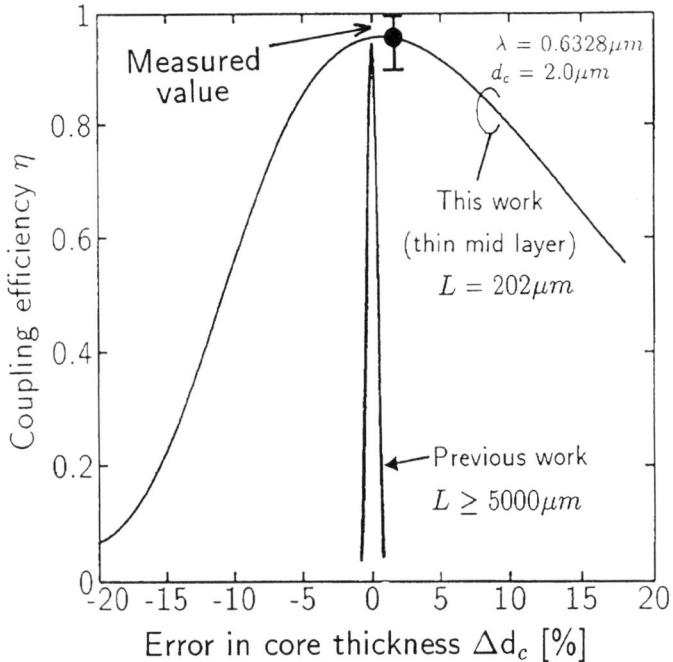

Figure 4.3.9
Calculated coupling efficiencies versus fabrication error of the core thickness

Figure 4.3.9 shows the calculated coupling efficiencies of the previous and proposed configurations versus the error in one of core thicknesses. In this analysis the field of the incident mode was expanded into two coupled modes and a continuous range of radiation modes. Thus the effect of radiation loss was taken into account. In the new configuration the tolerance to obtain the efficiency over 90% is $-3-5\%$ and is much larger than that of the previous structure, which is narrower than $-1-1\%$. In addition, the coupling length is 202 μm and is much shorter than that of the previous one, which is longer than 5000 μm.

A device which has a stripe width of 8 μm and a coupling length of 200 μm was fabricated. The reactive ion etching (RIE) technique was used to etch the lateral confinement layers with a Cr/photoresist two-layer mask. The bias RF sputtering technique was used to obtain the smooth surface of a SLC structure when the upper half core of the SLC structure was deposited after the etching of the SLC layers. The propagation loss characteristics were measured by using the cut-back method.

Figure 4.3.10 shows the measured results of propagation loss when the light was incident on port #1 and was observed from port #4. In this case the light does not experience any curved waveguides. The least squares fit of the propagation losses was evaluated to be 6.88 dB/cm ($\sigma = 0.00$ dB) before the crossconnect region and 5.81 dB/cm ($\sigma = 0.97$ dB) after the crossconnect region, respectively. The somewhat large propagation

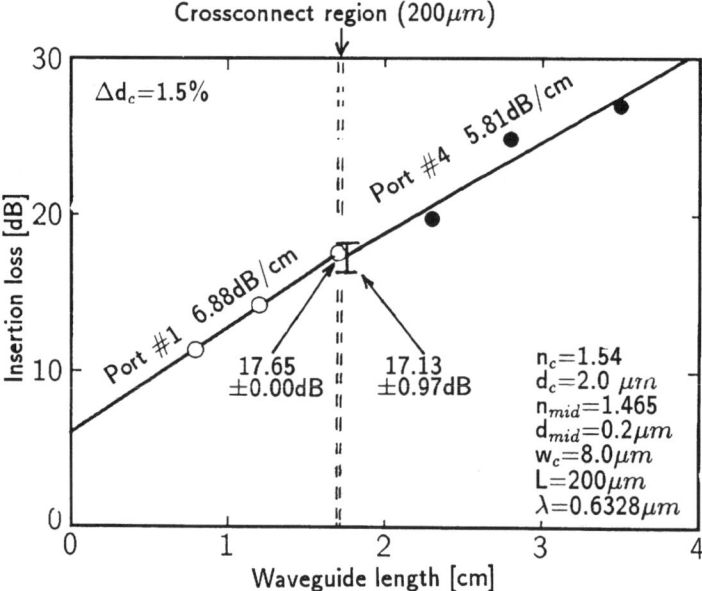

Figure 4.3.10
Measured results of propagation loss using the cut-back method

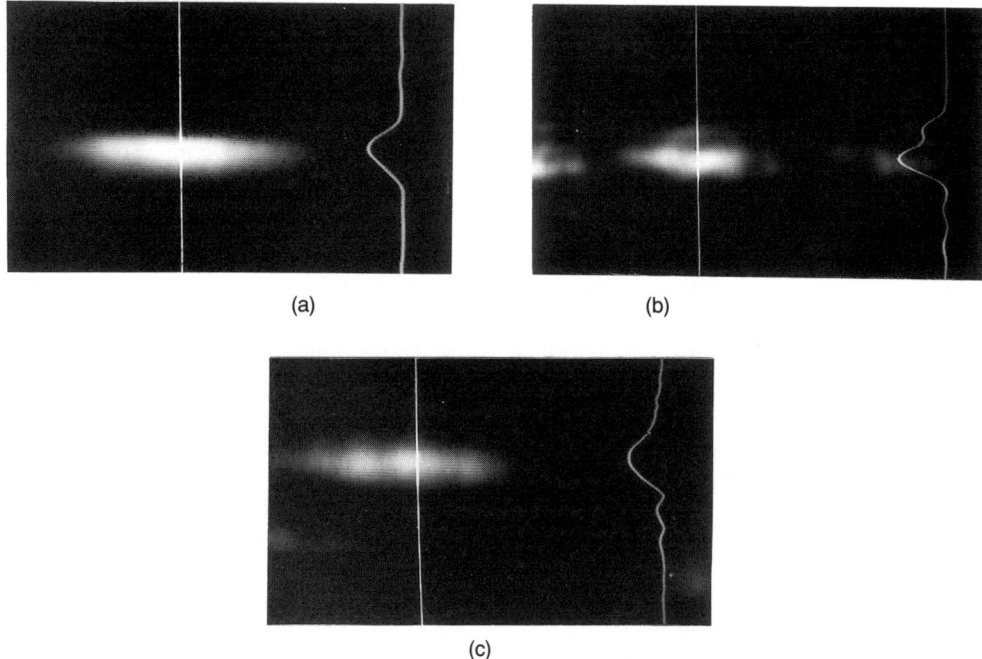

Figure 4.3.11
Observed near field patterns. (a) Before crossconnect (port #1). (b) Inside cross-connect. (c) After cross-connect (port #4)

loss is caused by the irregularity of the edge of the stripe patterns. The coupling efficiency was calculated to be 90–100% from the extrapolated values of insertion loss just before and after the coupling region, taking into account the measurement error. The fabrication error of the core thickness was measured to be 1.5% by SEM observation of the cross-section. The measured value of the coupling efficiency coincides with the theoretical value as shown in Figure 4.3.9, coupling efficiency of stacked crossconnects.

Figure 4.3.11(a)–(c) shows the near-field patterns, before, in, and after the crossconnect region. The transfer of optical field from the lower core to the upper core was observed.

5 Summary

In this study we proposed and developed a new stacked configuration for mutilevel interconnects by using ARROW-type waveguides. The stacked configuration will be effective not only for dense optical interconnects but also for the basic structure of functional waveguide devices such as polarization splitters and wavelength filters utilizing the multilayer cladding waveguides.

References

[1] M.A. Duguay, Y. Kokubun, T.L. Koch and L. Pfeiffer, *Appl. Phys. Lett.*, **49**(1), 13–15 (1986).
[2] Y. Kokubun, T. Baba, T. Sakaki and K. Iga, *Electron. Lett.*, **22**(8), 892–893 (1986).
[3] T. Baba and Y. Kokubun, *J. Quantum Electron.*, **28**(7), 1689–1700 (1992).
[4] T. Baba and Y. Kokubun, *Photon. Technol. Lett.*, **1**(8), 232–234 (1989).
[5] Y. Kokubun and S. Asakawa, *Photon. Technol. Lett.*, **5**(12), 1418–1420 (1993).
[6] T. Baba, Y. Kokubun and H. Watanabe, *J. Lightwave Technol.*, **8**(1), 99–104 (1990).
[7] T. Baba, Y. Kokubun, T. Sakaki and K. Iga, *J. Lightwave Technol.*, **LT-6** (9), 1440–1445, (1988).
[8] S. Asakawa, Y. Kokubun, M. Ohyama and T. Baba, *Electron. Lett.*, **29**(16), 1485–1486 (1993).
[9] M. Mann, U. Trutschel, C. Wächter, L. Leine and F. Lederer, *Opt. Lett.*, **16**(11), 805–807 (1991).
[10] T. Baba, Y. Kokubun and Y. Mera, *Topical Meeting on Integrated and Guided-Wave Optics (IGWO'89)*, Houston, 1989, TuBB5, Digest pp. 163–167.
[11] Y. Suematsu and K. Furuya, *IEEE Trans. Microwave Theory Technol.*, **MTT-23**, 170–175 (1975).
[12] M. Ohyama, Y. Kokubun and E. Ohta, *Electron. Lett.*, **30**(12), 951–952 (1994).

4.4

Design and Modeling of Optical Waveguide Devices

Masanori Koshiba

Abstract

A simple scalar finite element method is introduced for the design and modeling of optical waveguide devices and is utilized as a solver of an analysis/design package for step-index and graded-index optical waveguides with arbitrary cross-section. In this package isoparametric quadratic elements which fit the curved boundaries of a waveguide and can model accurately the refractive index change in each element are used. The scalar finite element method has as its main advantages: a smaller matrix dimension, less computer time, and no spurious solutions. Calculations are performed for a Ti-indiffused $LiNbO_3$ optical waveguide and a D-shaped optical fiber. The present package is incorporated into an optical CAD system named ELM (expert lightwave modeling system).

1 Introduction

As optical waveguide devices penetrate into many photonic systems, there is a demand for user-friendly design and modeling techniques. For this purpose, various numerical and approximate techniques have been developed [1]. In particular, the finite element method (FEM) is a powerful and efficient tool for the general (i.e. arbitrarily-shaped, inhomogeneous, dissipative, anisotropic, and nonlinear) optical waveguiding problems [2], and recently the vector FEM (VFEM) has been utilized as a waveguide solver of CAD packages [3,4]. Although the VFEM enables one to compute accurately the mode spectrum of a waveguide with arbitrary cross-section, it has been known to include nonphysical spurious solutions and to require a long computer time and/or large computer memory.

In this section a simple scalar FEM (SFEM) [2] is introduced and is utilized as a solver of an analysis/design package for step-index and graded-index optical waveguides with arbitrary cross-sections. It has been implemented on engineering workstations and has been incorporated into an optical CAD system which we called ELM (expert lightwave modeling system).

2 Basic Equations

An optical rib waveguide will support the propagation of waves that have two possible field configurations, classified as the E^x and E^y modes. The main field components of the E^x_{mn} modes are E_x and H_y, while those of the E^y_{mn} modes are E_y and H_x. The subscripts m and n designate the number of maxima of the dominant field in the x and y directions, respectively.

Noting that the E^x and E^y modes are well approximated by the TE^y ($E_y \equiv 0$, the leading function is E_x) and TM^y ($H_y \equiv 0$, the leading function is H_x) modes, respectively, the following Helmholtz equation can be derived from the Maxwell equation [2]

$$p_x \frac{\partial^2 \phi}{\partial x^2} + p_y \frac{\partial^2 \phi}{\partial y^2} - p_z \beta^2 \phi + q k_0^2 \phi = 0 \tag{1}$$

where k_0 is the free-space wavenumber, and β is the propagation constant in the z (axial) direction. The main field ϕ, the coefficients p_x, p_y, p_z, and q, and the remaining main field are given by

$$\phi = E_x \tag{2}$$

$$p_x = n_x^2/n_z^2, \qquad p_y = 1, \qquad p_z = 1, \qquad q = n_x^2 \tag{3}$$

$$H_y = \frac{1}{Z_0 \beta k_0} \left(\beta^2 \phi - \frac{n_x^2}{n_z^2} \frac{\partial^2 \phi}{\partial x^2} \right) \simeq \frac{n_{\text{eff}}}{Z_0} \phi \tag{4}$$

for the E^x modes, and

$$\phi = H_x \tag{5}$$

$$p_x = 1/n_y^2, \qquad p_y = 1/n_z^2, \qquad p_z = 1/n_y^2, \qquad q = 1 \tag{6}$$

$$E_y = -\frac{Z_0}{n_y^2 \beta k_0} \left(\beta^2 \phi - \frac{\partial^2 \phi}{\partial x^2} \right) \simeq -\frac{n_{\text{eff}} Z_0}{n_y^2} \phi \tag{7}$$

where n_x, n_y, and n_z are the refractive indices, Z_0 is the free-space impedance, and n_{eff} is the effective index and is given by $n_{\text{eff}} = \beta/k_0$.

3 Finite Element Approach

Dividing the waveguide cross-section Ω into a number of linear or quadratic triangular elements [2] and applying the standard finite element technique to (1), we obtain

$$[K]\{\phi\} - n_{\text{eff}}^2 [M]\{\phi\} = \{0\} \quad (8)$$

with

$$[K] = \sum_e \int\int_e \left[q\{N\}\{N\}^T - p_x \frac{\partial\{N\}}{\partial \bar{x}} \frac{\{N\}^T}{\partial \bar{x}} - p_y \frac{\partial\{N\}}{\partial \bar{y}} \frac{\{N\}^T}{\partial \bar{y}} \right] d\bar{x} d\bar{y} \quad (9)$$

$$[M] = \sum_e \int\int_e p_z \{N\}\{N\}^T d\bar{x} d\bar{y} \quad (10)$$

where the components of the $\{\phi\}$ vector are the values of ϕ at all nodes, $\{N\}$ is the shape function vector, $\{0\}$ is a null vector, the summation Σ_e extends over all different elements, $\bar{x}$ and $\bar{y}$ are the Cartesian co-ordinates normalized by the free-space wavenumber, namely $\bar{x} = k_0 x$ and $\bar{y} = k_0 y$, and T denotes a transpose. The form of (8) is a standard eigenvalue problem whose eigenvalue and eigenvector directly correspond to the effective index and modal field, respectively.

The SFEM has been successfully utilized for the analysis and design of various optical waveguide devices [5–9].

4 CAD Package Using SFEM

The CAD package described here consists of three modules: a pre-processor, a solver, and a post-processor.

The pre-processor is used to define the guide geometry and specify the optical wavelength in microns, the refractive indices, the boundary conditions (Neumann and/or Dirichlet conditions on artificial boundaries far from the core region, and conditions on the planes of symmetry), the desired mode (E_{mn}^x or E_{mn}^y modes, $m, n = 1.2$), and the type of elements (linear or quadratic triangular elements). In this package isoparametric quadratic elements that fit the curved boundaries of a waveguide are used. Furthermore, the numerical integration formula [2] is introduced, and therefore, for graded-index waveguides, the refractive index change in each element can be modeled accurately. Figure 4.4.1 shows an input menu popup window for the graded-index waveguides, where the refractive index is defined as

$$n_i(x, y) = n_{si} + \Delta n_i [f(x) g(y)]^{\alpha_i}, \quad i = x, y, z \quad (11)$$

Here n_{si} is the substrate index, Δn_i is the maximum index change, and $f(x)$ and $g(y)$ are the refractive index profile functions in the x and y directions, respectively. In the ELM, for the profile in the x direction, $f(x)$, uniform, Gaussian, modified error, generalized

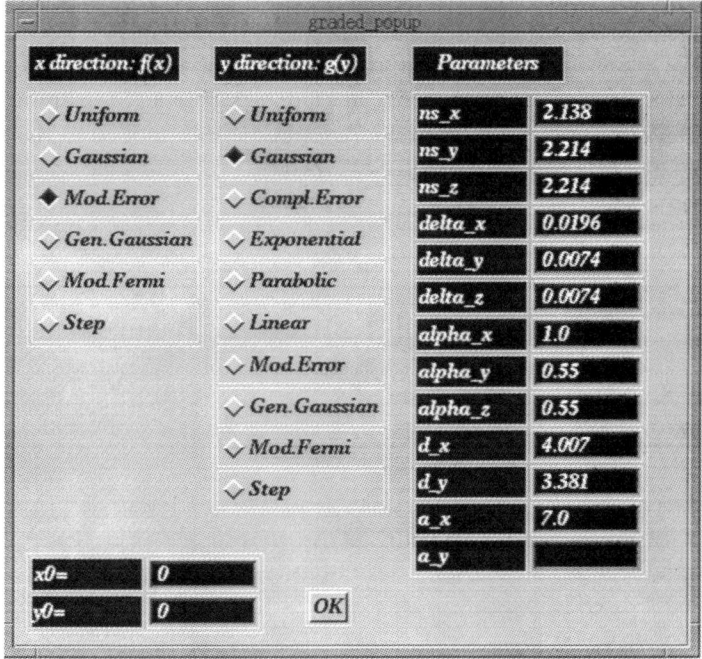

Figure 4.4.1
Input menu for graded-index optical waveguides

Gaussian, modified Fermi, and step functions are available, and for the profile in the y direction, $g(y)$, uniform, Gaussian, complementary error, exponential, parabolic, linear, modified error, generalized Gaussian, modified Fermi, and step functions are available. In the earlier CAD packages based on VFEM [3,4] only linear elements are used, and graded-index waveguides and waveguides with curved boundaries have not been treated.

As mentioned above, the solver is constructed by using SFEM [2]. It evaluates the phase constant, attenuation constant, confinement factor, spot size, and beam divergence angle (far-field pattern) given by

$$I(\theta_x, \theta_y) \propto \cos^2\theta \left| \int\int_\Omega \phi(x,y) \exp[j(k_x x + k_y y)] \, dx dy \right|^2 \quad (12)$$

with

$$k_x = k_0 \sin\theta_x = k_0 \sin\theta \cos\theta_a \quad (13)$$

$$k_y = k_0 \sin\theta_y = k_0 \sin\theta \sin\theta_a \quad (14)$$

$$\cos^2\theta = 1 - (\sin^2\theta_x + \sin^2\theta_y) \quad (15)$$

where θ is the polar angle to the z direction, θ_a is the corresponding azimuthal angle, θ_x and θ_y are the components of θ resolved along x and y, and $\cos\theta$ is the obliquity factor.

The spot size is defined by two method: one is defined as the beam waist at which the field amplitude is $1/e$ of its peak value, and the other is defined by using the quadratic moment of the field intensity [10]. The quadratic moment is applied to the waveguides

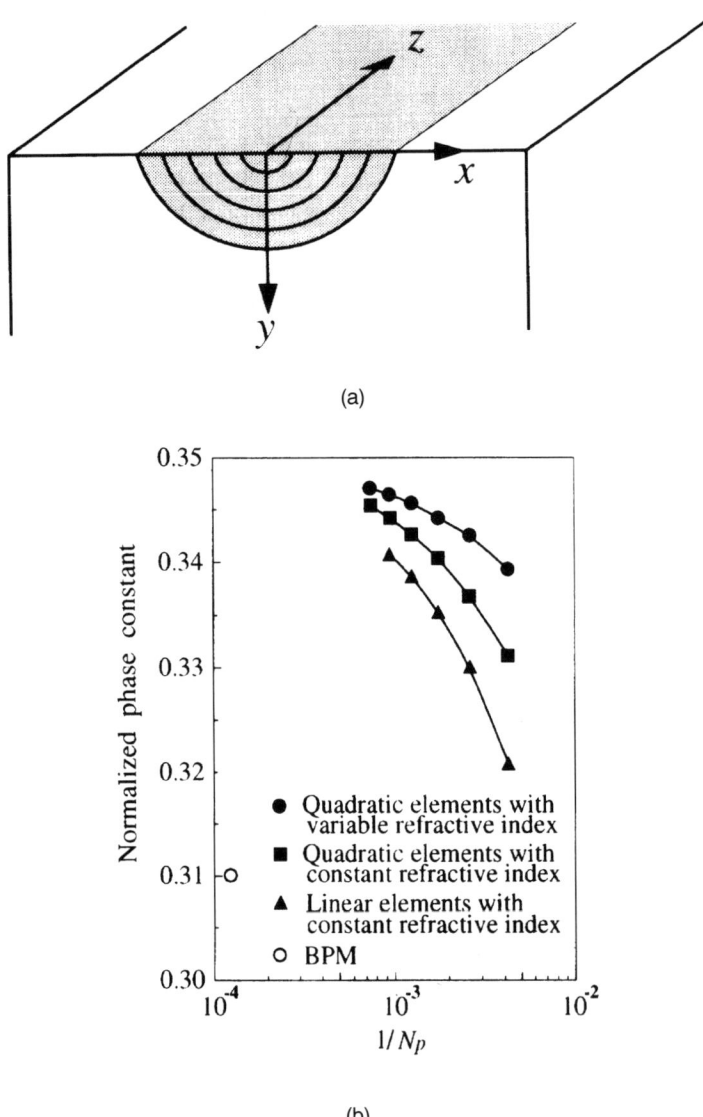

Figure 4.4.2
Anisotropic graded-index channel waveguide. (a) Waveguide structure. (b) Convergence of solution

with twofold symmetry. Although the phase and attenuation constants can be obtained directly by solving the complex eigenvalue problem, the attenuation constant can also be evaluated with the perturbation techniques [11].

The post-processor is used to display the near-field pattern and the far-field pattern. Results can be displayed as contour lines, color contours, or three-dimensional views.

5 Practical Work

First, we consider a Ti-indiffused X-cut Y-propagation LiNbO$_3$ optical waveguide as shown in Figure 4.4.2(a), where the refractive index profile functions in (1) are given by

$$f(x) = \frac{\text{erf}[(x+a_x)/d_x] - \text{erf}[(x-a_x)/d_x]}{2\text{erf}(a_x/d_x)} \quad (16a)$$

$$g(y) = \exp[-(y/d_y)^2] \quad (16b)$$

Here $2a_x$ is the Ti pattern width before diffusion, and d_x and d_y are the diffusion depths in the x and y directions, respectively.

Figure 4.4.2(b) shows the convergence of normalized phase constants of the fundamental E^x (E_{11}^x) mode, where N_p is the number of nodal points, and the normalized phase constant is defined as

$$b = \frac{n_{\text{eff}}^2 - n_{\text{sx}}^2}{(n_{\text{sx}} + \Delta n_x)^2 - n_{\text{sx}}^2} \quad (17)$$

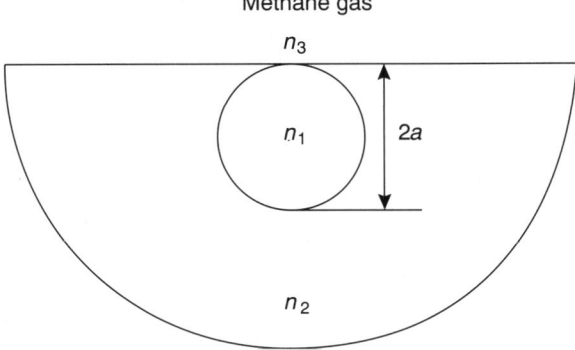

Figure 4.4.3
D-shaped optical fiber for methane gas sensing

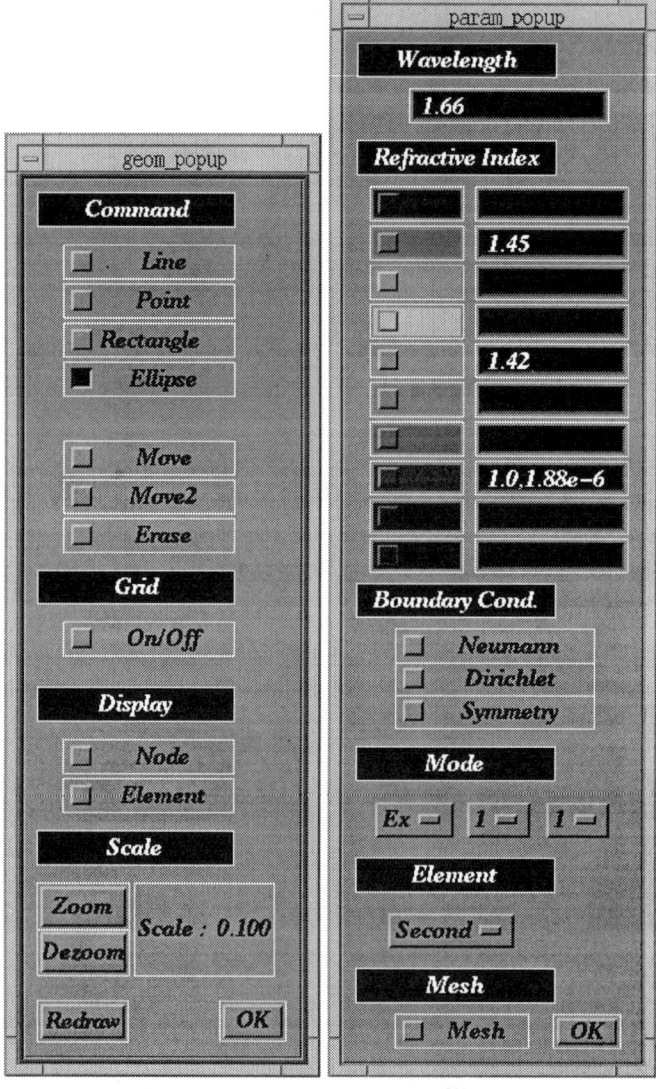

Figure 4.4.4
Main input menu

Design and Modeling of Optical Waveguide Devices

Here n_{eff} is the effective index. The operating wavelength, $\lambda = 1.52$ μm, and the waveguide parameters are assumed to be [5]

$$a_x = 7 \ \mu\text{m}, \qquad d_x = 4.007 \ \mu\text{m}, \qquad d_y = 3.381 \ \mu\text{m}$$
$$n_{sx} = 2.138, \qquad n_{sy} = n_{sz} = 2.214$$
$$\Delta n_x = 0.0196, \qquad \Delta n_y = \Delta n_z = 0.0074$$
$$\alpha_x = 1, \qquad \alpha_y = \alpha_z = 0.55$$

The input data are entered on the popup window as shown in Figure 4.4.1. The present approach using quadratic elements in which the refractive index change is taken into

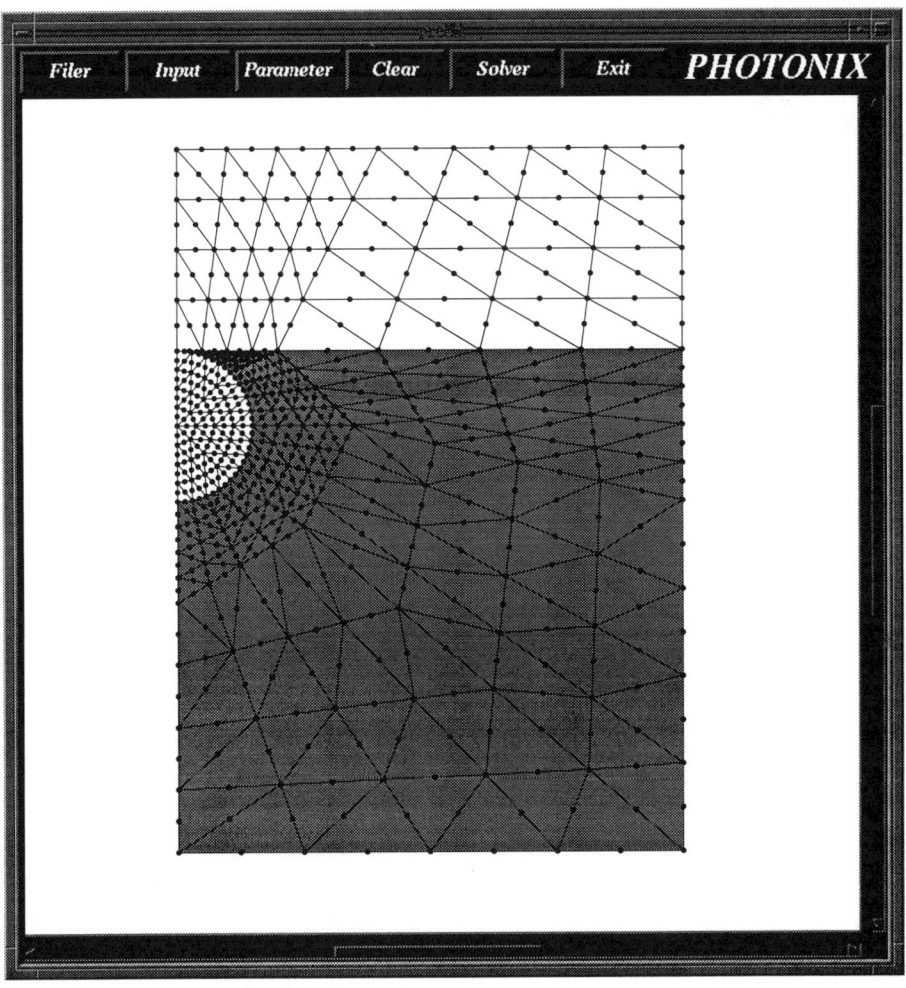

Figure 4.4.5
Mesh generation

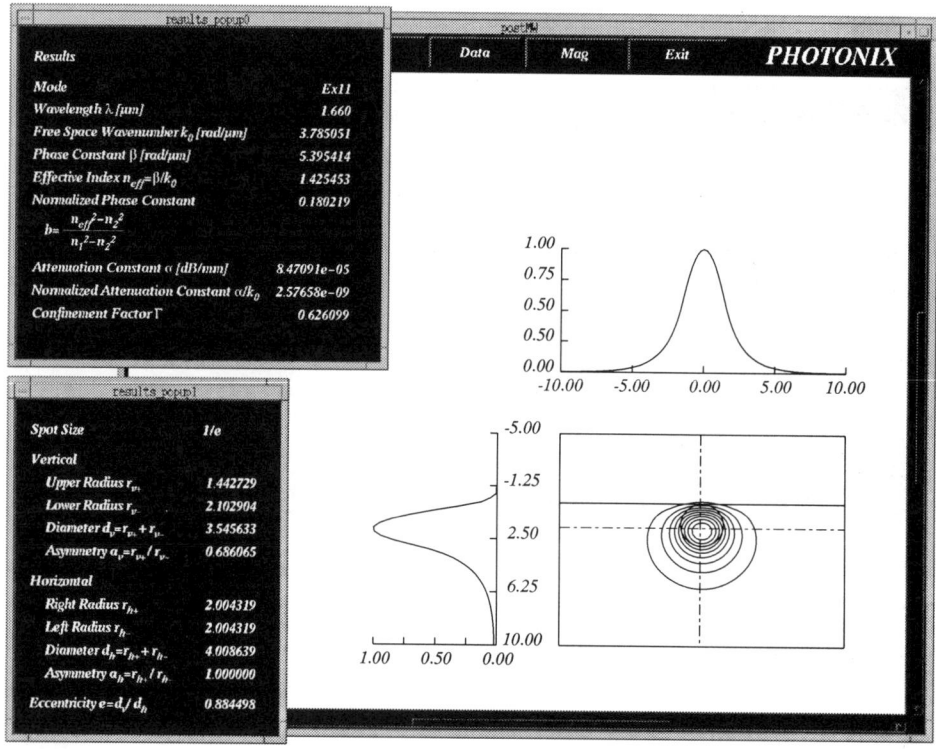

Figure 4.4.6
Window showing numerical results and near-field patterns

account is better in accuracy and in computing time than the earlier approach using linear or quadratic elements with a constant refractive index. The result of the beam propagation method (BPM) [5] is also plotted in Figure 4.4.2(b), where N_p is the number of sampling points. It seems that the BPM solution may not be fully convergent.

Next, we consider a D-shaped optical fiber for a methane gas sensor [12], as shown in Figure 4.4.3, where $\lambda = 1.66$ μm, and the waveguide parameters are assumed to be [12]

$$n_1 = 1.45, \qquad n_2 = 1.42, \qquad n_3 = 1.0 - j1.88 \times 10^{-6}$$

The input data are entered on the popup window as shown in Figure 4.4.4. After inputting all the required data, the finite element mesh is generated automatically. A sample mesh is shown in Figure 4.4.5.

Figures 4.4.6 and 4.4.7 show the near-field pattern and the far-field pattern for the E_{11}^x mode in the case of core diameter $2a = 3$ μm, respectively. Numerical results related to the effective index (1.425453), attenuation constant (8.47091 × 10^{-5} dB/mm), confinement factor (0.626099), vertical and horizontal spot sizes (3.545633 μm and 4.008639 μm), and vertical and horizontal beam divergence angles (FWHMs; 12.728288° and 10.510850°) are also shown.

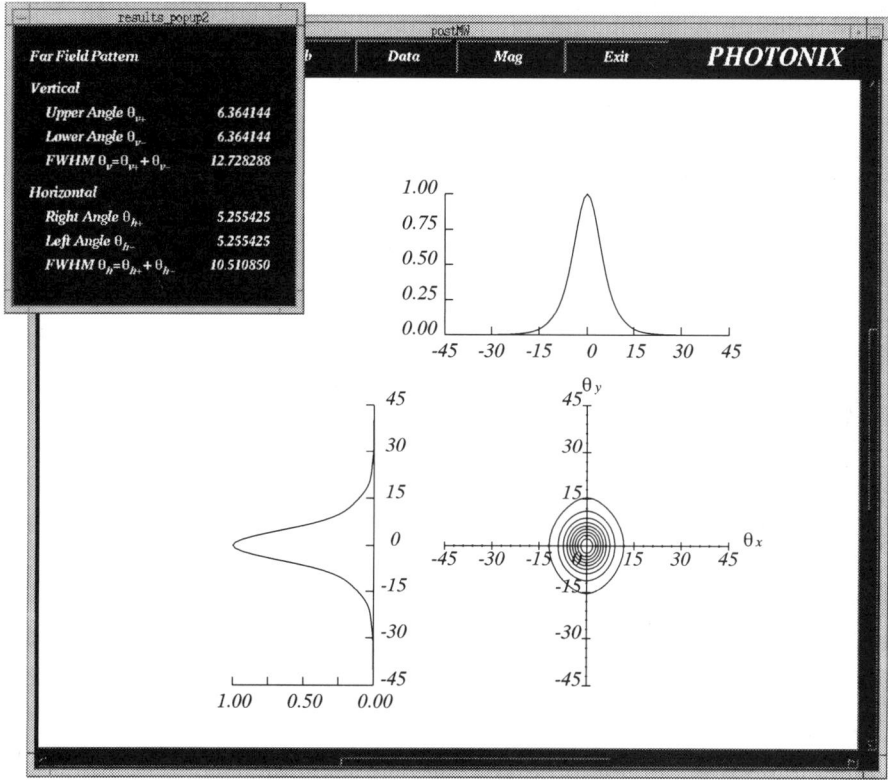

Figure 4.4.7
Window showing far-field patterns

Figures 4.4.8(a) and (b) show the effective index and the attenuation constant for the E_{11}^x mode, respectively. Our results are slightly different from those of the scalar finite difference method (SFDM) [12]. It is interesting to note that the cutoff core diameter is revealed in the present SFEM analysis. It seems that this is due to the waveguide asymmetry. The attenuation constant obtained by using the perturbation method [11] is in good agreement with that obtained by solving directly the complex eigenvalue problem. The CPU time in the former case is approximately one-tenth of that in the latter case.

6 Conclusions

We have developed an efficient analysis/design package for optical waveguides using a simple SFEM, which has been incorporated into ELM. The SFEM has as its main

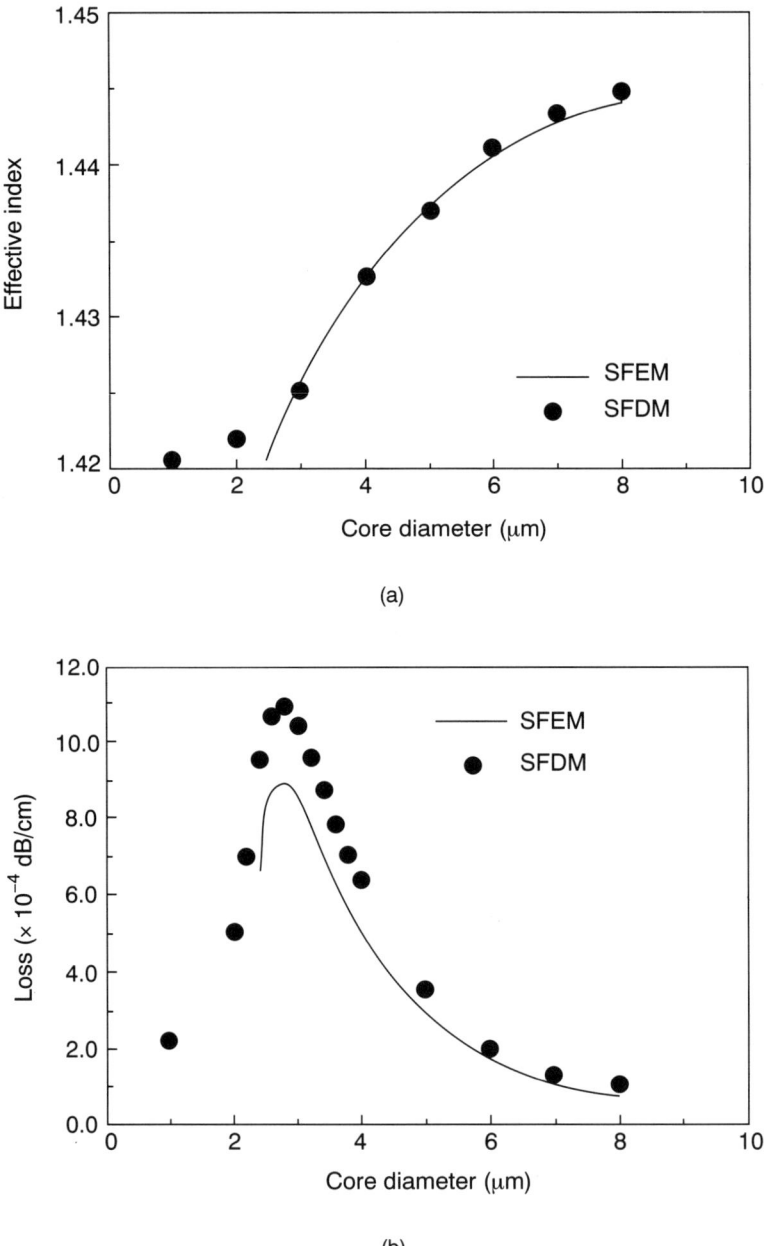

Figure 4.4.8
Propagation characteristics for the E_{11}^x mode of a D-shaped optical fiber for methane gas sensing. (a) Effective index. (b) Attenuation constant

advantages: smaller matrix dimensions, less computer time, and no spurious solutions. We are now working on extending this package to magneto-optic waveguides, nonlinear optical waveguides, and longitudinally varying waveguides.

References

[1] K.S. Chiang, *Opt. Quantum Electron.*, **26**, S113–S134 (1994).
[2] M. Koshiba, *Optical Waveguide Theory by the Finite Element Method*, KTK Scientific Publishers/Kluwer Academic Publishers, Tokyo/Dordrecht, 1992.
[3] T.P. Young and P. Smith, *GEC J. Res.*, **4**, 249–255 (1986).
[4] G. Sewell and S. Cvetkovic, *Adv. Eng. Software*, **11**, 169–175 (1989).
[5] L. Bersiner, U. Hempelmann and E. Strake, *J. Opt. Soc. Am. B*, **8**, 422–433 (1991).
[6] M. Koshiba, H. Saitoh, M. Eguchi and K. Hirayama, *IEE Proc.*, pt J, **139**, 166–171 (1992).
[7] M. Koshiba and X.-P. Zhuang, *J. Lightwave Technol.*, **11**, 1453–1458 (1993).
[8] M. Eguchi and M. Koshiba, *J. Lightware Technol.*, **12**, 607–613 (1994).
[9] N. Osman, M. Koshiba and R. Kaji, *J. Lightwave Technol.*, **12**, 821–826 (1994).
[10] K. Hayata, M. Koshiba and M. Suzuki, *Electron. Lett.*, **22**, 127–129 (1986).
[11] K. Hayata, M. Koshiba and M. Suzuki, *IEEE J. Quantum Electron.*, **QE-22**, 781–788 (1988).
[12] F.A. Muhammad and G. Stewant, *Electron. Lett.*, **28**, 1205–1206 (1992).

4.5

Guided Wave Solid State Laser and Optical Amplifier

Yasumitsu Miyazaki

Abstract

This section discusses integrated crystal thin-film optical waveguide devices with heavily doped ions to achieve large optical amplification and laser. When an Nd^{3+} ion is used as a laser active ion in crystal garnet films, the amplifier can amplify light signals of 1.06 μm and 1.3 μm bands. In this section we propose a garnet crystalline thin-film slab and channel waveguide type optical amplifier with heavily doped rare-earth ions, and report a gain of 6.5 dB over a length of 4 mm, a gain per unit length of 16 dB/cm, and a pump efficiency of 1.1 dB/mW. For convenient optical wavelengths of 1.5 μm and 1.3 μm, channel waveguide optical amplifiers are investigated using Er^{3+} and Pr^{3+} ions in garnet crystal films.

1 Introduction

Lasers are a key technology for optical industries, and optical amplifiers are important devices for optical fiber communication and optical signal processing [1,2]. Recently, optical fiber amplifiers and semiconductor amplifiers have been extensively studied. However, integrated optical amplifiers and lasers using optical pumping are rarely discussed. Fiber amplifiers and lasers have a drawback in that they need a length of over a few meters [3–5], and semiconductor amplifiers have noise problems. On the other hand, solid-state optical amplifiers and lasers constructed using thin films with optical pumping have advantages such as a combination of different optical thin-film waveguide devices (modulators, switches, and isolators) in monolithic optical integrated circuits [6]. Solid-state lasers of small size pumped by laser diodes have been studied [7] and developed for waveguide type lasers and amplifiers using optical film waveguides [8-12]. A few glass film amplifiers were studied and less than 1 dB/cm gain properties are reported. We consider integration using different crystal thin-film optical waveguide devices with heavily doped ions to achieve large amplification.

Garnet or $LiNbO_3$ films are used mainly as materials for the optical thin-film waveguide devices and optical integrated circuits. Therefore we also investigate the materials for optical amplifiers. When an Nd^{3+} ion is used as a laser active ion in crystal garnet films, the amplifier can amplify light signals of 1.06 μm and 1.3 μm bands.

In this section we propose a garnet crystalline thin-film slab and channel waveguide type optical amplifier, and report a gain of 6.5 dB over a length of 4 mm, a gain per unit length of 16 dB/cm, and a pump efficiency of 1.1 dB/mW. For convenient optical wavelengths of 1.5 μm and 1.3 μm, channel waveguide optical amplifiers are investigated using Er^{3+} and Pr^{3+} ions in garnet crystal films.

2 Structure of a Waveguide Optical Amplifier and Laser

Guided wave solid state lasers and optical amplifiers using optical waveguides on the amorphous and crystal substrates pumped by laser diodes are compatible with other optical guided wave devices. Guided wave solid-state lasers and optical amplifiers have the same fundamental structure, and guided wave solid-state lasers can be fabricated by using a mirror or cavity structure at the input and output in guided wave solid-state optical amplifiers.

Figure 4.5.1 shows the energy level diagrams of laser active ions of rare-earth ions for solid-state lasers and optical amplifiers and Table 4.5.1 shows the optical absorption and fluorescence characteristics of laser active ions. Garnet crystal films may have

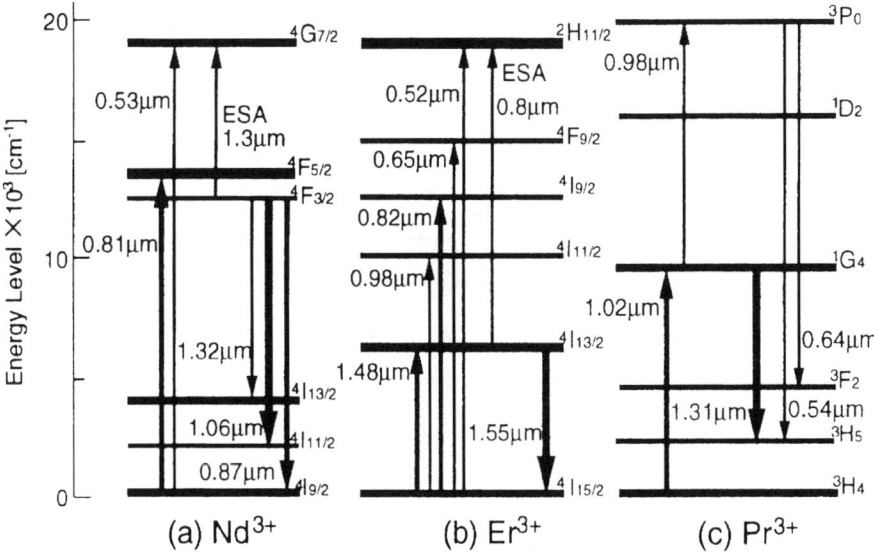

Figure 4.5.1
Energy levels

Table 4.5.1
Optical characteristics of laser active ions

Active ions	Pumping wavelength (μm)	Signal wavelength (μm)	Energy level
Neodymium (Nd^{3+})	0.8	1.06	four-level system
		1.3	
Erbium (Er^{3+})	1.48	1.55	three-level system
	0.98		
Praseodymium (Pr^{3+})	1.02	1.3	four-level system

heavy doped active ions, compared with glass films and fibers, and can yield high amplification gains and highly efficient laser oscillations.

Figure 4.5.2 shows integrated circuits with different optical crystal film waveguide devices that have a mode dispersion to confine light power in the waveguides. Figure 4.5.3 shows the structure of the slab waveguide type lasers and optical amplifiers using crystal films.

Nd:YGG ($Nd_{0.04}$:$Y_{2.96}$ Ga_5O_{12}) thin garnet crystal films were prepared by RF sputtering onto YAG ($Y_3Al_5O_{12}$) crystal substrates which had been cut and polished on the (1,1,1) plane. The lattice constants of Nd:YGG and YAG are 12.2 Å and 12.01 Å, and

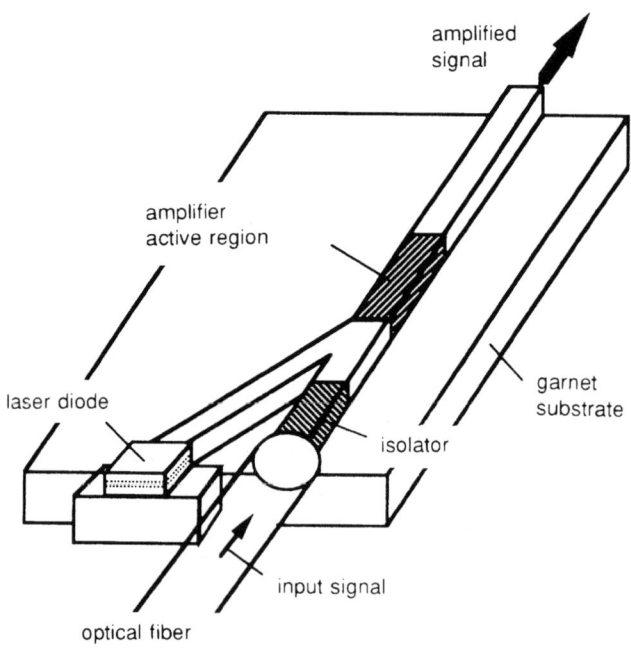

Figure 4.5.2
Structure of optical IC

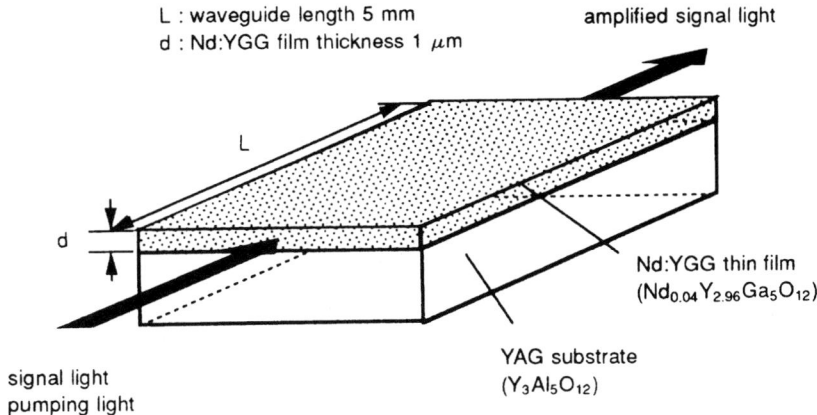

Figure 4.5.3
Structure of the waveguide type laser optical amplifier using Nd-doped garnet thin film

the Nd:YGG thin films can grow on the YAG substrate. The refractive index of 1.94 at a wavelength of 1.06 μm of the Nd:YGG thin film is greater than that of the YAG substrate (1.82), and the thickness of the Nd:YGG thin film is 1 μm for the single-mode condition. For active ion concentration the neodymium ion Nd^{3+} of active ion is contained 1.3 at.% in Nd:YGG film. The propagation loss of Nd:YGG thin film was 2.7 dB/cm at 633 nm and 2.1 dB/cm at 1061.5 nm, respectively.

A high gain can be achieved by increasing the active ion concentrations. The active ion concentrations in garnet crystals are about ten times greater than that in fiber amplifiers. In Nd:YGG, Y^{3+} is substituted with Nd^{3+}, and the radii of the two rare-earth ions differ by about 3%. Therefore high active ion concentrations cause quenching and attenuate the fluorescence intensity. For example, Nd concentrations in bulk Nd:YAG, which is a typical solid-state laser medium, are limited to less than 1.5 at.%. Hence the Nd concentration in the Nd:YGG thin film is designed to be 1.3 at.%. The pump light and the signal light are guided collinearly, and then the signal light obtains optical gain because of stimulated emission when it passes through the thin film which is maintained in a state of population inversion by the pump light.

There are two methods of optical pumping. One is longitudinal pumping (end pumping), which is done parallel to the surface of the waveguide. The other is transverse pumping (side pumping), which is done perpendicular to the surface. We adopt the former end pumping mechanism because end pumping has a high efficiency for film thicknesses of the order of 1 μm.

The efficient optical amplifier is essentially of three-dimensional channel waveguide type shown in Figure 4.5.4, formed by loading a layer of neodymium doped yttrium gallium garnet (Nd:YGG) on yttrium aluminum garnet (YAG) with a strip of zinc oxide (ZnO). Nd:YGG film and ZnO strip are fabricated by RF sputtering. Strip-type active channel waveguide devices may be fabricated by the direct etching method. A Nd:YGG film is deposited at 700–800 °C and crystallized by annealing at 950 °C. The refractive

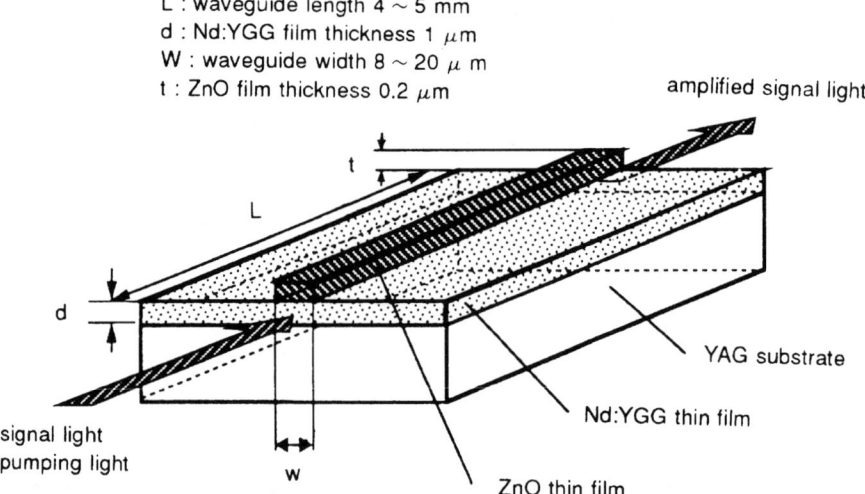

Figure 4.5.4
Structure of channel waveguide type optical amplifier

index of ZnO is 1.94 at the 1.06 μm wavelength. The thicknesses of the Nd:YGG film and the ZnO strip are 1.0 μm and 0.2 μm. The waveguide length is less than 1 cm. The signal light and the pumping light of coherent beams are confined in the Nd:YGG layer at a strip of ZnO region. The propagation losses of the waveguide are also 2.7 dB/cm and 2.1 dB/cm at 1.06 μm, respectively.

3 Amplification Theory

We discuss here optical amplifications using rate equations for the case of Nd^{3+} as the active ion. The active ions in the thin-film waveguide are the Nd^{3+} ions, which are a four-level laser medium. The photon density and stimulated emission density are given by four-level rate equations

$$\frac{\partial s(x, y, z)}{\partial t} + c\frac{\partial s(x, y, z)}{\partial z} = c\sigma n(x, y, z)s(x, y, z) - \alpha_s c s(x, y, z) + \frac{Fn(x, y, z)}{2\tau_f} \quad (1)$$

$$\frac{\partial n(x, y, z)}{\partial t} = -c\sigma n(x, y, z)s(x, y, z) - \frac{n(x, y, z)}{\tau_f} + \tau(x, y, z)\left(1 - \frac{n(x, y, z)}{N_T}\right) \quad (2)$$

where $s(x, y, z)$ is the photon density of the signal light, $n(x, y, z)$ is the population inversion density, $\tau(x, y, z)$ is the pumping rate, c is the velocity of light in the active medium, σ is the stimulated emission cross-section, α_s is the propagation loss of the signal light, τ_f is the lifetime of spontaneous emission, F is the coefficient factor of spontaneous emission for the guided mode, and N_T is the concentration of the Nd^{3+} ion.

The normalized distribution functions of the photon density and the pumping rate are defined as follows

$$s(x, y, z) = S(z)s_0(x, y) \qquad (3)$$

$$r(x, y, z) = R(z)r_0(x, y) \qquad (4)$$

and

$$\int_{-\infty}^{\infty}\int_{-\infty}^{\infty} s_0(x, y)dxdy = 1 \qquad (5)$$

$$\int_{-\infty}^{\infty}\int_{-\infty}^{\infty} r_0(x, y)dxdy = 1 \qquad (6)$$

where $S(z)$ and $R(z)$ are the photon density and the pumping rate per unit length, respectively.

The static solution is found by solving equations (1) and (2) under steady-state conditions with respect to time. By substituting equation (2) into equation (1) and integrating (1) over the x-y plane, we obtain the differential equation

$$\frac{dS(z)}{dz} = \frac{R(z)}{cS_a A(z)} S(z) - \alpha_s S(z) + F \frac{R(z)}{2cB(z)} \qquad (7)$$

where $S_3^{-1} = c\sigma\tau_f$ is the saturation density defined as the power density in an active medium at which the gain coefficient is reduced to one-half, and

$$A(z)^{-1} = \int_{-\infty}^{\infty}\int_{-\infty}^{\infty} \frac{s_0 r_0}{1 + S(z)s_0(x, y)/S_a + \tau_f R(z)r_0(x, y)/N_T} dxdy \qquad (8)$$

$$B(z)^{-1} = \int_{-\infty}^{\infty}\int_{-\infty}^{\infty} \frac{r_0}{1 + S(z)s_0(x, y)/S_a + \tau_f R(z)r_0(x, y)/N_T} dxdy \qquad (9)$$

The net gain can be obtained by solving equation (7) at a low level photon density $(s(x, y, z) \ll S_a)$ and a high-level photon density $(s(x, y, z) \gg S_a)$. Figure 4.5.5 shows the net gain versus signal power with an amplifier length of 1 cm with a waveguide film thickness of 2.5 μm calculated using $N_T = 10^{20}$ cm^{-3}, $\sigma = 5 \times 10^{-19}$ cm^{-2}, $\tau_f = 3 \times 10^{-4}$ s, and $\alpha_s = 0.1$ cm^{-1}. The signal wavelength is 1.06 μm. In the range of small signal power, the net gain is independent of the signal power. In the range of large signal power, the net gain decreases as the signal power increases. It shows that a constant gain can be obtained in the range of signal power $P_s < 10^{-6}$ W. The saturation signal power decreases as the pump power increases for small signal power; a 20 dB gain can be obtained with a pumping power of 0.5 mW. Figure 4.5.6 shows the characteristic gain versus input signal power of a channel waveguide type amplifier of waveguide length $L = 5$ mm and thickness 1 μm, which calculated for a pumping power of $P_p = 10$ mW and $\sigma = 5 \times 10^{-19}$ cm^2, $\alpha_s = 0.1$ cm^{-1}, $\tau_f = 3 \times 10^{-4}$ s, and $N_T = 10^{20}$ cm^{-3}. For channel widths smaller than 25 μm, a gain greater than 20 dB can be obtained in the small signal region.

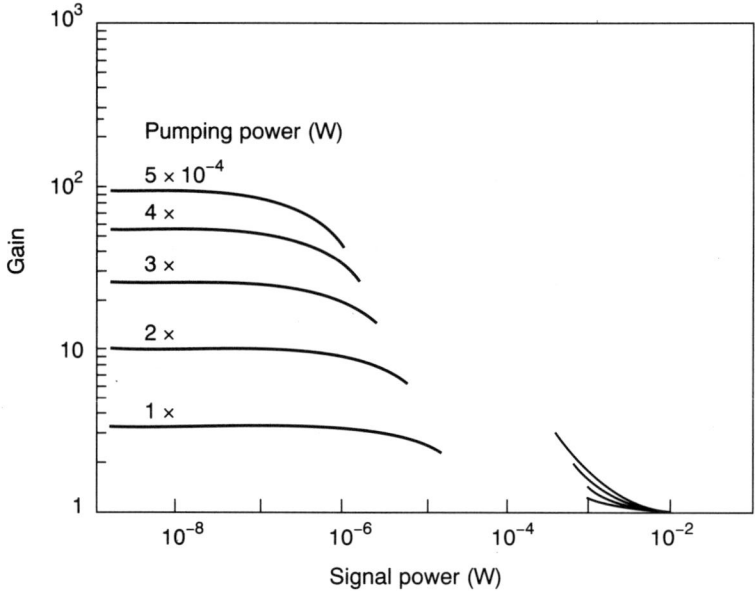

Figure 4.5.5
Net gain versus signal power

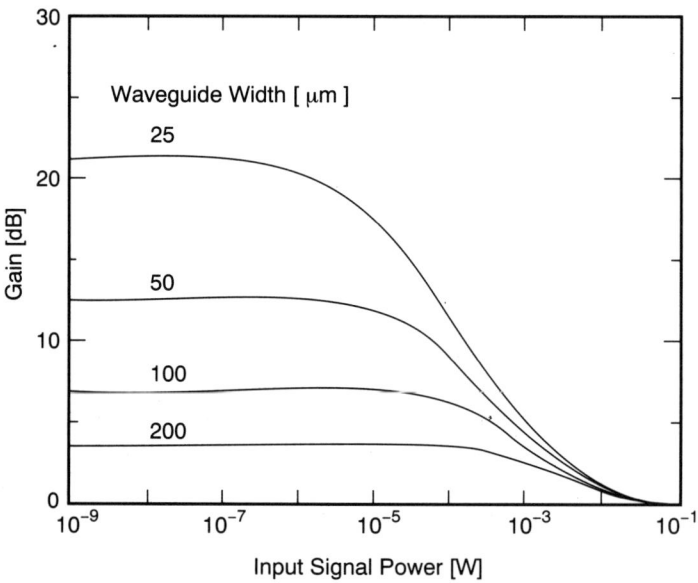

Figure 4.5.6
Gain characteristics versus the input signal power of the channel waveguide

4 Optical Characteristics and Spectroscopy

The sputtering system used to prepare the thin-film waveguides is of the quadruple type which can heat the substrate at more than 750 °C. The sputtering target is made by sintering a stoichiometric mixture of Y_2O_3 and Ga_2O_3 with 1.3 mole of Nd_2O_3 relative to Y_2O_3. The propagation loss is 2.8 dB/cm at $\lambda = 633$ nm and 2.2 dB/cm at $\lambda = 1061.5$ nm, and is caused by absorption and scattering.

The absorption and fluorescence spectra depend on the host material. Detailed measurements are required for high gain.

The absorption spectrum of the Nd film waveguide was measured for the pumping excitation as in Figure 4.5.7. LED, which has a luminescent center wavelength at 800 nm, was utilized for the experiment. As a result, the peak was located at 808 nm.

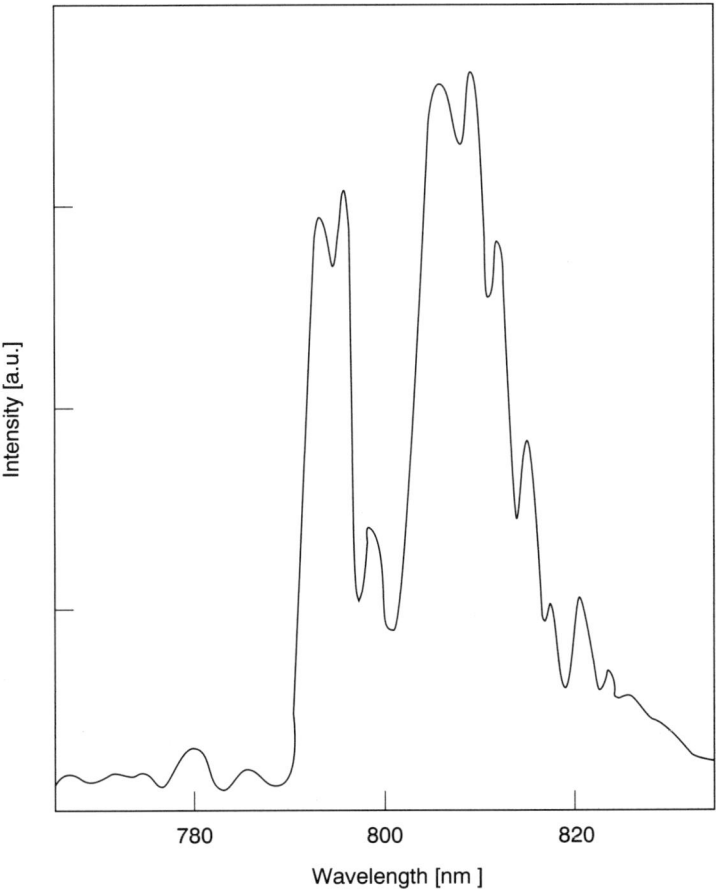

Figure 4.5.7
Absorption spectrum of the thin film around 0.8 μm. Transition $^4I_{9/2} \rightarrow {}^4F_{5/2} + {}^4H_{9/2}$

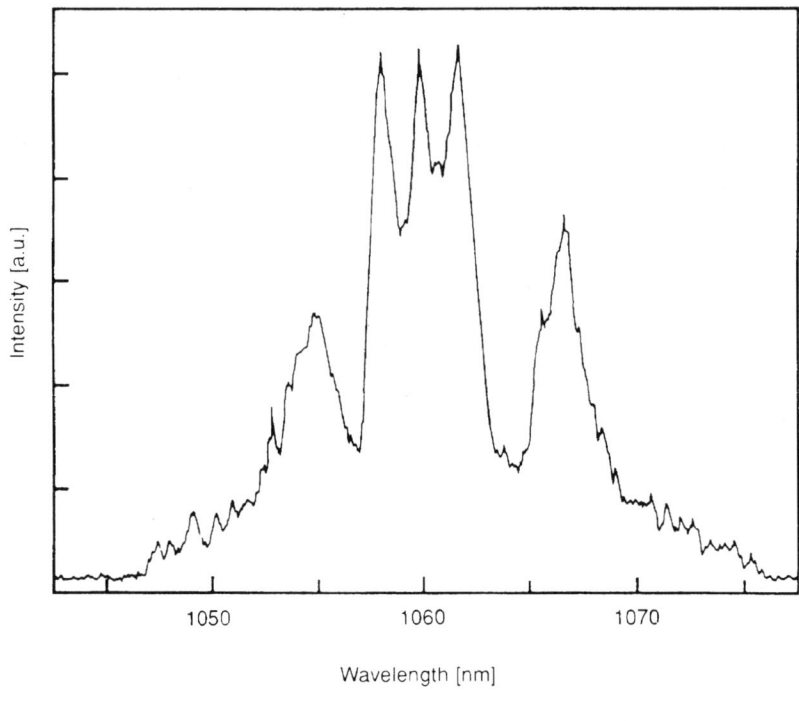

(a)

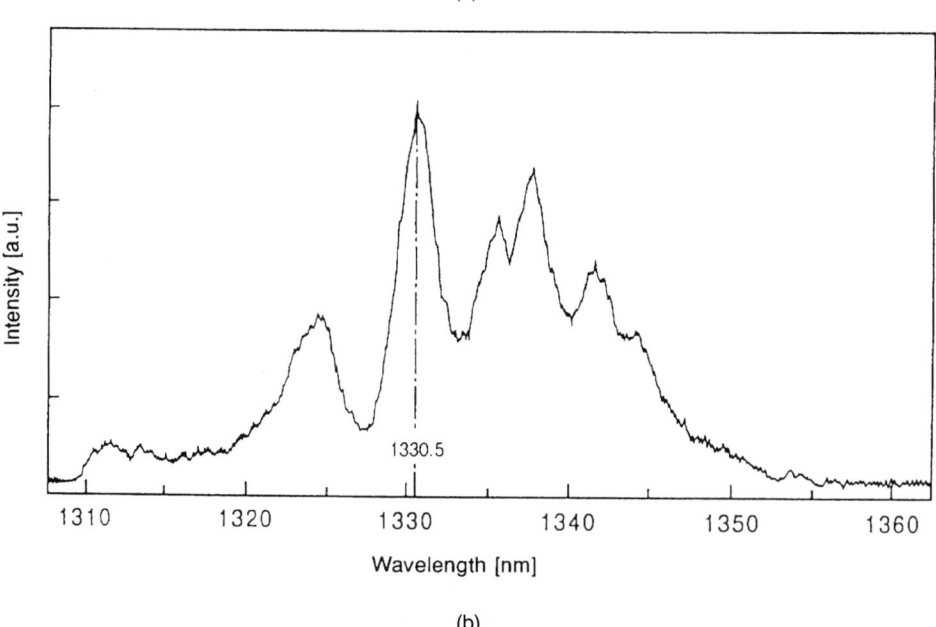

(b)

Figure 4.5.8
Fluorescence spectra of the thin film in (a) the 1.06 μm region and (b) the 1.3 μm region

Figure 4.5.8 shows the fluorescence spectrum of the Nd:YGG waveguide at the 1.06 μm and 1.3 μm regions. The peaks of the fluorescence spectrum for the waveguide are located at 1057.75 nm, 1060.0 nm, and 1061.87 nm. Therefore, a high gain is obtained for the signal light at $\lambda = (1060 \pm 2.5)$ nm.

The full width at half maximum for fluorescence is about 5 nm, which is a several times larger than that of the Nd:YAG single crystal. The fluorescence lifetime is 230 μm for the $^4F_{3/2}$–$^4I_{11/2}$ transition.

5 Experiment

In the amplifier experiment we measured the single-pass gain at the 1.06 μm region. The experimental setup for the gain characteristics is shown in Figure 4.5.9.

As the signal light source for the 1.06 μm region, we prepared a tunable Nd:YAG laser using an etalon plate pumped with a GaAlAs laser diode. The laser rod was flat at both ends and had a thickness of 3 mm and a cavity length of 8 cm. The Nd:YAG laser oscillated at two wavelengths, 1064.1 nm and 1061.5 nm, by tuning the angle of the etalon. For the tuned output at 1061.5 nm, the lasing threshold and the slope efficiency were 10 mW and 5%, respectively. We tuned at 1061.5 nm for the experiment.

For the pumping light we used a high-power GaAlAs laser diode of 40 mW tuned to $\lambda = 808$ nm. The maximum laser output power at 1061.5 nm was 0.5 mW with the absorbed pump power at 20 mW. The signal and pumping light beams were combined coaxially using a dichroic mirror. Next, their beams were coupled into the waveguide by the prism coupling method. After propagation, the beam from the output prism was separated from the pumping light with an IR filter which was detected with a Pin-Si photo diode through a monochromator.

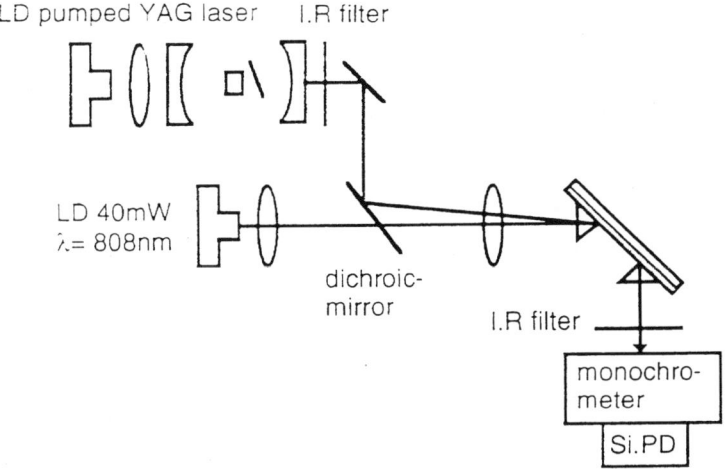

Figure 4.5.9
Experimental set-up for gain measurement

Figure 4.5.10 shows the single-pass gain versus pumping power, where the signal and pump power are the launched power into the waveguide of 1 μm thickness which are converted at prism coupling efficiency. The signal and pump light are modulated directly at frequencies of 100 Hz and 50 Hz, respectively. The gain and the S/N ratio are calculated from the following

$$G = 10 \log \frac{I_{02} - I'_{02}}{I_{01}} \qquad (10)$$

$$\frac{S}{N} = 10 \log \frac{I_{02} - I'_{02}}{I_{02}} \qquad (11)$$

where I_{01} and I_{02} are the signal levels with the pump light off and on, respectively. I'_{02} is the spontaneous emission level corresponding to the noise level. As the signal power increases, the gain decreases owing to gain saturation. Approximately linear gain properties were obtained for the small signal power of 10 μW. A maximum gain of 4.4 dB was obtained for a pumping power of about 14 mW by the slab waveguide. From the slope of the line, a pump efficiency of 0.31 dB/mW is obtained. A gain of

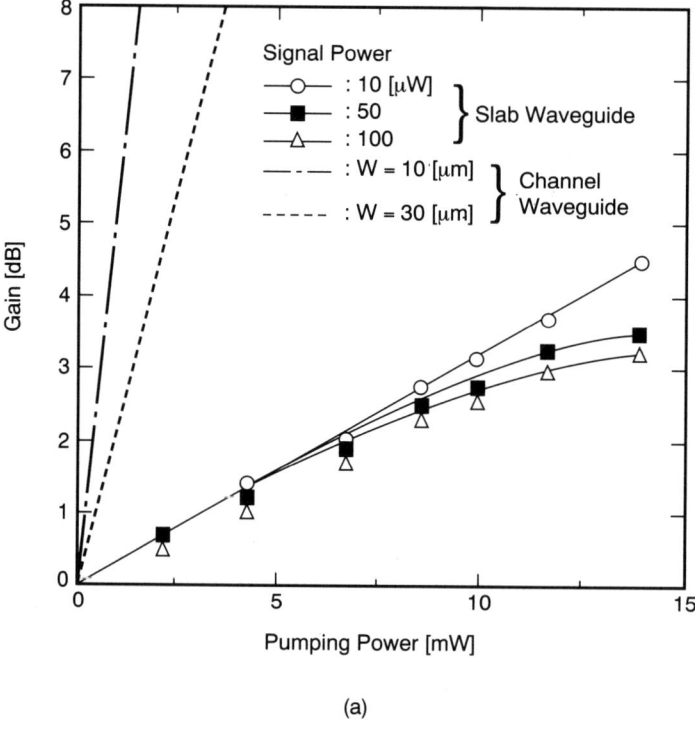

(a)

Figure 4.5.10
Gain characteristics of the optical slab waveguide amplifier. (a) The single-pass gain and (b) the S/N ratio versus the pump power. (c) The single-pass gain versus the waveguide length corresponding to the propagation length

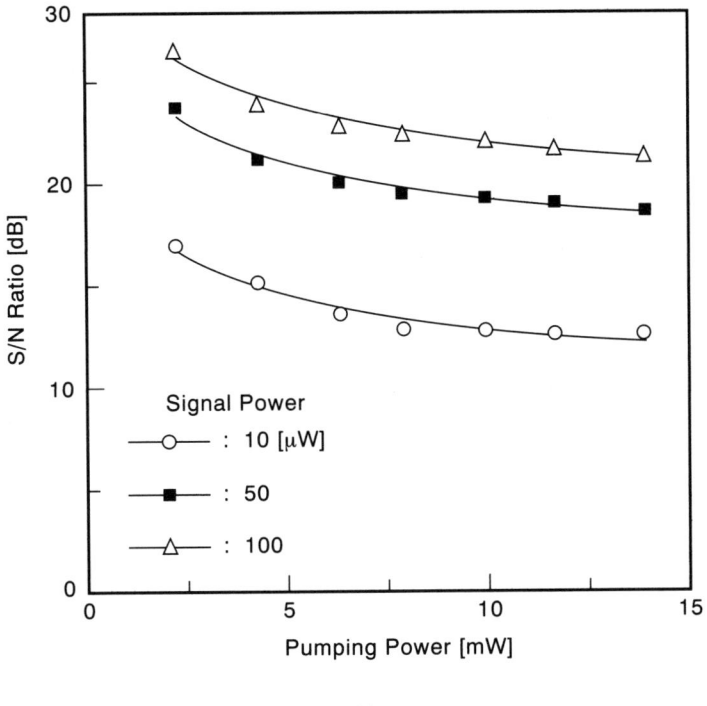

(b)

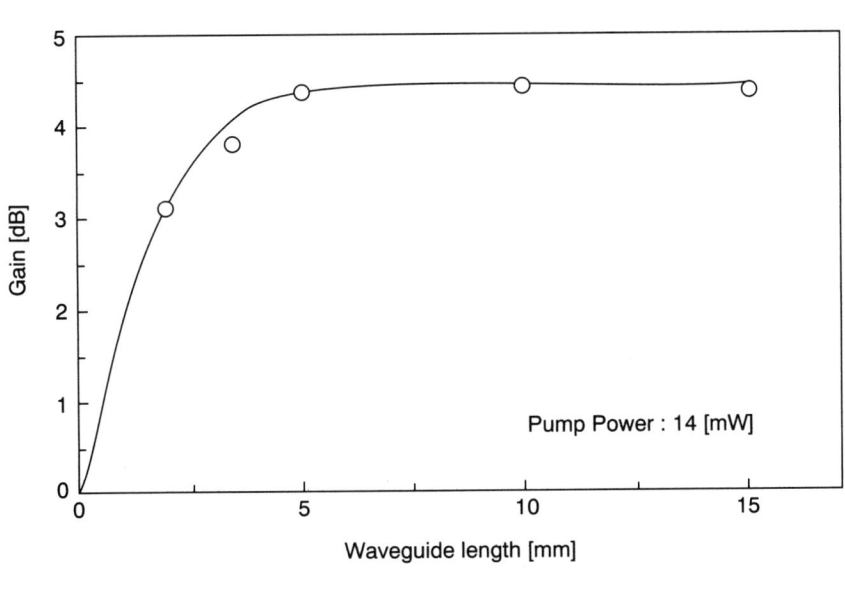

(c)

Figure 4.5.10 (*continued*)

about 10 dB can be expected with a small pump power of 30 mW in this case. The S/N is 13–20 dB. Furthermore, the dashed line shows the characteristic of the channel waveguide whose thickness and width are 1 μm and 30 μm, respectively.

As the propagation length is increased, the pump power is decreased due to pump absorption, and the pump power density also decreases because the slab waveguide confines light only in one direction. An improvement in amplification can be expected with channel type waveguides.

For improvement, we developed a three-dimensional channel waveguide amplifier. In case of the channel type, the signal light power, the waveguide length, and the waveguide width were 5 μW, 4 mm, and 20 μm, respectively. As shown in Figure 4.5.11, the gain depended linearly on pumping powers less than 4 mW. A gain of 6.5 dB was obtained at a pumping power of about 10 mW for a waveguide length of 4 mm. The pump efficiency was 1.1 dB/mW and the gain per unit length was 16 dB/cm. The dashed line is the theoretical result which is calculated using the rate equations. To explain the difference between experiment and theory, we consider that the Nd:YGG crystalline film is not a perfect crystal. The half line width of the Nd:YGG film is 5 mm, but that

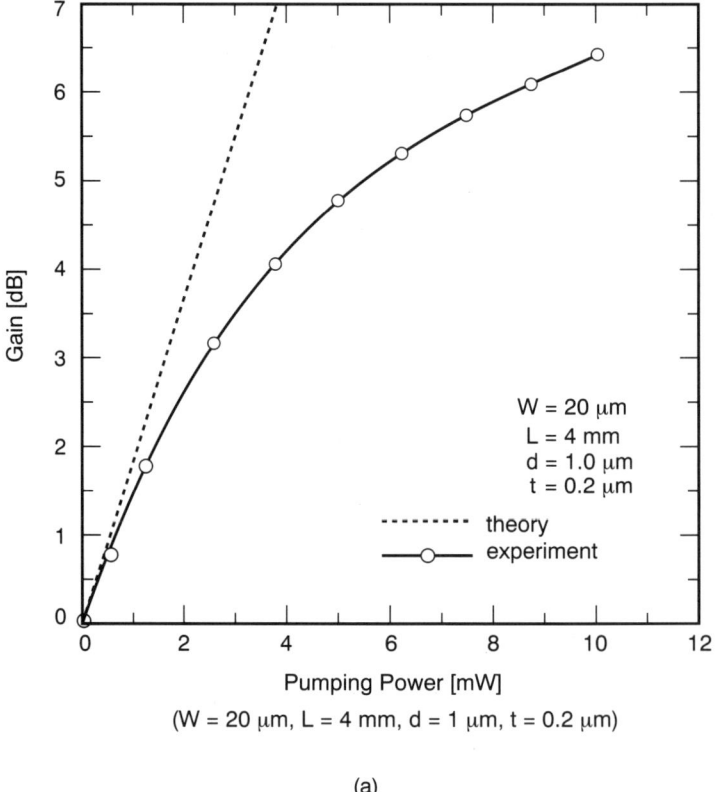

(a)

Figure 4.5.11
Gain characteristics of the optical channel waveguide type amplifier

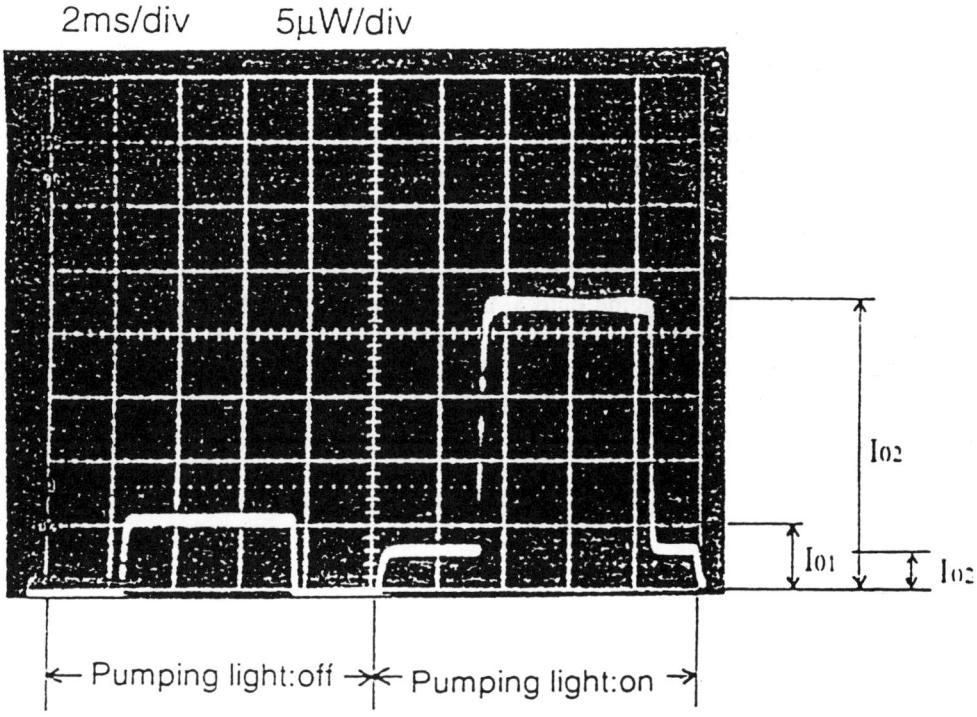

Figure 4.5.11 (*continued*)

of Nd:YAG used for the gain calculations using the rate equations is 2 nm, and that for the stimulated emission cross-section is 5×10^{-19} cm^2.

6 LNP Crystal Film Waveguide and Er, Pr Garnet Film Waveguides

LNP (LiNdP$_4$O$_{12}$) is an efficient crystal for use as a laser oscillator and optical amplifier with a high concentration of active ions and very weak concentration quenching. LNP crystals are very attractive because of their high absorption efficiency for pump light. Figure 4.5.12(a) shows the absorption spectrum. The Nd concentration of LNP is 42.1×10^{20} cm^{-3} (thirty times that of Nd:YAG [1 at.%]); however, the stimulated emission cross-section and lifetime are the same order as Nd:YAG. Figures 4.5.12(b) and (c) show the fluorescence spectra of LNP for 1.03–1.08 μm and 1.31–1.36 μm, respectively.

For optical amplification of the convenient wavelength 1.55 μm in optical communication, an Er:YGG 1.3 at.% film waveguide is fabricated by RF sputtering as shown in Figure 4.5.13. The optical characteristics of Er:YGG film waveguide amplification have been studied. On the basis of these studies, high gain and high efficient waveguide

224   *Chapter 4 Functional Photonics Devices and Integration*

type lasers and optical amplifiers of Er^{3+} and Pr^{3+} heavily doped crystal films on garnet films can be obtained for wavelengths of 1.55 μm and 1.3 μm.

7 Conclusion

We proposed optical amplifiers and lasers constructed with ion doped garnet crystal thin-film slab and channel waveguide and optical pumps, for miniaturization and combination of different optical waveguide devices into optical integrated circuits. Rare-earth-doped crystal thin films have a high active ion density, a high stimulated emission cross-section and a high density of light in the active crystal film region. We obtained Nd:YGG thin-film slab and channel waveguide onto a YAG single crystal substrate. For the gain of the Nd:YGG thin film slab waveguide a maximum gain of 4.4 dB was obtained for a signal light of 10 μW and a pumping light of about 14 mW; the propagation length was 5 mm. Furthermore, we discussed Nd:YGG thin film channel waveguide, calculated gain characteristic of channel waveguide, and showed possibility of achieving a high

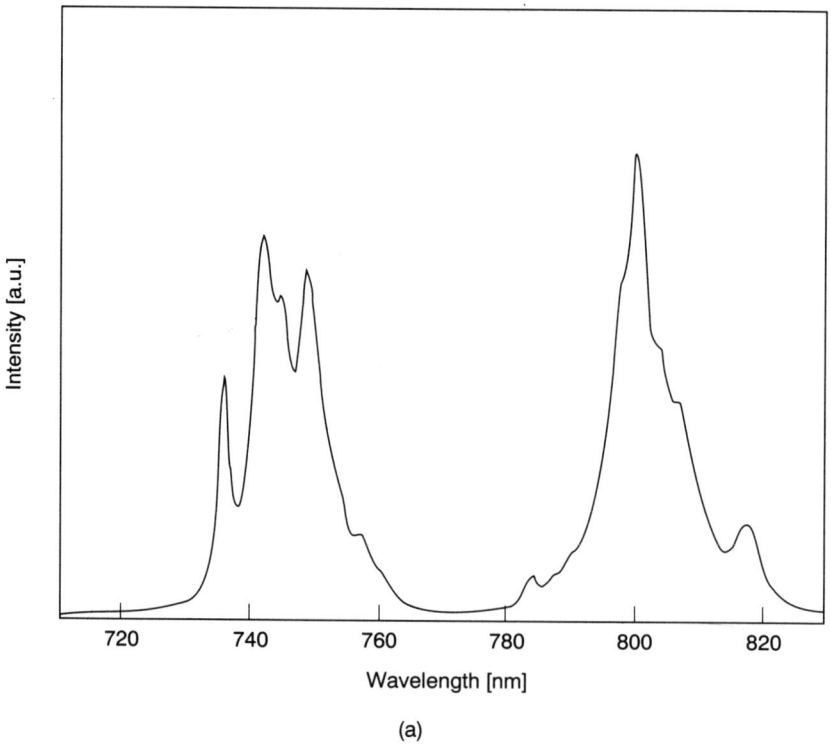

(a)

Figure 4.5.12
Optical characteristics of LNP crystal. (a) Absorption spectrum of LNP: 720–820 nm; (b) fluorescence spectrum of LNP and Nd:YAG: 1030–1080 nm; and (c) fluorescence spectrum of LNP and Nd:YAG: 1310–1360 nm

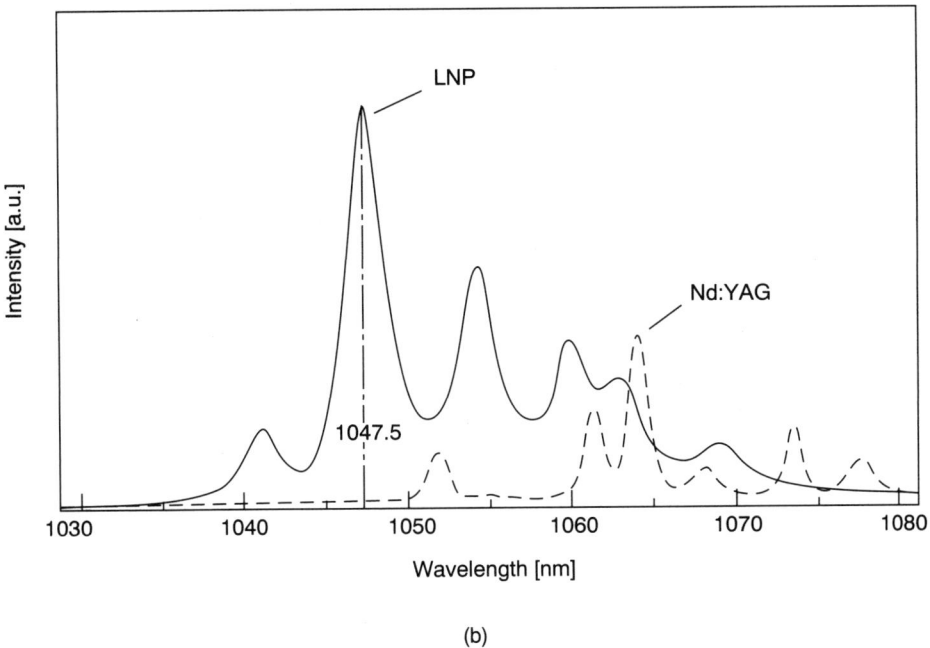

(b)

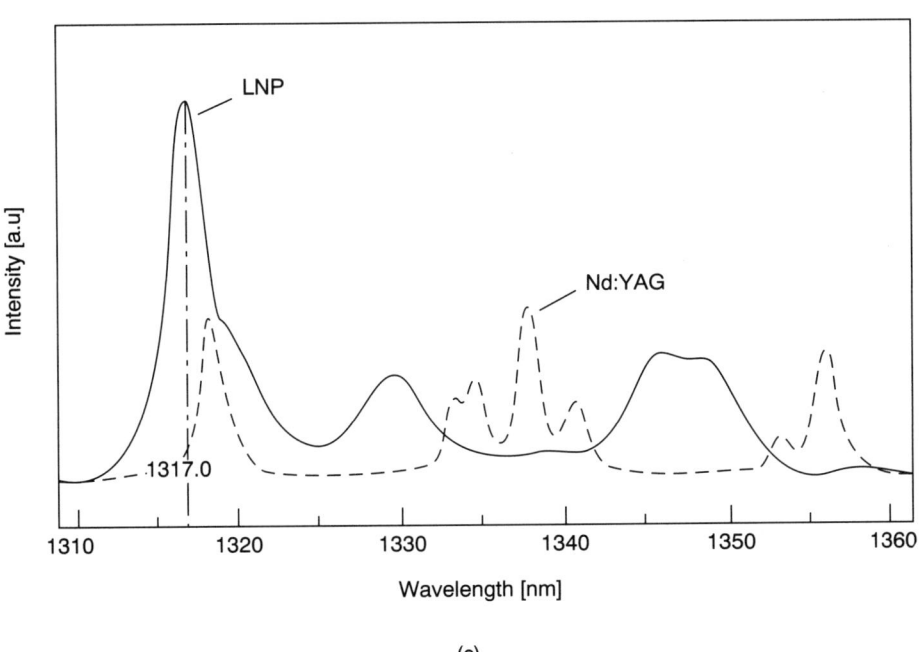

(c)

Figure 4.5.12 (*continued*)

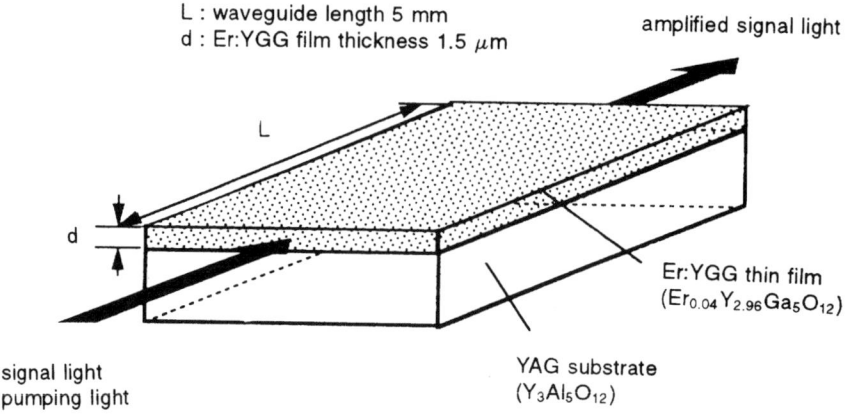

Figure 4.5.13
Optical waveguide type amplifier and laser of Er:YGG crystal film

gain and miniature optical amplifier. A channel waveguide optical amplifier and laser using Nd-doped garnet thin film was studied for the gain characteristics at the 1.06 μm region as the fundamental property. We obtained the result that the gain is linearly dependent on a pumping power less than 4 mW, and a gain of 6.5 dB is obtained at a pumping power of about 10 mW for a waveguide length of 4 mm. The pump efficiency was 1.1 dB/mW and the gain per unit length was 16 dB/cm. On the basis of these results, in the near future we can optimize the structure of the amplifier and laser of waveguide type and investigate doping of Pr^{3+} and Er^{3+} for amplification and optical oscillation at the 1.3 μm and 1.55 μm regions.

References

[1] W. Koechner, *Solid-State Laser Engineering*, Springer-Verlag, 1982.
[2] Y. Miyazaki, F. Tsuchiya and Y. Akao, Paper of Technical Group, EMT81-5, 1 (1981) [in Japanese].
[3] M. Shimizu, M. Yamada, M. Horiguchi, T. Takeshita and M. Okayasu, *Electron. Lett.*, **26**(20), 1641 (1990).
[4] M. Yamada, M. Shimizu, Y. Ohishi, J. Temmyo, M. Wada, T. Kanamori, M. Horiguchi and S. Takahashi, *IEEE Photon. Technol. Lett.*, **4**(9), 994 (1992).
[5] M.J.F. Digonnet and C.J. Gaeta, *Appl. Opt.*, **24**(3), 333 (1985).
[6] M. Yamaga and Y. Miyazaki, *Jpn J. Appl. Phys.*, **23**(3), 312 (1984).
[7] B. Zhou, T.J. Kane, G.J. Dixon and R.L. Byer, *Opt. Lett.*, **10**(2), 62 (1985).
[8] M. Yamaga, K. Yusa and Y. Miyazaki, *Jpn J. Appl. Phys.*, **25**(2), 194 (1986).
[9] M. Yamaga, K. Yusa and Y. Miyazaki, *Trans. IECE Jpn*, **E69**(9), 956 (1986).
[10] Y. Miyazaki, M. Wada and T. Matsushita, *Optical Amplifiers and Their Applications Technical Digest*, **13**, WA6, 226 (1990).
[11] K. Arakane, H. Murotani and Y. Miyazaki, *Optical Amplifiers and Their Applications Technical Digest*, **14**, MD15, 206 (1993).
[12] G. Nykolak, M. Haner, P.C. Becker, J. Shmulovich and Y.H. Wong, *IEEE Photon. Technol. Lett.*, **5**(10), 1185 (1993).

4.6

Optical Isolator with a Waveguide Structure

Yoshiyuki Naito

Abstract

The design and characteristics of a waveguide optical isolator employing a nonreciprocal phase shift are described. Compared with a conventional waveguide isolator based on the TE-TM mode conversion, the isolator employing the nonreciprocal phase shift has the advantages of no need of phase matching and complicated magnetization control. It is shown that the degradation of characteristics, due to deviations in the waveguide parameters from the design values, can be retained within an acceptable range with current technologies. Also, the characteristics of LPE-grown iron garnet suitable for the isolator are described together with the fabrication of magnetooptic rib waveguides.

1 Introduction to Waveguide Optical Isolators

Optical nonreciprocal devices are indispensable in protecting optical active devices from unwanted reflected light. Waveguide isolators have been studied in various configurations. The first proposal of a waveguide isolator based on the mode conversion was made by Wang et al. [1]. In the isolator utilizing the TE-TM mode conversion, nonreciprocal and reciprocal mode converters were installed for unidirectional mode conversion. The mode converters can be constructed making use of the Faraday and Cotton-Mouton effects in a magnetooptic crystal. In this case the magnetization in each converter is aligned orthogonally to each other, as shown in Figure 4.6.1(a) by the arrows. That is, the Faraday effect occurs for the magnetization aligned along the light propagation direction, while the Cotton-Mouton effect is obtainable in the section where the magnetization is transverse to the propagation direction. Castera and Hepner reported

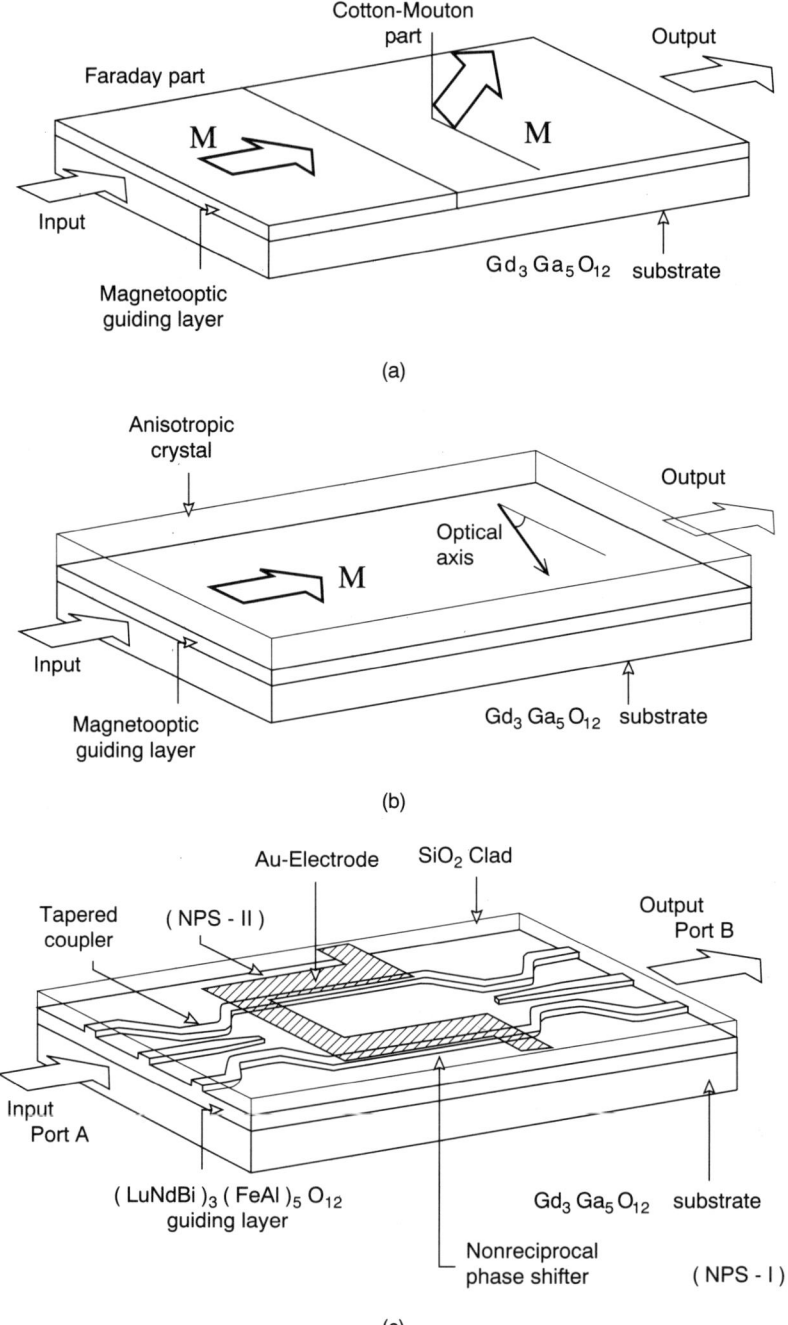

Figure 4.6.1
Structures of waveguide optical isolators. (a) TE–TM mode conversion type, (b) semileaky type, and (c) interferometric isolator employing the nonreciprocal phase shift. Arrows indicate the direction of magnetization

an experimental study on unidirectional mode conversion applying external magnetic fields [2]. Ando *et al.* demonstrated the feasibility of complicated magnetization control with a laser annealing technique, and reported an isolation ratio of 12.5 dB at a wavelength of 1.152 μm [3]. Besides the complicated and precise magnetization control, the phase matching between the modes concerned is essential in the TE-TM mode conversion isolator, which results in the necessity to control the waveguide parameters strictly.

In contrast with the mode conversion isolator, a semileaky waveguide isolator (Figure 4.6.1(b)) does not need the complicated magnetization control [4,5]. It is, however, needed to make a stable and uniform optical contact of an anisotropic crystal such as $LiNbO_3$ with the magnetooptic material of different crystal structure. Since the anisotropic crystal cannot be grown on to the magnetooptic one, we are forced into pressing the crystals to obtain the required optical contact along a mm-long device length. Because of this a sufficient isolation cannot be obtained reproducibly in the semi-leaky isolator.

To circumvent the difficulties associated with these isolators, it is effective to use the nonreciprocal phase shift which the TM mode experiences while traveling in a magnetooptic waveguide where the magnetization is aligned transversely to the light propagation direction in the film plane. Here, the nonreciprocal phase shift means that the phase change which the lightwave experiences differs depending on the propagation direction. Measurement of the nonreciprocal phase shift is reported in $Y_3Fe_5O_{12}$ and $(GdBi)_3Fe_5O_{12}$ slab waveguides [6,7]. Okamura *et al.* utilized this phase shift to obtain the nonreciprocal intensity difference in an $Y_3Fe_5O_{12}$ waveguide Mach-Zehnder interferometer [8].

Since only the TM_0 mode takes part in the operation, no phase matching is required in the isolator employing the nonreciprocal phase shift. It will be shown that, by virtue of this, the degradation of characteristics due to deviations in the waveguide parameters from designed values, such as the thickness of the guiding layer and the width of the channel waveguide, can be retained within an acceptable range with current technologies. In the following sections, the design of a waveguide isolator composed of a nonreciprocal phase shifter and branching couplers is described, and the characteristics of the isolator are discussed.

2 Design

The structure of the isolator employing the nonreciprocal phase shift is schematically shown in Figure 4.6.1(c). An optical interferometer is composed of branching devices, nonreciprocal phase shifters in two arms, and a reciprocal phase shifter in one arm. Although it is not shown explicitly in the figure, the reciprocal phase shift is achieved by an optical path difference between two arms. For the interferometer to work as an ideal isolator, a 90° nonreciprocal phase difference between two arms and a 90° reciprocal phase shift should be installed. A forward traveling lightwave launched into the central waveguide of the left branching device is divided into two waves with equal amplitude and in-phase. While traveling in two arms, the −90° phase difference produced in the nonreciprocal phase shifter is canceled by the 90° reciprocal phase change. Consequently, two waves are coupled into the central waveguide of the right branching device. For a backward traveling wave, the nonreciprocal phase shift changes its sign, i.e. +90°. This phase shift is added to the reciprocal one, and the whole phase difference between the two arms amounts to 180°. No output is coupled from the central waveguide of the branching device when two waves are in phase opposition.

In order to align the magnetization in-plane and produce the nonreciprocal phase shift effectively in a practical device, the magnetic field produced by an electric current flowing along a Au electrode is applied in the opposite direction in the two nonreciprocal phase shifters transversely to the propagation direction.

2.1 Nonreciprocal phase shifter

To realize a practical device using the nonreciprocal phase shift, a magnetooptic layer needs to possess in-plane magnetization together with a large Faraday rotation and low absorption. The in-plane magnetization characteristic reduces the required magnetic field, and consequently the electric current. A large Faraday rotation enables a short device length. It is well known that the substitution of Bi for rare-earth elements increases the Faraday rotation of the iron garnet. However, as Bi-substitution increases in the liquid phase epitaxy (LPE) growth, it tends to create a perpendicular magnetization.

It has been found that a $(LuNdBi)_3(FeAl)_5O_{12}$ film LPE-grown on a (111) $Gd_3Ga_5O_{12}$ substrate is suitable for the isolator employing the nonreciprocal phase shift [9]. The melt composition used in the growth is defined by the molar fractions given in Table 4.6.1. The samples were prepared using a conventional LPE technique of isothermal dipping with horizontal substrate rotation of 70 rev/min. The dependence of the characteristics on the growth temperature is described in what follows.

The external magnetic field intensity required to saturate the magnetization in-plane was measured by VSM. Results are shown in Figure 4.6.2. A minimum field of 5 Oe was obtained in the film grown at 755 °C. The electric current required to produce a

Table 4.6.1
The melt composition used in the LPE growth of the in-plane magnetized iron garnet

$$R_1 = \frac{[Fe_2O_3]}{[Lu_2O_3] + [Nd_2O_3]} = 11.0$$

$$R_2 = \frac{[PbO]}{[B_2O_3]} = 7.25$$

$$R_3 = \frac{[PbO]}{[Bi_2O_3]} = 5.0$$

$$R_4 = \frac{[Lu_2O_3] + [Nd_2O_3] + [Fe_2O_3] + [Al_2O_3]}{Total} = 0.10$$

$$R_5 = \frac{[Nd_2O_3]}{[Lu_2O_3]} = 0.206$$

$$R_6 = \frac{[Al_2O_3]}{[Fe_2O_3]} = 0.011$$

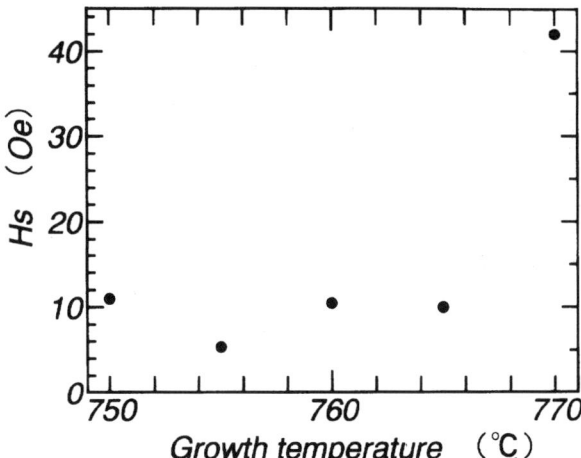

Figure 4.6.2
External magnetic field required to saturate magnetization in-plane H_s versus growth temperature

magnetic field of 5 Oe was calculated to be 16 mA using a 20 μm wide electrode in the isolator shown in Figure 4.6.1(c). A typical optical loss of the film grown at this temperature was 5 dB/cm at a wavelength of $\lambda = 1.31$ μm. The loss value includes a scattering loss as well as the optical absorption.

The nonreciprocal phase shift is determined by $\psi = (\beta_+ - \beta_-)L$, where L is the propagation distance. β_+ and β_- indicate the propagation constants of the lightwave propagating in the forward and backward directions, respectively, which are given as solutions of the following characteristic equation

$$\tan(\kappa d) = \frac{\frac{\kappa}{\varepsilon'}\left(\frac{p}{n_1^2} + \frac{q}{n_3^2}\right)}{\left(\frac{\kappa}{\varepsilon'}\right)^2 - \frac{pq}{n_1^2 n_3^2} + \left(\frac{q}{n_3^2} - \frac{p}{n_1^2}\right)\frac{\alpha}{n_2^2 \varepsilon'}\beta + \left(\frac{\alpha}{n_2^2 \varepsilon'}\beta\right)^2} \tag{1}$$

where

$$\beta^2 = n_1^2 k_0^2 + p^2 = \varepsilon'^2 k_0^2 - \kappa^2 = n_3^2 k_0^2 + q^2 \tag{2}$$

$$\varepsilon' = n_2^2 - \left(\frac{\alpha}{n_2}\right)^2 \tag{3}$$

k_0 is the free space wavenumber of the lightwave, and α is related to the Faraday rotation per unit length θ_F(deg/cm) by $\alpha = n_2 \lambda \theta_F / 1.8$. n_1, n_2, and n_3 indicate the refractive indices of a clad, a guiding layer, and a substrate, respectively, and d denotes the thickness of the guiding layer.

The Faraday rotation of samples grown at each temperature is shown in Figure 4.6.3. The Faraday rotation of the film grown at 755 °C is 600°/cm at $\lambda = 1.31$ μm. If we use the $(LuNdBi)_3(FeAl)_5O_{12}$ film grown at 755 °C as the guiding layer and SiO_2 as the clad,

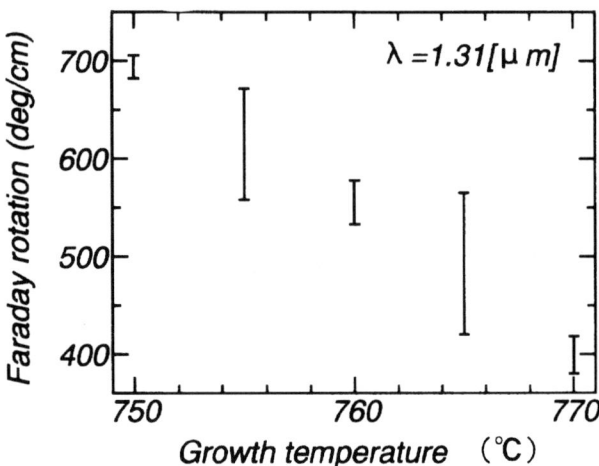

Figure 4.6.3
Faraday rotation at $\lambda = 1.31$ μm versus growth temperature

the maximum nonreciprocal phase shift is obtained for a 350 nm thick guiding layer, and a propagation distance of 6.21 mm is needed as the nonreciprocal phase shifter to achieve the ideal isolator operation. From a single-mode condition, the height of the rib and the width of the waveguide are determined to be 60 nm and 2 μm, respectively.

2.2 Tapered branching coupler

The branching device, the characteristics of which are immune to deviations in the waveguide parameters, is designed for constructing the isolator. It is composed of three

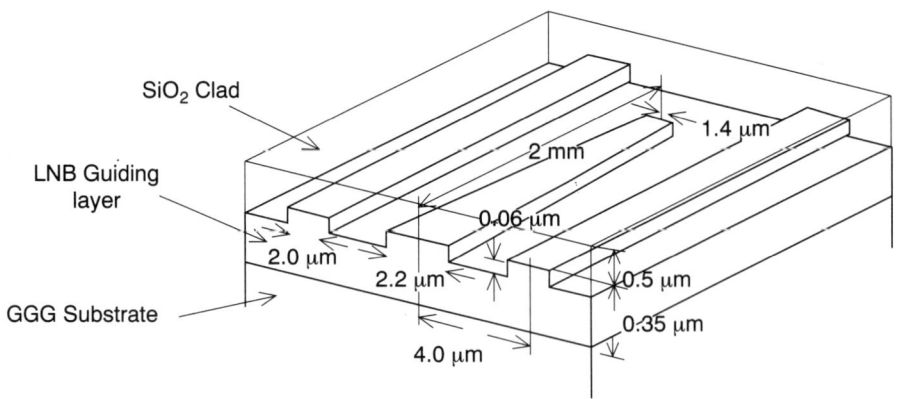

Figure 4.6.4
Branching device composed of tapered coupled waveguides

coupled waveguides, as shown in Figure 4.6.4. The width of the central waveguide varies linearly along the propagation direction. The lightwave launched into the central waveguide is divided into two waves with equal amplitude and in-phase. When the lightwaves launched into two waveguides placed on both sides of the central waveguide are of equal amplitude and 180° phase difference, no lightwave is coupled out from the central waveguide [10]. Its performance is similar to a conventional Y-branch composed of single-mode waveguides. From numerical calculations, we find the branching device structure which is matched to the above-mentioned nonreciprocal phase shifter. Its geometric dimensions are indicated in Figure 4.6.4.

3 Characteristics of the Isolator

The characteristics of the isolator are calculated at $\lambda = 1.31$ μm. Calculated forward and backward losses are shown in Figure 4.6.5, where material losses are excluded.

Figure 4.6.5(a) shows the dependence on the rib height. The thickness of the magnetooptic layer, 350 nm, is held constant. According to this figure, if the deviation is within ± 10 nm, a backward loss ≥ 50 dB is realizable. Also, an increase in the forward loss is ≤ 1 dB. Since we can make use of a dry etching process with an etching rate $\simeq 5$ nm/min, the rib height is controllable within this range of error. As shown in Figure 4.6.5(b), the dependence on the thickness of the magnetooptic guiding layer is of the same order, where a rib height of 60 nm has been kept constant.

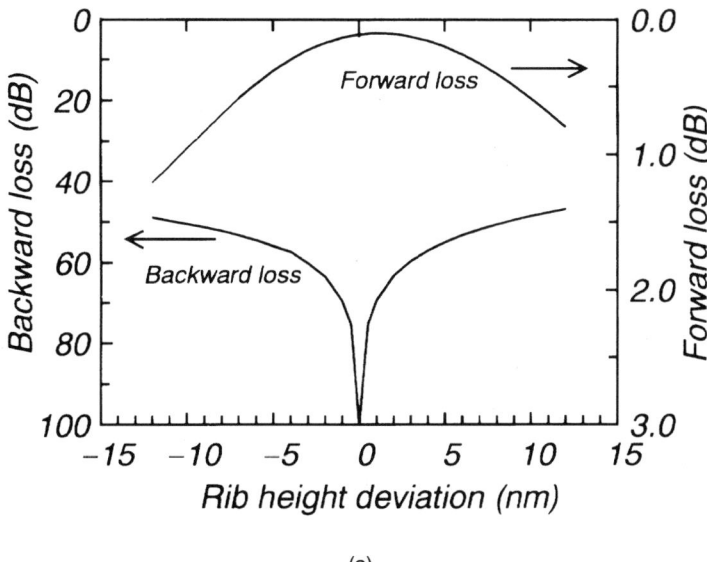

(a)

Figure 4.6.5
Dependence of forward and backward losses on (a) the rib height, (b) the guiding layer thickness, and (c) the waveguide width

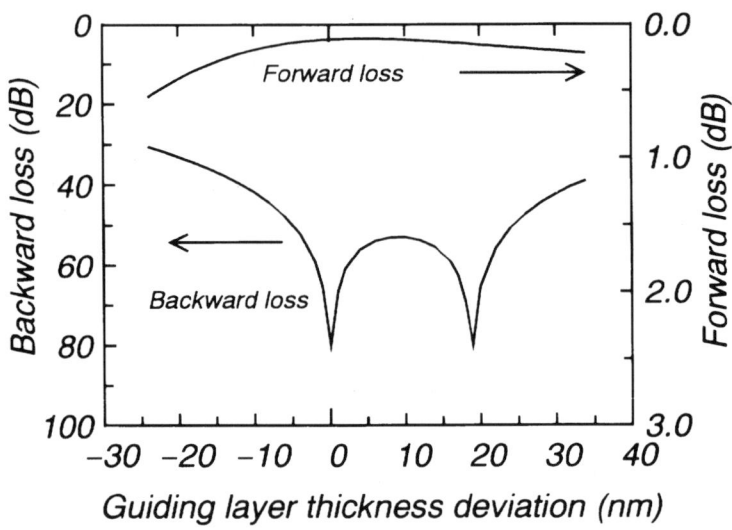

(b)

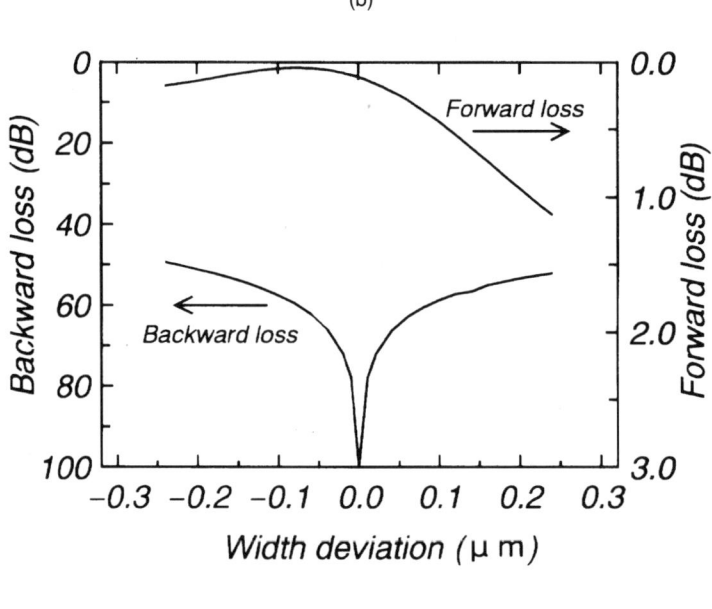

(c)

Figure 4.6.5 (*continued*)

The dependence on the width of the waveguide is shown in Figure 4.6.5(c). It is assumed that the width varies uniformly in every waveguide, while the other parameters are held constant. When the deviation is within ±0.2 μm, a backward loss ≥ 50 dB is obtainable together with a loss increase ≤ 1 dB in the forward direction. Considering the state of the art in the lithography and etching processes for patterning the waveguide, the width deviation can be retained within this range.

The relatively gentle degradation of characteristics is mainly attributed to the fact that the isolator operates in a single polarization.

4 Fabrication Technology

A magnetooptic rib waveguide can be formed using an Ar dry-etching process with the mask patterned through electron beam lithography. It was found that etching the $(LuNdBi)_3(FeAl)_5$ layer with a Ti mask deposited directly on the layer as shown in Figure 4.6.6(a) brought about an increase in the optical absorption. This can be attributed to the large absorption associated with Fe^{2+} ions in the near infrared region. In the rare-earth iron garnet, the iron exists normally as Fe^{3+}. However, if Ti^{4+} is incorporated through the bombardment of Ar^+ ions during the etching process, part of the Fe^{3+} ion changes into Fe^{2+} to compensate for the charge imbalance in the garnet. In order to circumvent this, the etching process shown in Figure 4.6.6(b) can be applied. SiO_2 is used as the etch resistant mask instead of Ti. In order to define the SiO_2 mask pattern, a CHF_3 reactive ion etching process is adopted with an Al mask.

The dependence of excess loss measured in straight rib waveguides fabricated employing the etching process of Figure 4.6.6(b) is shown in Figure 4.6.7. The indicated

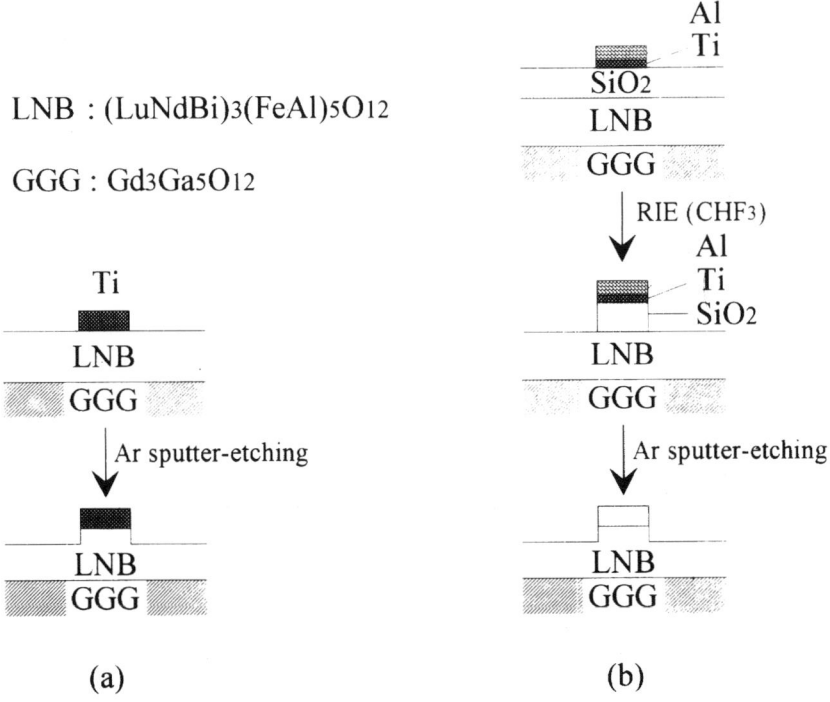

Figure 4.6.6
Etching processes for the fabrication of magnetooptic rib waveguides

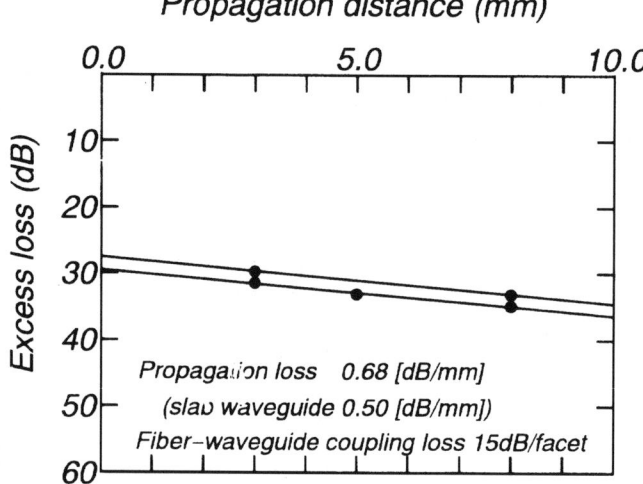

Figure 4.6.7
Propagation loss measured at $\lambda = 1.31$ μm

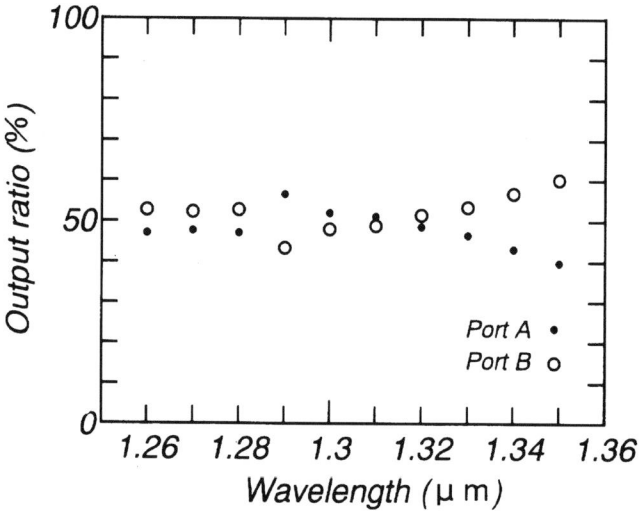

Figure 4.6.8
Measured wavelength dependence of branching ratio

value includes a coupling loss between a single mode fiber and a waveguide at two facets. By extracting the coupling loss, the propagation loss of the waveguide is determined to be approximately 7 dB/cm at $\lambda = 1.31$ μm. The measured imbalance of the dividing ratio in the branching device was $\leq 2\%$ in the 1.30–1.32 μm wavelength range, as shown in Figure 4.6.8.

5 Concluding Remarks

The design and characteristics of a waveguide optical isolator employing the nonreciprocal phase shift have been described. It has been shown that the degradation of characteristics, due to deviations in the waveguide parameters from design values, can be retained within an acceptable range with current technologies.

References

[1] S. Wang, M. Shah and J.D. Crow, *J. Appl. Phys.*, **43** (4), 1861–1875 (1972).
[2] J.P. Castera and G. Hepner, *IEEE Trans. Magnetics*, **MAG-13** (5), 1583–1585 (1977).
[3] K. Ando, T. Okoshi and N. Koshizuka, *Appl. Phys. Lett.*, **53** (1), 4–6 (1988).
[4] S. Yamamoto, Y. Okamura and T. Makimoto, *IEEE J. Quantum Electron.*, **QE-12** (12), 764–770 (1976).
[5] S.T. Kirsch, W.A. Biolsi, S.L. Blank, P.K. Tien, R.J. Martin, P.M. Bridenbaugh and P. Grabbe, *J. Appl. Phys.*, **52** (5), 3190–3199 (1981).
[6] T. Mizumoto and Y. Naito, *IEEE Trans. Microwave Theory Technol.*, **MTT-30** (6), 922–925 (1982).
[7] T. Mizumoto, K. Oochi, T. Harada and Y. Naito, *J. Lightwave Technol.*, **LT-4** (3), 347–352 (1986).
[8] Y. Okamura, H. Inuzuka, T. Kikuchi and S. Yamamoto, *J. Lightwave Technol.*, **LT-4** (7), 711–714 (1986).
[9] T. Mizumoto, S. Mashimo, T. Ida and Y. Naito, *IEEE Trans. on Magnetics*, **29** (6), 3417–3419 (1993).
[10] Y. Cai, T. Mizumoto and Y. Naito, *J. Lightwave Technol.*, **8** (10), 1621–1629 (1990).

4.7

Integrated Photonic Functional Devices Using Grating Couplers

Hiroshi Nishihara, Toshiaki Suhara and **Shogo Ura**

Abstract

Integrated optic (IO) devices of various functions for application to optical information processing systems can be constructed by integrating several types of grating components, *e.g.* grating couplers, focusing grating couplers, and grating beam splitters in planar waveguides. Device configurations, working principles, fabrication and experimental results are presented on integrated optic devices studied in the UUO project, i.e. IO pickup devices for serial and parallel data readout, an IO scanning optical microscope, an IO focal spot intensity modulator, and IO position/displacement sensors.

1 Introduction

Periodic structures or gratings having a period of the same order of magnitude as optical wavelengths fabricated in a wave-guiding structure play an important role in constructing integrated optical devices [1,2], since they can perform various functions, *e.g.* input/output coupling, phase matching, wavelength dispersion, and wavefront conversion. There has been much interest in such components since the earliest stage of integrated optics research. One of the most important grating components is a grating coupler which serves as an interface to couple the optical wave in free space to the guided-mode field. A focusing function can be incorporated into a grating coupler by modifying the grating pattern into a curved and chirped grating which performs a wavefront conversion based upon the principle of holography. Recent research and development activities in integrated optics have been directed toward the implementation of fully integrated complex devices incorporating several functions to perform a specific operation. Also, from the integration point of view the periodic structures offer many advantages. Gratings have a planar structure within a thin layer and therefore they can be fabricated by the well-established planar microfabrication processes suitable for mass production, i.e. photo and electron-beam lithographies associated with dry etching techniques. The many and complex functions realizable with gratings of the same or similar structure minimize the number of components and the fabrication process. Several kinds of grating components can be fabricated simultaneously. The characteristics of the periodic elements are insensitive to point defects, and this ensures good yields, even in complex integrated devices. Along with the versatility and the integration compatibilities, the application of grating components is an attractive subarea of increasing importance.

We have been conducting theoretical and experimental studies on such grating components and the implementation of various functional devices based on their integration. As for the fabrication technique, electron beam writing techniques have been applied. Various grating components were fabricated in glass, polymer, and $LiNbO_3$ waveguides. A typical example of the integrated devices proposed and demonstrated by us is an integrated-optic disc pickup device (IODPU) using a focusing grating coupler (FGC) as the key component. As a part of the UUO project, we have studied the implementation of new integrated photonic functional devices evolving from the IODPU or resulting from a new combination of grating components, as well as the design and fabrication of (focusing) grating couplers in AlGaAs waveguides [3,4] for future monolithic integration with laser diodes. In what follows the IODPU is presented first as a prototype device, and then examples of integrated photonic devices developed in the UUO project are presented.

2 *Integrated Optic Disc Pickup Devices*

Figure 4.7.1 shows the schematic illustration of the integrated optic disc pickup device (IODPU) [5,6], which is capable of the readout of disc data and focusing/tracking error detection. The readout head for optical disc systems is currently constructed with bulk microoptics and needs complex and time-consuming fabrication processes. The IODPU was proposed to solve these problems and to improve producibility, reduce size and weight, and enhance application flexibility. The device is constructed by integrating a focusing grating coupler (FGC), a twin grating focusing beam splitter (TGFBS), and a photodiode array in a thin glass film waveguide on an SiO_2 buffer layer on an Si substrate. The guided wave diverging from a butt-coupled laser diode (LD) is focused by the FGC onto a point on the disc. The wave reflected by the disc is collected and coupled back into the waveguide by the same FGC. The TGFBS divides the reflected wavefront in half, deflects and simultaneously focuses onto two detection points at a distance slightly shorter than that of the input point (LD). The complex functions of the TGFBS minimize the number of components. At the detecting points the guided waves are transferred to four PD elements, and the disc data are read out and the focusing/tracking error signals are also obtained by simple processing of the photocurrents based on Foucault/push-pull methods.

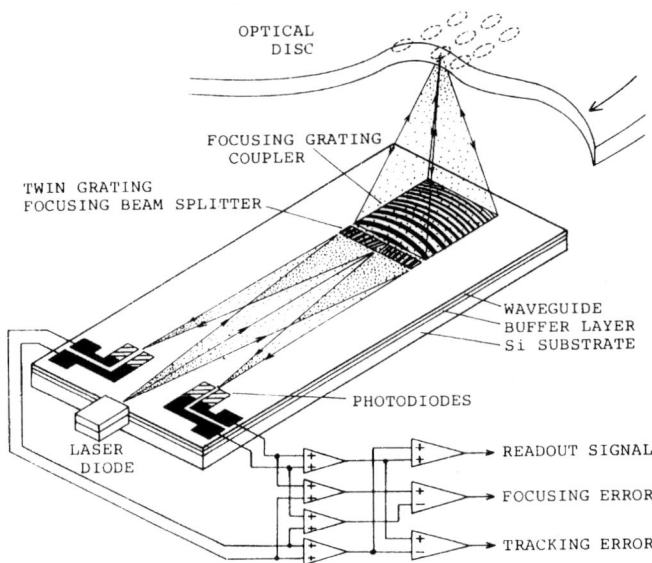

Figure 4.7.1
Integrated optic disc pickup device (IODPU)

Prototype devices having a FGC of 1 mm aperture and 1.3–2.0 mm focal length have been fabricated and tested. The waveguides were fabricated by depositing Corning 7059 glass onto an oxidized Si substrate with integrated PDs. The FGC/TGFBS gratings were fabricated by EB writing and reactive ion etching simultaneously in an Si-N layer deposited by PCVD onto the guiding layer. A LD of 790 nm wavelength was butt coupled to the waveguide. The measured focus spot widths were 1.1–1.5 times the theoretical diffraction-limited values. The minimum 3 dB full width was 1.4 μm. Although the performances obtained thus far are not sufficient for practical application, the fundamental operation, i.e. detection of the pit readout and error signals and focus servo operation, has been experimentally demonstrated.

A series of modifications to IODPU have been proposed and demonstrated as IO devices for disc pickup. They include IO devices for magnetooptic disc pickup. Theoretical analysis of the readout response showed that IODPUs exhibit inherent super-resolution characteristics. A review of these earlier work is given in Reference [6].

3 Integrated Optic Parallel Pickup for Optical Cards

We proposed and demonstrated an integrated optic device [7] for parallel pickup of optical data used in read-only optical cards. The proposed integrated optic parallel pickup device (IOPP) was constructed by integrating a linearly focusing grating coupler (LFGC), a beam-splitting/imaging grating (BSIG), and a PD array in a single mode waveguide on an Si substrate, as illustrated in Figure 4.7.2. The guided wave diverging

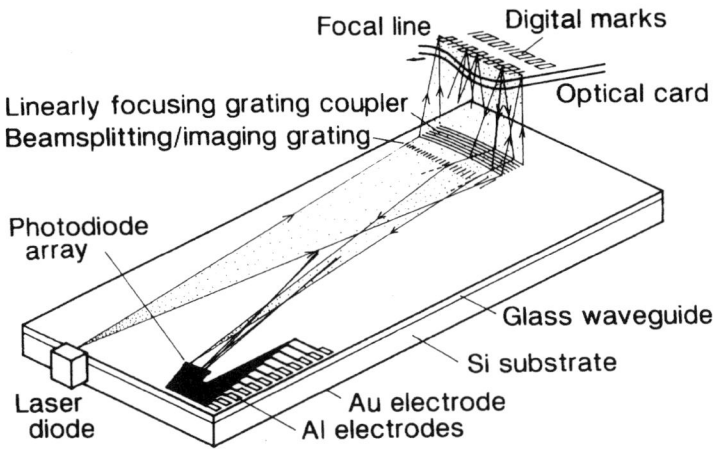

Figure 4.7.2
Integrated optic parallel pickup device

from a LD is diffracted by the LFGC to a focal line onto an optical card. The wave reflected by multiple digital marks on the focal line is coupled again by the same LFGC into the waveguide, and imaged by the BSIG onto the PD array.

The device was designed to detect 9 μm size optical reflection marks, and a prototype IOPP was fabricated by planar processes. The chip size is 10 mm × 17 mm. A glass plate on which the reflection marks were patterned, was set and moved on the LFGC focal plane, and output photocurrents were monitored. By using digital marks of 9 μm size and 18 μm pitch, it was confirmed experimentally that the multiple marks were detected separately and simultaneously.

4 Integrated Optic Scanning Optical Microscope

There are several advantages offered by scanning microscopy over conventional optical microscopy, i.e. the higher resolution obtained by the confocal optical configuration, the sequential nature suitable for electronic image processing, consistent resolution maintained over a large area, and the possibility of taking advantage of the various methods available for improving image resolution and contrast. Additional advantages can be gained by using a compact configuration for the optical system: it is less sensitive to mechanical shock and it is easier to scan the microscope head instead of the object. We have proposed and demonstrated an implementation of such compact optics for a scanning optical microscope (SOM) using the integrated optics technology [8].

The SOM sensing head is shown schematically in Figure 4.7.3. The device structure is similar to IODPU. Connections to the SO sensing head, however, are made via one input and two output optical fibers. There are two grating regions etched into the waveguide surface: the focusing grating coupler (FGC) and a twin grating

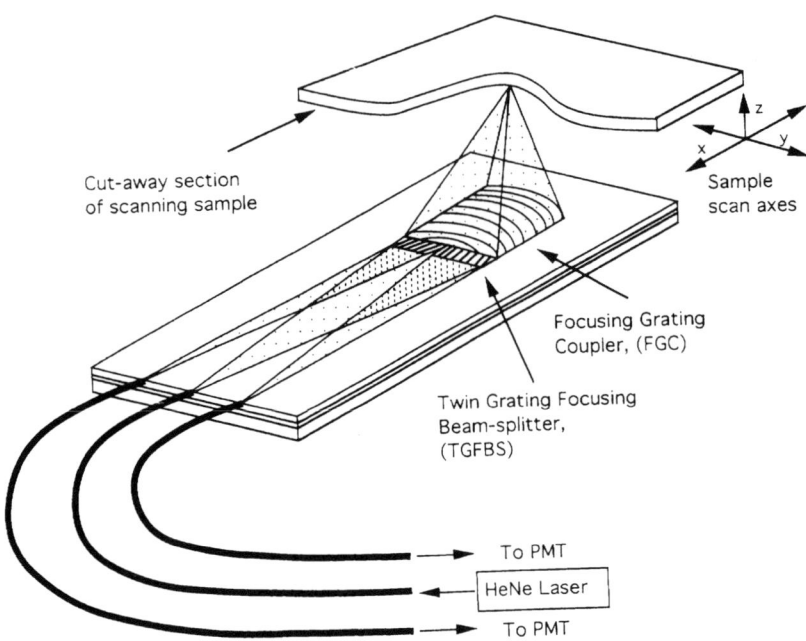

Figure 4.7.3
Integrated optic scanning optical microscope sensing head

focusing beam splitter (TGFBS). The FGC is used to convert the guided wave into a focus, some distance away from the waveguide. The focused light is then used to image an object scanned point-by-point in the focal plane of the FGC; the same FGC being used to convert the reflected light back into the waveguide mode. By using two output fibers, both intensity and differential images of the object can be produced by the SOM. In addition, owing to the properties of the waveguide and fiber modes, the images exhibit characteristics of confocal scanning microscopy which give rise to resolution improvement.

An IO SOM head of 2.0 mm focal length and 1 mm aperture for use at the He–Ne laser wavelength was designed and fabricated by a procedure similar to that of IODPU. A polarization-preserving fiber was used to ensure efficient coupling to the TE fundamental waveguide mode. Since the fiber–fiber spacing and the optic axis alignment are crucial to obtaining good SOM performance, the fibers were first located and fixed in a V-groove assembly, and the end face of the assembly was polished and coupled to the waveguide.

The performance of the fabricated head was tested by using an IC as an object and scanning it using a moving coil actuator and a dc motor stage. The images were displayed in a CRT by modulating the brightness of the signal from the photomultiplier tubes (PMTs) connected to the output fibers. It was experimentally confirmed that both intensity and differential images of the confocal features were obtained by using the sum and difference of the two PMT signals.

5 Integrated Optic Focal Spot Intensity Modulator

Poled polymers showing electrooptic (EO) effects are very attractive in constructing functional integrated optic devices, since they can be spin-coated to form waveguides on various substrates. The freedom of substrate choice with the easy coating method would result in a great improvement in design and application flexibility. We proposed and demonstrated [9] an integrated optic device for modulating the intensity of a focused spot in free space.

A schematic diagram of the proposed device is shown in Figure 4.7.4. The guided wave is excited by end-fire coupling to become a diverging TM mode in the single mode waveguide, it then propagates through the EO phase-distribution modulator and is coupled out by the FGC to be a focusing beam in the air. The waveguide consists of an upper-cladding layer, an EO polymer core layer, and a lower cladding layer on the substrate. A planar metal film below the waveguide serves as a bottom electrode for the phase distribution modulator and as a reflection film for the FGC to give high output coupling efficiency. The top electrodes of the phase distribution modulator are patterned to have three fingers of fan-out shape and circular input/output boundaries matching the wavefront. When a modulation signal is applied to the center finger of the top electrodes against the bottom electrode while the side fingers are connected to the bottom electrode, only the refractive index of the EO polymer core below the center finger is modulated. The resultant phase distribution variation of the guided wave in the FGC aperture causes a variation in the intensity distribution of the focus spot of the FGC.

A device was fabricated and the characteristics were measured. The modulator length was 3 mm. The focal length and aperture of the FGC were 5 mm and 1×1 mm^2, respectively. A light beam from an LD ($\lambda = 0.82$ mm) was end-fire coupled to be a TM$_0$ mode. The full width at half-maximum of the focal spot was 4.2 μm, while the diffraction limited value was 3.7 μm. The measured half-wave voltage was 54 V and

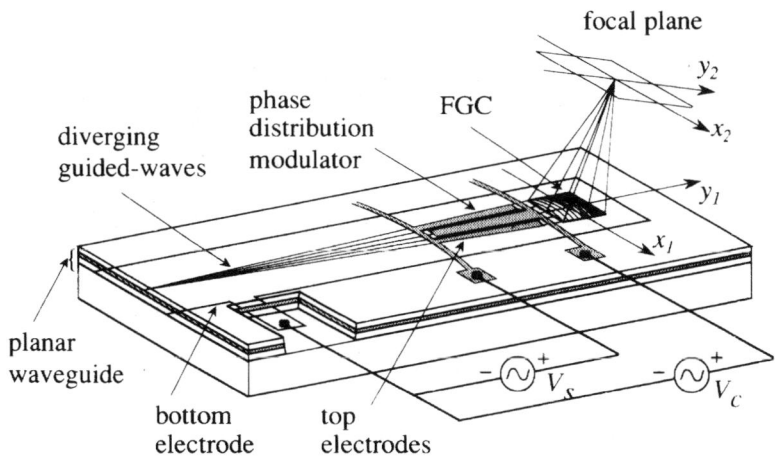

Figure 4.7.4
Integrated optic focal spot intensity modulator

the EO coefficient r_{33} was estimated to be 6.1 pm/V. A flat frequency response was measured up to 2 MHz. The device was left in a nonconditioned room without any applied voltage and the half-wave voltage was measured several times. More than 80% of the poled r_{33} was sustained over 300 h.

6 Integrated Optic Position/Displacement Sensors

Integration of optical interferometers for displacement sensing is very attractive because of its high resolution, high stability, device compactness, etc. and there have been several reports of integrated optic devices for sensing mirror vibration/displacement along the direction of a sensing beam propagation with a resolution of nanometers. In an integrated optic displacement/position sensor [10], an FGC has been integrated and a displacement measurement with a long working distance can be achieved without an external collimating lens. We have recently proposed and demonstrated a new integrated optic device [11] using a pair of LFGCs for sensing the displacement of a grating scale, and introduce the device here although another integrated optic device for sensing two-dimensional displacement of a grating scale is under investigation [12].

A schematic view of the proposed integrated optic sensor is illustrated in Figure 4.7.5. The waveguide consists of a glass core layer and a SiO_2 buffer layer on a Si substrate. The linearly focusing grating coupler (LFGC) pair and the PD pair are monolithically integrated. The guided wave diverging from a butt-coupled laser diode is diffracted by the LFGCs to become two beams in the air and overlap with each other on a grating scale. Both overlapped beams are diffracted by the grating scale to the same direction normal to the scale plane, focused to a line and interfere on the PDs. Thus the sensor head constructs an optical interferometer by utilizing positive or negative first-order diffraction of the grating scale. The phases of the two beams returning to the PDs vary in the opposite way to each other when the grating scale moves along the grating

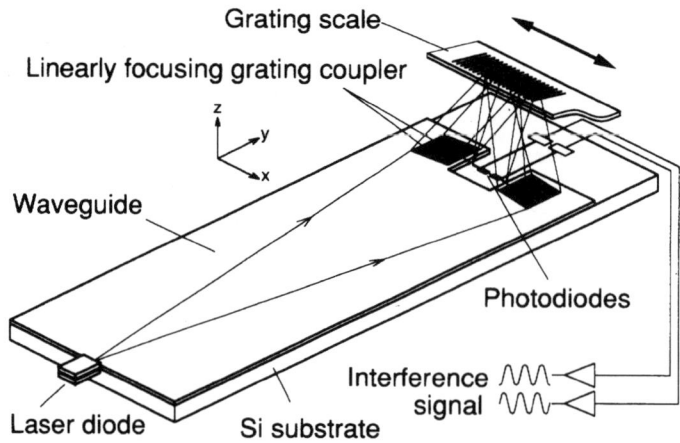

Figure 4.7.5
Integrated optic grating scale displacement sensor

vector direction (x-direction), and the displacement is determined by the variation in the interference signal. When a phase retardation has been introduced in the photocurrent variation against the other, the displacement direction can be discriminated by the phase relation.

The sensor was fabricated and the operation principle was confirmed. A grating scale of 4.6 μm period was set and moved in the x-direction as shown in Figure 4.7.5 while the output photocurrents were measured. It was confirmed that the photocurrents varied sinusoidally with a period of half of the grating scale period.

7 Conclusion

It has been demonstrated theoretically and experimentally that various integrated photonic functional devices can be implemented by integrating waveguide grating components. Subjects for future work include improvement of device performances for practical applications, monolithic integration of grating components and laser diode using semiconductor waveguides, and application of grating components to nonlinear IO devices.

References

[1] T. Suhara and H. Nishihara, *IEEE J. Quantum Electron.*, **OE-22**, 845–867 (1986).
[2] H. Nishihara, M. Haruna and T.Suhara, *Optical integrated circuit*, (McGraw-Hill Publishing, 1989).
[3] T. Suhara, K. Okada, T. Saso and H. Nishihara, *IEEE photon. Technol. Lett.*, **4**, 903–905 (1992).
[4] S. Ura, S. Kido, T. Suhara and H. Nishihara, *Design of high efficiency 3rd order grating coupler in semiconductor waveguide*, Optoelectronics Conference, Chiba, July 12–15, 1994.
[5] S. Ura, T. Suhara, H. Nishihara and J. Koyama, *J. Lightwave Technol.*, **LT-4**, 913–918 (1986).
[6] T. Suhara and H. Nishihara, *Proc. SPIE*, **1136**, 92–99 (1989).
[7] S. Ura, M. Shinohara, T. Suhara and H. Nishihara, *An integrated-optic parallel data pickup device*, Technical Digest of International Symposium on Optical Memory, Sapporo, October 1–4, 1991, pp. 27–28 (paper 1C-4).
[8] S. Sheard, T. Suhara and H. Nishihara, *J. Lightwave Technol.*, **11**, 1400–1403 (1993).
[9] M. Oh, S. Ura, T. Suhara and H. Nishihara, *Focal spot intensity modulator using electrooptic polymer waveguide*, Technical Digest (Post Deadline Papers) of Fourth Microoptics Conference and Eleventh Topical Meeting on Gradient-Index Optical Systems, Kawasaki, October 20–22, 1993, pp. 22–25 (paper PD6).
[10] S. Ura, T. Suhara and H. Nishihara, *IEEE J. Lightwave Technol.*, **1**, 270–273 (1989).
[11] S. Ura, M. Shinohara, T. Suhara and H. Nishihara, *IEEE Photon. Technol. Lett.*, **6**, 239–241 (1994).
[12] S. Ura, T. Endoh, T. Suhara and H. Nishihara, *Linearly focusing grating couplers for sensing two-dimensional grating-scale displacement*, Technical Digest of Topical Meeting of the International Commission for Optics, Kyoto, April 4–8, 1994, p. 164 (paper 7A-5).

5
ACTIVE PHOTONIC DEVICES

5.1

Multifunctional Optoelectronic Integrated Devices

Susumu Noda and **Akio Sasaki**

1 Introduction

There is great interest in optical signal processing and optical computing. Various types of optical devices have been proposed and demonstrated for these fields [1–4]. Among them, we have proposed and demonstrated the approach of vertical and direct integration of optoelectronic devices [1,2]. In this approach the constituent devices are integrated without the conducting elements used for the electrical connections between them. They can be considered to form a single device. Thus, we have called it an optoelectronic integrated device (OEID) rather than an optoelectronic integrated circuit (OEIC). The integrated device shows not only the simple combined characteristics of constituent devices but also the more elevated characteristics due to the mutual interaction among them.

The OEID was first demonstrated by integrating a heterojunction phototransistor (HPT) and a light-emitting diode (LED), and the optical amplification, switching, and bistable functions were successfully demonstrated [1,5,6]. The LED part was then replaced by a laser diode (LD) and the optical gain was greatly increased; the additional functions due to the lasing oscillation have been demonstrated [2,7,8]. At the same time, the degree of integration has been increased step by step, and it has been shown that the greater the degree of integration, the higher the functionality of the device [2,9–11]. In the following, the structures, fabrication, and characteristics of the OEIDs with various combinations of HPTs and LDs are described. A trial to make the LD part have a surface-emitting function is also briefly described in section 6.

2 OEID-I Composed of One HPT and One LD

The structure of OEID-I, composed of one HPT and one LD, is shown in Figure 5.1.1. The device was fabricated by three-step liquid phase epitaxy (LPE). In the first step,

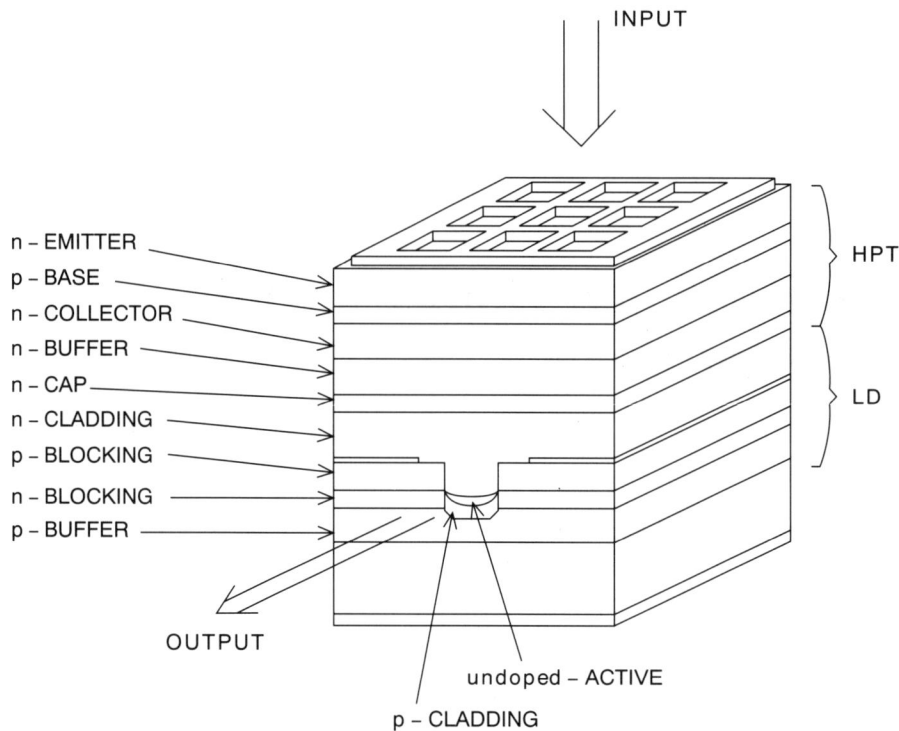

Figure 5.1.1
Schematic structure of OEID-I composed of one HPT and one LD

the current blocking layers of p-InP, n-InP, and p-InP were grown on a p-InP substrate. After the formation of a groove with 2 μm width and 300 μm separation on the wafer, the layers of a p-InP cladding, an undoped InGaAsP active ($\lambda = 1.3$ μm), an n-InP cladding, and an n-InGaAsP layers were grown at the second LPE. A stripe geometry LD was formed by these growth processes. Then the HPT composed of an n-InP buffer, an undoped n-InP collector, a p-InGaAsP base ($\lambda = 1.2$ μm), and an n-InP emitter was grown at the third LPE. After the growth, a AuGe/Ni/Au mesh electrode was deposited on the epilayer side and an AuZn/Au electrode was deposited on the reverse side. The device dimension was 270×300 μm^2.

Since the HPT is integrated just above the LD, the spontaneous emission is incident from the LD to the HPT. It induces the positive feedback mechanism for the current increase in the device. Figure 5.1.2 shows the output-input light powers characteristic of the device. It is seen that the device was switched with a very low input power of 56 nW and emitted an output power of 4 mW. Therefore, the optical switching gain is as large as 1×10^5. The large gain and small switching power are due to the achievement of the lasing oscillation and the effective utilization of the internal optical feedback [7]. The energy required for the switching was measured by using an optical pulse input. The minimum switching energy was as small as 80 fJ.

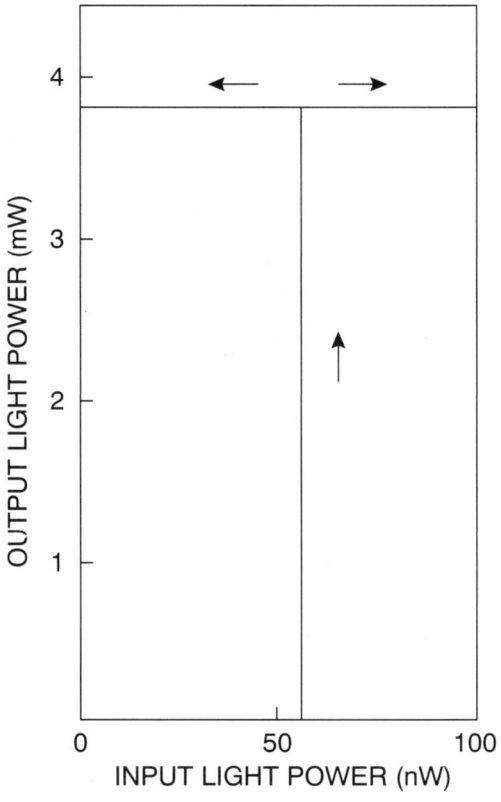

Figure 5.1.2
Optical switching function

3 OEID-II Composed of Three HPTs and One LD

When linearity between the output and input light powers is required, the optical feedback inside the device should be controlled appropriately. For this purpose, we have developed OEID-II, composed of three HPTs and one LD [8]. The schematic structure of the device is shown in Figure 5.1.3. The positions of HPT-1 and HPT-3 are shifted slightly away from the stripe of the LD to weaken the optical feedback. Two optical absorptive layers, AB1 and AB2, are inserted between the HPTs and the LD for additional suppression of the optical feedback. The device fabrication processes were almost the same as in OEID-I. The dimension of each HPT was about 210×290 μm^2.

When HPT-1 was utilized as the HPT part, the output and input light powers characteristics became as shown in Figure 5.1.4, where the relative separation of the HPT-1 and the LD was about 180–200 μm. It is seen in Figure 5.1.4 that a fairly good linear relationship between the input and output power characteristics has been obtained, which suggests that optical feedback does not substantially occur in part of the HPT-1 and the LD. When the load resistance was selected as 10 Ω, the optical gain was

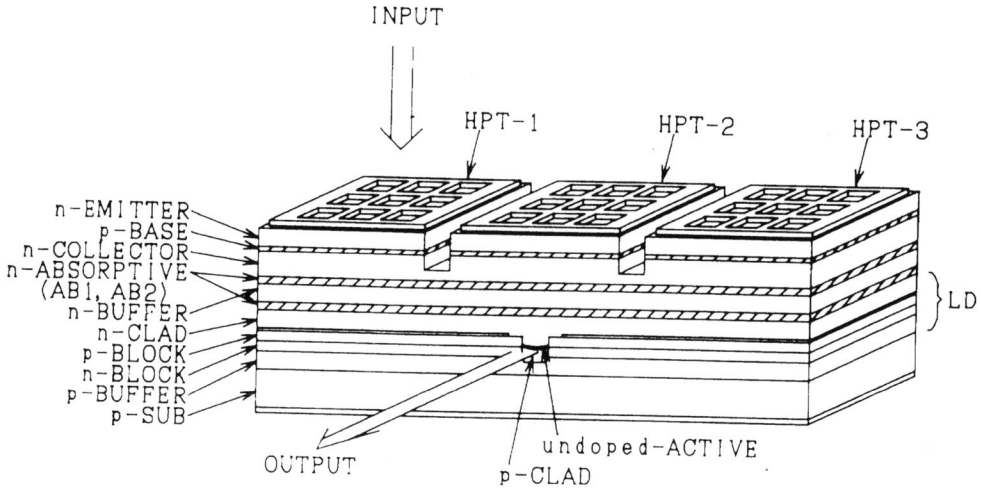

Figure 5.1.3
Schematic structure of OEID-II composed of three HPTs and one LD

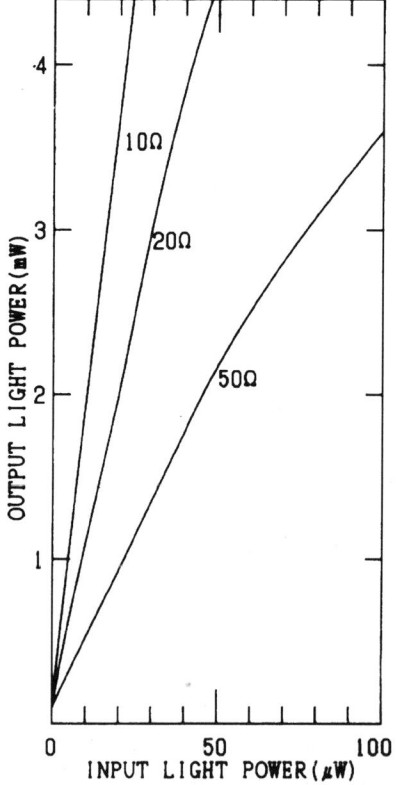

Figure 5.1.4
Optical amplification function

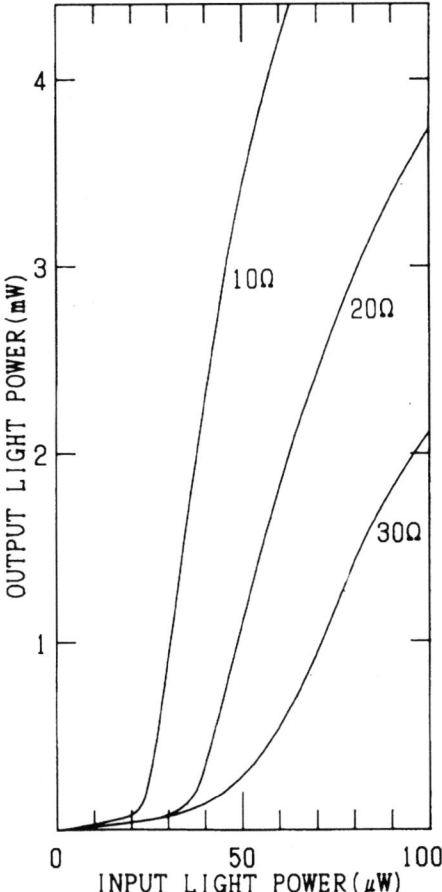

Figure 5.1.5
Optical thresholding function

as high as 200 and highest among those of the integrated devices ever reported. From the spectral response measurement of the device, it was also found that the optical gain is over 100 in the wide wavelength range from 0.75 to 1.25 μm. When the operational condition was slightly changed, the optical threshold function shown in Figure 5.1.5 was achieved, where the optical threshold function means that the device can emit a light output when the input light power exceeds a threshold level and is very important for optical neural network systems.

When HPT-3 was utilized as the HPT part, the optical feedback becomes moderate and the optical bistable function has been obtained, where the relative separation of the HPT-1 and the LD was about 150–160 μm. A detailed calculation of the optical feedback inside the device is shown in Reference [8].

4 OEID-III Composed of Four HPTs and One LD

We have shown the optical switching, amplification, thresholding, and bistable functions by the OEID-I and II. The functions come from a simple combination of one HPT and one LD. To achieve more elaborate functions, the simultaneous utilization of multiple HPTs and an LD is considered effective. Here we describe OEID-III, composed of four HPTs and one LD [9,11]. The schematic structure of OEID-III is shown in Figure 5.1.6. The fabrication procedure is almost the same as for OEID-II except for a slight modification of the doping levels of some layers. The dimension of each HPT was 100×120 μm^2 and the cavity length of the LD was about 400 μm. The periphery of the HPTs was coated with a polyimide film to avoid the leakage of current due to the formation of the surface states.

By careful experiments it has been found that there are three optical coupling modes between any two HPTs through the LD stripe: (1) bidirectional coupling; (2) unidirectional coupling; and (3) no coupling [9]. Here, bidirectional, unidirectional, and no coupling mean that the two HPTs affect each other, one HPT is affected by the other but the reverse case does not hold, and two HPTs do not affect each other respectively. Figure 5.1.7 shows a schematic drawing of the bidirectional optical coupling between HPT-A and HPT-B through the LD [10]. When the optical input is irradiated to HPT-A, a photo-induced current flows through the HPT-A + LD part. In this case the light emitted from the LD stripe under HPT-A is fed back not only to HPT-A but also to HPT-B, as shown in the figure. The current–voltage characteristics of the HPT-B + LD part

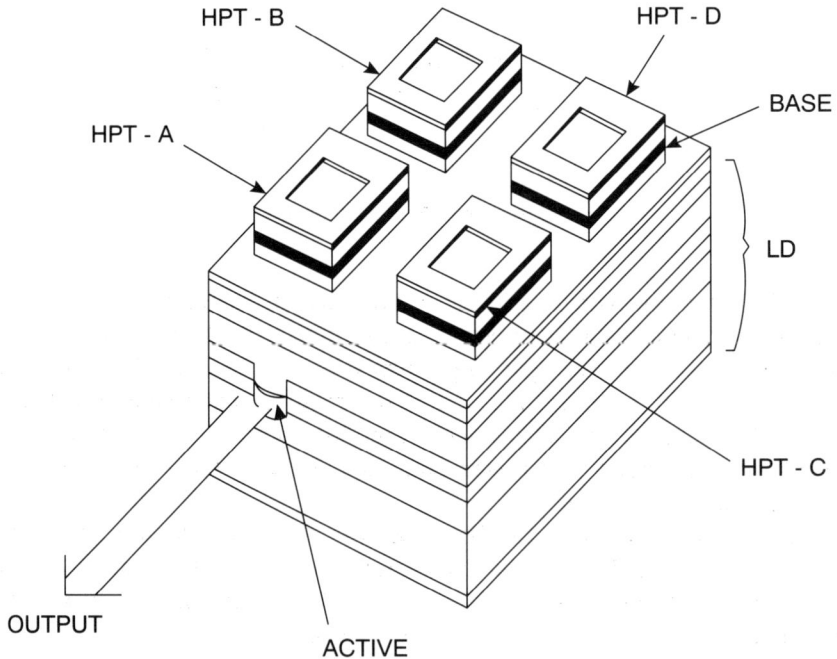

Figure 5.1.6
Schematic structure of OEID-III composed of four HPTs and one LD

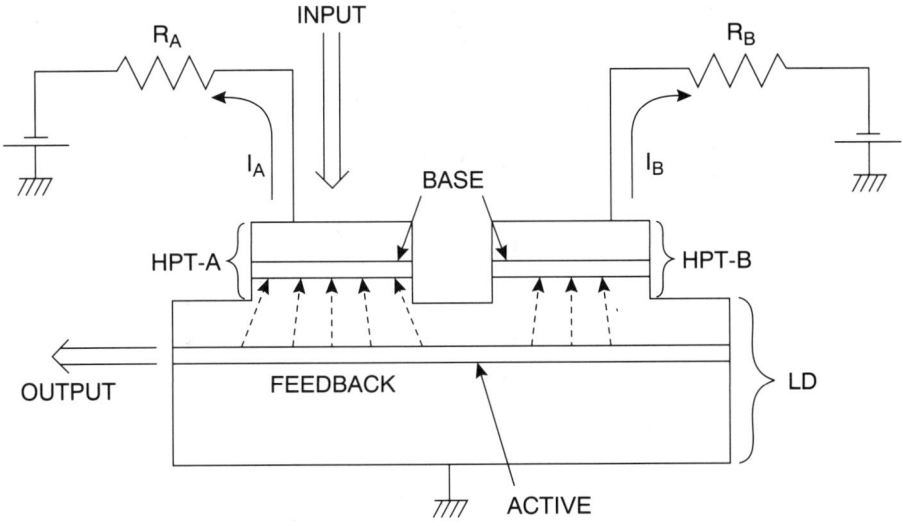

Figure 5.1.7
Schematic drawing to show bidirectional optical coupling between HPT-A and HPT-B through an LD stripe

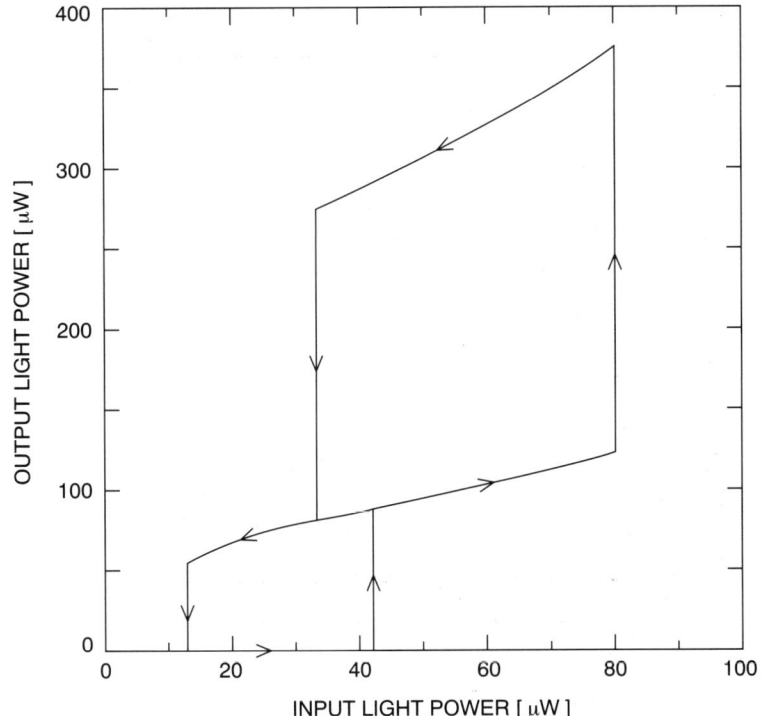

Figure 5.1.8
Optical tristable function

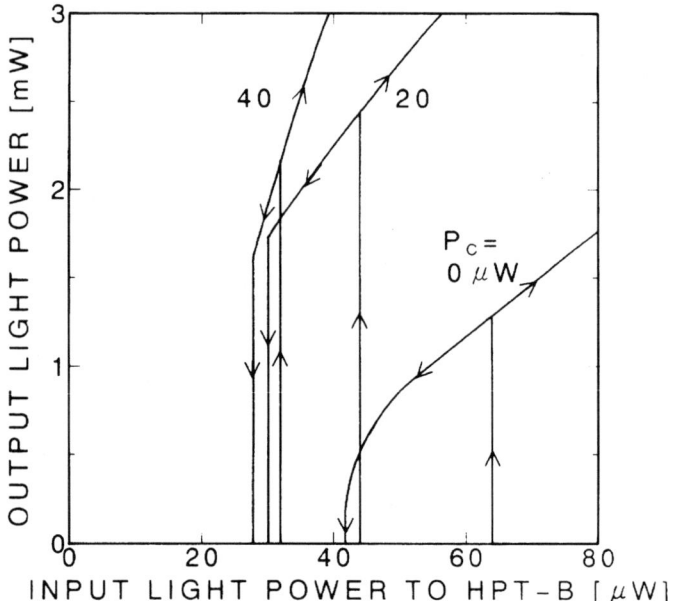

Figure 5.1.9
Light-controlled optical bistable function

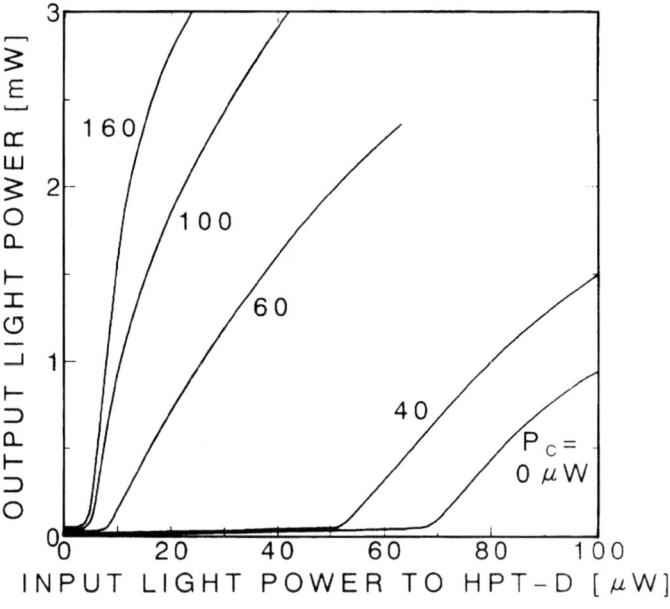

Figure 5.1.10
Light-controlled optical threshold function

are thus significantly affected by the HPT-A + LD part. Similarly, the current–voltage characteristics of the HPT-A + LD part are strongly affected by the HPT-B + LD part. This means that HPT-A and HPT-B are affected bidirectionally through the LD stripe. By utilizing these optical coupling phenomena, the following three functions have been successfully achieved: (1) an optical tristable function by bidirectional coupling between HPT-A and HPT-B through the LD; (2) a light-controlled optical bistable function by unidirectional coupling between HPT-B and HPT-C; and (3) a light-controlled optical thresholding function by no coupling between HPT-C and HPT-D. Figures 5.1.8–5.1.10 show the functions obtained experimentally.

5 OEID-IV Composed of Six HPT and Two LD

In previous sections we have shown that the greater the degree of integration, the higher the functionality of the device. To achieve more elaborate functions, we have developed OEID-IV, composed of six HPTs and two LDs, as shown in Figure 5.1.11. The fabrication processes were almost the same as those of OEID-III, but the number of constituent devices were increased. The dimension of each HPT was 100×120 μm^2, and the cavity length and the separation length of the LDs was 400 μm and 300 μm, respectively.

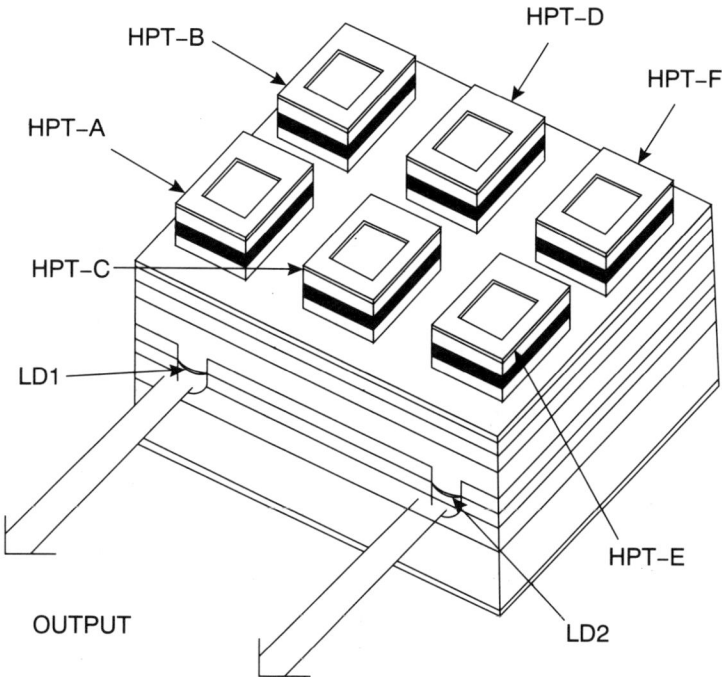

Figure 5.1.11
Schematic structure of OEID-IV composed of six HPTs and two LDs

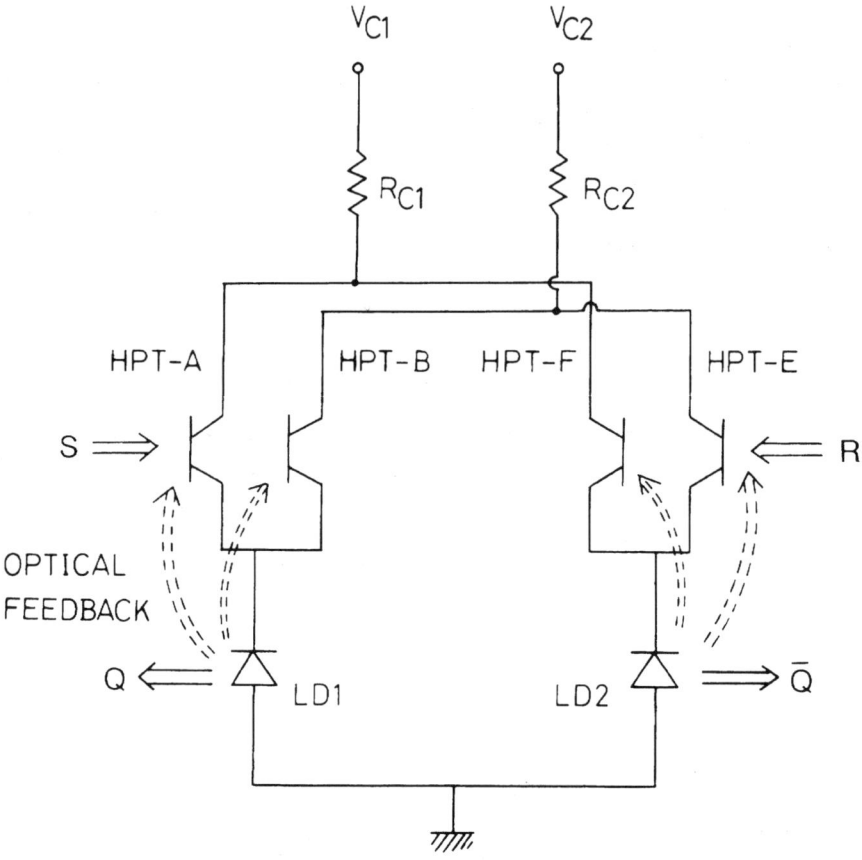

Figure 5.1.12
Equivalent circuit for the tristable flip-flop function

Since the device has two LDs as the output portions, the most interesting function is to switch the output portion from LD1 to LD2 or from LD2 to LD1 by the corresponding optical inputs. This function is the so-called bistable set–reset flip-flop, and can be achieved by simultaneous utilization of the HPT-A + LD1 and the HPT-E + LD2 parts which are connected in parallel through a common load resistance. Due to the parallel connection, only one part of the HPT-A + LD1 and the HPT-E + LD2 can be switched on, and the sequential input to the HPT-A and to the HPT-E enable the flip-flop operation. We have achieved not only bistable flip-flop but also tristable flip-flop functions by utilizing additional HPTs (HPT-B and HPT-F) and their internal optical couplings as described in section 4. Figures 5.1.12 and 5.1.13 show the equivalent circuit of the tristable flip-flop function and the operation obtained experimentally. This multistable logic is considered very important in reducing the device numbers inside the system. The HPT-C (or HPT-D) which is integrated between the LD1 and the LD2 can be utilized for the reset operation of the device by connecting it in parallel to the LDs [11].

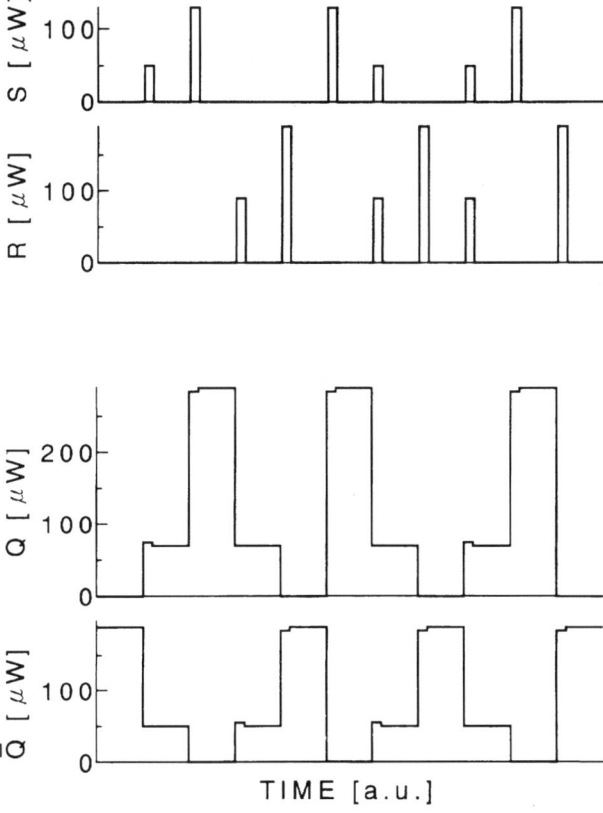

Figure 5.1.13
Optical tristable flip-flop function

6 Summary and Extension to an OEID with a Surface-emitting Function

We have developed various optoelectronic integrated devices composed of multiple HPTs and LDs. It has been shown that the greater the degree of integration, the higher the functionality of the device due to the increase of the internal optical and/or electronic couplings among constituent devices. More than 13 functions have been achieved: optical switching, amplification, thresholding, bistability, tristability, light-controlled bistability, light-controlled threshold, set and reset functions in bistable, tristable, and tetrastable states, flip-flop functions in bistable, tristable, and astable states, and so on, including functions that have not been described here.

When considering two-dimensional arrays of the OEIDs, we should utilize a surface-emitting LD (SEL) instead of an edge-emitting LD. Among various types of SELs, a circular grating-coupled SEL (CGSEL) [12] has the advantages of a narrow beam divergence, a symmetric circular far-field pattern, and a stable single-mode oscillation, although a CW oscillation has not yet been achieved. The HPT can be integrated on

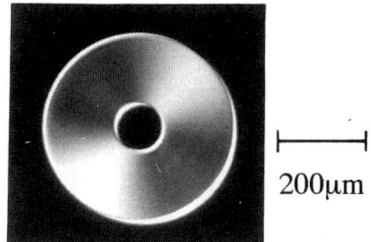

Figure 5.1.14
Circular gratings formed on SiN_x deposited on InP substrate for CGSEL

the center part of the CGSEL where no grating coupler is present. This means that the optical feedback from the CGSEL to the HPT can be only spontaneous emission and thus it is very easy to control the optical feedback inside the device. We have recently started to develop an OEID using a CGSEL. Figure 5.1.14 shows the circular gratings of SiN_x deposited on InP substrate with a period of 0.4 μm, which has been formed by using the electron beam lithography technique. Details of the device characteristics as well as the fabrication processes will be reported elsewhere [13].

References

[1] A. Sasaki and M. Kuzuhara, *Jpn J. Appl. Phys.*, **20**, L283 (1981).
[2] S. Noda and A. Sasaki, *J. Fiber Integrated Optics*, **12**, 319 (1993).
[3] D.A.B. Miller, D.S. Chemla, T.C. Damen, T.H. Wood, C.A. Burrus, A.C. Gossard and W. Weighmann, *IEEE J. Quantum Electron.*, **QE-21**, 1462 (1985).
[4] K. Kasahara, Appl. Phys. Lett., **52,** 679, (1988).
[5] A. Sasaki, K. Taneya, H. Yano and S. Fujita, *IEEE Trans. Electron Devices*, **ED-31**, 805 (1984).
[6] A. Sasaki, S. Metavikul, M. Itoh and Y. Takeda, *IEEE Trans. Electron Devices*, **35**, 679 (1988).
[7] S. Noda, T. Takayama, K. Shibata and A. Sasaki, *IEEE Trans. Electron Devices*, **39**, 305 (1992).
[8] S. Noda, T. Takayama, K. Shibata and A. Sasaki, *IEEE J. Lightwave Technol.*, **10**, 2023 (1992).
[9] S. Noda, Y. Kobayashi and A. Sasaki, *IEEE Photon. Technol. Lett.*, **34**, 1142 (1992).
[10] S. Noda, Y. Kobayashi, T. Takayama, K. Shibata and A. Sasaki, *IEEE J. Quantum Electron.*, **28**, 2714 (1992).
[11] S. Noda, Y. Kobayashi, K. Shibata and A. Sasaki, *IEEE J. Quantum Electron.* to be published in IEEE J. Quantum Electron. (August 1995).
[12] M. Toda, *IEEE J. Quantum Electron.*, **26**, 473 (1990).
[13] S. Noda, T. Ishikawa, M. Imada and A. Sasaki, Tenth International Conference on Integrated Optics and Optical Fiber Communication (IOOC'95), June 26–30, 1995, Hong Kong, paper TuBl-5.

5.2

Strained Layer Superlattices for High-speed Optical Devices

Ikuo Suemune

Abstract

The intrinsic strain effects on semiconductor band structures and on lasing characteristics are described. In addition, the high-speed modulation characteristics of strained quantum well (QW) semiconductor lasers are shown to be critically dependent on the band offsets in separate-confinement heterostructures (SCH). Especially, lower band offsets result in a higher thermionic emission of carriers from QWs to the SCH layers, and this reduces the maximum modulation frequency due to the transport limit in the SCH layers. It is shown that the above transport limit is reduced by employing SCH lasers grown on InGaAs ternary substrates and that the maximum modulation frequency near 100 GHz is expected by the reasonable selection of the optical confinement factor.

1 Introduction

The application of strain onto semiconductors modifies the band structures, and a reduction of the lasing threshold currents in strained quantum well (QW) lasers was theoretically predicted [1,2]. The capability of high-speed performance of semiconductor lasers was also predicted theoretically in strained QW lasers [3].

The strain effects on semiconductors will be briefly reviewed in the following section. The lasing threshold current is calculated based on theoretically calculated band structures considering the strain effect. In 1.3 μm and 1.55 μm semiconductor lasers, which are the key light sources for long-distance optical communication systems, Auger recombinations contribute significantly to the recombination processes. The strain effect on the Auger recombination is briefly discussed.

In addition to the intrinsic properties of the strained QW active layers, the design of heterostructures also modifies the performance of semiconductor lasers. Especially the barrier heights of the QWs, which are given by the band offsets at the heterointerfaces, influence the laser performance significantly, *e.g.* the temperature dependence of the lasing threshold currents and the laser modulation bandwidth. It is shown that the performance of 1.3 μm lasers is drastically improved by the introduction of the heterostructures grown on InGaAs substrates. A maximum modulation frequency as high as 100 GHz is theoretically predicted.

2 *Strain Effects on Semiconductor Band Structures*

The fundamental concept of the strain effect on semiconductors is shown in Figure 5.2.1. The details of the analysis is given in Reference [4]. When a semiconductor film with a lattice constant larger than the others is sandwiched, the semiconductor film is under biaxial compression. The top of the valence band near the zone center is four times degenerate in zinc blende semiconductors, but the biaxial strain lifts the degeneracy. When the (001) semiconductor film is under biaxial compression, the bands designated

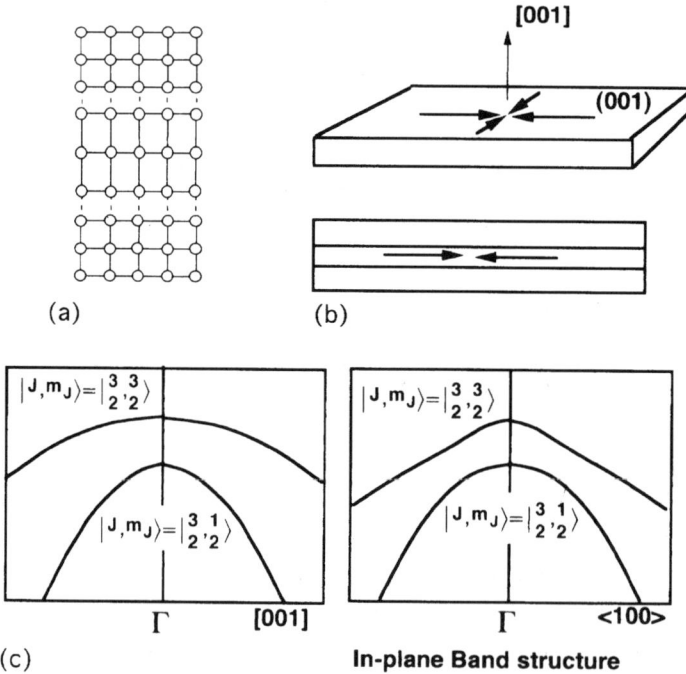

Figure 5.2.1
(a) Schematic lattice image where a semiconductor film with a larger lattice constant is coherently sandwiched between two semiconductors having a smaller lattice constant. (b) The sandwiched film experiences biaxial compression. The (001) plane is shown as an example. (c) Bulk band structures under biaxial compression. Left is in the [001] direction and right is in the $\langle 100 \rangle$ directions in the (001) plane

as $|J, m_J\rangle = |\frac{3}{2}, \frac{3}{2}\rangle$ (corresponding to the bulk heavy-hole (HH) band) and $|J, m_J\rangle = |\frac{3}{2}, \frac{1}{2}\rangle$ (corresponding to the bulk light-hole (LH) band) are split in energy at the zone center. The valence band structure near the zone center becomes anisotropic under this circumstance.

The band designated as $|J, m_J\rangle = |\frac{3}{2}, \frac{3}{2}\rangle$ shows a weaker energy dependence in the [001] direction perpendicular to the (001) plane, suggesting a heavy-hole mass in this direction. However, the corresponding band structure in the (001) plane shows a higher energy dependence, suggesting a lighter hole mass. Therefore in strained QWs, the quantum state energy is determined by the effective mass in the [001] direction, while the density of states (DOS) is mainly determined by the effective mass in the (001) plane. In strained QWs under biaxial compression, the DOS near the top of the valence band is reduced by the lighter-effective mass in the (001) plane.

In strained QWs under biaxial tension, the band designated $|J, m_J\rangle = |\frac{3}{2}, \frac{1}{2}\rangle$ is lifted upward. In this case, the effective mass in the (001) plane is heavier than that in the [001] direction. Therefore, the DOS in the valence band is not much reduced in this case, but the energy separations between the quantum state energies get larger by the lighter effective mass in the [001] direction. These anisotropic valence bands are caused by the strain change in the performance of the semiconductor lasers.

3 Lasing Characteristics of Strained Semiconductor Lasers

The fundamental lasing characteristics will be discussed on the strained QW under biaxial compression. The relations of the optical gain and the Auger recombinations to the injection current are compared between the strained and unstrained QWs.

3.1 Injection current and optical gain

The detailed methods for calculating the optical gain and the current density considering the nonparabolic conduction and valence bands are given in Reference [5]. Figure 5.2.2 shows a comparison of the characteristics calculated on a 5 nm GaAs QW and a 5 nm $In_{0.37}Ga_{0.63}As$ QW. In both cases, $Al_{0.4}Ga_{0.6}As$ barriers are assumed, and the GaAs QW is unstrained, while the InGaAs QW is under a biaxial compression of 2.6%.

In the biaxially compressed $In_{0.37}Ga_{0.63}As$ QW, the band near the top of the valence band is dominated by the $|J, m_J\rangle = |\frac{3}{2}, \frac{3}{2}\rangle$ band and is nearly parabolic. The in-plane mass near the zone center in the (001) plane is nearly isotropic [4] and is reduced, as schematically shown in Figure 5.2.1. In the lattice-matched GaAs QW, the valence bands designated $|J, m_J\rangle = |\frac{3}{2}, \frac{3}{2}\rangle$ and $|J, m_J\rangle = |\frac{3}{2}, \frac{1}{2}\rangle$ lie very close in energy near the zone center. This causes significant mixing of the two bands and the resultant band near the top of the valence band is nonparabolic and highly anisotropic in the (001) plane.

In the actual calculation, the valence band structure in the (001) plane was represented by the one in the principal [100] and [110] directions. The two curves for the GaAs QW shown in Figure 5.2.2 were calculated by the [100] and [110] band structures. The two curves for the $In_{0.37}Ga_{0.63}As$ QW almost coincide due to the nearly isotropic

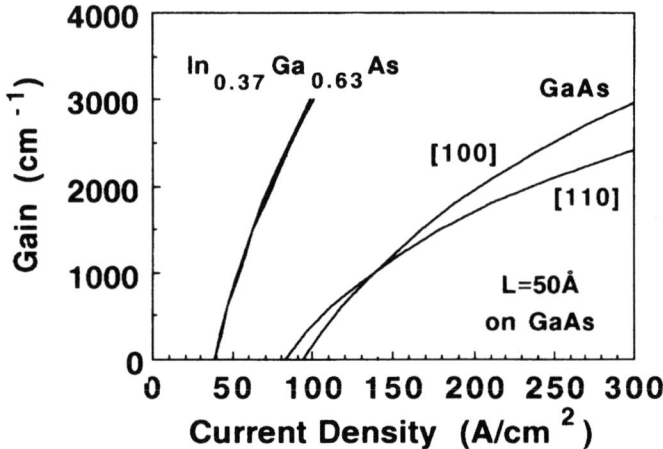

Figure 5.2.2
Current density dependence of the optical gain calculated for a strained $In_{0.37}Ga_{0.63}As$ QW and an unstrained GaAs QW both with 5 nm well width

band structure in the (001) plane. The current density for the positive gain, J_0, in the $In_{0.37}Ga_{0.63}As$ QW shown in Figure 5.2.2 is much reduced in comparison with the GaAs QW. This is because of the reduced mass in the (001) plane in the biaxially compressed $In_{0.37}Ga_{0.63}As$ QW and the corresponding reduction of the DOS. The measured J_0 value is as low as 25 A/cm^2 [6] and is in reasonable agreement with the calculated J_0 value of 38 A/cm^2 considering the uncertainty of the parameters.

The difference in the slope for the two cases in the optical gain versus the current characteristics shown in Figure 5.2.2 also originates from the difference of the valence band DOS. The high-speed capability of semiconductor lasers is closely related to the differential gain given by the slope of the gain versus carrier concentration characteristics, i.e. $g' = \partial g/\partial n$. The differential gain is also improved for the reduced DOS, and the value for the $In_{0.37}Ga_{0.63}As$ QW was more than three times larger than that for the GaAs QW [5]. When the heterostructure is properly designed, as discussed later, this leads to a significantly expanded modulation bandwidth in strained QW lasers.

3.2 Auger recombination

High-speed modulation capability is important especially in optical fiber communication systems. 1.3 μm and 1.55 μm semiconductor lasers designed for optical communications show a higher temperature sensitivity of the lasing threshold currents. The main factor in temperature sensitivity is Auger recombination. Auger recombination is the collision process where two transitions are involved and the overall energy and momentum are conserved. The major Auger processes are shown in Figure 5.2.3. One is called the CCCH process, where an electron (1) in the conduction band recombines with a hole (1') in the valence band and another electron (2) is excited to higher energy in the conduction band. The other is called the CHHS process, where an electron (1')

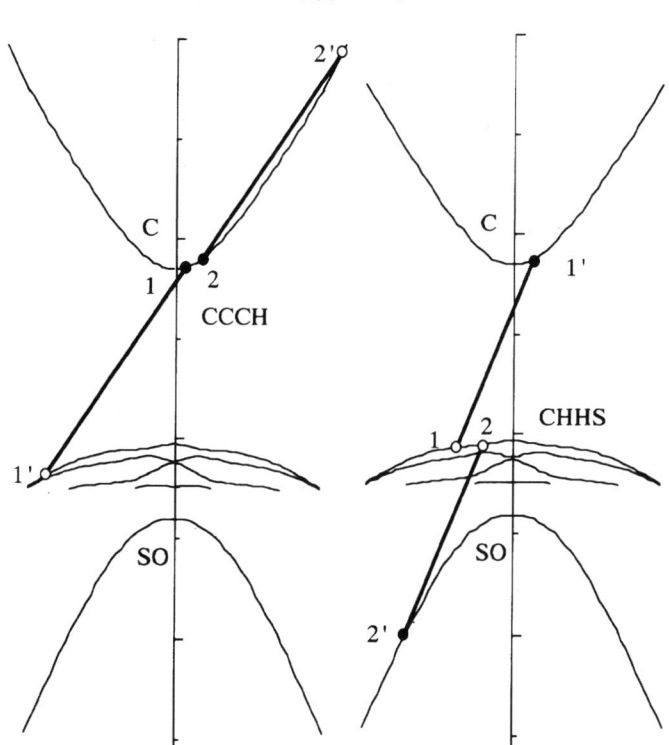

Figure 5.2.3
CCCH and CHHS Auger recombination processes near the transition threshold. The band structure was calculated for an $In_{0.53}Ga_{0.47}As$ (5 nm)/$In_{0.52}Al_{0.48}As$ QW

recombines with a hole (1) and another hole (2) near the band top is excited to higher energy in the spin–orbit split-off (SO) band.

The details of the calculations and the numerical results are given in Reference [7], and the main points are summarized in what follows. In unstrained QWs the CHHS process dominates which is enhanced with the higher hole concentration near the top of the valence band. The biaxial compression reduces the DOS near the top of the valence band, as discussed above, and the hole concentration for the lasing threshold is also reduced. This, as well as the modified valence band structure near the zone center, significantly reduces the contribution of the CHHS process in strained QWs.

In the CCCH process, on the other hand, electron (1) transits to hole (1′) deep in the valence band. The valence band structure, remote from the zone center, is not much modified by the strain effect. Therefore a reduction in the CCCH transition rate with the strain is not significant. The overall lasing threshold current was still dominated by the Auger processes even in strained QWs, but the absolute current density was reduced to ∼1/4 in the strained QW compared with the unstrained QW [7]. Therefore the problem of the Auger process and the temperature sensitivity of the threshold current are still an open question.

4 Heterostructures Grown on InP and InGaAs Substrates

1.3 μm and 1.55 μm semiconductor lasers are usually grown on InP substrates. However, especially 1.3 μm separate-confinement heterostructure (SCH) lasers grown on InP substrates have relatively small band offsets, i.e. ΔE_c (conduction band offset) = 179 meV and ΔE_v (valence band offset) = 40 meV. Another possibility of selecting the substrates was proposed recently [8]. When a similar SCH laser is designed on an InGaAs substrate (with a mole fraction of 26%), the band offsets can be increased to $\Delta E_c = 347$ meV and $\Delta E_v = 180$ meV.

To study the influence of the band offset, the relation between the injected carrier concentration and the maximum optical gain was calculated for the SCH with 8 nm QWs corresponding to a 1.3 μm lasing wavelength. A compressive strain of 1.4% in the QW layer was assumed, and the SCH layer widths on both sides of the well layer were set at 50 nm. The results are shown in Figure 5.2.4 for the SCHs grown on InP and InGaAs substrates. The sheet carrier density is the carrier concentration integrated across the SCH layers. It can be seen that the optical gain for the InP substrate is much lower than that for the InGaAs substrate. This is because of the carrier population to the SCH layers due to the lower band offsets in the SCH grown on InP substrates. This significant carrier population to the SCH layers also results in a higher temperature sensitivity of the optical gain, and the characteristic temperature, T_0, of 173 K for the InGaAs substrate is reduced to 74 K for the InP substrate [9].

The modulation characteristics of semiconductor lasers are closely related to the differential gain, $g' = \partial g / \partial n$, as discussed above. The calculation of the differential gain

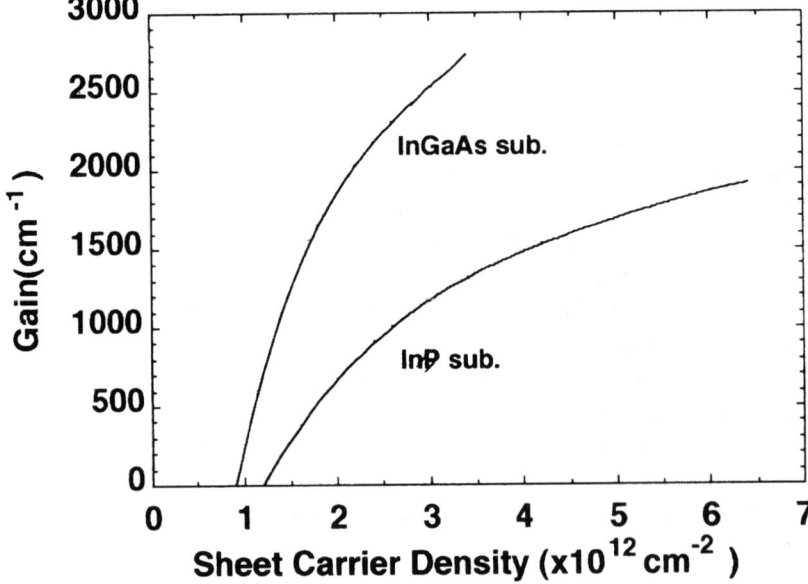

Figure 5.2.4
Optical gain calculated for the sheet carrier density integrated over the quantum well and the SCH layers for 1.3 μm SCH lasers grown on InP and InGaAs substrates

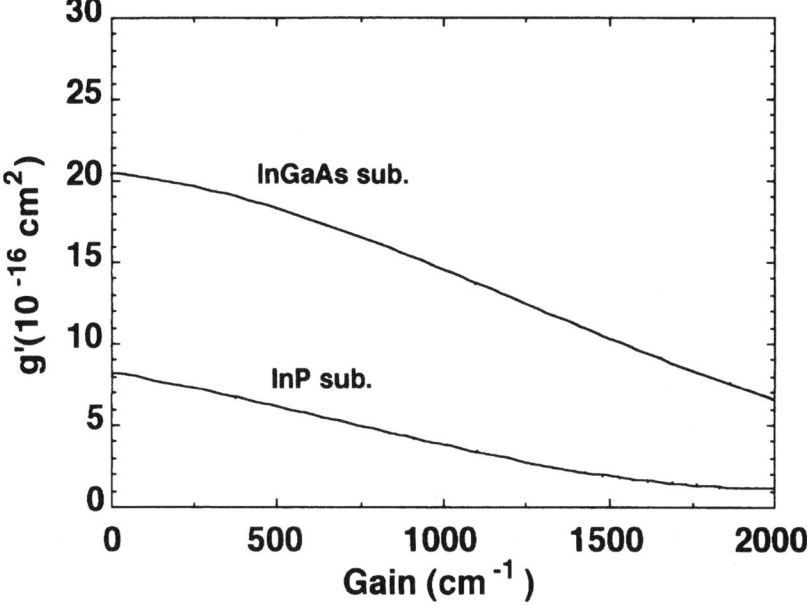

Figure 5.2.5
Differential gain versus optical gain for 1.3 μm SCH lasers grown on InP and InGaAs substrates, calculated from Figure 5.2.4

for a given optical gain is shown in Figure 5.2.5. In the case of the SCH on the InGaAs substrate, the calculated differential gain is dominated by the intrinsic characteristics of the QW. The much lower differential gain for the SCH on the InP substrate is due to the carrier population to the SCH layers.

5 Band Offsets and Modulation Characteristics

A close relation between the band offset and the differential gain was clarified in the previous section. The relation between the laser modulation dynamics and the band offset will be discussed in this section. In Reference [10] it was discussed that the carrier transport across the SCH region influences the high-speed modulation characteristics of semiconductor lasers. The main points are summarized as follows:

(1) When the relaxation time of the carriers into the QW after transportation into the SCH region is given by τ_r and the thermionic emission time of carriers from the QW to the SCH region is given by τ_e, as shown in Figure 5.2.6, the effective differential gain is reduced from g'_{QW} to g'_{QW}/χ (where $\chi = 1 + \tau_r/\tau_e$).

(2) When the carrier transport limit is taken into account, measured modulation characteristics can be explained by the bulk value for the nonlinear saturation coefficient of the optical gain, ε, which is usually assumed to be larger in QWs than in bulk active layers.

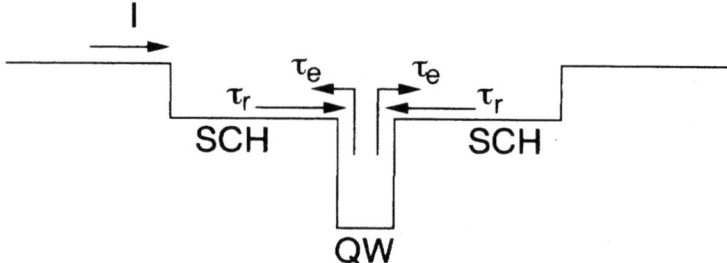

Figure 5.2.6
Schematic to show the transport across the SCH layers into the QW layer and thermionic emission out of the QW layer

When the effect of the finite band offsets is taken into account during these considerations, the lower band offsets in the SCH structure will modify the thermionic emission time, τ_e. The values estimated for the SCHs grown on the InP and InGaAs substrates are shown in Figure 5.2.7. It was several picoseconds for the InP substrate, while it was more than 100 ps on the InGaAs substrates due to a larger band offset. Therefore the thermionic emission will be much smaller in the latter case. The other parameter, τ_r, is estimated to be ~1.8 ps assuming hole diffusion across the 50 nm thick SCH layer.

When the photon lifetime of the laser cavity is given by τ_p, the maximum modulation frequency is given by $f_{max} = 2^{3/2}\pi/K$, where $K = 4\pi^2(\tau_p + \varepsilon/(g'_{QW}/\chi))$. From the lasing threshold condition, $1/\tau_p = (c/n)\Gamma g$ holds, where Γ is the optical confinement

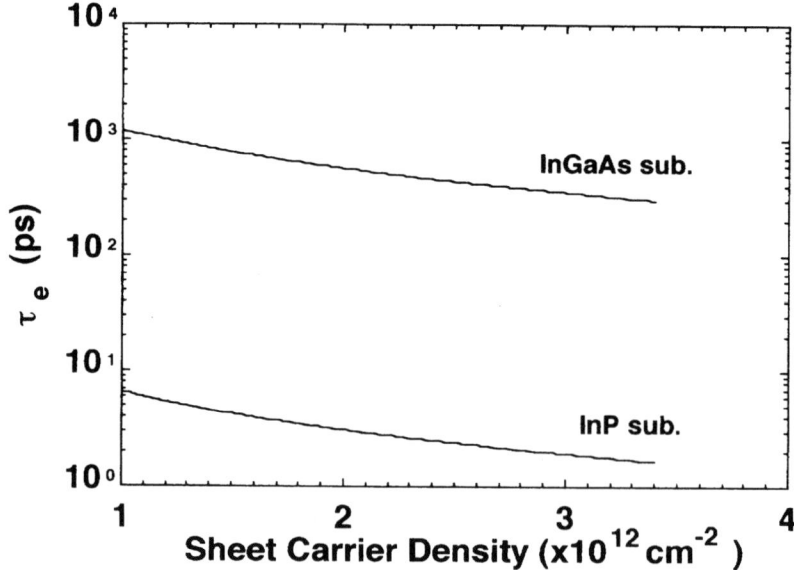

Figure 5.2.7
Thermionic emission time for holes from the quantum well to the SCH layers calculated for 1.3 μm SCH lasers grown on InP and InGaAs substrates

factor of the active layer. One way to calculate the maximum modulation frequency is to employ the above estimated values for τ_r and τ_e. Another method was found from a steady-state rate equation analysis, and this showed that the effective differential gain given by g'_{QW}/χ is equal to the differential gain directly estimated from Figure 5.2.4. This comes from the fact that the sheet carrier concentration in the QW, $N_{s,QW}$, is balanced with the total sheet carrier concentration integrated over the whole SCH region, N_s, with the relation of $N_s = \chi N_{s,QW}$.

Using this relation the maximum modulation frequency was calculated for $\Gamma = 0.01$ and $\varepsilon = 1 \times 10^{-17}$ cm^{-3} (bulk value) and is shown in Figure 5.2.8. The difference in the differential gain is reflected in the modulation characteristics and nearly twice higher modulation frequency is expected for the SCH on the InGaAs substrate. The reduction in the maximum frequency for the lower optical gain for both cases in Figure 5.2.8 is due to the increase in the corresponding photon lifetime, τ_p, which increases the K factor. This term is independent of the differential gain and is minimized for a larger optical confinement factor.

This is evidenced in Figure 5.2.9. The optical confinement factor can be increased by multiple quantum well (MQW) active layers, and the maximum modulation frequency will be as high as 100 GHz by increasing the optical confinement factor up to 5%. A higher optical confinement factor will reduce the contribution of the first term in the calculation of the K factor, and the influence of the band offset through the carrier population to the SCH layers will be more significant on the modulation characteristics. An optical confinement factor of 5% is realized by a MQW structure with the four wells.

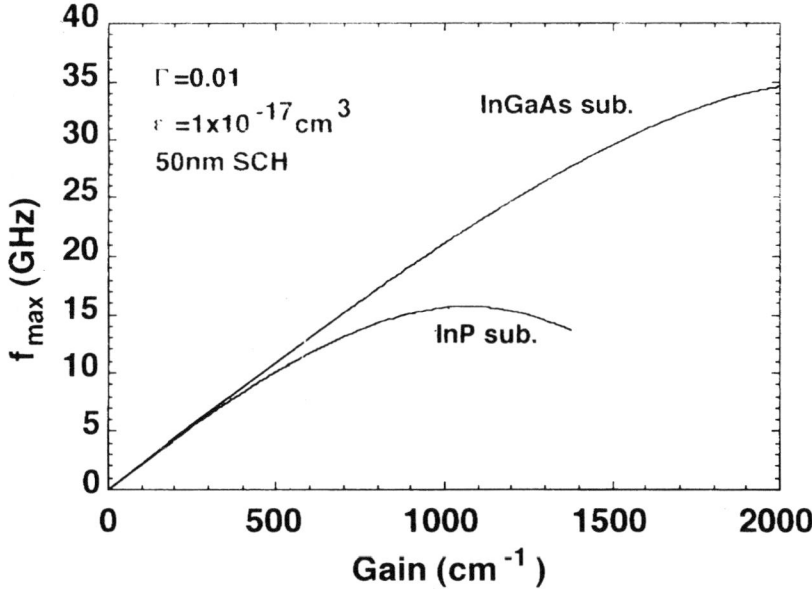

Figure 5.2.8
Maximum modulation frequency versus optical gain calculated for 1.3 μm SCH lasers grown on InP and InGaAs substrates. Lasers designed on InP substrates show a narrower bandwidth due to the carrier population to the SCH layers

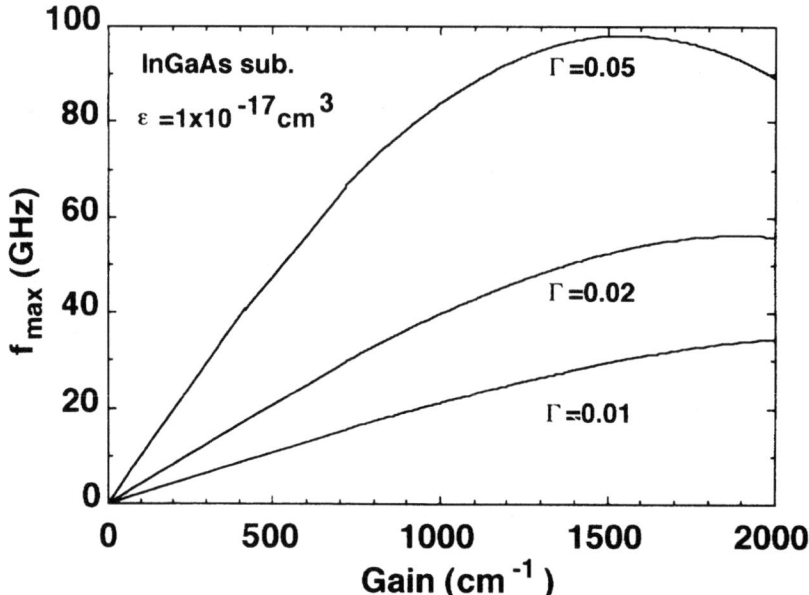

Figure 5.2.9
Dependence of the maximum modulation frequency on the optical confinement factor for SCH lasers grown on an InGaAs substrate. With the optimum design of the laser cavity to have a threshold optical gain of ~ 1500 cm^{-1}, the bandwidth will be as large as 100 GHz

6 Conclusions

After a brief review of the strain effects on semiconductors, the strain effects on the intrinsic lasing characteristics, such as the threshold currents and Auger recombinations in long-wavelength lasers, were discussed. Significant reductions in the threshold currents and the Auger recombination rates were shown by the introduction of the stain.

In SCH heterostructures, commonly used for semiconductor lasers, the band offsets were shown to play an intrinsic role in determining the maximum modulation frequency. The main issues to realize the wide modulation bandwidth are the following two points: (1) The band offsets should be large enough to suppress the carrier population to the SCH layers. (2) The optical confinement factor should be large enough to reduce the K factor through the photon lifetime term. MQW structures will be preferable for this purpose. Larger band offsets, on the other hand, may cause problems for the carrier transport between QWs in the MQW structures. The design of barrier layers in the MQW having a large enough tunneling probability will become important from this viewpoint.

References

[1] E. Yablonovitch and E.O. Kane, *IEEE J. Lightwave Technol.*, **LT-4**, 504 (1986).
[2] A.R. Adams, *Electron. Lett.*, **22**, 249 (1986).

[3] I. Suemune, L.A. Coldren, M. Yamanishi and Y. Kan, *Appl. Phys. Lett.*, **53**, 1378 (1988).
[4] I. Suemune, *Phys. Rev.*, **B43** (17), 14099 (1991).
[5] I. Suemune, *IEEE J. Quantum Electron.*, **27** (5), 1149 (1991).
[6] H.K. Choi and C.A. Wang, *Appl. Phys. Lett.*, **57**, 321 (1990).
[7] I. Suemune, *Appl. Phys. Lett.*, **55** (25), 2579 (1989).
[8] H. Ishikawa, *Appl. Phys. Lett.*, **63**, 712 (1993).
[9] H. Ishikawa and I. Suemune, *IEEE Photon. Technol. Lett.*, **6**, 344 (1994).
[10] R. Nagarajan, T. Fukushima, M. Ishikawa, J.E. Bowers, R.S. Geels and L.A. Coldren, *IEEE Photon Technol. Lett.*, **4**, 121 (1992).

# 5.3

Functional Integration in Distributed Feedback Lasers

Yoshiaki Nakano and **Kunio Tada**

Abstract

This section describes two new types of semiconductor lasers which are characterized by wavelength tunability and chirping control capability. These extended functions are incorporated by integrating different optical and electrical elements with distributed feedback (DFB) lasers. The lasing wavelength tunability is brought about through the incorporation of a linearly chirped grating. A single-mode quasi-continuous tuning range of 2.2 nm at around 868 nm has been achieved. The control of wavelength chirping due to longitudinal spatial hole burning is carried out by a tailored current density distribution along the laser axis. Integrated multiple distributed electrodes are used to produce such a current density profile. A chirping reduction in an actual phase-adjusted DFB laser is demonstrated.

1 Introduction

Distributed feedback (DFB) semiconductor lasers are important light sources for present and future optoelectronic applications since they give one of the best spectral purities among all types of diode lasers. There are, however, several points that need to be improved toward the next generation applications. Their wavelength tunability is rather limited compared with other tunable semiconductor lasers, and their lasing wavelength is not very stable when they are directly modulated.

In the meantime, integration of different optical and electrical elements with existing optoelectronic devices has been shown to result in a new class of devices with extended functions and higher performances. Such an approach could be used together with semiconductor DFB lasers in order to solve the above-mentioned problems. In this section we describe two new types of DFB lasers obtained by the integration technique. One is characterized by its wavelength tunability while the other is characterised by its chirping control capability.

2 Chirped-grating-tunable DFB Laser

Wavelength-tunable semiconductor lasers are key devices in optical communication systems and other optoelectronic applications employing a wavelength division multiplex (WDM) scheme. A variety of tunable lasers have been studied and developed. Among monolithic ones, tunable DFB lasers are advantageous since they do not have the problem of spectral linewidth broadening during wavelength tuning. However, the tuning range of the tunable DFB laser is relatively limited, as mentioned earlier, because of the difficulty in changing the refractive index from the value at threshold. In order to extend its tunability, we proposed the integration of chirped gratings (diffraction gratings with nonuniform period) with DFB lasers.

Chirped gratings have been utilized in semiconductor DFB lasers for the purpose of improving single-mode selectivity, suppressing longitudinal spatial hole burning, and reducing spectral linewidth [1-3]. We intended to achieve a larger tunability by making use of the chirped grating since lasing wavelength tuning may be brought about not only by the refractive index change, which has been the tuning mechanism of most conventional tunable DFB lasers, but also by the chirped grating effect. That is to say, when the photons are localized in the cavity, they feel a pitch of the chirped grating at that particular position. If the position of the localization changes by any means, the effective pitch changes accordingly.

This section describes the analysis, fabrication, and characteristics of such chirped-grating-tunable (CGT) DFB lasers [4,5].

2.1 Theoretical analysis

We consider the simple case of a two-section CGT DFB laser with the rear facet as cleaved (CL) and the front facet being anti-reflection coated (AR). The laser has a linearly-chirped grating which is approximated by a stepped chirped grating in the numerical analysis, as shown in Figure 5.3.1. From computed photon density profiles, we understood that the photon localization along the cavity was controllable through the ratio of injection currents into the front (I_F) and the rear (I_R) sections. This should result in an additional wavelength shift to the conventional index effect.

Simulated wavelength tuning characteristics are shown in Figure 5.3.2. The tuning range of 2.7 nm for a conventional uniform-grating 1.55 μm DFB laser (dotted curve in the figure) agrees well with the reported experimental value [6]. It is also understood that the chirp rate D (as defined in Figure 5.3.1) should be chosen appropriately in order to ensure a large tuning range. A negative D is apparently preferable in this particular example.

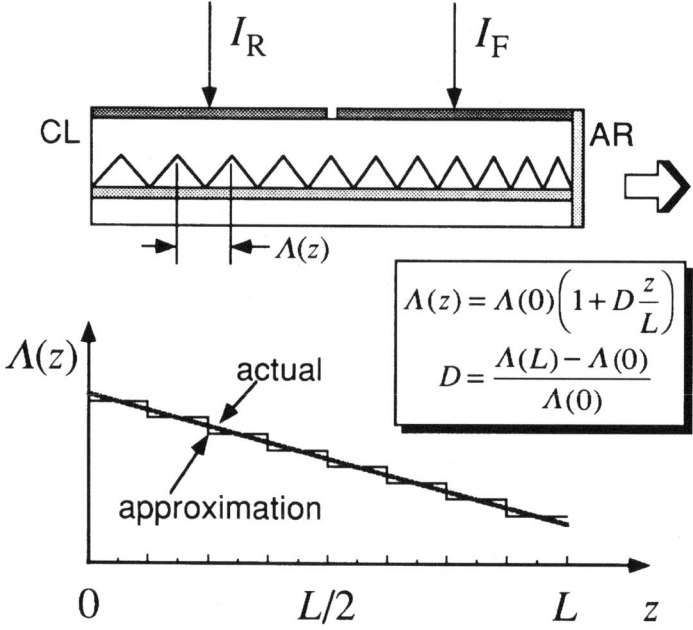

Figure 5.3.1
Conceptual drawing of the two-section CGT DFB laser with a linearly-chirped grating, and definition of the parameters

2.2 Fabrication

Figure 5.3.3 is a schematic drawing of a CGT DFB laser with three sections fabricated. The grating pitch becomes longer toward the farther facet (linear chirping). The average pitch, Λ_0, is 0.255 μm (second order), and the chirp per unit length, $d\Lambda/dz$, is 3.4×10^{-7}. This linearly chirped grating was fabricated by spherical wave holographic exposure [7]. Two-step liquid phase epitaxy [LPE] was carried out to form the double channel planar buried heterostructure (DCPBH) [8]. Tilted end facets that suppress reflection were fabricated using reactive ion etching [9]. The top electrode was divided into three sections with the central-section length being equal to the sum of the side-section lengths.

Here we focus on the characteristics of two-section CGT DFB lasers with one of the facets having a high reflectivity as discussed in subsection 2.1 since the localization of photon density occurs more easily than three-section devices with both facets having low reflectivity. The two-section CGT DFB laser was obtained by cleaving the illustrated three-section device in Figure 5.3.3 into two at the center. Therefore, one of the facets (rear) had cleaved facet reflectivity while the other (front) had very low reflectivity of the tilted-etched facet. The lengths of the two electrodes were identical, i.e. 200 μm. The chirping rate D is calculated to be -0.53×10^{-3}.

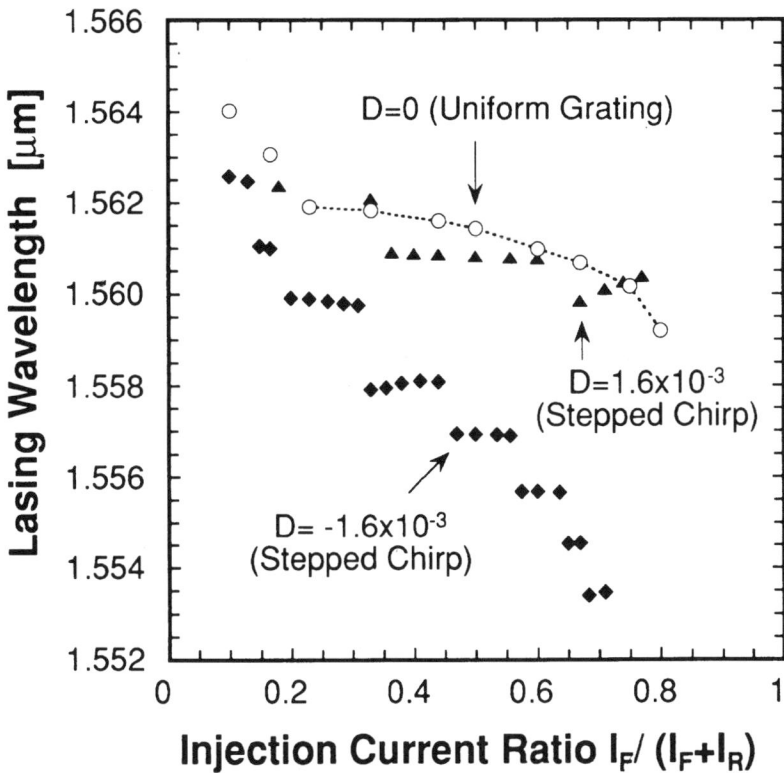

Figure 5.3.2
Simulated tuning characteristics of two-section DFB lasers with different gratings. The output power from the front facet is kept at 1 mW

2.3 Characteristics

Wavelength tuning obtained by changing the ratio of the front current I_F to the rear current I_R is shown in Figure 5.3.4. Throughout the tuning the output power from the tilted front facet was kept constant at 1 mW. A good heat sink and pulsed driving currents were used in the measurement for the reduction of the thermal tuning. In the figure, a single-mode tuning range of 2.2 nm is achieved with one small mode hopping around the injection current ratio of 0.5. This range, corresponding to approximately 4 nm in the 1.55 μm band, is large as that of (quasi-) continuous tuning. In our GaAlAs/GaAs short-wavelength lasers, the effect of an index change is small compared with InGaAsP/InP long-wavelength lasers because the amount of index variation due to the free carrier plasma effect is proportional to the square of the wavelength. Therefore, wavelength tuning in our device is mainly given by the chirped grating effect.

For the sake of comparison, we also fabricated two-section DFB lasers with uniform and reverse linear chirp (positive D) gratings. Besides the grating, they have identical structures (including ARCL facets) to the previous one. The tuning characteristics of

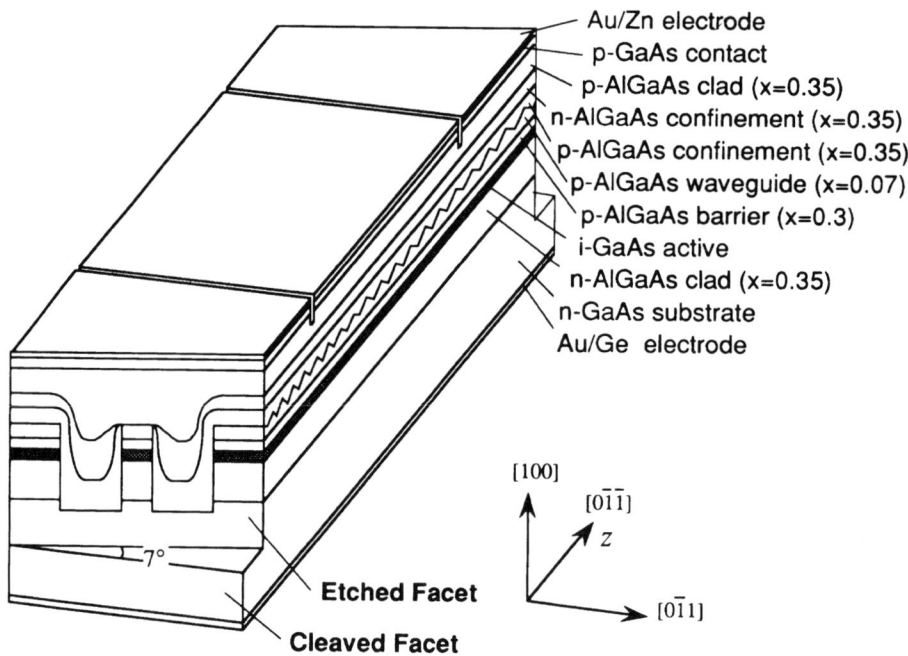

Figure 5.3.3
Structure of the GaAlAs/GaAs three-section CGT DFB laser fabricated. The pitch of the grating becomes longer toward the farther facet. 7°-tilted end facets are formed by reactive ion etching

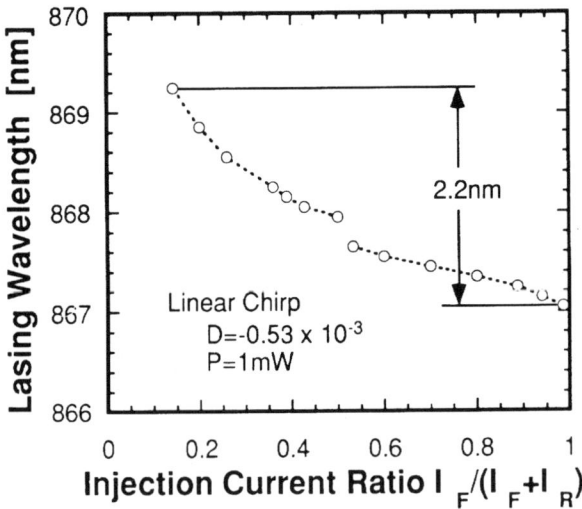

Figure 5.3.4
Measured tuning characteristics of the GaAs two-section CGT DFB laser with a negative value of the chirping rate D. The output power from the tilted front facet was kept at 1 mW

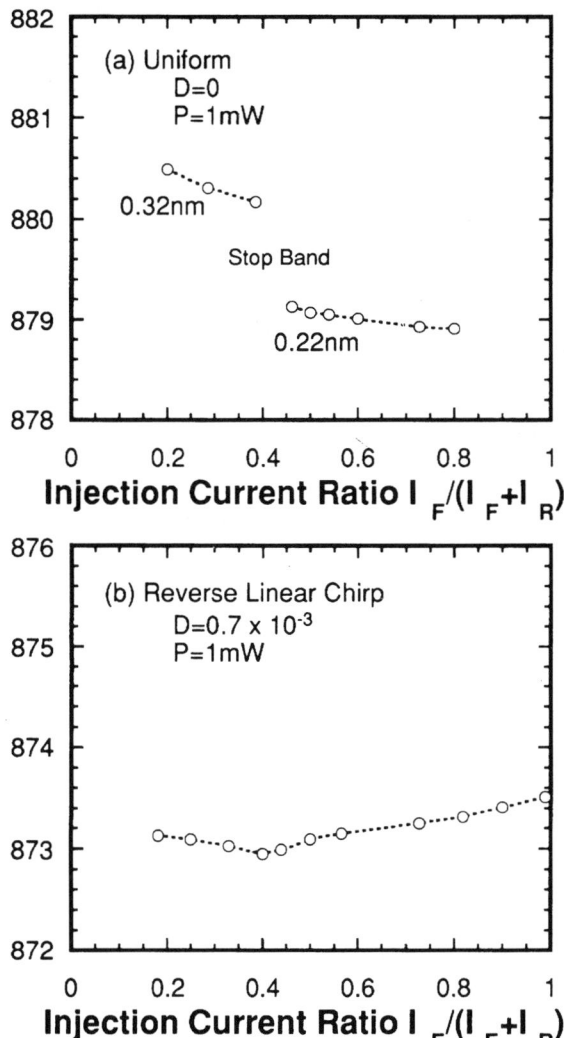

Figure 5.3.5
Measured tuning characteristics of GaAs two-section DFB lasers with different gratings: (a) uniform and (b) reverse linear chirp (positive D). The output power from the tilted front facets was kept at 1 mW

these lasers are shown in Figures 5.3.5(a) and (b). The index tuning effect in GaAs lasers is small, as mentioned earlier. This is proved by the very small continuous tuning range of the uniform grating device in Figure 5.3.5(a). Therefore the wavelength tuning in the CGT DFB lasers could mainly be due to the chirped grating effect. The reverse linear chirp in Figure 5.3.5(b) gives rise to a net wavelength shift toward the opposite direction after canceling the shift due to the index change. This is consistent with the results shown in Figure 5.3.2. Therefore, an appropriate choice of chirp direction is essential in achieving a wide tuning range.

3 Chirping-controlled DFB Laser with Distributed Electrode

Wavelength chirping in directly-modulated DFB laser diodes is a serious problem in high-bit-rate long-haul optical transmission systems. Longitudinal spatial hole burning (LSHB) is a major cause of chirping at modulation frequencies from a few megahertz (after the thermal effect vanishes) to a few gigahertz (before relaxation oscillation takes over). In order to manage the LSHB, the DFB laser often has to be segmented into two or three parts and driven by a different current to each part. In this case, however, external circuitry for driving becomes complicated and therefore is not compatible with high-speed modulation. Here we describe a different approach to managing the LSHB and hence to controlling the wavelength chirping in DFB lasers, namely a distributed electrode (DE) structure. With this structure incorporated, the chirping can be controlled without using more than one current source.

3.1 Device structure and simulation

Figure 5.3.6 shows a schematic of the laser structure we have fabricated. The electrode is divided into small pieces with different lengths and placed at proper intervals along the laser axis. Such a distributed electrode scheme causes an effective resistance profile. Under a constant applied voltage, there arises an injection current profile in accordance with the electrode duty cycle. As a result, the carrier density distribution along the laser axis can be tailored. Another feature of the device shown in Figure 5.3.6 is the modulated stripe width (MSW) structure which gives rise to an effective index profile along the cavity to match the phase condition. This scheme is more beneficial than the quarter-wave phase-shifted structure in the sense that the fabrication is much simpler and the LSHB is less significant [10].

We carried out a numerical analysis from the view point of chirping suppression, and designed a symmetric DE structure consisting of 19 electrodes where the electrodes were densely allocated near the center, as in Figure 5.3.6. By solving the coupled-wave equations and the rate equations concurrently, we simulated chirping characteristics the results of which are depicted in Figure 5.3.7. We assumed a two-step drive current as shown in the inset. The first abrupt current step is to turn the laser on, and the lasers exhibit a damping oscillation in the wavelength. This is due to the relaxation resonance of the lasers.

The lasing wavelength shift accompanying the second gradual current step with a 3 ns rise time is the one associated with the LSHB. In the case of the MSW DFB laser without DE in Figure 5.3.7(a), one observes a substantial shift toward the shorter wavelength side (blue shift). This is consistent with the standard theory of LSHB. In contrast, in the MSW DFB laser with DE in Figure 5.3.7(b), that particular shift becomes very small. It is clear that the DE structure is effective in compensating the blue shift of the lasing wavelength due to LSHB.

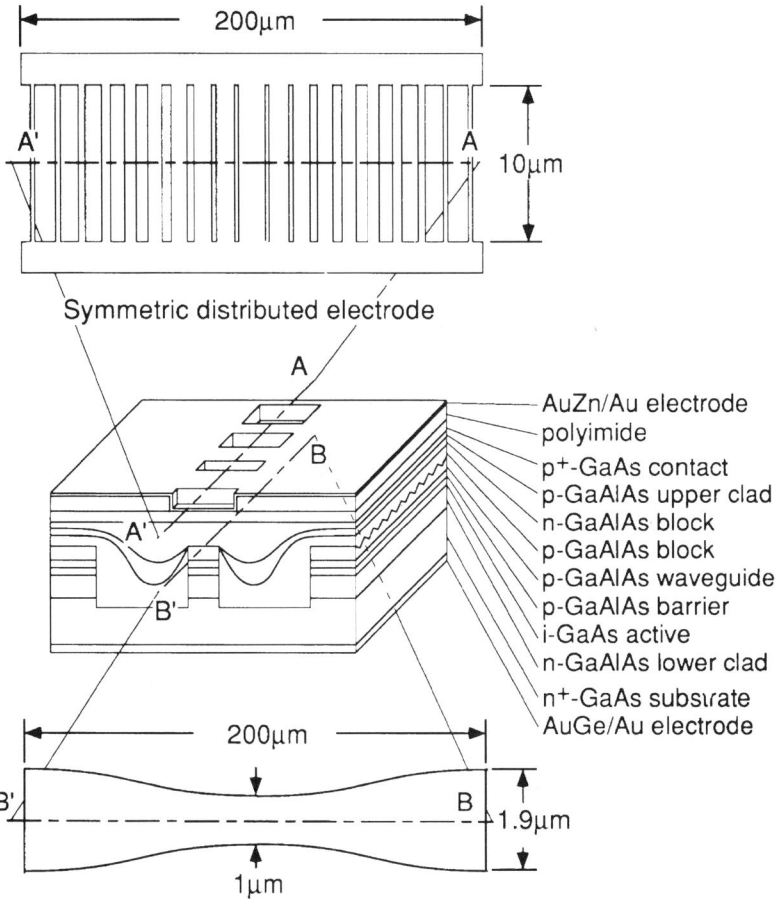

Figure 5.3.6
Schematic drawing of the GaAlAs/GaAs DCPBH DFB laser with the distributed electrode (DE) and modulated stripe width (MSW) structures

3.2 Fabrication and chirping characterization

The DCPBH structure was fabricated by two-step LPE. The DE pattern was formed on the regrown surface using photosensitive polyimide. Its 200 μm long cavity was defined by the etched tilted end facets, as described in subsection 2.2. Figure 5.3.8 is a micrograph showing the top view of the completed wafer. A unit device was obtained by cutting the wafer along the broken line.

The laser devices were driven by current pulses with a 3 ns rise time. The light output was detected by an avalanche photodiode (APD) through a monochromator. The signal

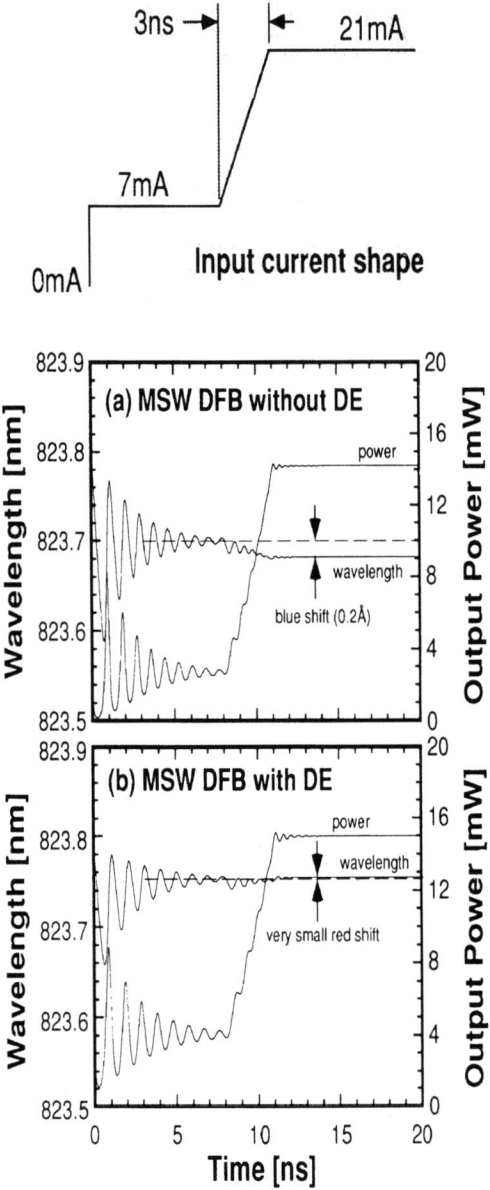

Figure 5.3.7
Simulated chirping characteristics for MSW DFB lasers without (a) and with (b) DE. The assumed input current shape is shown on the top

during the 3 ns rise time was sampled by a high-speed head and a Boxcar averager. The time evolution of the intensity was taken at different wavelengths, and finally, by exchanging the time and wavelength axes in a data processor, we obtained shifts of the lasing wavelength during the 3 ns period. MSW DFB lasers without the DE structure from the same wafer were measured for the sake of comparison.

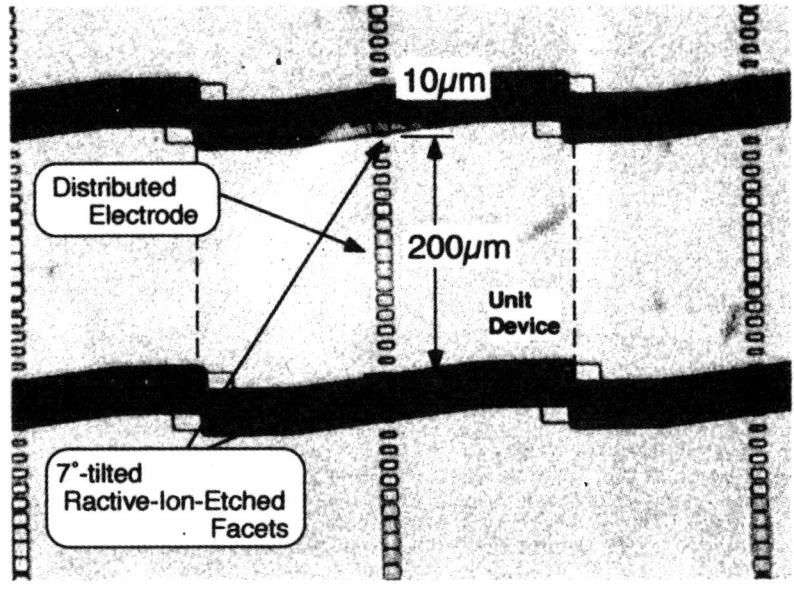

Figure 5.3.8
Micrograph showing the top view of the completed wafer. A unit device is obtained by cutting the wafer along the broken line

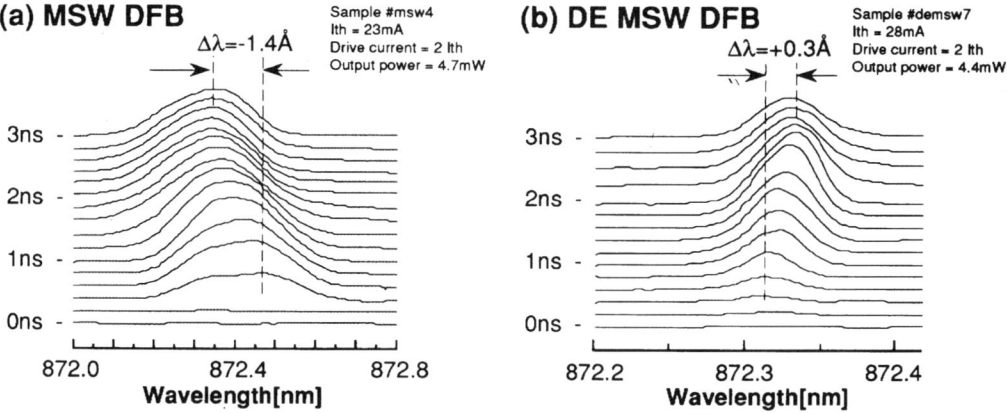

Figure 5.3.9
Measured time-resolved spectra for the MSW DFB laser (a) and for the DE MSW DFB laser (b) obtained from the same wafer. The former shows a large blue shift of the wavelength during the 3 ns rise time of the current pulse while the latter shows a very small red shift

Figure 5.3.9 shows the observed time-resolved spectra for the MSW and DE-MSW DFB lasers. The amplitude of the modulation current pulse was chosen to be twice the threshold ($2I_{th}$), which corresponds to output power of 4.7 mW and 4.4 mW, respectively. In the MSW DFB laser, the oscillation wavelength shifted by 1.4 Å toward the shorter wavelength side (blue shift). This chirping behavior is normal and agrees with the simulation. On the other hand, the DE-MSW DFB laser showed a reverse but much smaller wavelength shift toward the longer side (red shift) by 0.3Å. Obviously, the incorporation of the DE structure has compensated the blue shift by curtailing excess carrier injection that would otherwise have caused large LSHB. This DE structure could also be used to enlarge chirping for an efficient frequency modulation.

4 Conclusions

In this section we have shown two new types of semiconductor DFB lasers with extended functions which are incorporated by integrating different optical and electrical elements. One of them is a tunable DFB laser with a linearly-chirped grating. A single-mode quasi-continuous tuning range of 2.2 nm at around 868 nm wavelength has been achieved in a two-section device. This CGT DFB laser will be useful in various WDM systems due to its broader continuous tuning range and ease of fabrication.

The other is a DFB laser integrated with multiple distributed electrodes (DE) which control wavelength chirping due to longitudinal spatial hole burning by making use of tailored current density distribution. The DE DFB laser fabricated has shown much smaller chirping than the laser without DE. High frequency-modulation sensitivity may be achieved if hole burning is enhanced by the DE.

References

[1] A. Suzuki and K. Tada, *Proc. SPIE*, **239**, 10–18 (1981).
[2] P. Zhou and G. S. Lee, *Appl. Phys. Lett.*, **56** (15), 1400–1402 (1990).
[3] X. Pan, H. Olesen and B. Tromborg, *Technical Digest, Conference on Lasers and Electrooptics (CLEO)*, May 1991, CWF15, p. 252.
[4] N. Chen, A. Kitamoto, Y. Nakano and K. Tada, *Proc. SPIE*, **1979**, 347–350 (1993).
[5] N. Chen, Y. Nakano and K. Tada, *Jpn J. Appl. Phys.*, **33**, pt 1 (1B), 856–858 (1994).
[6] M. Okai, S. Sakano and N. Chinone, *Proceedings of the 15th European Conference on Optical Communication (ECOC)*, 1989, pp. 122–125.
[7] A. Suzuki and K. Tada, *Thin Solid Films*, **72**, 419–426 (1980).
[8] Y. Nakano and K. Tada, *Appl. Phys. Lett.*, **49** (18), 1145–1147 (1986).
[9] Y. Nakano, Y. Hayashi, N. Chen, Y. Sakaguchi and K. Tada, *Jpn J. Appl. Phys.*, **29** (12), L2430–L2433 (1990).
[10] Y. Nakano and K. Tada, *IEEE J. Quantum Electron.*, **24** (10), 2017–2033 (1988).

5.4

Light-emitting Devices by GaAs-on-Si Technology

Masayoshi Umeno and **Takashi Egawa**

Abstract

A room-temperature pulsed AlGaAs/GaAs vertical-cavity surface-emitting laser (VCSEL) has been grown on a Si substrate using metalorganic chemical vapor deposition (MOCVD). The VCSEL on Si exhibits a threshold current of 79 mA and a threshold current density of 4.9 kA/cm^2 under pulsed conditions at 300 K. We have also demonstrated the first successful fabrication of the monolithic integration of a reliable AlGaAs/InGaAs laser, a p–n photodetector, and GaAs metal–semiconductor field-effect transistors (MESFETs) grown on a Si substrate using selective regrowth by MOCVD. The reliability of the laser can be improved by a strain-relieved AlGaAs/InGaAs laser with an InGaAs intermediate layer.

1 Introduction

Demands for functionality and capability are increasing in future devices such as optoelectronic integrated circuits (OEICs). GaAs/Si technology has the potential to realize OEICs because the GaAs/Si technology allows the integration of GaAs and Si circuits on the same chip, which can take advantage of high-density Si technology and high-speed or optoelectronic GaAs technology. The GaAs/Si technology is particularly attractive for optical interconnections in a Si VLSI. For example, interconnections between Si VLSI subsystems at high data rates could be achieved by using GaAs-based optoelectronic interface units instead of wire interconnections [1]. However, GaAs/Si has a high dislocation density ($> 10^6$ cm^{-2}) and stress ($\sim 10^9$ dyn/cm^2) because of the different lattice parameters and thermal expansion coefficients in the two materials [2,3].

In spite of these problems key devices such as lasers, light-emitting diodes (LEDs), metal–semiconductor field-effect transistors (MESFETs), and GaAs ICs have been grown on Si substrates [4–6]. However, the rapid degradation of the GaAs-based laser is an obstacle to the fabrication of OEICs on a Si substrate because the GaAs-based laser on Si has a rapid degradation due to the formation of dark-line defects (DLDs) [7]. The formation of DLDs is caused by the high density of dislocations and the large stress in the GaAs/Si [2]. Thermal cycle annealing, strained layer superlattice, selective-area growth, and a combination of these techniques have been proposed to reduce the dislocation and stress in the GaAs/Si, but crystallinity is not still good enough for fabrication of a reliable GaAs-based laser on Si [8,9].

Recently, vertical-cavity surface-emitting lasers (VCSELs) on Si have been attracting increasing interest because of their advantages over edge-emitting lasers, such as the potential for wafer scale testing, high-density two-dimensional array fabrication, ultrafast parallel optical information processing, and the possibility of monolithic integration with other optical or electronic devices. In addition, the strain and dislocations can be reduced in the VCSEL on Si with a small active volume, which results in the fabrication of reliable lasers on Si. To our knowledge, however, there has been only one demonstration of VCSEL on Si with a pulsed threshold current (I_{th}) of 125 mA, corresponding to a threshold current density (J_{th}) of 70.8 kA/cm^2 at 300 K. In this study we demonstrate the low-threshold pulsed operation of an AlGaAs/GaAs VCSEL on Si at 300 K [10]. We also demonstrate the first fabrication of a monolithically integrated reliable AlGaAs/InGaAs laser, p–n photodetector, and GaAs MESFET grown on a Si substrate using selective regrowth by metalorganic chemical vapor deposition (MOCVD) [11,12].

2 Experimental

The samples were grown on n$^+$–Si substrates oriented 2° off (100) toward [110] using MOCVD at atmospheric pressure. The samples were grown on Si at 750 °C by the conventional two-step growth technique. Figure 5.4.1 shows a schematic cross-sectional structure of the AlGaAs/GaAs VCSEL grown on Si [10]. The structure consists of a 0.85 μm thick thermal cycle annealed n$^+$-GaAs buffer layer, 20 pairs of quarter-wave n$^+$-AlAs/n$^+$-GaAs (71 nm/59 nm) multilayer distributed Bragg reflectors (DBRs), a 0.46 μm thick lower n-Al$_{0.7}$Ga$_{0.3}$As cladding layer, a 70 nm thick lower Al$_{0.3}$Ga$_{0.7}$As confining layer, a 9 nm thick GaAs active layer, a 70 nm thick upper Al$_{0.3}$Ga$_{0.7}$As confining layer, a 0.34 μm thick upper p-Al$_{0.7}$Ga$_{0.3}$As cladding layer, and an 80 nm thick p$^+$-GaAs contact layer. Thermal cycle annealing was performed five times by varying the substrate temperature between 350 °C and 850 °C in order to reduce the threading dislocation density. As shown in Figure 5.4.1, Au–Sb/Au was used for the contact on the n$^+$-Si substrate. Nonalloyed Au–Zn/Au of 40 × 40 μm^2, which is used for the top mirror and the electrical contact, was formed by the photoresist patterning and the lift-off technique. The devices were mounted junction up on a sample holder and tested under pulsed conditions (100 ns, 1 MHz) at 300 K. Light output characteristics were measured by detecting the light emitted around the edges of the Au–Zn/Au metallization.

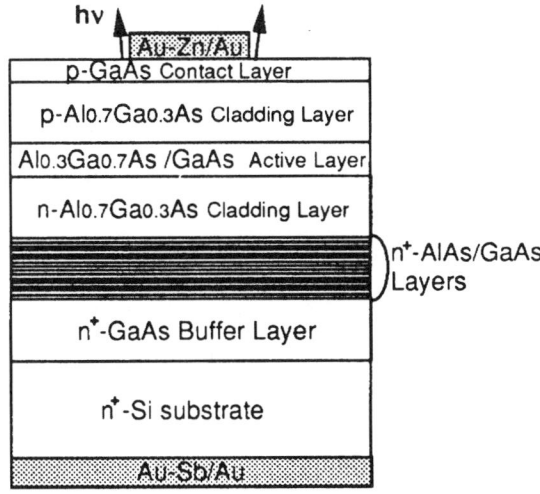

Figure 5.4.1
Schematic cross-sectional structure of the AlGaAs/GaAs VCSEL grown on Si by MOCVD

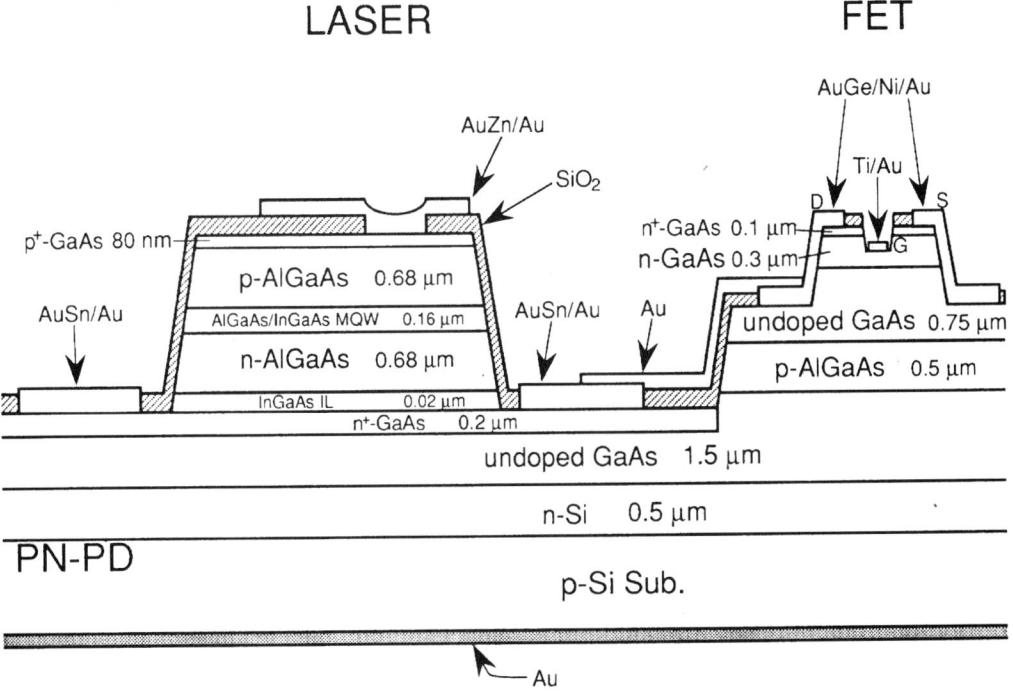

Figure 5.4.2
Schematic cross-sectional structure of the monolithically integrated laser diode, photodetector, and MESFET grown on a Si substrate

Transmission electron microscopy (TEM) was used to study the effect of AlAs/GaAs DBR on the dislocation bending. The dark spot density (DSD) was measured by the electron-beam-induced current (EBIC) method. Surface morphology observation was made in air at room temperature with a Seiko Instruments SPI 3700/SPA-300 atomic force microscope (AFM).

Figure 5.4.2 shows the schematic structure of the monolithically integrated device grown on a SiO_2 back-coated p-Si substrate at 750 °C using selective regrowth by MOCVD [11,12]. After growth of the GaAs MESFET on a p-Si substrate using the two-step growth technique, a SiO_2 film was deposited on the n^+-GaAs layer and openings were etched in the SiO_2 film to expose the undoped GaAs layer. By using a SiO_2 film as a mask, selective regrowth was performed for the AlGaAs/InGaAs MQW laser with $In_{0.08}Ga_{0.92}As$ as the intermediate layer (InGaAs IL). During the GaAs layer growth, a p–n photodetector was formed near the surface of the p-Si substrate by diffusing the As atoms.

3 Results and Discussion

Previous studies have indicated that the AlAs/GaAs DBR can be effective in bending the dislocations. Figure 5.4.3 shows a cross-sectional TEM micrograph of the VCSEL

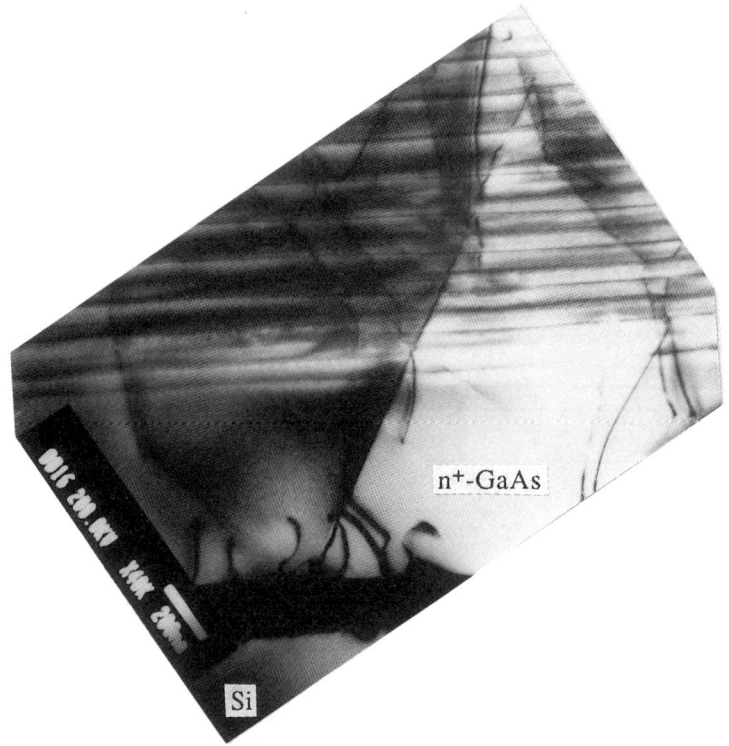

Figure 5.4.3
Cross-sectional TEM micrographs of a AlAs (71 nm)/GaAs (59 nm) DBR grown on Si

grown on Si by MOCVD. As shown in Figure 5.4.3, some dislocations, which originate at the GaAs/Si interface, can be effectively confined into the thermal cycle annealed n$^+$-GaAs layer. However, dislocation bending is not observed in the AlAs/GaAs DBR. Many dislocations originating at the GaAs/Si interface propagate into the active layer. This VCSEL grown on Si by MOCVD, which was grown under thermal cycle annealing and the AlAs/GaAs DBR, shows a DSD of 2.5×10^7 cm^{-2}. This value is almost the same as the sample without the AlAs/GaAs DBR. These results also indicate that the AlAs/GaAs DBR cannot be effective in bending the threading dislocations because there is less mismatch in the lattice constant for the AlAs/GaAs layers.

The calculated and measured reflectivities of the 20 pairs of AlAs/GaAs DBRs are shown in Figure 5.4.4. The measured reflectivity is greater than 90% at the wavelength region between 820 nm and 870 nm, and 93% at the wavelength of 860 nm. Figure 5.4.5 shows the AFM image of the surface morphology for the VCSEL grown on Si by the two-step growth technique. The microroughness consisting of hillocks and depressions of different size is observed in the surface morphology, and is caused by the three-dimensional growth at the initial growth stage, which also causes interfacial roughness in the AlAs/GaAs DBR. The peak-to-valley height is about 42 nm. The relatively lower reflectivity is probably due to the three-dimensional growth at the GaAs/Si interface. The reflectivity can be increased by the use of AlGaAs/AlGaP intermediate layers, which contribute to the specular surface morphology in the GaAs/Si [3,4]. A higher reflectivity will be also expected by increasing the number of pairs. However, a thicker epitaxial layer cannot be grown on Si due to the formation of microcracks. The top mirror has a reflectivity of 60% for the alloyed and 96% for the nonalloyed Au–Zn/Au. The higher reflectivity of the nonalloyed Au–Zn/Au is caused by a smooth morphology due to the lack of thermal annealing.

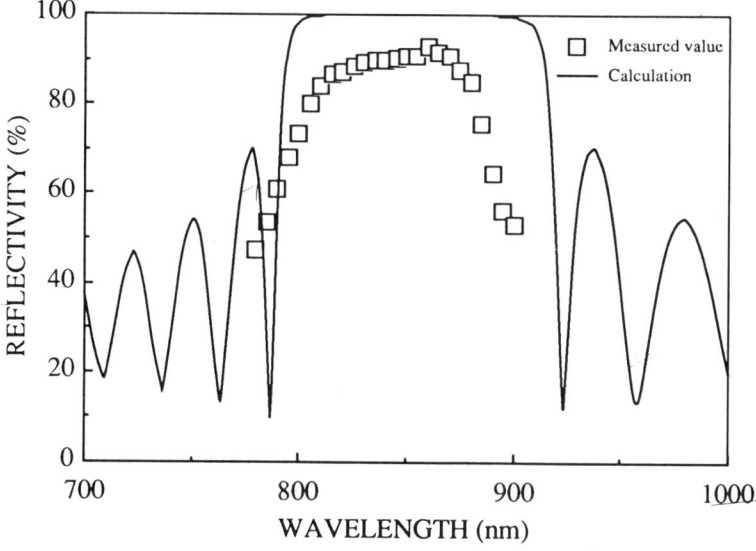

Figure 5.4.4
Calculated and measured reflectivities of the 20 pairs of AlAs (71 nm)/GaAs (59 nm) DBRs on Si

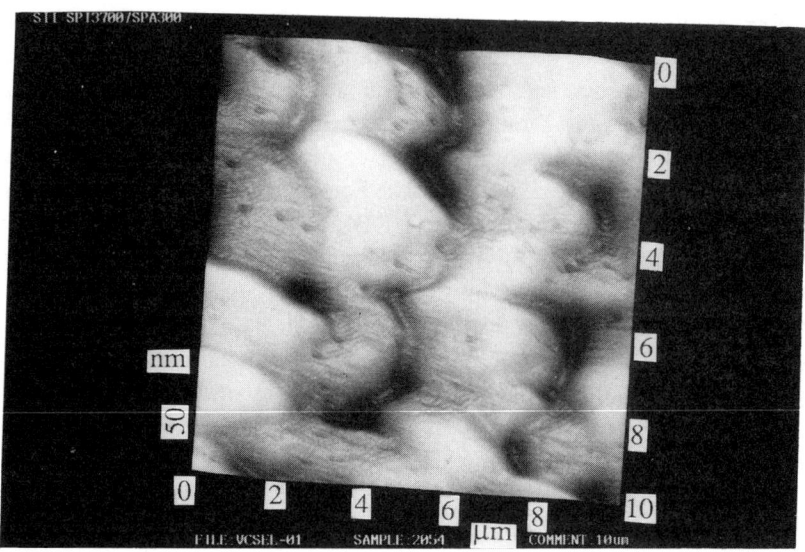

Figure 5.4.5
AFM image of the surface morphology of the VCSEL grown on Si by the two-step growth technique

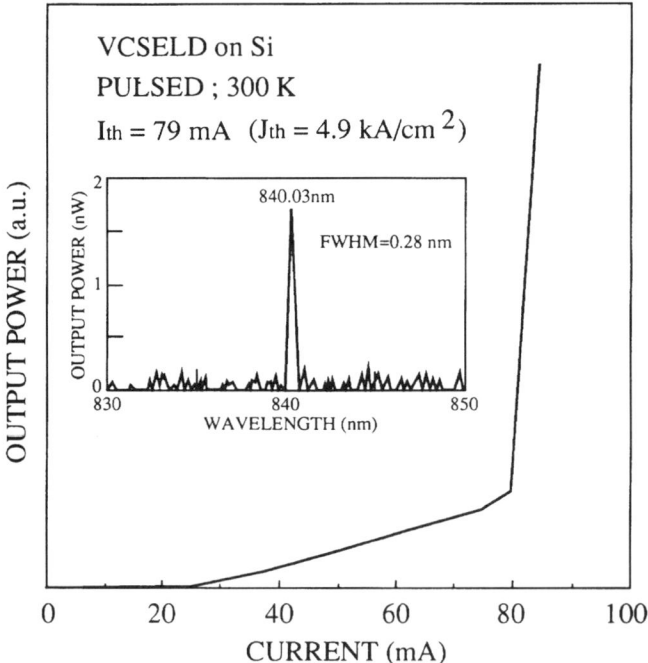

Figure 5.4.6
Room-temperature pulsed L–I characteristic and emission spectrum of the VCSEL on Si

The turn-on voltage and the series resistance are 1.3 V and 26 Ω, respectively. These values are comparable with those of the conventional edge-emitting lasers on Si. Figure 5.4.6 shows the pulsed light–current (L–I) characteristic of the VCSEL on Si at 300 K. This VCSEL exhibits a threshold current (I_{th}) of 79 mA, a threshold current density (J_{th}) of 4.9 kA/cm^2, and a lasing wavelength of 840.03 nm with the full width at half maximum (FWHM) of 0.28 nm. Compared with previously reported results, this VCSEL on Si has remarkable improvements, such as a lower threshold and a narrower lasing spectrum, which are probably due to a lower dislocation density and laser structure. A cw operation at 300 K might be expected by increasing the reflectivity of the DBR and reducing the threading dislocation density.

Figure 5.4.7 shows the results from the aging tests at 300 K. The conventional AlGaAs/GaAs laser has a rapid degradation because the laser has an etch pit density (EPD) of 2×10^7 cm^{-2} and a larger tensile stress. Although the stress in the active layer is relieved for the AlGaAs/InGaAs laser without the InGaAs IL, this laser also shows a rapid degradation because of the EPD of 2×10^7 cm^{-2}. In contrast, the AlGaAs/In$_{0.02}$Ga$_{0.98}$As laser with InGaAs IL has operated for 24 h. These results indicate that the stress relief by the InGaAs active layer and the reduction of the formation of DLDs by the InGaAs IL are required to fabricate a reliable laser on Si. The MESFET with a 2.5 μm gate length exhibits a transconductance of 90 mS/mm and a good pinch-off characteristic, which result from the use of SiO$_2$ back-coated Si substrate [11,12]. The cw L–I characteristic of the laser diode using an external

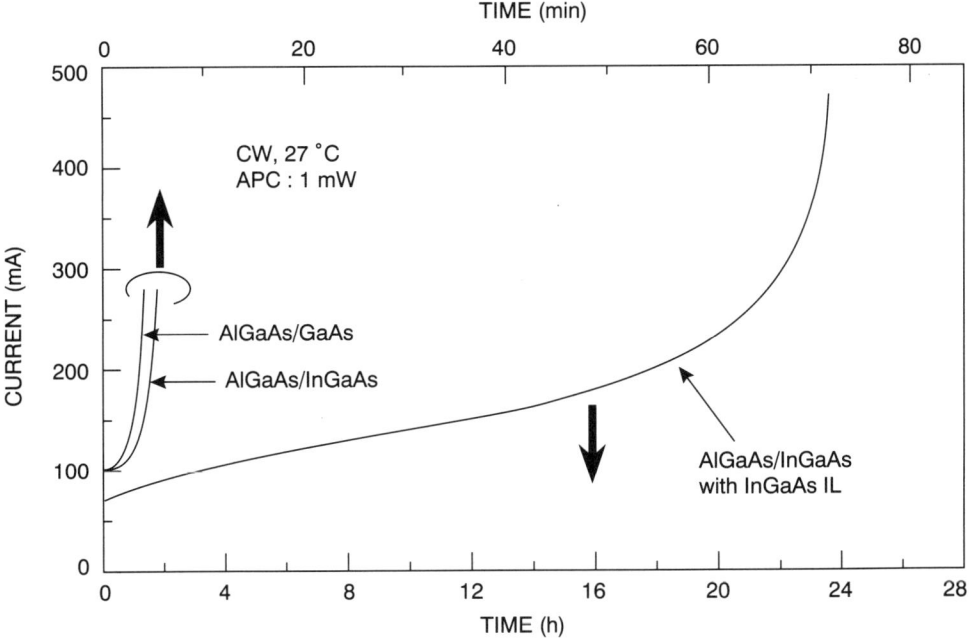

Figure 5.4.7
Results from the aging tests at 300 K

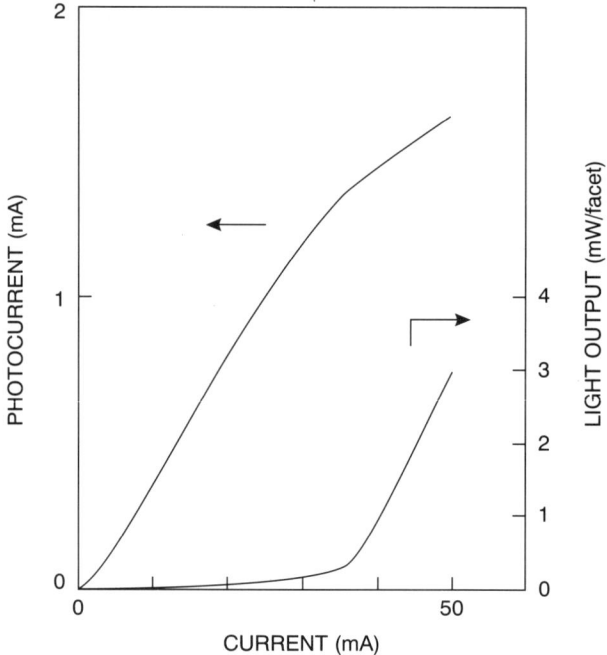

Figure 5.4.8
L–I characteristic of the laser and the photocurrent monitored by the photodetector

detector and a photocurrent monitored by the internal p–n photodetector are shown in Figure 5.4.8. The laser diode has a cw threshold current of 34 mA, which is small enough to be modulated by the MESFET. The photocurrent curve slope efficiency is 4% below the injection current of 34 mA. Above the injection current of 34 mA, the photocurrent curve slope efficiency is 1.9% because the lasing emission occurs at 34 mA.

4 Conclusions

The room-temperature pulsed operation of a AlGaAs/GaAs VCSEL with 20 pairs of AlAs/GaAs DBRs has been grown on Si using the MOCVD technique. The measured reflectivity of the 20 pairs of AlAs/GaAs DBRs was 93% at a wavelength of 860 nm. TEM observations showed that the AlAs/GaAs DBR cannot be effective in bending the threading dislocations because there is less mismatch in the lattice constant for the AlAs/GaAs layers. The threshold current and the threshold current density were 79 mA and 4.9 kA/cm^2, respectively, which are remarkable improvements over previously reported results. We have also demonstrated the first successful fabrication of the monolithic integration of a reliable AlGaAs/InGaAs laser, a p–n photodetector, and a GaAs MESFET grown on Si substrates. This new type of OEICs is promising in future applications such as optical interconnections and optical computing.

References

[1] T. Egawa, T. Jimbo and M. Umeno, *IEICE Trans. Electron.*, **E76-C**, 106–111 (1993).
[2] T. Egawa, Y. Hasegawa, T. Jimbo and M. Umeno, *Jpn J. Appl. Phys.*, **31**, 791–797 (1992).
[3] T. Egawa, T. George, T. Jimbo and M. Umeno, *IEEE Photon. Technol. Lett.*, **6**, 150–152 (1994).
[4] T. Egawa, T. Soga, T. Jimbo and M. Umeno, *IEEE J. Quantum Electron.*, **27**, 1798–1803 (1991).
[5] Y. Hasegawa, T. Egawa, T. Jimbo and M. Umeno, *Jpn J. Appl. Phys.*, **32**, L997–L999 (1993).
[6] T. Egawa, S. Nozaki, T. Soga, T. Jimbo and M. Umeno, *Appl. Phys. Lett.*, **58**, 1265–1267 (1991).
[7] T. Egawa, T. Jimbo, Y. Hasegawa and M. Umeno, *Appl. Phys. Lett.*, **64**, 1401–1403 (1994).
[8] T. Soga, T. Jimbo and M. Umeno, *Appl. Phys. Lett.*, **56**, 1433–1435 (1990).
[9] Y. Kobayashi, T. Egawa, T. Jimbo and M. Umeno, *Jpn J. Appl. Phys.*, **30**, L1781–L1783 (1991).
[10] T. Egawa, T. Jimbo and M. Umeno, *Technical Digest of International Electron Devices Meeting*, Washington, DC, 1993, pp. 597–600.
[11] T. Egawa, T. Jimbo and M. Umeno, *IEEE Photon. Technol. Lett.*, **4**, 612–614 (1992).
[12] T. Egawa, T. Jimbo and M. Umeno, *Jpn J. Appl. Phys.*, **32**, 650–653 (1993).

5.5

Organic Light-emitting Diodes with Microcavity Structures

Tetsuo Tsutsui

Abstract

Organic electroluminescent diodes with an optical microcavity structure were fabricated. The device structure consisted of a dielectric reflector composed of SiO_2/TiO_2 bilayers, an indium–tin–oxide electrode, a hole transport layer, a europium complex as an emission layer, an electron transport layer, and a MgAg electrode. The dielectric reflector and the MgAg metal electrode constituted a planer microcavity. Sharply directed emission from the europium complex was proved to be produced in organic electroluminescent diodes with an optical microcavity, when operated under a dc drive voltage.

1 Introduction

Recently much progress has been made in the research and development of organic thin-film electroluminescent (EL) diodes. Bright EL emissions, covering the entire visible spectral regions spanning from blue to red, under low dc applied voltages less than 10 V, have been reported. Maximum brightness exceeding 70 000 cd/m^2 and an external quantum efficiency of more than 4% have been attained [1]. A variety of organic materials, such as vacuum-sublimed dye films, conjugated polymers, and dye-polymer composites have been found to be useful for the fabrication of organic EL diodes [2–4]. Organic EL diodes are promising for future applications for high-quality, full-color flat panel displays. The most important issue to be resolved for display applications is a relatively short working lifetime of organic EL diodes, although continuous operations of more than 5000 h have been reported.

Progress in high-performance organic EL diodes that exhibit very high luminance and can be driven with low dc voltages have provided a promising means to develop EL diodes with optical microcavity structures. When multilayers of sublimed dye films are used for the fabrication of EL diodes, one can easily introduce an optical microcavity structure into EL diodes: sublimed-dye EL diodes are composed of two or three thin dye films, which are called the hole transport layer (HTL), an emission layer (EML) and an electron transport layer (ETL). For an anode and a cathode, a transparent indium–tin–oxide (ITO) electrode and a reflective vacuum-deposited metal film, respectively, have been used. The total thickness of the organic layers sandwiched between two electrodes is around 100 nm, which is shorter than the wavelength of visible emissive light. By replacing the ITO electrode with highly reflective mirrors, a Fabry–Perot microcavity can be introduced into usual thin-film EL diodes.

Several experimental studies on optical microcavities using organic materials have been reported so far. Kuhn studied the effect of mirrors on the emission lifetime of excited states of an europium complex placed in front of a mirror and gave clear evidence of the effect of a confined radiation field on the spontaneous emission of excited states [5]. Emission from planer microcavities made of fluorescent dye solutions or Langmuir–Blodgett films was examined by means of optical pumping [6,7]. We reported the effect of a confined radiation field on the spontaneous emission lifetime in vacuum-sublimed dye films [8]. In those reports, however, excited states of organic materials have been produced via optical pumping and no experimental work on the emission from microcavities produced by electric pumping, in other words, electroluminescent diodes with microcavity structures, has been performed, except for our preliminary study on organic multilayer electroluminescent diodes [9].

In 1993 we published one of the first reports on the emission produced by electric pumping from microcavities made of organic dye films [10]. Both spectral narrowing and a large alteration in the spacial distribution of the intensity of emitted light were actually observed in EL diodes with two reflective metal electrodes. In that paper we pointed out two major difficulties with regard to substantiation of EL diodes with sharply-directed emission: One is insufficient stability of the devices under continuous operation due to surface roughness of a vacuum-deposited silver electrode, and the other was smearing out of the microcavity effect owing to the intrinsically broad bandwidth of the emission spectrum of a dye used. The first difficulty can be overcome by the use of an ITO coated dielectric reflector in place of the Ag electrode proposed by Nakayama and his coworkers [11], and the other is perfectly solved by adopting an Eu complex which forms stable homogeneous films with vacuum sublimation as an EML material. Thus, strongly directed emission is observed in a quite simple organic EL diode under not optical but electric pumping at room temperature.

2 Design of EL Diodes with Microcavity Structure

Figure 5.5.1 shows the device structure of the EL diode with the microcavity and the molecular structures of the materials used. The device consists of a dielectric reflector/ITO layer/HTL/EML/ETL/MgAg mirror. We assume a phase shift of π when emitted light is reflected at the HTL/MgAg interface. The dielectric reflector used is

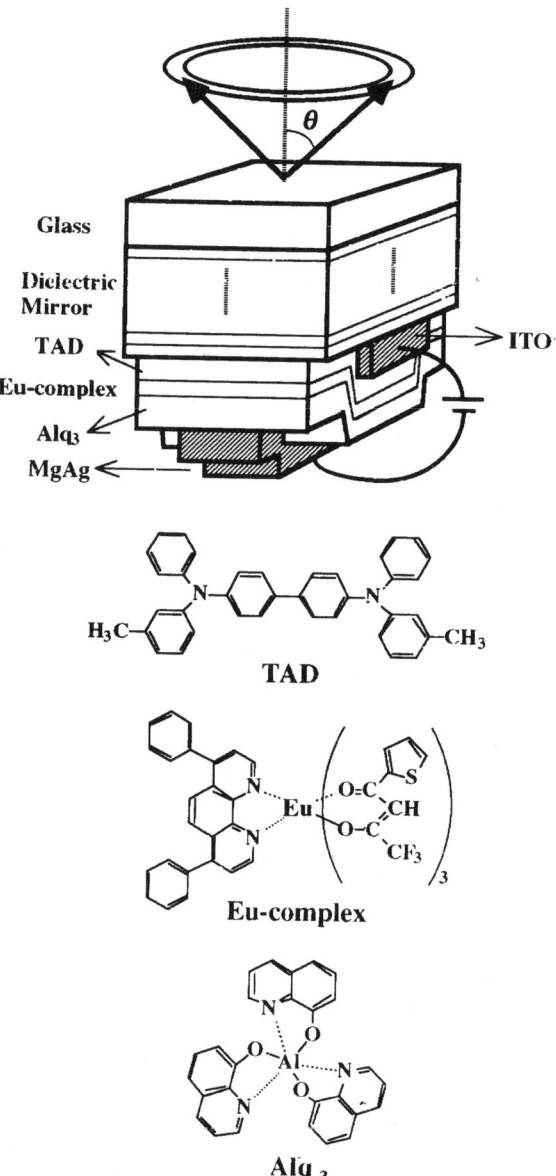

Figure 5.5.1
Molecular structures of dyes used and the device structure of the EL diode with a microcavity

a quarter-wave stack composed of four pairs of SiO_2/TiO_2 layers. For the microcavity structures, a detailed description of the reflection of emitted light at the quarter-wave stack is extremely complicated and thus we roughly estimate the phase shift at the dielectric reflector using an effective penetration depth [12].

A Fabry–Perot cavity sandwiched between the metal mirror and the dielectric reflector consists of three organic layers with thicknesses L_1, L_2 and L_3, the ITO layer with thickness L_4 with the equivalent optical path length of $\lambda/2$, and a top low-refractive-index SiO$_2$ layer of thickness L_5 with an equivalent optical path length of $\lambda/4$. Moreover, an additional optical path length due to the effective penetration depth of the dielectric stack reflector should be taken into account. This term is described as the effective optical path length n_6L_6. The resonance condition for emitted light with wavelength λ in the microcavity is expressed by equation (1)

$$\sum_{i=1}^{6} n_i L_i \cos\left\{\sin^{-1}\left(\frac{\sin\theta}{n_i}\right)\right\} = m\frac{\lambda}{2} \qquad (1)$$

where n_i and L_i represent the refractive index and the thickness of the ith layer, respectively, and θ represents the outer emission angle. m is a nonzero integer.

We now examine the simplest case, where emission is vertically directed to the device surface. The equivalent optical path lengths n_4L_4 (ITO) and n_5L_5 (SiO$_2$) are $\lambda/2$ and $\lambda/4$, respectively, and that of the effective penetration depth of the dielectric stack mirror is calculated as corresponding to the effective optical length of about λ using the refractive indices of SiO$_2$ and TiO$_2$ of 1.45 and 2.3, respectively. Thus the shortest value of the total optical path length of the three organic layers, $n_1L_1 + n_2L_2 + n_3L_3$, is determined to be $\lambda/4$, with $m = 4$. Assuming the refractive indices of all organic layers to be 1.7 and taking λ to be 617 nm, the total thickness of the organic layers, L_t, is calculated to be about 90 nm. One should note that this value only gives a rough measure for the design of the thicknesses of organic layers because the refractive indices and the thicknesses of the layers are roughly assumed values. In practice, the thicknesses of the organic layers must be carefully tuned based on experimental results. These rough estimations indicate that the $\lambda/4$ portion of the standing wave of 2λ within the effective cavity is located at the organic layers, and the node of the standing wave is not located at the thin EML placed between the organic layers.

3 Emission Characteristics of EL Diodes with Microcavity

EL diodes with a microcavity structure were fabricated on a glass substrate with a dielectric reflector coated with an ITO conductive layer (158 nm). The reflectance of the substrate at around 617 nm was 85%. On a substrate with a patterned ITO electrode, a TAD layer for HTL, a 12 nm thick Eu complex layer for EML, an Alq$_3$ layer for ETL, and a top MgAg alloy cathode (150 nm) were formed. The thicknesses of the HTL and ETL varied from 35 nm to 65 nm.

The EL diodes were driven by dc voltage as large as 15 V, and sharp red emission was observed. The intensity of emitted light was proportional to the input current. Observed emission from the diodes was sharply directed along fixed outer emission angles which form cone surfaces. Figure 5.5.2 shows the emission pattern of the device with $L_t = 142$ nm. In the device with $L_t = 142$ nm, the emission was only observable at a specific viewing angle of about 40° off the direction normal to the device surface, and no emission was observed at different viewing angles, indicating that the emission is strongly directed along the cone surface. The angle at a maximum emission intensity θ_{max} measured from the direction normal to an emission surface decreased with an decrease

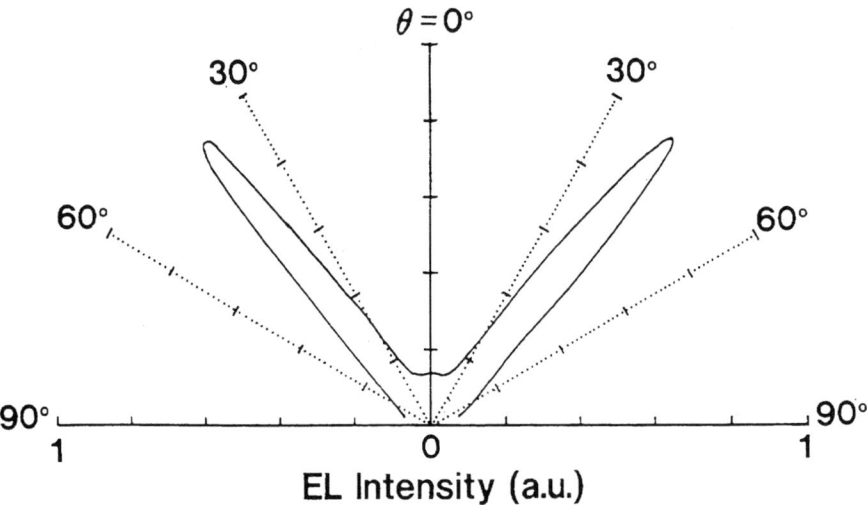

Figure 5.5.2
Angular dependence of intensity of emitted light in the EL diode with a microcavity ($L_t = 142$ nm). θ is the outer emission angle

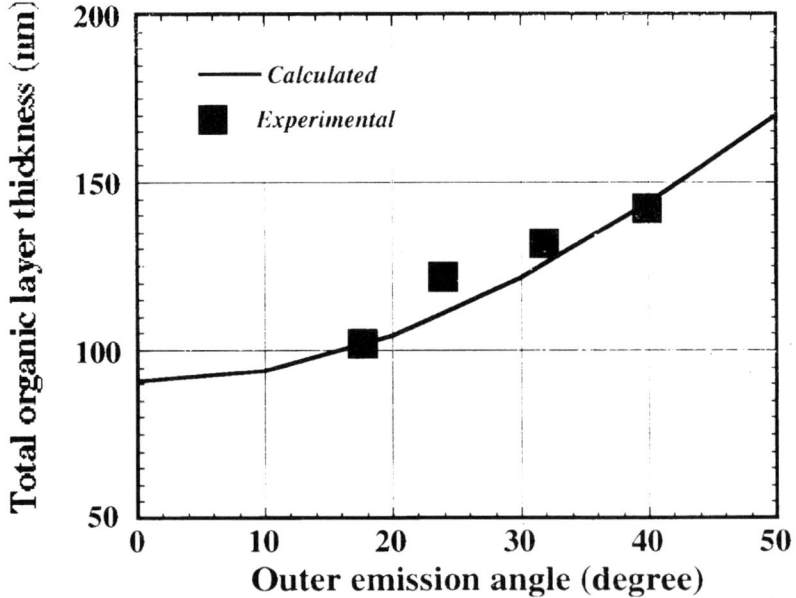

Figure 5.5.3
Relationship between the total organic layer thickness L_t and the outer emission angle θ. The solid line represents the curve calculated using equation (1)

in L_t. Figure 5.5.3 shows the relationship between the total thickness of the organic layer L_t and the outer emission angle θ. The filled squares and the solid line in the figure show the experimentally obtained values and the curve calculated using equation (1), respectively. The agreement between the experimental values and the calculated ones is fairly good, considering that the estimations of the effective optical path lengths are very rough. Thus this confirms the validity of our microcavity design concept.

On the basis of the relationship between the thickness of the organic layers and the emission angle, we succeeded in designing and fabricating a device with the

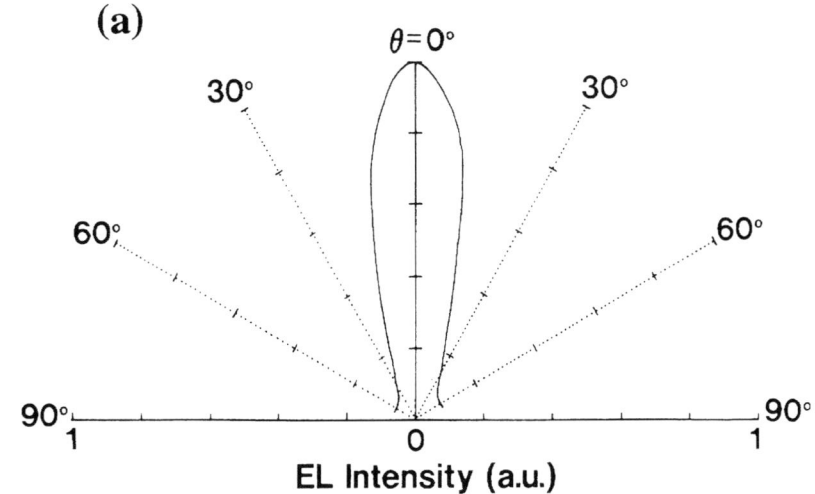

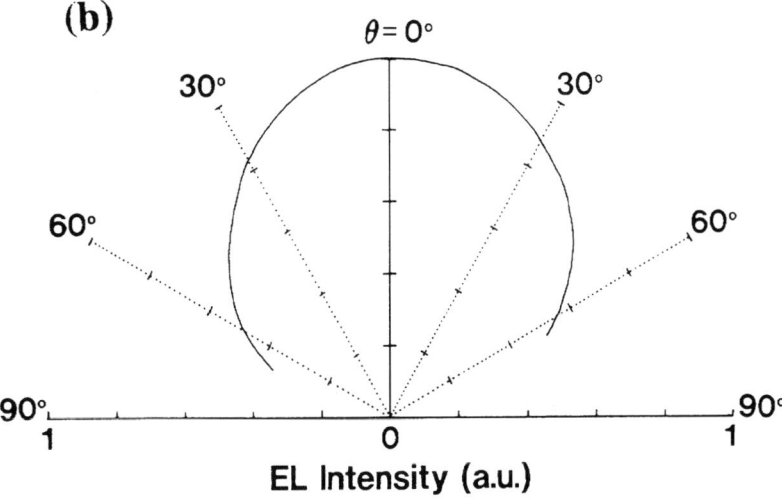

Figure 5.5.4
Angular dependences of intensity of emitted light in the EL diodes (a) with and (b) without a microcavity. The total organic layer thickness L_t is 87 nm. θ is the outer emission angle

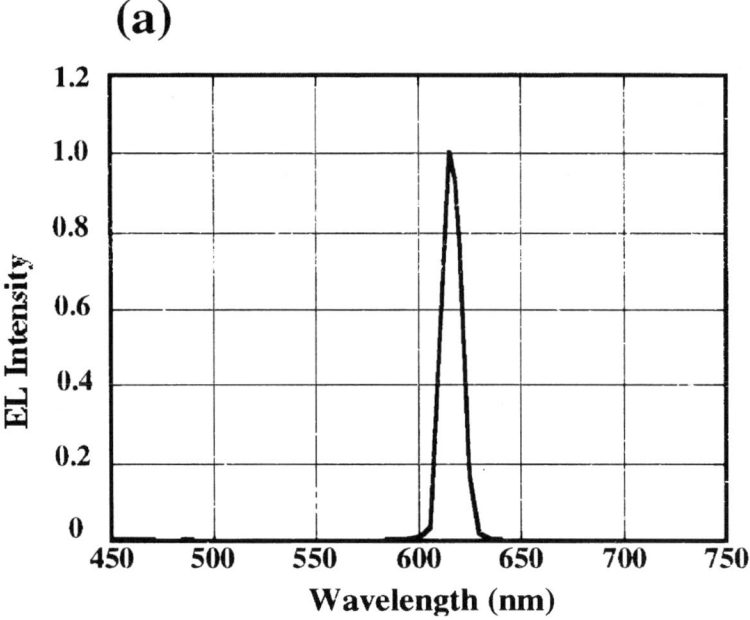

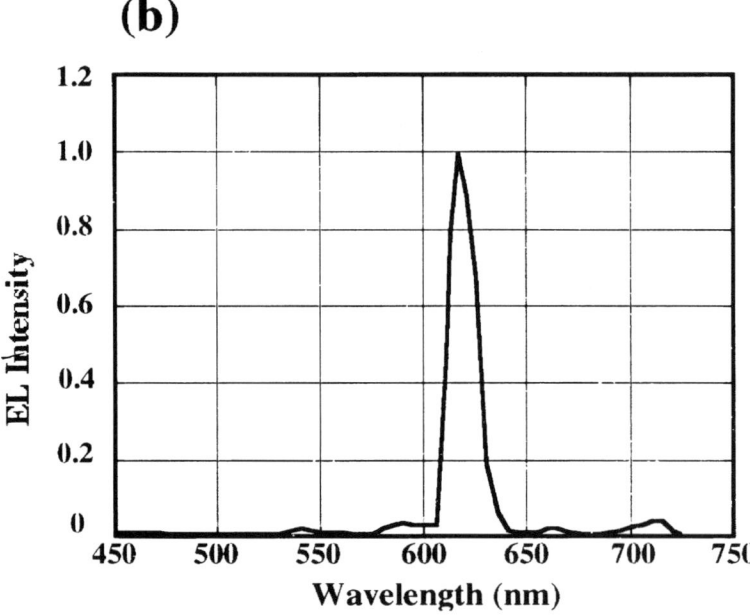

Figure 5.5.5
Emission spectra of EL diodes (a) with and (b) without a microcavity taken at the direction perpendicular to the device surface. The total organic layer thickness L_t is 87 nm

sharply-directed emission vertical to the emission surface. Figure 5.5.4 shows the emission pattern of the microcavity device with $L_t = 87$ nm. For comparison, the emission pattern of the conventional device without a microcavity with the same organic layer thickness is shown. The emission sharply directed vertical to the device surface is observed in the device with a microcavity. In contrast, the emission with a uniform spacial distribution nearly Lambertian is observed in the conventional device without a microcavity structure.

Figure 5.5.5 compares the EL emission spectra between the devices with and without a microcavity structure. The EL spectrum from the device without a cavity contained several small peaks characteristic of the emission from an Eu ion in free space. These small peaks completely disappeared in the device with microcavity. This is more clear evidence that the microcavity structure is surely effective in attaining a single resonance in the cavity.

The sharp spacial distribution of emission patterns in the EL devices with a microcavity is similar to those in laser diodes. The question arises whether the emission we observed is due to spontaneous emission or stimulated emission. It is well known that the transition from spontaneous emission to stimulated emission in microcavities is not clearly determined, when the excitation energy is increased. Detailed analyses of device characteristics, such as input–output energy relations, emission spectra, and time responses, are needed. Thus the question remains open for future investigations.

4 Conclusions

By incorporating a simple planer optical microcavity into high-efficiency organic electroluminescent diodes, we have succeeded in fabricating unique light-emitting devices with strongly directed emission. In other words, we have shown one of the promising ways to realize controlled-spontaneous-emission diodes which are expected to be a kind of novel emission devices that exhibit emission characteristics similar to laser diodes. We should stress that the device structure is quite simple and easy to fabricate. The sizes of the devices can be minimized; for example, the order of tens of micrometers is not so difficult to fabricate and are easy to integrate on a substrate. Thus these devices are promising for a variety of applications such as light sources for signal processing, holography, optical interconnections, and so on.

References

[1] S. Saito, T. Tsutsui, M. Era, N. Takada, C. Adachi, Y. Hamada and T. Wakimoto, *SPIE Proc.*, **1910**, 212 (1993).
[2] T. Tsutsui and S. Saito, in *Intrinsically Conducting Polymers: An Emerging Technology*, M. Aldissi, ed., Kluwer Academic Publishers, 1993, pp. 123.
[3] J.H. Burroughes, D.D.C. Bradley, A.R. Brown, R.N. Marks K. Mackay, R.H. Friend P.L. Burns and A.B. Holmes, *Nature*, **347**, 539 (1990).
[4] J. Kido, M. Kohda, K. Okuyama and K. Nagai, *Appl. Phys. Lett.*, **61**, 761 (1992).
[5] H. Kuhn, *J. Chem. Phys.*, **53**, 101 (1970).

[6] F. De Martini, G. Innocenti, G.R. Jacobivitz and P. Mataloni, *Phys. Rev. Lett.*, **59**, 2955 (1987).
[7] M. Suzuki, H. Yokoyama, S.D. Brosen and E.P. Ippen, *Appl. Phys. Lett.*, **58**, 998 (1991).
[8] T. Tsutsui, C. Adachi, S. Saito, M. Watanabe and M. Koishi, *Chem. Phys. Lett.*, **182**, 143 (1991).
[9] T. Tsutsui, C. Adachi and S. Saito, in *Photochemical Processes in Organized Molecular Systems*, K. Honda ed., Elsevior Science Publishers B.V., 1991, pp. 437.
[10] N. Takada, T. Tsutsui and S. Saito, *Appl. Phys. Lett.*, **63**, 2032 (1993).
[11] T. Nakayama, Y. Itoh and A. Kakuta, *Appl. Phys. Lett.*, **63**, 594 (1993).
[12] N.E. Hunt, E.F. Schbert, R.A. Logan and G.J. Zydzik, *Appl. Phys. Lett.*, **61**, 2287 (1992).

6
HIGH SPEED DEVICE TECHNOLOGIES

6.1

Quantum Wire and Quantum Box Lasers: GaAs Systems

Yasuhiko Arakawa

Abstract

Various issues concerning quantum wire and quantum box lasers are discussed. First, we theoretically discuss the fundamental properties of quantum wire and box lasers together with band structures. Then we show recent progress in the fabrication technique for quantum wires and boxes. Finally, we discuss fabrication of quantum wire lasers with a microcavity in which the interaction of 1D electrons with 2D photons is realized.

1 Introduction

With the progress in semiconductor microfabrication technology, quantum nanostructures have received great attention, impacting not only on electronic devices, but also on optical devices. Particularly, quantum well lasers utilizing 2D electron gas, and optical modulators with quantum confinement effect are now used in practice. Moreover, optical devices with quantum wires and boxes are now widely investigated, focusing on application to lasers and infrared detectors.

Since the first fabrication of quantum well lasers in 1975 by molecular beam epitaxy (MBE) [1], much effort has been devoted to clarifying the physics in quantum well lasers, including the suppressed temperature dependence of the threshold current and possible lasing oscillation at a photon energy lower than the bandgap by a LO phonon [2]. It was in 1982 that quantum well lasers really received great attention by industry, when a threshold current as low as 0.25 kA/cm^2 was achieved in Bell Labs [3]. This triggered the intensive research on quantum well lasers at many research places in universities and industries. In addition, it was also predicted that dynamic behaviors, which are sometimes more important in practical optical communication systems, are significantly enhanced with quantum effects [4]. A threshold current as low as 0.2 mA, a modulation bandwidth as wide as 30 GHz, and a spectral linewidth as narrow as 20 kHz have been so far obtained, which demonstrates the superior nature of quantum well lasers.

On the other hand, quantum wire and box lasers were proposed in 1982, predicting a significant reduction in the temperature dependance of the threshold current [5]. In addition, quantum wire effects in semiconductor lasers were demonstrated using high magnetic fields where the electrons are confined by the Lorentzian force. Since then, detailed analysis of the threshold current and the prediction of enhanced dynamic properties have been investigated theoretically [6]. In parallel, issues on the carrier injection and nonlinear gain effect were also discussed [7].

In this section, we first theoretically discuss the fundamental properties of quantum wire and box lasers together with band structures. Then we show recent progress in the fabrication technique for nano-structures. Finally, we also discuss the fabrication of quantum wire lasers with a microcavity in which the interaction of 1D electrons with 2D photons is realized.

In sections 2 and 3 we describe the fundamental properties of the quantum box lasers based on a simple laser model. In section 4, several issues related to quantum box lasers are briefly discussed. In section 5 we discuss strain effects in quantum wire lasers, including band structure and lasing properties and, in section 6 and 7, recent progress in the fabrication technique of the quantum wires by MOCVD selective growth and their optical properties. On the basis of these results, in section 8 we show the fabrication of a microcavity quantum wire laser as the first step toward fully confined photonic and electronic systems. Finally, in section 9, the fabrication of quantum box structures is also discussed.

2 Lasing Properties of Quantum Box Lasers [8]

With the reduction in dimensionality of the electron motion in quantum wells, quantum wires, and quantum box structures, the density of states changes from a parabolic function to a step-like function, a reciprocal of the square-root function, and a delta function, respectively, which leads to a narrower density of states.

In quantum boxes, the density of states (per unit energy and box) of electrons is given by

$$\rho_C(E) = 2 \sum_{k,m,n}^{\infty} \delta(E - E_k - E_m - E_n) \tag{1}$$

where E_k, E_m, and E_n are the quantized energy levels of the quantum boxes in x, y, and z directions, respectively. If we assume a structure with cubic GaAs/AlGaAs quantum boxes with a dimension of L_z, the energy levels $\{E_i\}$, $i = k, m, n$, are simply expressed by the following equations, if the barrier height in the potential potential profile is assumed to be infinite:

$$E_i = \frac{h^2}{2m_{C,i}} \left(\frac{\pi}{L_z}\right)^2 \tag{2}$$

where $m_{c,i}$ is the effective mass of electrons along the i-axis. When higher subband energy states in both conduction bands are ignored, the electronic state is equivalent to that of two-level atomic systems. Therefore, the basic properties of quantum box lasers should be quite similar to those of atomic lasers. For more detailed discussions, the band mixing effects must be included to study the quantized states in quantum boxes.

In real devices, distribution of carriers in all quantum boxes is important. If the quantum boxes are completely isolated, the uniform distribution of carriers in the quantum boxes is difficult. Therefore, in order to realize a highly uniform distribution of carriers, the tunneling effect between each quantum box should be utilized, while maintaining the quantum box effects. For this purpose, the barrier thickness must be designed carefully so that the tunneling time, t_T, satisfies the following relations:

$$\tau_r \gg \tau_T \gg \tau_{in} \qquad (3)$$

where τ_r and τ_{in} are the recombination lifetime and the intraband relaxation time, respectively. The first unequal relationship implies that the tunneling time should be short compared with the recombination time so that carrier injection occurs sufficiently before carrier recombination. On the other hand, the second unequal relationship implies that the tunneling time should be long enough so that the broadening effect in the gain profile due to this tunneling effect can be negligibly small compared with the intraband relaxation effect. The issue related to the bottleneck problem will be discussed in section 4.

Since the density of states has sharp structures, the gain profile $g(E)$ of semiconductor lasers having quantum box structures is extremely narrow. The expression of $g(E)$ for the quantum box laser is given by the following equation if Lorentzian energy broadening due to the intraband relaxation is assumed:

$$g(E) = 2 \sum_{k,m,n}^{\infty} A \frac{h/t_{in}}{(h/t_{in})^2 + (E - E_k - E_m - E_n)^2} \qquad (4)$$

where A is a constant proportional to the matrix element. If higher subbands can be ignored, the gain profile of this laser is very narrow. Therefore, the importance of Fermi statistics in shaping the gain profile is reduced in quantum box structures. This property results in an extremely low threshold current and an extremely wide modulation bandwidth. Note that the dipole matrix element of the electron–hole pairs in the quantum boxes is equal to that in double heterostructures since both structures are isotropic structures.

Figure 6.1.1 shows the modal gain (i.e. the bulk gain multiplied by the optical confinement factor Γ), plotted as a function of the injected current for a various number of quantum boxes. In the calculation it is assumed that the quantum box structure is made of 5 nm wide active GaAs cubes, sandwiched between AlGaAs barriers of 5 nm. Since higher subband effects are almost suppressed in quantum box structures with 5 nm, the lasing characteristics in this laser exhibit achievable limits. In fact, the lasing characteristics of quantum box lasers with larger dimensions are inferior to those with 5 nm dimensions. The cavity length of the laser is 300 μm, and as a related problem, the relaxation of carriers from barrier regions into the quantum boxes should also be noticed. This issue will be discussed later. τ_{in} is 0.1 ps. It is assumed that all carriers

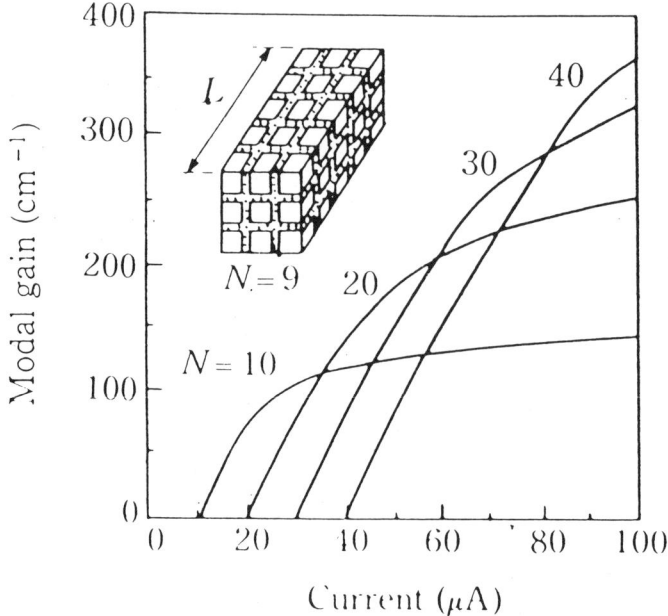

Figure 6.1.1
The modal gain at the lasing wavelength plotted as a function of the injected current for various numbers of quantum dots. The dimension L_z of the cubic quantum dot is 5 nm

are injected into the active region, ignoring nonradiative effects such as carrier leakage. For simplicity, in the direction of the optical cavity, the active layer is filled with quantum boxes sandwiched between the barriers. Therefore, the controllable parameter concerning the total number of quantum boxes in this case is the number of the quantum box N in a plane which is perpendicular to the direction of the cavity length. It is also assumed that the laser has a separate confinement structure which has optical confinement regions surrounding the quantum box active region, so that Γ can be assumed to be proportional to N. Since the gain saturates with an increase in the injected current, as shown in Figure 6.1.1, there exists an optimum N for minimizing the threshold current. If, for instance, the modal gain is 100 cm^{-1}, the optimum N is about 10, which results in a threshold current of about 25 μA. Note that the value of the threshold current strongly depends on the intraband relaxation time and the configuration of the broadening function. If the broadening function is a Gaussian function, the gain value is enhanced, leading to a lower threshold current.

Figure 6.1.2 shows the threshold current plotted as a function of the cavity length for various reflection coefficients R on the basis of the gain properties shown in Figure 6.1.1. As shown in the figure, there exists an optimum cavity length to minimize the threshold current for each R. The result indicates that an extremely low threshold current of sub-μA can be obtained with $R = 0.9$. This value is very small compared with quantum well lasers in which the expected lowest threshold current is about 100 μA. As a related problem, the relaxation of carriers from the barrier regions into the quantum boxes should be also noticed. This issue will be discussed later.

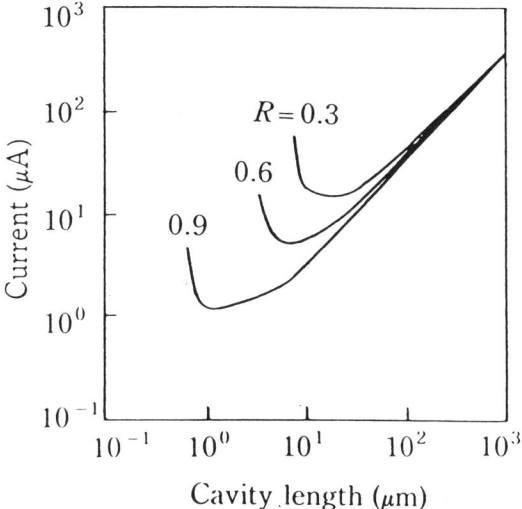

Figure 6.1.2
Threshold current plotted as a function of the cavity length for various mirror reflectivities R

3 Dynamic Properties of Quantum Box Lasers [8]

Modulation characteristics are very important for practical application of semiconductor lasers to high-speed optical communication systems. The important parameter for the modulation bandwidth is the relaxation resonant frequency f_r, which can be derived by conventional rate equations. The result is

$$f_r = \sqrt{\frac{g' P_0}{\tau_p}} \qquad (5)$$

where, P_0, τ_p, and g' are the stationary photon density in the cavity, the photon life time, and the differential gain (i.e. $g' = \partial g/\partial n$, where n is the carrier density), respectively. Equation (5) indicates that an increase in g', or P_0, and a decrease in τ_p leads to the enhancement of f_r. In the quantum box laser, we can expect g' to be much enhanced compared with double heterostructure lasers since the gain profile is extremely narrow. Therefore, use of the quantum box structure in semiconductor lasers should lead to ultra-high-speed modulation.

Figure 6.1.3 shows g' of the quantum box laser plotted as a function of the quasi-Fermi energy level e_{fc}, which is measured on the basis of the quasi-Fermi energy level ε_{trans} at which the transparent condition is satisfied. For comparison, g' values of the quantum well laser and the quantum wire laser are also plotted. The results indicate two important features. The first is that g' is enhanced in quantum box lasers compared with quantum wire lasers by a factor of three and quantum well lasers by a factor of ten, respectively, if low quasi-Fermi energy level is chosen by increasing the number of quantum boxes. In this case, the maximum value of g' in a quantum box laser reaches

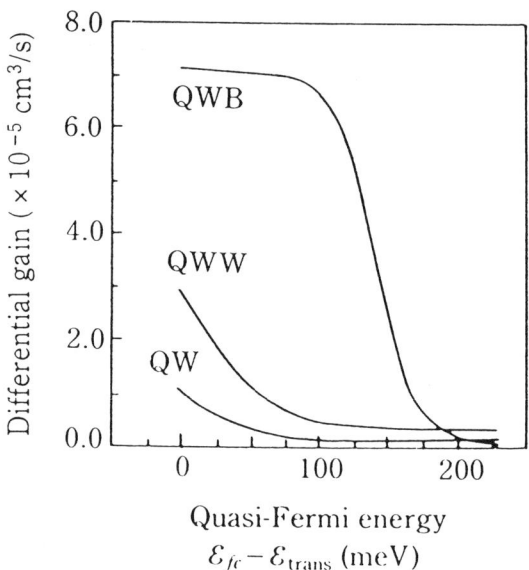

Figure 6.1.3
Differential gain of quantum-well, quantum-well wire, and quantum-well box lasers plotted as a function of the Fermi energy level of the conduction band

over 7×10^{-5} cm^3/s. The second feature is that g' is strongly dependent on the quasi-Fermi energy levels, which is more pronounced with a decrease in the dimensionality of electron motion. Due to this fact, a wide modulation bandwidth can be achieved by increasing N, since the quasi-Fermi energy is lower at the lasing condition with large N. As a result, the modulation bandwidth f_r of the quantum box laser is about three times as high as that of the quantum well laser and six times as high as that of a double heterostructure laser.

The spectral properties of semiconductor lasers with quantum boxes are also improved. It is now widely recognized that in semiconductor lasers the spectral linewidth $\Delta\nu$ is enhanced by a factor of $(1 + \alpha^2)$ compared with the Schawlow–Townes spectral linewidth $\Delta\nu_{ST}$ [7]:

$$\Delta\nu = \Delta\nu ST(1 + \alpha^2),$$

$$\Delta\nu ST = \frac{V^2 gh\nu \Gamma g R_m n_{sp}}{p P_0}, \qquad (6)$$

$$\alpha = \frac{\partial \chi_R/\partial n}{\partial \chi_I/\partial n},$$

where V_g, $h\nu$, Γ, g, R_m, P_0, and n_{sp} are the group velocity, photon energy, optical confinement factor, bulk gain coefficient, mirror loss, output power intensity, and spontaneous emission factor, respectively. χR and χI are the real and imaginary part of the complex susceptibility, respectively, and n is the carrier density. α is called the linewidth enhancement factor. From the above equations the reduction in $|\alpha|$ leads to a narrower spectral linewidth. Therefore, it is important to decrease $|\alpha|$ in order to achieve pure spectral properties in semiconductor lasers. In an atomic laser system α is almost zero since the photon energy E_0 at which α becomes zero is tuned to the photon energy E_{max} at which the gain is maximum. On the other hand, E_0 is usually detuned from E_{max} in the bulk quantum well lasers, which leads to an increase in α. In fact, the value of α has been measured in the range between 2.2 and 7.0 in double heterostructure lasers. However, in quantum box lasers in which the electronic state is similar to that of gas lasers, α is expected to become zero. This is a big advantage not only for laser devices but also for optical modulators with low chirping effects.

4 Problems Inherent in Quantum Box Lasers [9–13]

There have been some reports on the fabrication of quantum wire structures and quantum boxes structures. However, at present, fabrication still includes difficult problems to solve. The first one is to fabricate 'good' structures with small dimensions. In fact, structures smaller than 300 Å are very difficult to fabricate at present. In addition, even if this problem is overcome, the size fluctuation problem still exists. With a decrease in cubic dimensions, this factor becomes more important. The size fluctuation is equivalent to an inhomogeneous broadening of the gain profile, which reduces the high gain effects. In fact, the size fluctuation leads to energy broadening of the quantizing level.

Figure 6.1.4 shows g' plotted as a function of $x = \Delta Lz/Lz$ for $L_z = 5$ nm, where L_z is the average dimension of the cubic quantum boxes and Lz is the standard deviation of L_z due to the size fluctuation. In this calculation the size fluctuation effects are included

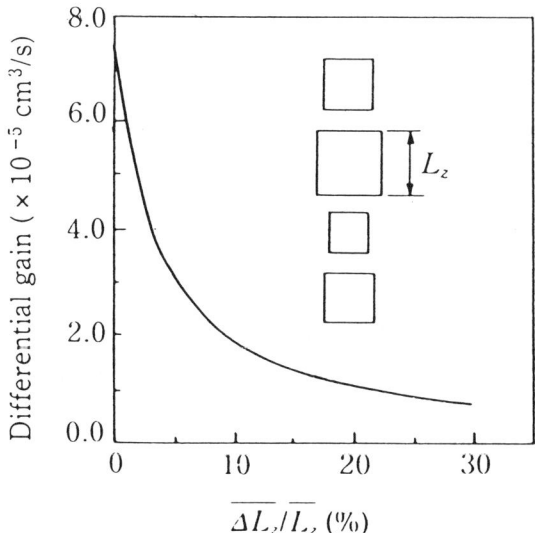

Figure 6.1.4
Effect of size fluctuation on the differential gain. The differential gain is plotted as a function of the standard deviation divided by the quantum dot dimension (5nm)

by considering a Lorentzian broadening in the gain profile due to both the intraband relaxation time and the size fluctuation. The result indicates that g' decreases drastically with an increase in x. In this calculation we assume all three cubic dimensions to be simultaneously fluctuated. If the fluctuation occurs in only one particular direction, a variation larger by a factor of three is allowed to achieve the same g'.

With an increase in optical power, the nonlinear gain effect becomes dominant, which results in degradation of the modulation dynamic as well as of the spectral characteristics. We predicted theoretically that in quantum well lasers this nonlinear gain effect is enhanced, which was also evidenced experimentally. This is due to the fact that the inhomogeneous broadening property is suppressed with the quantum confinement of electrons. Therefore, it is expected that in quantum box lasers in which the inhomogeneous broadening effect is most suppressed the gain nonlinearity is most enhanced.

For simplicity of our discussion, a single-mode lasing condition is considered. Therefore, only a symmetric nonlinear gain is taken into account. In this case, the nonlinear gain $g(e, n, I_{in})$ of the semiconductor lasers at the photon energy E can be expressed as follows:

$$g(E, n, I_{in}) = g_l(E, n) - \varepsilon(E, n)I_{in} \tag{7}$$

where $g_l(E, n)$, n, and I_{in} are the linear gain, carrier density, and light intensity inside the cavity (W/cm^2), respectively; $\varepsilon(E, n)$ is the nonlinear coefficient which can be derived using the density matrix formalism for the third-order perturbation theory.

Since $\varepsilon(E, n)$ includes the density of states, the nonlinear gain should be strongly affected by the dimensionality of the electron motion in the active layer. The calculation result shows that $\varepsilon(E, n)$ decreases with an increase in the dimensionality of the quantum confinement, where the saturation power density is defined as the power density at which the gain is half the linear gain. These results indicate that the nonlinear gain effect is enhanced in quantum box lasers.

Recently carrier relaxation phenomena from the barrier region into the quantum boxes have been discussed, predicting a significant reduction in the carrier relaxation rate due to increasing boxes. In a fully quantized system, the transition of electrons to lower energy levels can occur only when the phonon energy is resonant with the energy difference between the two energy levels. Benisty *et al.* [11] discussed this issue and attributed the decreased photoluminescence (PL) intensity with a reduction in the lateral size of the quantum boxes to this bottleneck. Moreover, they predicted that this bottleneck leads to a drastic increase in the threshold current in quantum box lasers. Although the simple model exhibits serious problems for quantum box lasers, the question whether the bottleneck problem really exists should be clarified by experiments. As discussed later, even from quantum boxes with a lateral size of 25 nm, PL can be observed. Moreover, even if the bottleneck exits, there are several ways to avoid this bottleneck in quantum box lasers by carefully designing the quantum box arrays, as discussed below.

To eliminate the bottleneck, two structures were discussed, as shown in Figures 6.1.5(a) and (b) [12]. In Figure 6.1.5(a) the quantum boxes are coupled to each other so that the carrier relaxes from the excited mini-band state to the ground level of

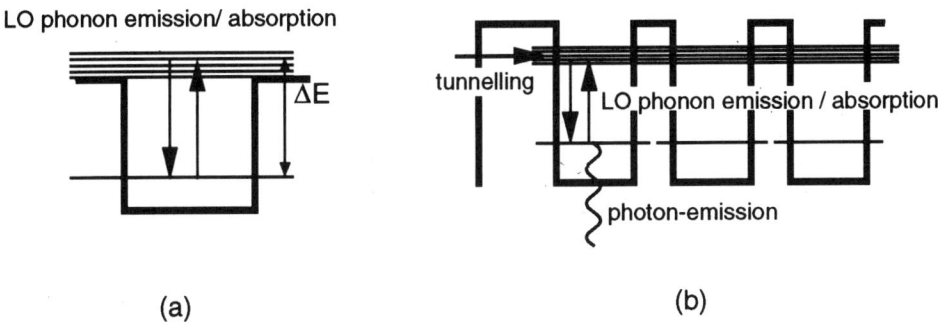

Figure 6.1.5
Examples of the structures where the bottleneck problem can be reduced

the quantum boxes by LO-phonon emission. On the other hand, in the second structure (Figure 6.1.5(b)), the ground energy level of the quantum boxes is sufficiently close to the potential barrier to realize direct carrier relaxation from the barrier to the ground level of the quantum boxes. Here we focus on the structure in Figure 6.1.5(b). We calculated the capture rate into the quantum boxes plotted as a function of ΔE. In this calculation the energy dispersion of the LO-phonon and the wavefunctions of the initial and final states are considered. Using this result, the threshold current of quantum box lasers is discussed [13]. We assumed that the carrier escape time from the ground state is determined by absorption of the LO phonon and is equal to the dephasing time of the optical dipole moment. We also assumed a Lorentzian-type lineshape function of the gain and ignored inhomogeneous broadening. The laser structure is a vertical cavity with a mirror reflectivity of 0.99. The rate equations are solved in the system where the energy level inside the quantum box and miniband levels in the barrier region are formed, as indicated. The calculation shows the threshold current and the dephasing time as a function of the size of the quantum boxes. The result shows that a sub-μA threshold current can be obtained if the size of the quantum boxes is appropriately chosen. The total number of the quantum boxes is optimized so that the threshold current is a minimum.

5 Band Structures of InGaAs Strained Quantum Wires and Lasing Characteristics [14]

5.1 Concept of InGaAs strained quantum wires

In this section we investigate whether the combination of strain and quantum wire effects can enhance the performance characteristics of quantum wire lasers. The results indicate that strained quantum wires take advantage of both the strain effect and the quantum wire effects. A strained InGaAs quantum wire completely surrounded by GaAs barriers has the possibility of being a type II device. In addition, the stronger 3D compression would decrease the maximum critical width of a wire compared with a well, making dislocation-free fabrication difficult for all but the thinnest wires. Therefore, we suggest

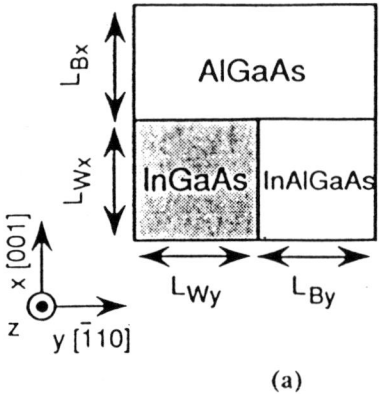

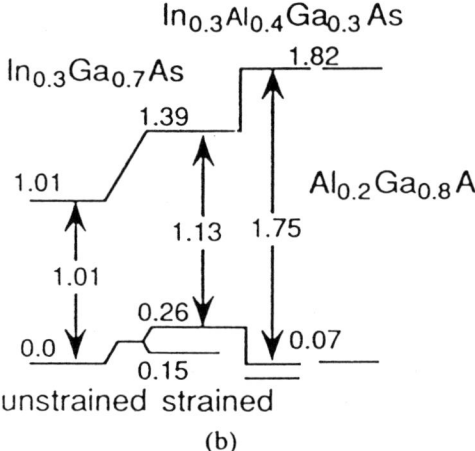

Figure 6.1.6
(a) Schematic cross-sectional illustration of the modified strained quantum wire array. (b) The bulk band edge of the novel strained $In_{0.3}Ga_{0.7}As-(Al_{0.2}Ga_{0.8}As, In_{0.3}Al_{0.4}Ga_{0.3}As)$ system. The energies are in eV

another structure for strained quantum wires, as shown in Figure 6.1.6(a). In what follows we call these 'modified' strained structures. The direction of the wire is taken to be parallel to [110], with the directions of confinement therefore being [001] and [$\bar{1}$10]. The coordinates are defined so that the x-, y-, and z-directions are parallel to [001], [$\bar{1}$10], and [110], respectively. Here, the thicknesses of the wells are equal to L_{Wx} and L_{Wy}, and the barriers thickness are equal to L_{Bx} and L_{By}. The unit cell is periodically repeated in the two quantized directions with periods L_x and L_y.

The quantum wires shown in Figure 6.1.6(a) are assumed to be grown on GaAs substrates. The $In_{0.3}Ga_{0.7}As$ wire region is surrounded by two kinds of barrier layers, $Al_{0.2}Ga_{0.8}As$ and $In_{0.3}Al_{0.4}Ga_{0.3}As$. The indium concentration of the InAlGaAs barrier is taken to be the same as that in the InGaAs wire region. Therefore the lattice constant of the InGaAs is the same as that of the InAlGaAs and these two layers together feel the induced biaxial stress from the AlGaAs layers in the x-direction [001]. In terms of the strain, the situation is similar to that of a strained quantum well in which a 2D layer of one lattice constant is sandwiched between layers of another lattice constant. In the following subsections we show that the biaxial compression in these quantum wires could reduce the strong band mixing and nonparabolicity.

5.2 Valence band structure of biaxially strained InGaAs–AlGaAs quantum wires

First we discuss the valence band structures of a GaAs–AlGaAs and a strained InGaAs–AlGaAs quantum well in order to clarify the strain and the quantum confinement effect as compared with a quantum wire. The results for the [110] direction in-plane dispersion of the valence bands of a GaAs(5 nm)–$Al_{0.4}Ga_{0.6}As$(5 nm) and a strained $In_{0.3}Ga_{0.7}As$(5 nm)–$Al_{0.2}Ga_{0.8}As$(5 nm) quantum well whose plane is normal to (001) (i.e. the direction of quantization is parallel to (001)) is shown in Figures 6.1.7(a) and (b). The zero of energy in the case of the GaAs–AlGaAs quantum well is taken as the valence band maximum of the GaAs bulk. A value for the valence band offset of 43% of the direct band gap difference is used, i.e. Miller's discontinuity. The zero of energy of the strained InGaAs–AlGaAs quantum wells is taken to be the valence band maximum of the unstrained $In_{0.3}Ga_{0.7}As$ bulk.

As can be seen, in the GaAs–AlGaAs quantum well structure strong nonparabolicities of the dispersion curve are observable. Moreover, the effective masses of the heavy holes 1 and 2 (VB1 and VB3) in the in-plane directions are much smaller than that of the light hole 1 (VB2), in contrast to the quantized direction. In addition, the effective mass of light hole 1 (VB2) in the in-plane direction is slightly negative. These properties are well-known typical features of GaAs–AlGaAs quantum wells.

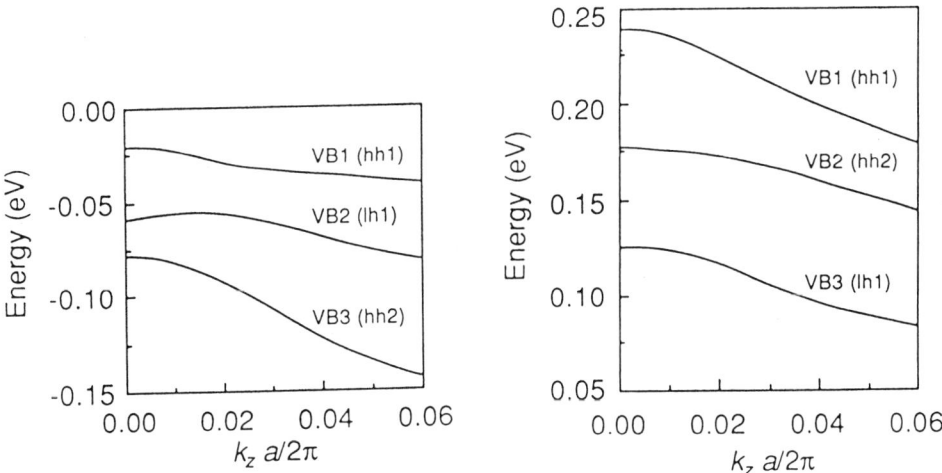

Figure 6.1.7
(a) Valence band structures in the [110] direction for (a) a GaAs (5 nm)–$Al_{0.4}Ga_{0.6}As$ (5 nm) quantum well whose orientation is (001), and (b) a $In_{0.3}Ga_{0.7}As$(5 nm)–$Al_{0.2}Ga_{0.8}As$(5 nm) quantum well

Next we discuss the valence band structures of a GaAs–Al$_{0.4}$Ga$_{0.6}$As quantum wire with the same dimensions, 5 nm × 5 nm, as the biaxially strained wire in order to clarify the strain effects. Also, $L_{Bx} = L_{By} = 5$ nm. The band calculation for the GaAs–AlGaAs quantum wires shown in Figure 6.1.9(a) below indicates that both the nonparabolicities and the negative effective mass nature are more enhanced compared with that of quantum wells. The effective mass of the first uppermost valence band (VB1) is 0.148m_0, which is one-third of the effective mass of heavy holes in bulk GaAs (0.45m_0). In contrast, the effective mass of the conduction band is 0.085m_0, which is larger compared with the effective mass of electrons in bulk GaAs (0.066m_0).

The previous results indicate that the strain effect reduces both the nonparabolicities and the negative effective mass nature compared with that of the GaAs-AlGaAs quantum wells. On the other hand, the 2D confinement effects enhance both natures compared with the quantum wells, as described in what follows.

Figure 6.1.8 shows the valence band structure of the novel biaxially strained quantum wires shown in Figure 6.1.8(a). This example consists of In$_{0.3}$Ga$_{0.7}$As with the two kinds of barriers consisting of In$_{0.3}$Al$_{0.4}$Ga$_{0.3}$As and Al$_{0.2}$Ga$_{0.8}$As. The zero of energy is taken to be the valence band maximum of the unstrained In$_{0.3}$Ga$_{0.7}$As bulk. Here a wire region of 5 nm × 5 nm is used, which is considered to be below the critical thickness. L_{Bx} and L_{By} are both equal to 5 nm for the barriers. Figure 6.1.8(b) shows calculated results for the strained quantum wires, indicating that both the nonparabolicities and the negative effective mass nature are reduced compared with the unstrained GaAs-AlGaAs quantum wires. The improvement can also be seen by examining the effective masses in the wire direction. The effective mass of the uppermost valence band (VB1) is 0.113m_0, which is only 25% of the effective mass of the heavy holes in the unstrained In$_{0.3}$Ga$_{0.7}$As bulk. In contrast, the effective mass of the conduction band is 0.073m_0, which is 1.41 times as large as that of electrons in unstrained InGaAs bulk. The effective mass ratio of the

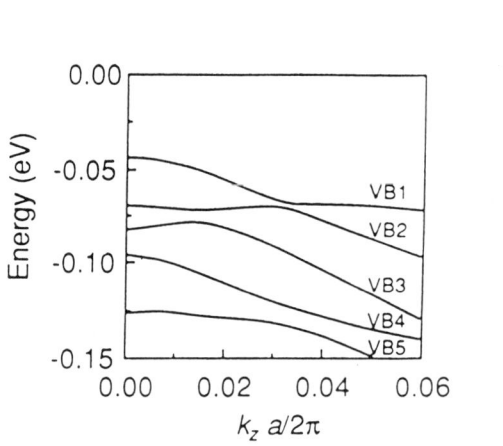
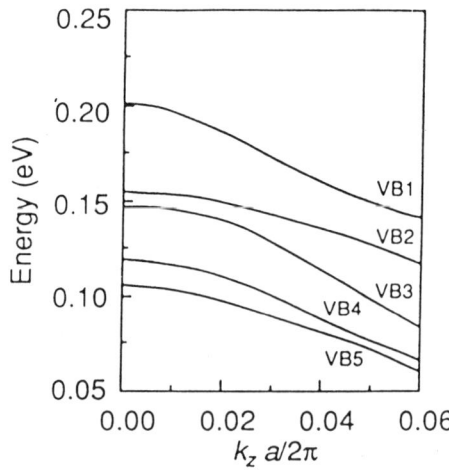

Figure 6.1.8
Valence band structure in the [110] direction for (a) a $L_{Wx} = L_{Bx} = L_{Wy} = L_{By} = 5$ nm GaAs–Al$_{0.4}$Ga$_{0.6}$As quantum wire, and (b) a $L_{Wx} = L_{Bx} = L_{Wy} = L_{By} = 5$ nm strained In$_{0.3}$Ga$_{0.7}$As–(Al$_{0.2}$Ga$_{0.8}$As, In$_{0.3}$Al$_{0.4}$Ga$_{0.3}$As) quantum wire

lowest conduction band to the uppermost valence band, $r_m = m_{VB1}/m_{CB1}$, is summarized for the GaAs bulk, the strained InGaAs–AlGaAs quantum well, and the GaAs–AlGaAs quantum wire clearly reveals an improvement in the ratio caused by introducing the strain effects. The strained quantum wire discussed here has the smallest r_m value, 1.55. Thus the strain effects in addition to the quantum wire confinement will reduce the asymmetry between the conduction and valence band densities of states. Note that in the structure discussed here the barrier layers are thick enough so that the wires are separate (i.e. the miniband width is small).

5.3 Lasing properties of the strained quantum wire lasers

On the basis of the results in the previous section, we discuss how the band structure affects the lasing characteristics. There are two features of band structures particularly significant for laser device performance: the nonparabolicities in the valence band and the asymmetry between the conduction and valence band densities of states. The increased presence of the nonparabolicities would affect the density of states and cause an increase in the threshold current. The asymmetry between the conduction and valence band densities of states is reduced for smaller ratios of the effective masses, r_m. A reduction in r_m would cause an increase in the hole occupation probability of the valence band, which would lead to a reduced threshold current.

Taking the above two features into consideration, we now discuss the modulation dynamics and spectral properties, and show that these properties are significantly improved.

Theories based on the Kronig–Penney model have shown that the differential gain is enhanced by a factor of four in quantum well lasers compared with double heterostructure lasers. In quantum wire lasers, additional improvement in the differential gain is expected [1,3]. For a discussion of the modulation dynamics, the damping effect and the nonlinear gain effect must be also taken into account. But here in order to clarify the effects of the strain on the quantum structures, we focus on investigating how the differential gain is changed.

The differential gain is calculated as a function of the modal gain for the strained quantum wire laser, the GaAs–AlGaAs quantum wire laser, and the GaAs–AlGaAs quantum well laser. The results are shown in Figure 6.1.9. As shown there, the differential gain of the GaAs–AlGaAs quantum wire laser is much higher than that of the GaAs–AlGaAs quantum well laser, even when the nonparabolicity is included.

For the strained quantum wire lasers, an additional enhancement ($\approx 30\%$) of the differential gain is predicted for the strained quantum wire laser. The main reason for the improvement in the differential gain for the strained quantum wire laser is the reduction of both the nonparabolicity and the asymmetry in the band dispersion curves of the lowest conduction band and the highest valence band.

Previous work has suggested that $|\alpha|$ can be significantly reduced by the introduction of quantum confinement effects [6]. As discussed in section 3, in quantum box lasers $|\alpha|$ can be equal to 0, if the structure is optimized. Here we will discuss whether the strained effect is useful or not in the quantum wire lasers in order to reduce $|\alpha|$. Figure 6.1.9 shows calculated $|\alpha|$ for the strained quantum well laser, the GaAs–AlGaAs

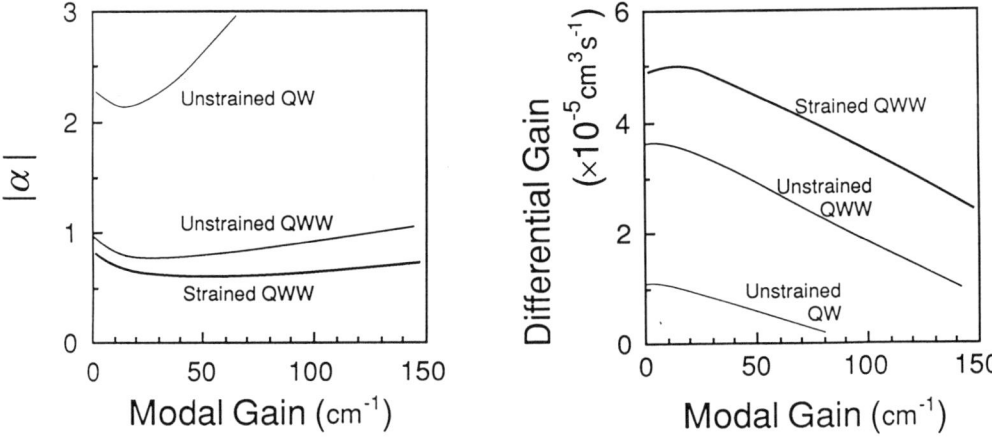

Figure 6.1.9
The calculated differential gain and linewidth enhancement factor plotted as a function of the modal gain for a strained quantum wire laser, a GaAs-AlGaAs quantum wire laser, and a GaAs-AlGaAs quantum well laser

quantum wire laser, and the GaAs-AlGaAs quantum well laser. The results shown in Figure 6.1.8 indicate that $|\alpha|$ can be additionally reduced if the strain effect is introduced in the quantum wires. Thus, the strain effect, in addition to suppressing valence band mixing and reducing the difference in the effective masses of the first conduction band relative to that of the first valence band, enhances the modulation dynamics and spectral properties. The advantages of both the quantum wire effect and the strain effect are successfully incorporated in this structure.

6 *Fabrication of Quantum Wires [15,16]*

In order to fabricate the quantum nano-structures, wet chemical etching, reactive ion etching [9], ion beam implantation, and ion beam milling have been investigated. These methods, however, suffer from free surface defects, the creation of a damage field during implantation, or a loss of interface control due to the random nature of the disordering mechanism. To avoid these problems, growth techniques on masked substrates and nonplanar substrates have been also investigated. Here, we discuss an *in situ* fabrication technique for quantum wires and quantum boxes by utilizing MOCVD selective growth on a SiO_2 patterned substrate on which the V-groove structures are formed by the growth.

The MOCVD growth was performed in a low-pressure, horizontal, rf-heated MOCVD reactor, using trimethylgallium (TMG), trimethylaluminum (TMA), and arsine (AsH_3) as group III and V sources, respectively. The ratio of group V to group III was 100. The growth temperature was 700 °C. Purified H_2 with a 6 l/min flow rate was used as a carrier gas. The growth pressure was 100 Torr.

The fabrication procedure for the quantum wires is as follows. First, a SiO_2 layer with a thickness of 20 nm is formed by plasma chemical vapor deposition on a semi-insulating (100) GaAs substrate. PMMA is then patterned on the SiO_2 layer by the electron beam lithography technique, followed by wet chemical etching to pattern the SiO_2. After this procedure a triangular-shaped GaAs structure with (111)A facet sidewalls is grown on the masked substrate by MOCVD growth. The formation of the triangular structure is due to the large growth rate difference between the (100) orientation and (111)B or (111A). Further continuation of the growth leads to the lateral growth above the SiO_2 mask, making the gap between the triangular prisms small. The gap between the triangular prisms is then filled up with an $Al_{0.4}Ga_{0.6}As$ layer by switching the growth from GaAs to $Al_{0.4}Ga_{0.6}As$. In contrast to GaAs, orientation dependence of the growth rate is weak in AlGaAs materials, although the dependence is strongly affected by temperature and Al content. As a result, a sharp corner at the bottom of the V-groove between the triangular prisms is formed by the growth of the AlGaAs layer. At this point the growing layer is again switched to GaAs. Since at the V-groove corner there is a small point which has a (100) orientation, triangular-shaped quantum wires can be grown at the V-groove corner. This quantum wire is connected to thin quantum wells on (111)A-sidewalls of large triangular GaAs prisms. Thus, GaAs quantum wires which are coupled to thin quantum wells are formed between the triangular prisms without being exposed to air. Figure 6.1.10 shows a high-resolution, cross-sectional SEM image of the quantum wire array with 20 nm period and its illustration. As shown in this photograph, even though there is ~30% lateral size fluctuation in the SiO_2 mask, the quantum wires are uniformly formed, which is due to a reduction in the size variation by the lateral selective growth of the triangular prisms.

By changing the growth time of the GaAs material for the quantum wires, we obtained quantum wires with various lateral widths. Figure 6.1.11 shows high-resolution SEMs of the quantum wire region with lateral widths of ~10, ~15, ~25, and ~35 nm, respectively. Each quantum wire smoothly connects to quantum well layers with ~2,

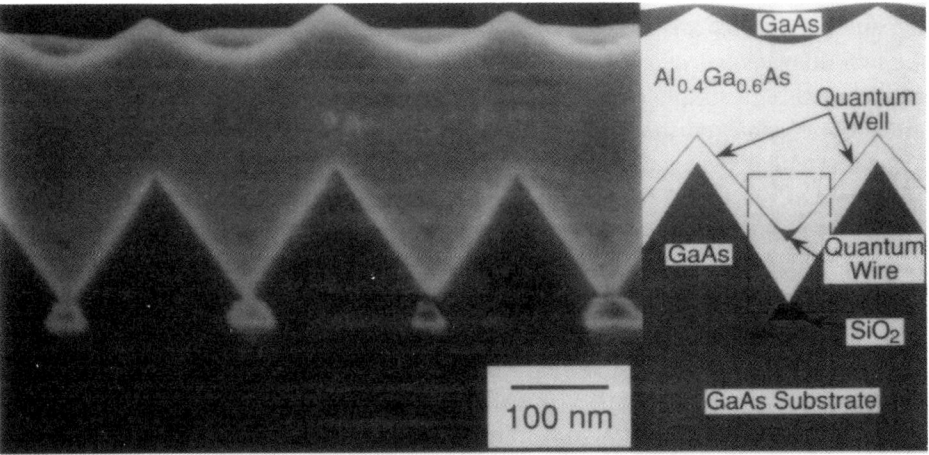

Figure 6.1.10
A high-resolution scanning electron micrograph of GaAs triangular-shaped quantum wire array structures with a lateral width of ~15 nm and its illustration

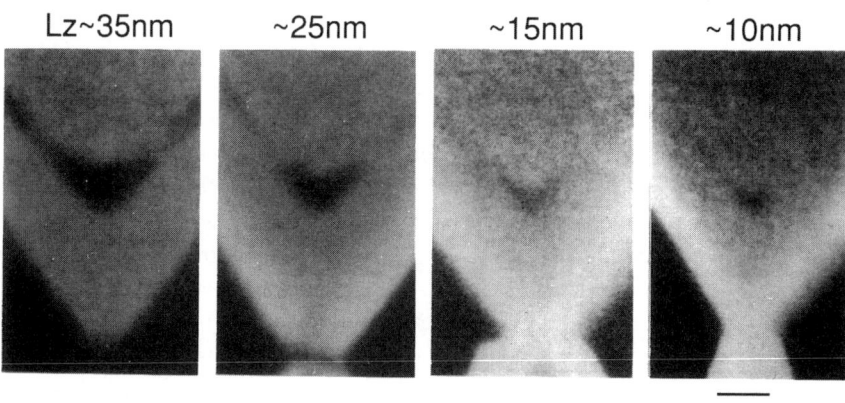

Figure 6.1.11
High-resolution scanning electron micrographs of part of the quantum wires with lateral widths of ~10, ~15, ~25, and ~35 nm, respectively

~3, ~5, and ~7 nm thickness. As shown in this photograph, the quantum wires with size in the 10 nm range were obtained by systematic change of the growth time.

7 Optical Properties of Quantum Wires [14–16]

Photoluminescence (PL) spectra from the quantum wire structures are measured at 20 K as a function of the lateral widths of 0, ~7, ~10, ~15, ~25, ~30, and ~35 mn, as shown in Figure 6.1.12(a). In this figure the hatched PL peaks correspond to the quantum wires. Figure 6.1.12(b) shows the energy shift ΔE of the PL peak of the quantum wires versus the lateral width Lx. ΔE is defined as the energy difference between the PL peaks of the GaAs bulk and the quantum wires. As shown in these figures, systematic blue shifts are observed with decreasing Lx, which is due to enhancement of the two-dimensional quantum confinement effect. The solid calculation curve in the figure is based on a simple one-band model. These results indicate that a strong lateral confinement is achieved in the present structures.

PL measurements are a useful tool to confirm the quantum wire effects in the structures. However, it is necessary to obtain additional clear evidences. When the lateral potential exists in the quantum wires, the behavior of the Landau shift should depend on the direction of the applied magnetic fields. Therefore measurements of magneto-PL spectra were performed for quantum wires with a lateral width of 20 nm using pulsed magnetic fields at 4.2 K. The pulse duration of the magnetic field was 10 ms and the maximum field was about 40 T. PL spectra were detected with an optical multi-channel analyzer system installed at the exit of the monochromator through the optical fiber.

The PL peak position from the quantum wires and the bulk at the three configurations (i.e. $B \parallel x$, $B \parallel y$, $B \parallel z$) are plotted as a function of applied magnetic fields in Figure 6.1.13. As shown in this figure, the behavior of the PL peak shift ΔE is different between the bulk and the quantum wires. The bulk PL peak shifts at various magnetic

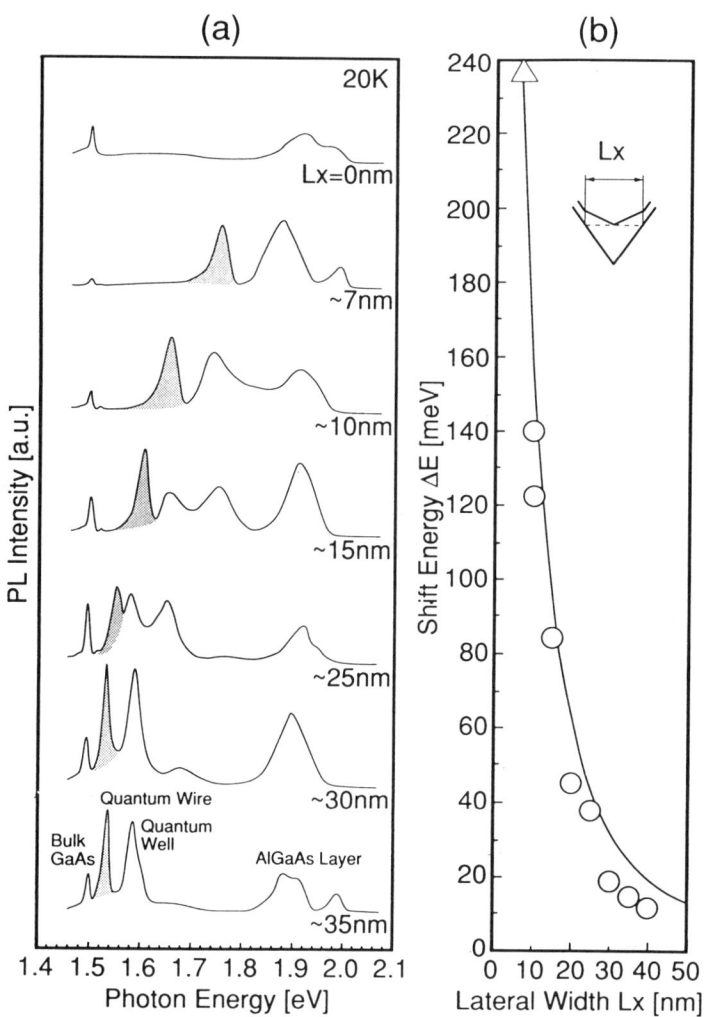

Figure 6.1.12
(a) Photoluminescence (PL) spectra from quantum wire structures measured as a function of the lateral wire widths at 20 K. (b) The energy shift ΔE of the PL peak of the quantum wires versus the lateral width Lx

fields are almost equally independent of the configurations throughout the magnetic field region. In contrast, PL peak shift of the quantum wires is clearly dependent on the configuration. When the magnetic field is applied in parallel with the quantum wires ($\boldsymbol{B} \parallel \boldsymbol{x}$), the energy shift with an increase in the magnetic field is the smallest. This can be explained by a classical picture in which the cyclotron motion on the plane perpendicular to the magnetic field is restricted by the two-dimensional lateral potentials of the quantum wires. On the other hand, there is also an anisotropic effect when the magnetic field is applied in the two directions perpendicular to the quantum wires. The results indicate that the cross-sectional shape of the quantum wires is not isotropic.

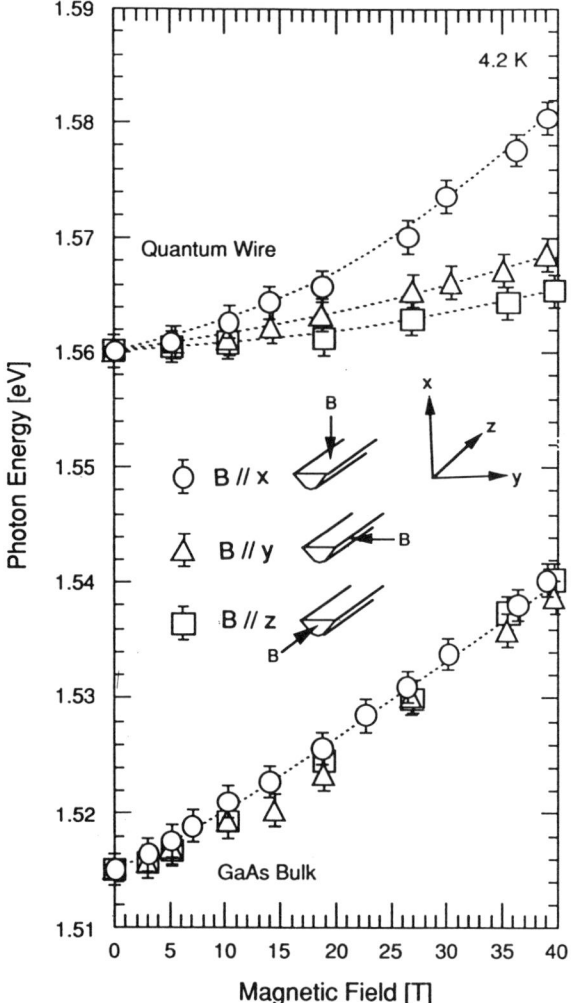

Figure 6.1.13
PL peak positions from quantum wires and the bulk as a function of applied magnetic field for various configurations. Circles, triangles, and squares represent the data for $\boldsymbol{B} \parallel \boldsymbol{x}$, $\boldsymbol{B} \parallel \boldsymbol{y}$, and $\boldsymbol{B} \parallel \boldsymbol{z}$, respectively

In order to analyze this behavior, we analyzed the experimental results assuming that the quantum wire potential is expressed by the following harmonic potential $f(x, y, z)$:

$$f(x, y, z) = \tfrac{1}{2} m^* \omega_{0y}^2 y^2 + \tfrac{1}{2} m^* \omega_{0x}^2 x^2 \qquad (13)$$

In this case the energy shift, ΔE, due to the magnetic filed applied perpendicularly to the quantum wires can be expressed as follows:

$$\begin{aligned}
\Delta E &= \tfrac{1}{2}\hbar\sqrt{\omega_c^2 + \omega_{0y}^2}, \quad \boldsymbol{B} \parallel \boldsymbol{y} \\
\Delta E &= \tfrac{1}{2}\hbar\sqrt{\omega_c^2 + \omega_{0x}^2}, \quad \boldsymbol{B} \parallel \boldsymbol{x}
\end{aligned} \qquad (14)$$

where ω_c is equal to eB/m. By fitting the curve to the measured data using the above relationships, the value of ω_{0x}/ω_{0y} is 2.9. It should be noted that the ω_i is approximately proportional to $1/L_i^2$ ($i = x$ or y), where L_i is the thickness of the quantum well. Since the SEM observation showed that Lx and Ly are about 12 nm and 20 nm, the value of $(Lx/Ly)^2$ is in good agreement with the value of ω_{0x}/ω_{0y}. To measure the exciton lifetime, samples were mounted in a cryostat and excited by frequency doubled output (532 nm) of a mode-locked Nd$^+$:YAG laser with a fiber-grating pulse compressor. The pulse width was less than 10 ps and the repetition rate was 82 MHz. Time-resolved PL spectra were obtained using a streak camera with a 0.25 m monochromator. The overall time resolution was less than 50 ps. In all the measurements the sample temperature was maintained at about 9 K and the excitation power was 10 mW, which correspond to a photon flux per pulse of about 4×10^{10} photons/cm^2.

8 Fabrication of InGaAs Vertical Microcavity Lasers [17–19]

It is important to combine confined electrons and confined photons [17]. As a first step towards such directions, we have fabricated a quantum wire laser with a microcavity. Figure 6.1.14 shows the fabrication process of the laser structure. In the microcavity, triangular-shaped In$_{0.3}$Ga$_{0.7}$As strained quantum wires are grown between (111)A facets of [011]-oriented GaAs triangular prisms by selective MOCVD growth on an SiO$_2$ masked DBR mirror region [18]. The quantum wires are compressively strained owing to the difference in the lattice constants of GaAs and In$_{0.3}$Ga$_{0.7}$As. A scanning electron micrograph and a schematic illustration of the cross section of the laser structure are shown in Figure 6.1.15. The top and bottom DBRs consist of 25 and 30.5 quarter-wave pairs of Al$_{0.4}$Ga$_{0.6}$As/AlAs, respectively. To obtain an upper DBR of high quality, the region above the quantum wires must be flat. Therefore, the quantum wires were buried in GaAs, instead of AlGaAs. Since the diffusion length of Ga is larger than that of Al, the growth of GaAs can flatten the uneven surface left after the quantum wire growth. In order to use GaAs as the barrier region on the quantum wires, InGaAs material of which the bandgap is smaller than that of GaAs was grown as the quantum wire region. In addition, as discussed before, the use of the strain effects leads to substantial improvements in the lasing characteristics.

The growth process of the In$_{0.3}$Ga$_{0.7}$As strained quantum wire region is the same as that of previously reported GaAs quantum wires. First, the triangular-shaped GaAs prisms with (111)A facet sidewalls were grown on the masked DBR region. The formation of the triangular structure is due to the large growth rate difference between (100) and (111)B or (111)A orientations. Further continuation of the growth leads to lateral growth above the SiO$_2$ mask, making the gap between the triangular prisms small. The gap between the triangular prisms was then filled with Al$_{0.4}$Ga$_{0.6}$As. In contrast to GaAs, the orientation dependence of the growth rate is weak in AlGaAs materials. As a result, a sharp corner at the bottom of the V-groove between the triangular prisms is formed by the isotropic growth of the AlGaAs layer. Then the growth layer is switched to InGaAs, which has anisotropic growth conditions similar to GaAs. As a result of

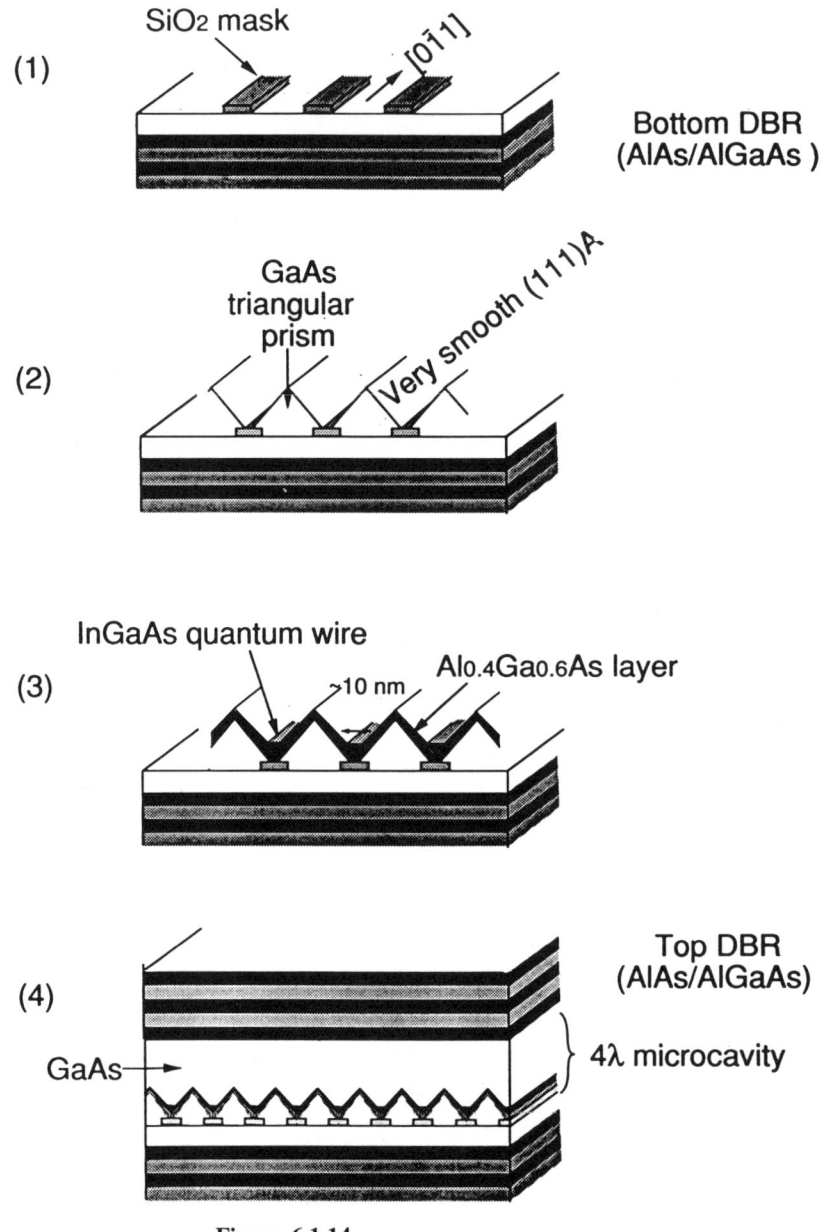

Figure 6.1.14
Fabrication process of the laser structure

rapid growth in the (100) direction, InGaAs quantum wires which are connected to thin quantum wells are formed between the triangular prisms. The essential feature of this fabrication technique is that sharp V-grooves can be obtained not by etching but by selective growth, without the quantum wires being exposed to air. The lateral width of the quantum wires was estimated to be about 10 nm with a period of 200 nm. The shift

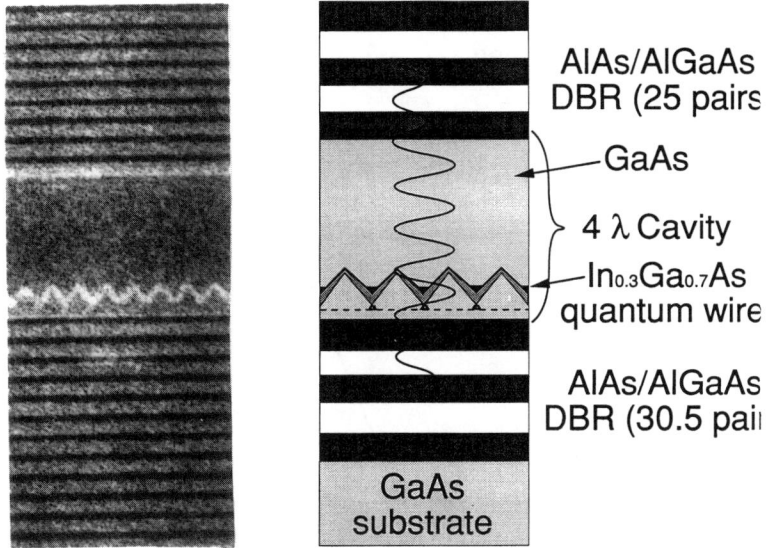

Figure 6.1.15
A scanning electron micrograph and a schematic illustration of the cross-section of the laser structure

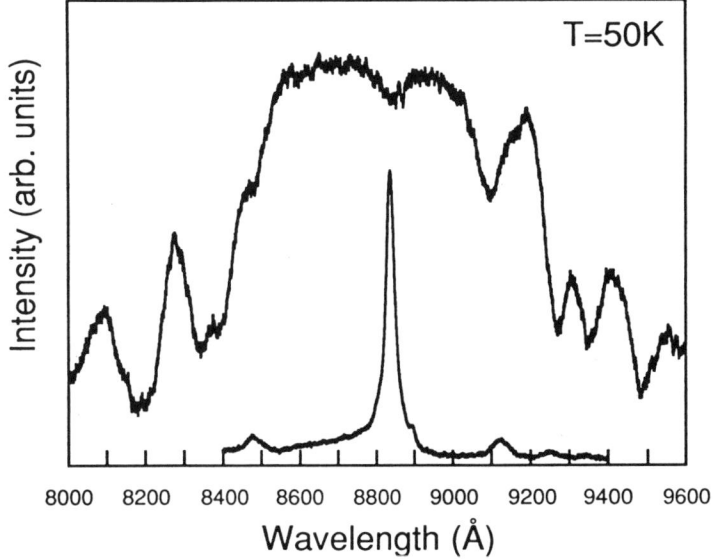

Figure 6.1.16
The reflectivity and PL spectrum of the sample measured at 14 K

of the photoluminescence (PL) peak of the InGaAs quantum wires is consistent with an increase in the In composition if both the quantum confinement and the strain effect are included.

Figure 6.1.16 shows the reflectivity and PL spectrum of the sample measured at 14 K. The reflectivity spectrum of the sample was measured using white light. The

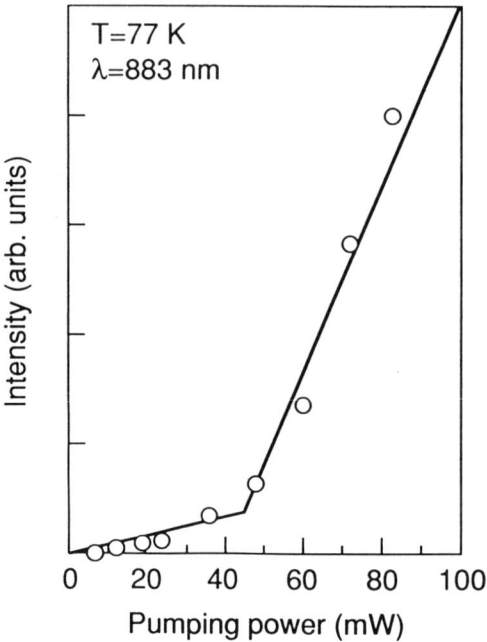

Figure 6.1.17
The output power of the light emitted from the sample as a function of pumping laser power. The beam diameter of the pump beam was about 20 mm

cavity resonance was designed so that the resonance photon energy coincides with the PL peak of the quantum wires. As shown in this figure, a sharp PL spectral line with a full width at half-maximum (FWHM) of than less 3 nm was observed from the quantum wires with the cavity. On the other hand, the FWHM of the PL spectra from the quantum wires without the cavity was about 16 nm. The narrowing of the spectrum due to the cavity clearly demonstrates the cavity effect.

The lasing properties of this sample were measured at 77 K, using optical pumping with a continuous wave Ti:sapphire laser with a wavelength of 733 nm. Figure 6.1.17 shows the output power of the light emitted from the sample as a function of pumping laser power. The beam diameter of the pump beam was about 20 μm. The measurement shows a clear threshold at a pump power of ~40 mW, demonstrating lasing oscillation at this temperature. This result demonstrates the possibility of an extremely low threshold current, less than 100 mA, by reducing the diameter of the cavity to 1 μm.

9 *Fabrication of Quantum Boxes [22–24]*

Fabrication trials of quantum box structures using a masked substrate have been intensively investigated by several groups. Here the GaAs quantum boxes were fabricated by the MOCVD selective growth technique on SiO_2 patterned (100) GaAs substrates. The masks consist of 100 nm × 100 nm mm^2 windows with a period of 140 nm. First,

$Al_{0.4}Ga_{0.6}As$ plinths are formed on the SiO_2 masks. Then, the GaAs is grown on top of the AlGaAs plinths, followed by growth of $Al_{0.4}Ga_{0.6}As$ so that the GaAs quantum boxes are embedded by $Al_{0.4}Ga_{0.6}As$. Figure 6.1.18 is a cross-sectional view of the GaAs surrounded by AlGaAs and its illustration. The photograph indicates that the lateral with of the quantum boxes is 25 nm × 25 nm. We believe that this lateral width is the smallest so far for GaAs quantum boxes embedded by AlGaAs materials.

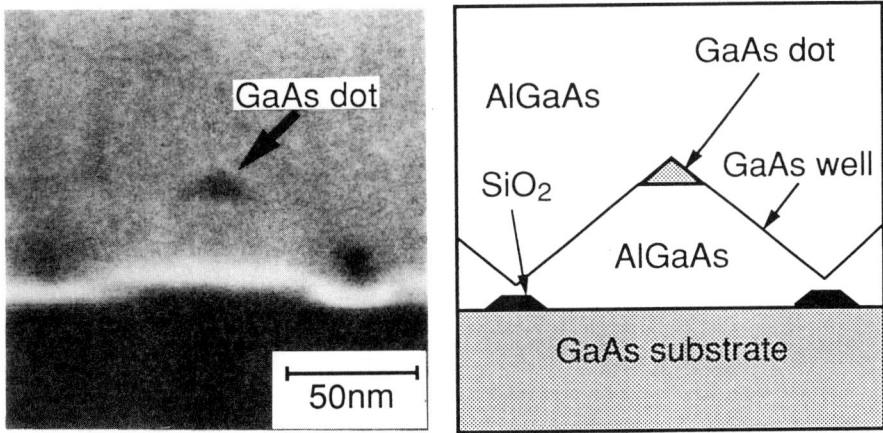

Figure 6.1.18
The Cross-sectional view of the GaAs surrounded by AlGaAs and its illustration

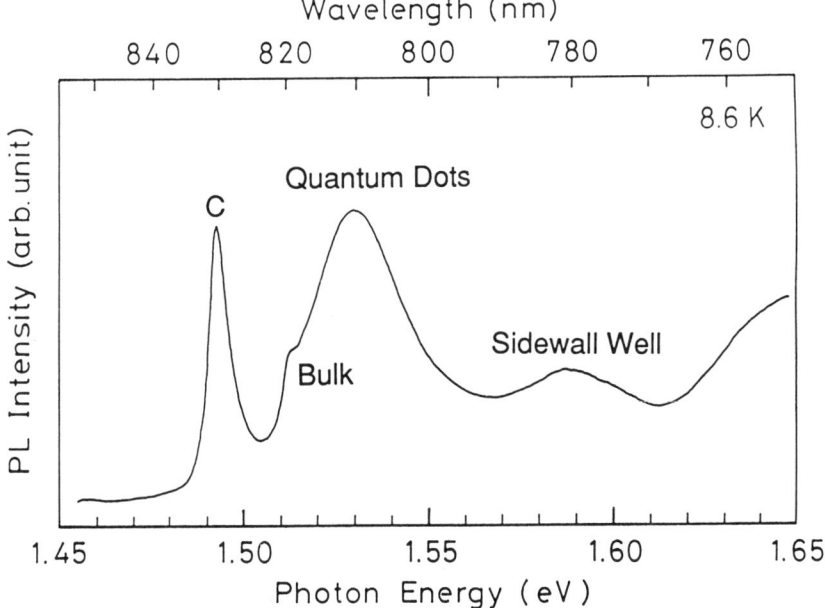

Figure 6.1.19
PL spectra of the quantum dot structures excited by a CW argon laser

Even from this small structure, PL can be observed. Figure 6.1.19 shows PL spectra of the quantum box structures excited by a CW argon laser. In this case electron-hole pairs are excited in the barrier region and then relaxed into the quantum box region. The energy shift of the PL peak from the quantum box region is about 18 meV. We believe that this energy shift results from lateral confinement of the electrons. The full width with half maximum (FWHM) of the PL peak is broad compared with those of the quantum wires. However, the tail of the luminescence on the low-energy side disappears at the photon energy of the bulk GaAs. This indicates that the broadening is not due to strain and defect effects but due to size variation of the quantum boxes. In the measurement, more than 100 000 quantum boxes are simultaneously excited. Excitation of a smaller number of quantum boxes is discussed elsewhere.

10 Conclusion

In this section we discussed nano-structure semiconductor lasers, focusing on confined electron systems for the lasers. For an ultimate light source, confined electrons and confined photons will play very important roles. However, to reach the goal, we have to develop nano-technologies for both photons and electrons.

References

[1] J.P. van der Ziel, R. Dingle, R.C. Miller, W. Wiegmann and W.A. Nordland Jr, *Appl. Phys. Lett.*, **26**, 463-465 (1975).
[2] N. Holonyak, Jr, R.M. Kolvas, R.D. Dupuis and P.D. Dapkus, *IEEE J. Quantum Electron.*, **QE-16**, 170-181 (1980).
[3] W.T. Tsang, *Appl. Phys. Lett.*, **39**, 786-788 (1981).
[4] Y. Arakawa, K. Vahara and A. Yariv, *Appl. Phys. Lett.*, **45**, 950-952 (1984).
[5] Y. Arakawa and H. Sakaki, *Appl. Phys. Lett.*, **40**, 939-941 (1982).
[6] Y. Arakawa, K. Vahala and A. Yariv, *Appl. Phys. Lett.*, **45**, 950-952 (1984).
[7] Y. Arakawa and A. Yariv, *IEEE J. Quantum Electron.*, **QE-22**, 1887-1899 (1986).
[8] Y. Arakawa and T. Takahashi, *Optoelectronics*, **3**, 155-162 (1988).
[9] Y. Arakawa and T. Takahashi, *Electron. Lett.*, **25**, 169 (1989).
[10] Y. Arakawa, *Solid State Electronics*, **37**, 523-528 (1994).
[11] H. Benisty, C.M. Sotomayor-Torres and C. Weisbuch, *Phys. Rev. B*, **44**, 1045 (1991).
[12] Y. Arakawa, *Solid State Electron.*, **37**, 523 (1994).
[13] H. Nakayama and Y. Arakawa, *14th IEEE Int. Semiconductor Laser Conf.*, Maui, 1994.
[14] Y. Yamauchi, T. Takahashi and Y. Arakawa, *IEEE J. Quantum Electron.*, **QE-22**, 1817-1823 (1991).
[16] S. Tsukamoto, Y. Nagamune, M. Nishioka and Y. Arakawa, *J. Appl. Phys.*, **71**, 533 (1992).
[14] S. Tsukamoto, Y. Nagamune, M. Nishioka and Y. Arakawa, *Appl. Phys. Lett.*, **63**, 310 (1993).
[17] Y. Nagamune, Y. Arakawa, S. Tsukamoto and M. Nishioka, *Phys. Rev. Lett.*, **69**, 2963 (1992).
[18] T. Kono, Y. Nagamune, M. Nishioka and Y. Arakawa, *Appl. Phys. Lett.*, **64**, 1564 (1994).
[19] For example, Y. Arakawa, *Extended Abstracts of 1990 International Conference on Solid State Devices and Materials*, Sendai, Japan, 1990, p. 745.

[20] T. Arakawa, S. Tsukamoto, Y. Nagamune, M. Nishioka, J. Lee and Y. Arakawa, *Jpn J. Appl. Phys.*, **32**(10A), 1377L–1379L (1993).
[21] T. Arakawa, T. Kono, M. Nishioka, Y. Nagamune and Y. Arakawa, *The Microoptics Conference and The Topical Meeting on Gradient-Index Optical Systems (MOC/GRIN'93)*, Japan, 1993.
[22] Y. Nagamune, S. Tsukamoto, M. Nishioka and Y. Nagamune, *J. Crystal Growth*, **126**, 707 (1992).
[23] Y. Nagamune, S. Tsukamoto, M. Nishioka and Y. Arakawa, *1992 Gallium Arsenide and Related Compounds, Karuizawa, Inst. Phys. Conf. Ser.*, **129**, 335 (1992).

6.2

GaInAsP/InP Quantum Well and Quantum Wire Semiconductor Laser Amplifiers

Kazuhiro Komori and **Masahiro Asada**

Abstract

Low noise characteristics of semiconductor laser amplifiers consisting of strained quantum well and low-dimensional quantum well structures were theoretically obtained using density matrix theory. Due to a sharper gain spectrum as well as a smaller population inversion parameter in these structures, predominant two-beat noises of travelling-wave semiconductor laser amplifiers were found to be reduced. The noise figure can be reduced to 3.3 dB in a quantum box structure. GaInAs/GaInAsP/InP quantum wire semiconductor lasers/amplifiers were fabricated using selective growth on grooved side walls, and the lasing action was achieved at 77 K.

1 Introduction

A semiconductor laser amplifier (SLA) [1-3] is very attractive for applications to photonic integrated circuits (PICs) as a booster amplifier as well as a preamplifier because of its high gain property, compact size, and feasibility of integration. However, the noise figure of SLAs reported up to now is inferior to that of a Er^{+3}-doped fiber amplifier [4] and that is a problem in the optical communication systems.

Recently, some improvements in the amplification characteristics of SLAs, such as large saturation power [5] and low noise [6] characteristics, have been achieved by introducing quantum well structures. Since these improvements resulted from a smaller value of the population inversion parameter due to smaller waveguide loss in respect of the medium gain, the lower dimensional quantum well structures, such as quantum wire and quantum box structures, could give superior amplification characteristics of travelling-wave-type semiconductor laser amplifiers (TW-SLAs).

In this section we describe the noise characteristics of SLAs consisting of low-dimensional (quantum wire and quantum box) structures analyzed using density matrix theory. The noise characteristics of strained quantum well (film) SLAs are also discussed.

Up to now, various technique for the fabrication of low-dimensional quantum wells have been widely studied, as described in the previous and next sections. Here we also show briefly a new fabrication method and lasing action of GaInAs/GaInAsP/InP quantum wire lasers, as a first step towards realizing low-noise SLAs.

2 Analysis of Noise Characteristics

In direct detection systems using a SLA and a photo detector, the noise is predominated by signal–spontaneous beat noise and spontaneous–spontaneous beat noise due to the beating between the amplified signal and the amplified spontaneous emission (ASE).

The signal-to-noise ratio (SNR) and the noise figure F, which are the important noise parameters in optical communication systems, are obtained by solving the master equation of photon density [7]. By neglecting shot noise terms, the SNR of the output and noise figure are given as follows for SLAs [3,7].

$$(\text{SNR})_{\text{out}} \sim \frac{2\langle S_{\text{in}}\rangle^2}{(2\gamma_{\text{sp}}\langle S_{\text{in}}\rangle + \gamma_{\text{sp}}^2 \Delta f)B} \tag{1}$$

$$F = \frac{(\text{SNR})_{\text{out}}}{(\text{SNR})_{\text{in}}} \sim 2\gamma_{\text{sp}} + \frac{\gamma_{\text{sp}}^2 \Delta f}{\langle S_{\text{in}}\rangle} \tag{2}$$

where $\langle S_{\text{in}}\rangle$ is the average input photon number, Δf is the effective noise bandwidth for spontaneous–spontaneous beat noise, B is the baseband width of the detection system, and γ_{sp} is the population inversion parameter. In bulk materials, γ_{sp} can be approximately expressed by using the carrier density N and the carrier density for transparency N_{g}, while in quantum well structures, it is difficult to express γ_{sp} by such an approximation since the gain is not linearly increasing with the carrier density.

Here, we precisely calculate γ_{sp} using the density matrix theory of semiconductor lasers [8]. The population inversion parameter γ_{sp} is defined as the spontaneous emission rate divided by the net stimulated emission rate per photon (i.e. stimulated emission subtracted by absorption) [7]:

$$\gamma_{\text{sp}} = \frac{\xi T_{\text{spon}}}{\xi T_{\text{stim}} - \alpha v_{\text{g}}} \tag{3a}$$

$$\alpha = \xi(K_0 N + \alpha_2) + (1 - \xi)\alpha_{\text{ex}} \tag{3b}$$

where α is the loss coefficient in the active waveguide, ξ is the optical confinement factor, K_0 is the proportional constant related to the intervalence band absorption, α_2 is the absorption for transitions between the split-off band and the acceptor level, α_{ex} is the loss in the cladding layers, and v_{g} is the group velocity of guided light.

T_{spon} is the rate of spontaneous emission and T_{stim} is that of the net stimulated emission given as follows [8]:

$$T_{\text{spon}} = \frac{\omega}{n_r^2 \varepsilon_0} \int \langle R_{cv}^2 \rangle \frac{f_c(1-f_v)g_{cv}(\hbar/\tau_{\text{in}})}{(E_{cv} - \hbar\omega)^2 + (\hbar/\tau_{\text{in}})^2} dE_{cv} \qquad (4)$$

$$T_{\text{stim}} = gv_g = \frac{\omega}{n_r^2 \varepsilon_0} \int \langle R_{cv}^2 \rangle \frac{(f_c - f_v)g_{cv}(\hbar/\tau_{\text{in}})}{(E_{cv} - \hbar\omega)^2 + (\hbar/\tau_{\text{in}})^2} dE_{cv} \qquad (5)$$

where E_{cv} is the transition energy between the levels c and v, ω is the angular frequency of the input light, ε_0 is the dielectric constant of the vacuum, τ_{in} is the intraband relaxation time, n_r is the refractive index, g is the power gain coefficient per unit length in the active medium, R_{cv} is the matrix element of the dipole moment, and f_c and f_v are the Fermi functions in conduction and valence bands, respectively.

g_{cv} is the density of state for electron–hole pairs expressed as follows for quantum film (QF), quantum wire (QW), and quantum box (QB) structures:

$$\text{QF}: \quad g_{cv} = \frac{1}{\pi \hbar^2 W_y} \frac{m_c m_v}{m_c + m_v} u(E_{cv} - E_{cn} - E_{vn} - E_g) \qquad (6)$$

$$\text{QW}: \quad g_{cv} = \frac{1}{2\pi \hbar W_x W_y} \left(\frac{m_c m_v}{m_c + m_v}\right)^{1/2} (E_{cv} - E_{clm} - E_{vlm} - E_g)^{-1/2} \qquad (7)$$

$$\text{QB}: \quad g_{cv} = \frac{2\delta(E_{cv} - E_{clmn} - E_{vlmn} - E_g)}{W_x W_y W_z} \qquad (8)$$

where m_c and m_v are the effective masses, E_{cn}, E_{clm}, and E_{clmn} are the quantized energies of the QF, QW, and QB structures, respectively, and E_{vn}, E_{vlm}, and E_{vlmn} are those of holes. The sizes of quantum well structures for the x-, y-, and z-directions are denoted by W_x, W_y, and W_z, respectively.

We assume an ideal travelling wave semiconductor laser amplifier (TW-SLA) with its facet reflectivity $R_f = 0$, in which the single-pass gain is expressed as follows:

$$G = \exp[(\xi g - \alpha)L] \qquad (9)$$

The effective bandwidth Δf is given by

$$\Delta f = \int \frac{[G(f) - 1]^2 \gamma_{\text{sp}}(f)^2}{[G(f_s) - 1]^2 \gamma_{\text{sp}}(f_s)^2} df \qquad (10)$$

where f and f_s are the light frequency and the input signal light frequency, respectively.

3 Gain and Noise Characteristics of Low-dimensional Quantum-well SLAs

Figure 6.2.1 shows the structural model of the SLA used in the calculation, where the active region consists of quantum film (QF), quantum wire (QW), and quantum box (QB) structures. The QW structures are assumed to consist of GaInAs well and

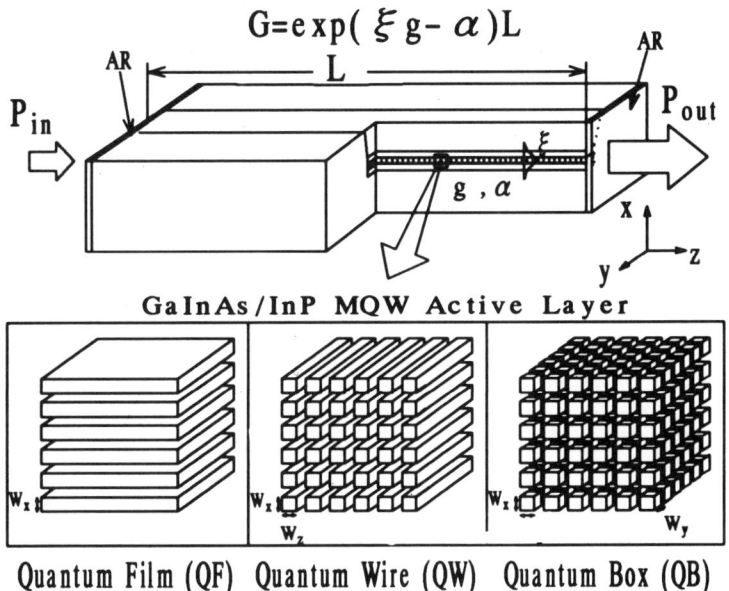

Figure 6.2.1
A model of a travelling wave semiconductor laser amplifier used for calculation © IEEE [11]

InP barrier layers. The SCH type waveguide structure, which consists of upper and lower waveguide layers with 0.2 μm thick GaInAsP ($\lambda_g = 1.10$ μm), is considered to increase the optical confinement factor. We assume an ideal TW-SLA with a facet reflectivity of $R_f = 0$.

The optical confinement factors used in the calculation of QF, QW, and QB structures are 9%, 4.5%, and 2.25%, respectively, considering a 5-period 10 nm thick GaInAs quantum well and 10 nm thick InP barrier layers, because we compare the noise characteristics with the same number of quantum well layers and assume the space filling factor of the well region to be $\frac{1}{2}$. The single-pass gain G in equation (9) was fixed at 20 dB by choosing an appropriate length of the amplifier L. Other parameter values used in the calculation are: the loss parameters $K_0 = 4 \times 10^{-17}$ cm^2, $\alpha_2 = 30$ cm^{-1}, $\alpha_{ex} = 5$ cm^{-1}; the intraband relaxation time $\tau_{in} = 1 \times 10^{-13}$ s; effective masses $m_c = 0.041 m_0$ and $m_v = 0.42 m_0$ for GaInAs, and $0.08 m_0$ and $0.45 m_0$ for InP, the band discontinuities at the GaInAs/InP heterointerface $\Delta E_c = 0.24$ eV and $\Delta E_v = 0.36$ eV.

Figure 6.2.2 shows the wavelength dependence of a single-pass gain of low-dimensional quantum well structures, where the 3 dB down bandwidth of the single-pass gain G becomes narrower in lower dimensional quantum well structures, i.e. the full width at half maximum (FWHM) is 68 nm, 14 nm, and 8 nm for quantum film (QF), quantum wire (QW), and quantum box (QB) structures, respectively, at a carrier density of $N = 5 \times 10^{18}$ cm^{-3}.

In the QB structure, the amplified spontaneous emission at the wavelength of the second quantized level is suppressed to a very small value, since the stimulated absorption rate is much higher (by several thousand cm^{-1}) than the stimulated emission rate at this level. Hence the QB structure behaves like an amplifier with a narrow bandpass

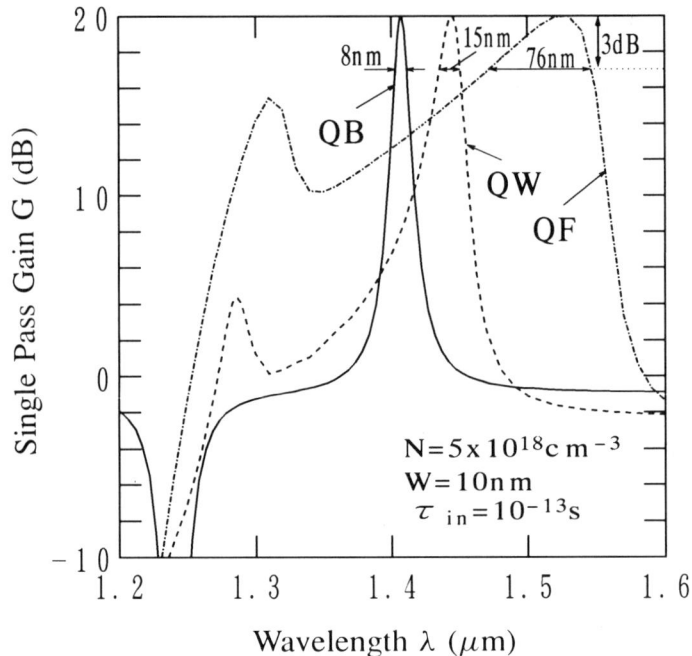

Figure 6.2.2
Single-pass gain G of quantum film (QF), quantum wire (QW), and quantum box (QB) structures. © IEEE [11]

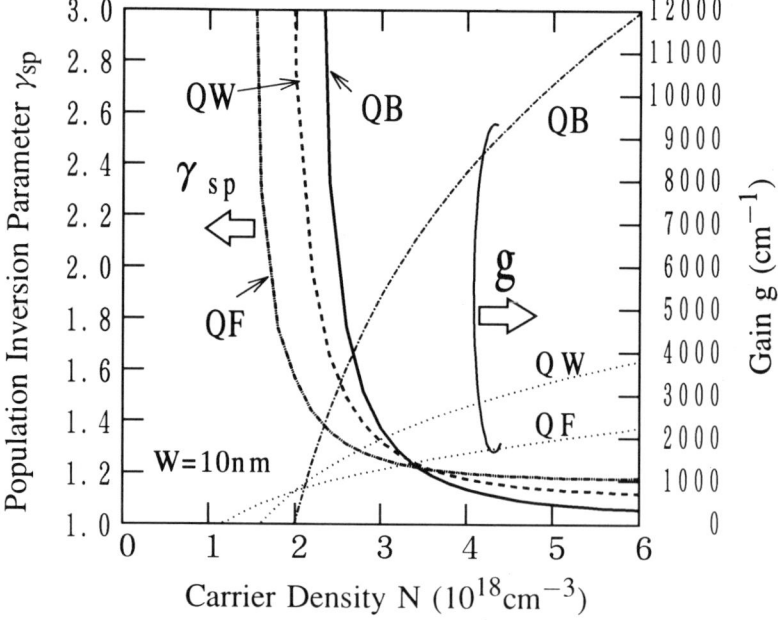

Figure 6.2.3
Gain coefficient g and population inversion parameter γ_{sp} as a function of the carrier density of low-dimensional quantum well structures. © IEEE [11]

filter which eliminates the spontaneous–spontaneous beat noise. For instance, the spectral width of the single-pass gain of the QB structure is only 5% of that of bulk medium (FWHM ~150 nm), and therefore it is possible to reduce the spontaneous–spontaneous beat noise without inserting a narrow bandpass filter by using the QB structure.

The population inversion parameter γ_{sp} and the gain coefficient g of low-dimensional quantum well structures as a function of carrier density N are shown in Figure 6.2.3. In the calculation, the size of the quantum well is set at 10 nm and wavelength is set at the gain peak wavelength. In each quantum well structure the population inversion parameter γ_{sp} decreases as the carrier density N increases and gradually approaches a certain value at a carrier density N larger than 5×10^{18} cm^{-3}. As can be seen, γ_{sp} is higher in lower dimensional QW structures at a carrier density N lower than 3.4×10^{18} cm^{-3}, since the carrier density for transparency, N_g, is larger in lower dimensional quantum well structures, while, at a higher carrier density region ($N > 3.4 \times 10^{18}$ cm^{-3}), γ_{sp} becomes smaller in the lower dimensional quantum well structures. Since the number of energy levels decreases and the quasi-Fermi level rapidly increases with an increase in the carrier density in lower dimensional quantum well structures, the population inversion is efficiently formed by increasing the carrier density.

Figure 6.2.4 shows the gain parameter g and population inversion parameter γ_{sp} as a function of the size of the quantum well ($W = W_x = W_y = W_z$). When W is large, the number of quantized levels increases and injected carriers are distributed over many quantized levels, hence the maximum gain at the lowest level is limited. Thus, the population inversion parameter increases with the size of the quantum well. On the other hand, when the size W is small, the carrier density for transparency, N_g, increases as the size of the quantum well decreases, hence the population inversion parameter at a fixed carrier density N increases. Consequently, there is an optimum size of the quantum well to minimize the population inversion parameter. For a carrier density of $N = 5 \times 10^{18}$ cm^{-3}, the optimum quantum well size is 6–10 nm, 7–10 nm, and 10–12 nm for QF, QW, and QB structures, respectively.

From Figures 6.2.3 and 6.2.4, it should be noted that the minimum population inversion parameter decreases in the lower dimensional quantum well structures, which is mainly due to a small loss and the efficient formation of population inversion. The reason for the latter can be explained as follows. In the lower dimensional quantum well structures, the energy difference between fundamental and second quantized levels increases, and this enhances the energy difference between the quasi-Fermi level and the energy level of the fundamental quantized level.

Figure 6.2.5 shows the noise figure F of low-dimensional quantum well structures as a function of average input power P_{in}. The size of quantum well $W(= W_x = W_y = W_z)$ was taken as 10 nm, which is almost the optimum size to minimize the population inversion parameter for QF, QW, and QB structures at a carrier density of $N = 5 \times 10^{18}$ cm^{-3}. In the relatively high input power region ($P_{in} > -20$ dB), F is mainly determined by the signal–spontaneous beat noise and F approaches twice the population inversion parameter $2\gamma_{sp}$, where γ_{sp} becomes smaller in lower dimensional quantum well structures. The noise figure of the QB structure can be reduced to 3.3 dB, which is close to the theoretical limit of 3 dB.

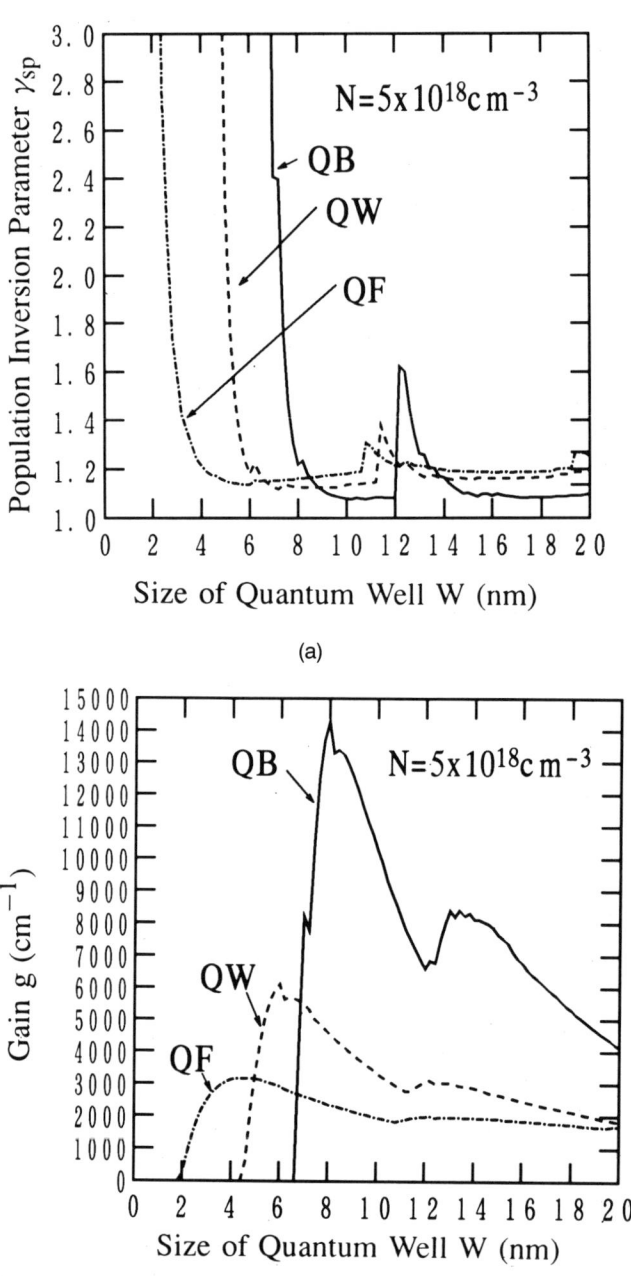

Figure 6.2.4
Gain coefficient g and population inversion parameter γ_{sp} as a function of the size of the quantum well. © IEEE [11]

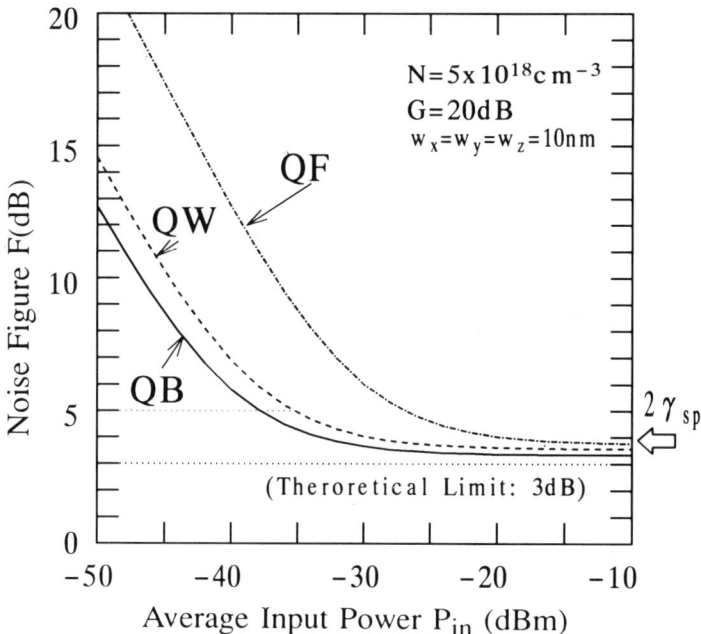

Figure 6.2.5
Noise figure of a semiconductor laser amplifier with low-dimensional quantum well structures. A carrier density of $N = 5 \times 10^{18}$ (cm^{-3}) is used in the calculation. © IEEE [11]

In the low-input power region ($P_{in} < -20$ dBm), the spontaneous–spontaneous beat noise becomes dominant and the noise figure increases with a decrease in input power. In an amplifier with a bulk active medium, a narrow bandpass optical filter is usually required when it is used in the low-input power region, while in low-dimensional quantum well structures, such as the QB and QW structures, the noise figure at the low-input power region becomes very small.

In the region of input power higher than -35 dBm the noise figure of the QB amplifier without a bandpass filter is less than 5 dB, which corresponds to the noise figure of a conventional amplifier using a bulk active medium with a bandpass filter. Thus we can eliminate the optical bandpass filter in a semiconductor laser amplifier by introducing a QB structure, which is very attractive for various applications of optical amplifiers.

The main reason for the low-noise characteristics in QW and QB structures is that the carrier density is concentrated at the gain peak wavelength, and total number of carriers required for amplification is much smaller than that in bulk and QF structures.

The effect of the size fluctuation on the noise figure can be approximately calculated for low-dimensional quantum wells by increasing the spectral broadening factor $\hbar/\tau_{in}$ due to fluctuation (9 meV/1 nm for a 10 nm cube). Considering an inhomogeneous size fluctuation (standard distribution) of 5% with the Gaussian distribution along the z- and y-directions in Figure 6.2.1, the noise figure of the quantum box degrades and becomes almost the same level as that of fluctuation-free quantum wire. However, this value is still lower than that of the quantum film.

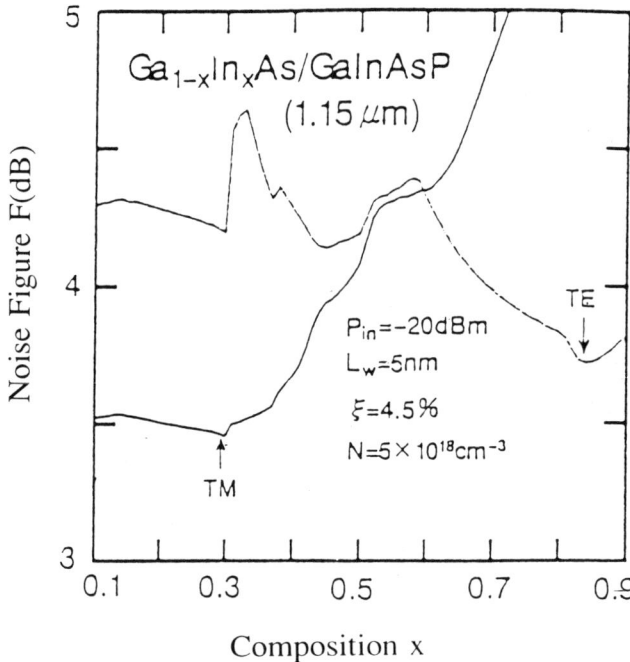

Figure 6.2.6
Noise figure F as a function of indium composition x for $Ga_{1-x}In_xAs/GaInAsP/InP$ strained quantum wells at a carrier density of 5×10^{18} cm^{-3}. © IEEE [10]

4 Noise Characteristics of Strained Quantum Well SLAs

The noise characteristics of a strained quantum well laser amplifier were analyzed taking into account the band-mixing effect in the gain calculation with the density-matrix theory [10]. In the analysis we assumed that the amplifier has a single-pass gain of 20 dB by choosing an appropriate amplifier length, reflection from the facet is neglected, and that the injected carrier density is 5×10^{18} cm^{-3}.

Figure 6.2.6 shows the calculation results of the noise figure of a strained quantum well SLA of a $Ga_{1-x}In_xAs/GaInAsP$ ($\lambda_g = 1.15$ μm) SCH structure as a function of indium composition x which determines the strain in the active region. The thickness of the quantum well is 5 nm, and the optical confinement in the waveguide is 4.5%. The input optical power is -20 dBm, where the signal–spontaneous beat noise is predominant. The noise figure is greater than 4 dB in the unstrained quantum well, while it is reduced to about 3.7 dB in compressive strain and to 3.5 dB in tensile strain. These results were obtained due to an increase in the gain and a reduction in the population inversion parameter in strained quantum well structures.

5 Fabrication of Quantum-wire Lasers/Amplifiers

A new fabrication method for high-density and small-size QW and QB structures, using selective growth on grooved side or corner walls of ultra-fine multilayers, was proposed

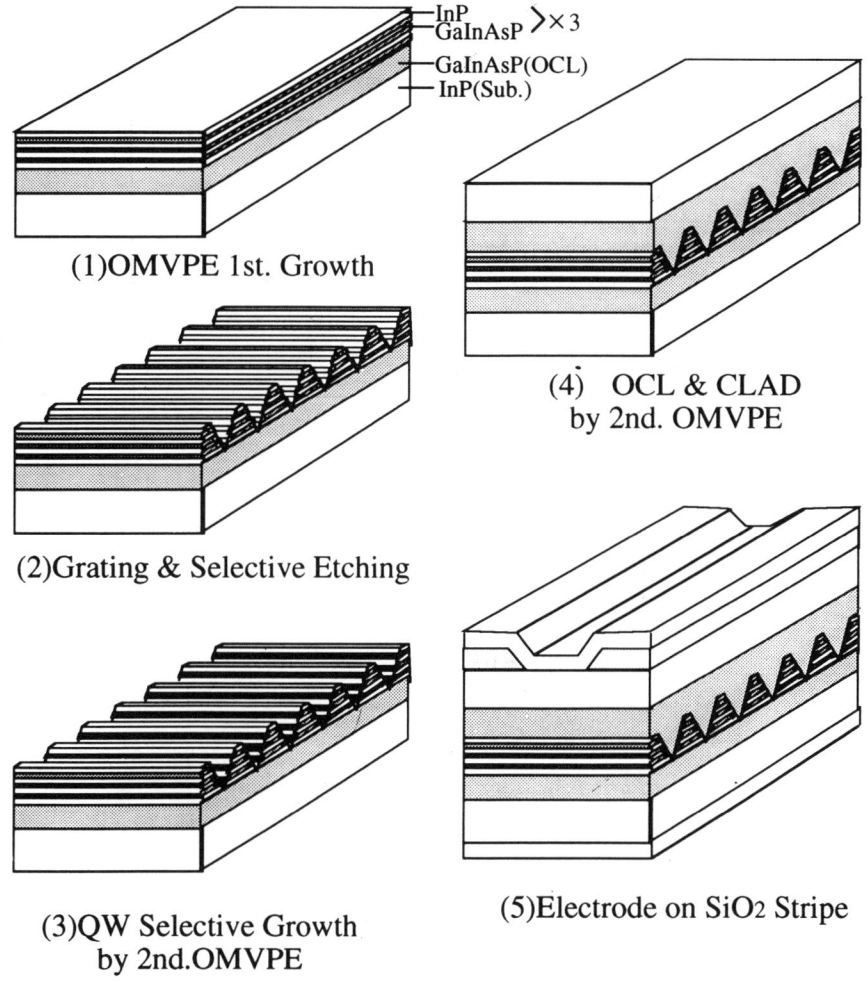

Figure 6.2.7
New fabrication process of a quantum wire structure

and demonstrated. Figure 6.2.7 shows the fabrication process of this method. An n-InP buffer layer and an n-GaInAsP ($\lambda_g \sim 1.2$ μm) optical confinement layer are grown on an InP substrate at first by OMVPE, and then ultra-thin multilayers of n-InP/n-GaInAsP ($\lambda_g \sim 1.2$ μm) are grown. A grating pattern is formed on the SiO_2 mask, put on the multilayers by holographic lithography, and mesa etching with a nonselective etchant is performed. Then, selective etching of GaInAsP in the multilayers is conducted to make side grooves. An n-InP buffer layer, a GaInAs active layer, and a GaInAsP protection layer are grown into the grooves, and then a GaInAsP ($\lambda_g \sim 1.2$ μm) optical confinement layer, a p-InP cladding layer, and a p^+-GaInAsP cap layer are grown. Finally, a SiO_2 stripe and an electrode are formed.

The schematic structure and cross-sectional SEM view are shown in Figure 6.2.8 for a device fabricated by this method. A GaInAs QW structure grown selectively on the

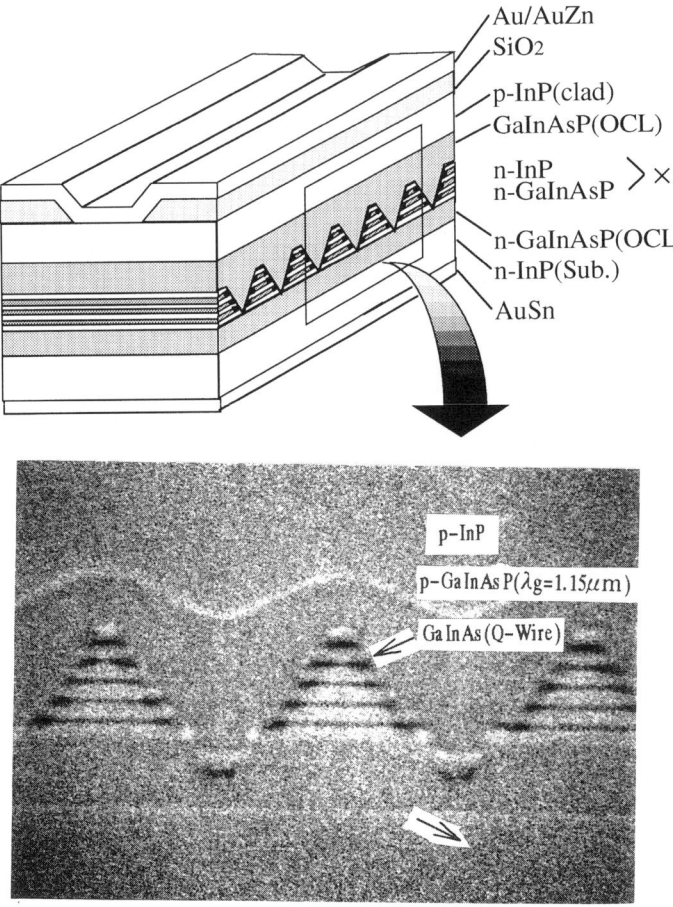

Figure 6.2.8
Structure of a fabricated device and its cross sectional SEM image

grooved side wall is clearly seen. In this method the total volume of the QW array can be comparable with that of a single-layer quantum film, which is advantageous in the device application.

To obtain highly efficient emission and lasing action, we used a $PH_3 + AsH_3$ atmosphere during the period of rising temperature at the growth process, and also a GaInAsP protection layer instead of the InP layer reported before [9]. These were effective for improvement of the overgrown interface and the current injection, respectively. By these improvements, we obtained a lasing action at 77 K of QW structure fabricated by this method for the first time. The wire was 30 nm wide and 15 nm thick, and the lasing wavelength was 1.45 μm, which is in agreement with the theoretical value estimated from the wire size and induced strain. The threshold current density was 12 kA/cm^2 at 77 K. We believe this high value is due to the yet remaining interface problem, and can be reduced by the improvement of the fabrication process.

6 Conclusion

The theoretical noise characteristics of semiconductor laser amplifiers with strained quantum well and low-dimensional quantum well structures are calculated using the density-matrix theory. It is clarified that the signal–spontaneous beat noise can be reduced by using low-dimensional quantum well structures due to the smaller population inversion parameter. In the quantum box structure, the minimum noise figure can be reduced to 3.3 dB, which is close to the theoretical limit of 3 dB. The bandwidth of spontaneous noise can be reduced in the lower dimensional quantum well structure, and this enables a significant reduction in the spontaneous–spontaneous beat noise. The noise figure of solitary semiconductor laser amplifiers consisting of quantum box and quantum wire structures without an optical bandpass filter is smaller than that consisting of the bulk medium with a filter at the small input power level (~ -35 dBm). Thus, we can eliminate the optical bandpass filter in lower-dimensional quantum well semiconductor laser amplifiers, which is very attractive for various applications of optical amplifiers.

As a first step toward realizing low-dimensional quantum well SLAs, GaInAs/GaInAsP/InP quantum-wire semiconductor lasers/amplifiers were fabricated using selective growth on grooved side walls, and a lasing action was achieved at 77 K.

References

[1] M.J. Coupland, K.G. Hambleton and C. Hilsum, *Phys. Lett.*, **7**, 231–232 (1963).
[2] T. Kambayashi and Y. Suematsu, *Trans. IEICE Jpn*, **E64**, 489–496 (1981).
[3] T. Mukai, Y. Yamamoto and T. Kimura, in *Semiconductor and Semimetals*, **22-E**, Academic Press, 1985, pp. 265–319.
[4] R.J. Mears, L. Reekie, I.M. Jauncey and D.N. Payne, *Conference on Optical Fiber Communication (OFC'87), Technical Digest*, Reno/Nevada, 1987, W12.
[5] S. Yamamoto, H. Taga, N. Edagawa, K. Mochizuki and H. Wakabayashi, *International Conference on Integrated Optics and Optical Fiber Communication (IOOC'89)*, Kobe, Japan, 1989, post-deadline paper, 20PDA-9.
[6] D. Personick, *Bell Syst. Tech. J.*, **52** (6), 843–874 (1973).
[7] K. Shimoda, H. Takahashi and C.H. Townes, *J. Phys. Soc. Jpn*, **12** (7), 686–700 (1957).
[8] M. Asada, Y. Miyamoto and Y. Suematsu, *IEEE J. Quantum Electron.*, **QE-22** (9), 1915–1921 (1986).
[9] K. Komori, A. Hamano, S. Arai, Y. Miyamoto and Y. Suematsu, *Jpn J. Appl. Phys.*, **31**, L535–L538 (1992).
[10] Y. Huang, K. Komori and S. Arai, *IEEE J. Quantum Electron.*, **29** (12), 2950–2956 (1993).
[11] K. Komori, S. Arai and Y. Suematsu, *IEEE J. Quantum Electron.*, **28** (9), 1894–1900 (1992).

6.3

Quantum Well and Wire Lasers: InP Systems

Shigehisa Arai, Yasuyuki Miyamoto, Kazuhito Furuya and **Yasuharu Suematsu**

Abstract

The fabrication process and the fundamental lasing properties of GaInAs/InP quantum well lasers with a narrow wire-like active region, namely quasi-quantum wire lasers, are described in this section. The fabrication process, consisting of the direct writing method by electron beam exposure followed by wet-chemical etching and two-step organo-metallic vapor-phase-epitaxial growth, has become a more or less mature technology. A comparison of the threshold current as well as that of the spontaneous emission peak–wavelength shift due to the lateral quantum-size effect exhibited advantages of adoption of strained quantum well structures into GaInAs/InP quantum wire lasers

1 Introduction

Low-dimensional quantum well structures, such as quantum wire and quantum box structures, are very attractive for applications to various optoelectronic devices because of their inherent advantages of high differential gain under carrier injection [1,2] for lasers and amplifiers, and large refractive index variation induced by an electric field [3] for compact optical switches [4]. As for semiconductor lasers, an extremely low threshold current operation [1,2], superior temperature dependence [1], and narrow linewidth or small chirping operations due to smaller linewidth enhancement factor [5] are expected.

Several methods to fabricate quantum wire and quantum box structures have been reported up to now. Special crystal growth technologies have been developed, such as fractional layer growth [6], selective growth on a fine patterned wafer [7] or on a special surface crystal orientation [8], and the strain-induced lateral-layer ordering scheme [9], and so on. On the other hand, an ultra-fine lithography technology is very important and attractive since it is more general and has various applications.

In this section the fabrication process of GaInAs/InP low-dimensional quantum well structure lasers utilizing electron beam exposure (EBX) direct writing and wet-chemical etching followed by organometallic-vapor-phase (OMVPE) regrowth is described. An improved fabrication process, which enabled a room temperature continuous wave (CW) operation of GaInAs/InP quasi-quantum wire lasers, is explained. Other problems of high performance operation are also discussed.

2 Fabrication Process

An outline of the fabrication process of GaInAs/GaInAsP/InP quantum wire lasers is shown in Figure 6.3.1 [10]. First, the initial quantum well wafer was grown by OMVPE

(a) DH-growth (OMVPE)

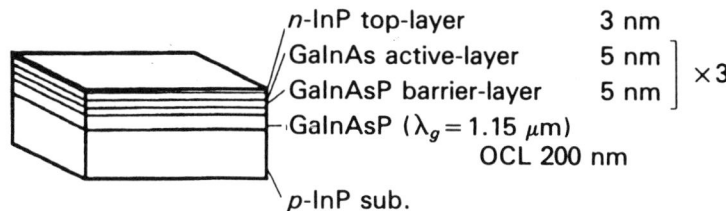

(b) EB lithography

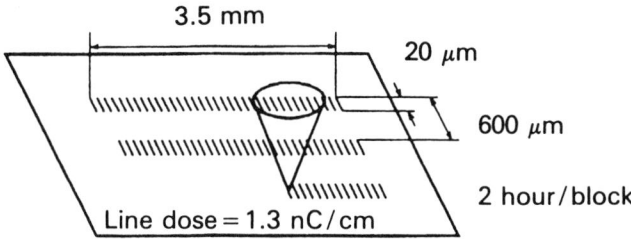

(c) Wet-etching

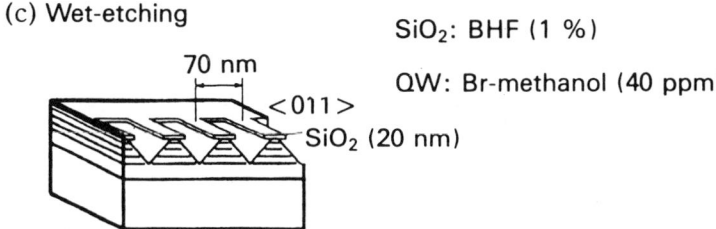

(d) Embedding growth (OMVPE)

(e) BH-growth (LPE)

Figure 6.3.1
Fabrication process of GaInAs/GaInAsP/InP SCH quantum wire lasers [10,12]

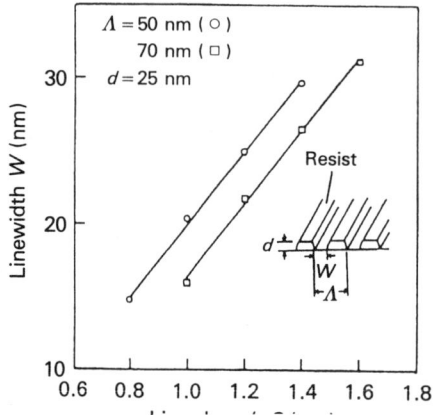 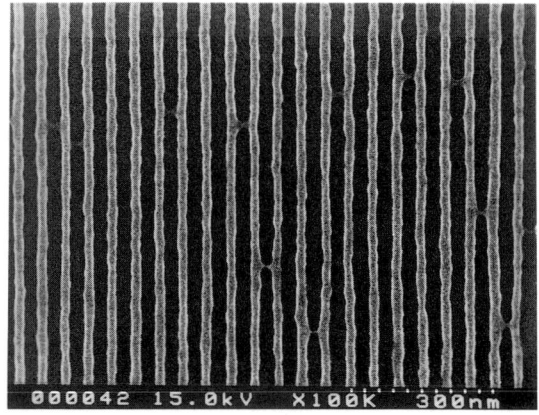

Figure 6.3.2
(a) Exposed linewidth of thin PMMA resist as a function of the line dose amount and (b) an SEM photograph of a 50 nm periodic line pattern

on an InP substrate (Figure 6.3.1(a)). Then 20–30 nm thick SiO$_2$ was deposited by CVD and a grating pattern along the [011] direction with a period of 50–100 nm was transferred to part of the wafer (20 μm × 3.5 mm block, 10 blocks with an interval of 600 μm) by an EBX system, as shown in Figure 6.3.1(b)). It took 2 h to expose each block to the line dose condition of 1.3 nC/cm. Figure 6.3.2 shows (a) the developed linewidth W as a function of the dose condition for grating periods of 50 nm and 70 nm and a PMMA thickness of 25 nm, and (b) an SEM photograph of a 50 nm period PMMA grating pattern (Pt is coated on for better contrast). The fluctuation of the wire width is around ±4 nm, which is still much larger than that required for high performance device application. After etching of the SiO$_2$ mask with 1% BHF for 13 s, wet chemical etching was done with 40 ppm (in volume) Br-methanol for 5 s. A grating depth of more than 30 nm was obtained.

The wafer then underwent an OMVPE regrowth process after removing the SiO$_2$ mask. We have found two important treatments, one is a preheating process with a pure H$_2$ atmosphere, i.e. the wafer was set at 400 °C for about 30 min, and the other is a thin (3 nm) InP layer growth prior to the growth of the GaInAsP optical confinement layer (OCL). The former process was effective for a reduction in the donor concentration at the regrowth interface, and the latter was effective for a reduction in the threshold current of GaInAs/InP lasers with a 70–140 nm wide wire-shaped active region, which were fabricated using conventional laser holographic lithography instead of the EBX lithography.

Figure 6.3.3 shows the threshold current density as a function of the inverse of the cavity length for three different groups: (a) one-step grown multiple quantum film (MQF) lasers; (b) two-step grown wire-like MQF lasers grown on a p-type substrate; and (c) two-step grown wire-like MQF lasers grown on an n-type substrate, where the initial wafer structure is the same (five periods of MQF with a well/barrier thickness of 8 nm/8 nm). As can be seen, the threshold current density of group (b) was $J_{th} = 0.9$ kA/cm^2 for a cavity length of 1 mm, and was almost half that of group (c) and only

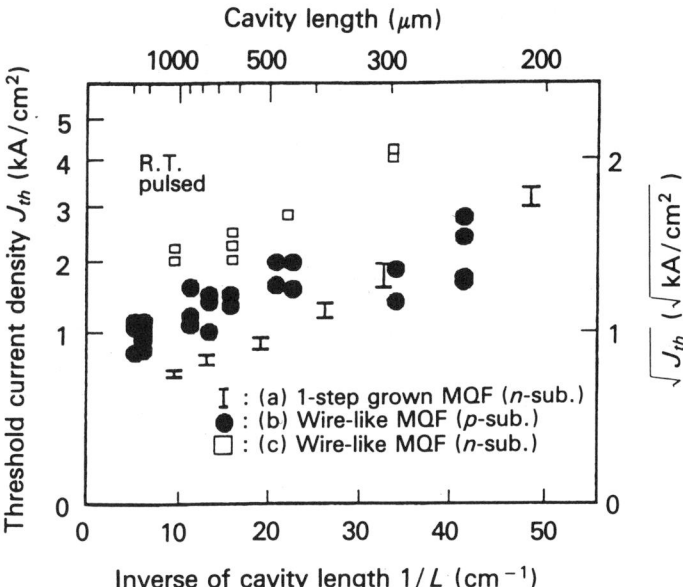

Figure 6.3.3
Threshold current densities of three different groups of lasers as a function of the inverse of the cavity length [10]

20–30% higher than that of one-step grown MQF lasers (group (a)). This difference may be attributed to less optical confinement of the active region of the wire-like MQF lasers ($\xi = 2\%$, almost half that of the original MQF wafer).

Finally, a BH structure with an active region width of 2 μm was fabricated by conventional LPE growth, where the stripe was formed in the direction perpendicular to the wire direction.

3 Lasing Properties of GaInAs/InP Quasi-quantum-wire Lasers

Since strained quantum well structures are very attractive for the high performance operation of semiconductor lasers, and the reduced effective mass of holes will enhance the difference between subband energy levels in quantum wires and quantum boxes, the introduction of a strained quantum well structure into low-dimensional systems is very attractive [11]. Hence, we fabricated three types of $Ga_{1-x}In_xAs$/GaInAsP/InP quasi quantum wire lasers from different quantum film structures, namely (a) a lattice-matched multiple quantum well wire (LM-MQW: $x = 0.53$) structure; (b) a compressively-strained multiple quantum well wire (CS-MQW: $x = 0.70$) structure; and (c) a tensile-strained single quantum well wire (TS-SQW: $x = 0.34$) structure. The thickness W_y and the width W_x, the period Λ, and the number of layers N_w of the quasi quantum wire structure are listed in Table 6.3.1, where the typical lasing wavelength λ, the optical confinement factor of the active medium ξ, the cavity length L, the threshold current

Table 6.3.1
Summary of GaInAs/InP quasi quantum wire lasers

Type	x of In	λ (μm)	n	W_y (nm)	W_x (nm)	Λ (nm)	ξ (%)	L (μm)	I_{th} (mA)	J_{th} (A/cm^2)
LM-MQW	0.53	1.51	3	8	10–30	70	0.6	700	82	5900
CS-MQW	0.70	1.54	5	3	30–60	100	0.9	910	29	1650
TS-SQW	0.34	1.46	1	12	30–40	70	0.95	980	16	816
(Film)	—	1.48	1	12	—	—	1.90	950	8	421

I_{th}, and the threshold current density J_{th} under room temperature CW conditions are also listed.

As for quasi quantum wire lasers with an LM-MQW active region, even though room temperature CW operation was obtained, the threshold current was very high due to poor optical confinement and a broad gain spectrum. We then tried to fabricate a slightly wider wire-like active region by introducing a compressively strained (CS) MQW structure, since the effective mass of heavy holes is expected to be reduced in the lateral direction, which will enhance the separation of quantized energy levels [11]. Even though the size of each wire was confirmed to vary from 30 nm at the top to 60 nm at the bottom, the threshold current was much reduced compared with that of the lattice-matched device. This lower threshold current operation of the CS-MQW wire laser seemed to be attributed to an increase in the optical confinement of the active medium and would be further improved by using a thicker quantum well structure.

The characteristic temperature T_0, by which the temperature dependence of the threshold current is empirically given as $I_{th}(T) = I_0 \exp(T/T_0)$, of quantum wire and quantum box lasers was predicted to be much higher than that of quantum film lasers [1]. T_0 of the CS-MQW wire laser was measured to be 119 K at the low-temperature range ($T < 220$ K) and was 70 K at around room temperature, which was slightly higher than that of the CS-MQW film laser [10]. Moreover, the lasing wavelength of the CS-MQW wire laser was approximately 40 nm shorter than that of the CS-MQW film laser fabricated on the same wafer. The difference in the spontaneous emission peak wavelength at a similar carrier injection level was observed to be 15–35 nm (8–18 meV), which was in good agreement with a theoretical energy level shift for the wire width of 60–30 nm [11]. These blue shifts may be attributed to the lateral quantum-size effect.

Because the wire width of each stacked quantum well differs a lot due to the mesa shape caused by wet-chemical etching, a single quantum well (SQW) structure is suitable for reducing the wire width deviation. A thicker quantum well is preferable for lowering the threshold carrier density, quasi quantum wire lasers consisting of a tensile-strained (TS) SQW (well/barrier thickness of 12 nm/12 nm) was fabricated [12]. Actually, there have been several reports that TS-QW lasers have improved lasing properties and such a TS-QW structure would have a reduced effective mass of light holes in the lateral direction. Figure 6.3.4 shows the schematic structure and its cross-sectional SEM view of a graded-index separate confinement heterostructure (GRIN-SCH) TS-SQW laser with a 70 nm periodic wire active region. The initial wafer consisted of a 12 nm thick active layer (+1.5% strain) sandwiched between 0.2 μm thick GaInAsP

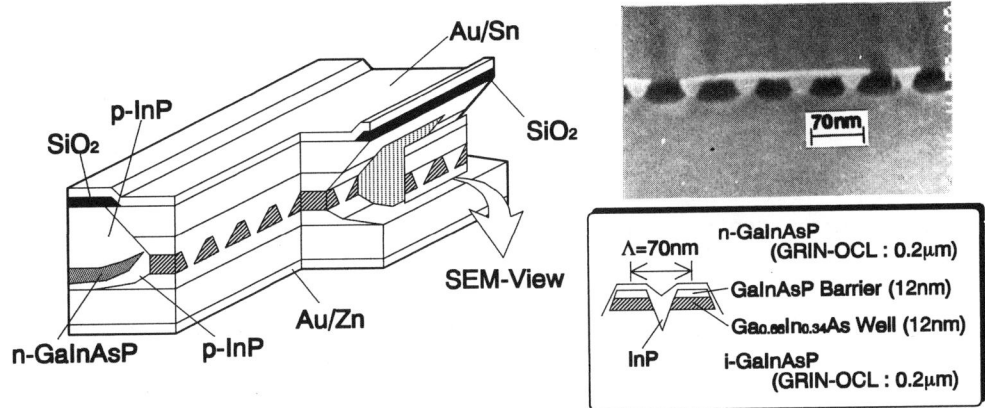

Figure 6.3.4
Schematic structure of a tensile-strained GaInAs/GaInAsP/InP quasi quantum wire laser and its cross-sectional SEM view [12]

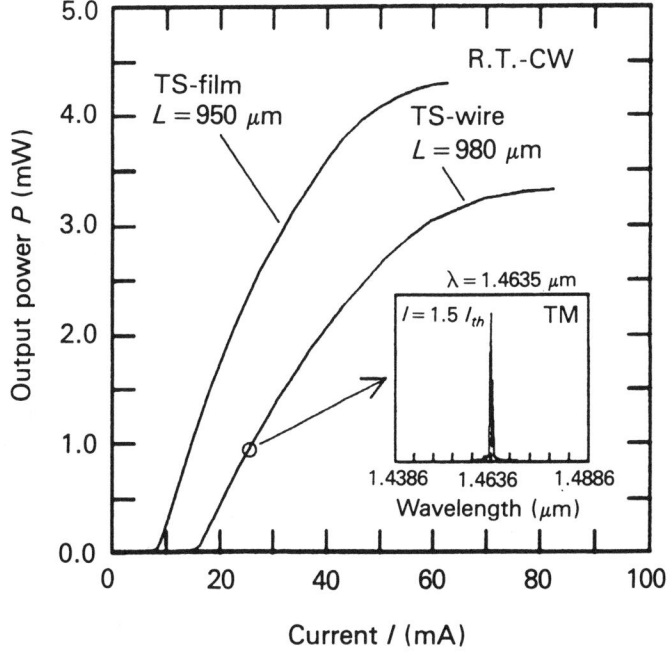

Figure 6.3.5
Light output characteristic of a tensile-strained (TS) GaInAs/GaInAsP/InP quasi quantum wire laser and that of a TS quantum film laser cut from the same wafer [12]

linearly graded ($\lambda_g = 1.05–1.25$ μm) optical confinement layers. As can be seen, the wire width was observed to be 30–40 nm.

Figure 6.3.5 shows the light output characteristic of a TS-SQW wire laser with a cavity length of 980 μm and its lasing spectrum at $I = 1.5 I_{th}$, along with the light output property of a 950 μm long TS-SQW film laser cut out from the same wafer.

Both lasers operated in TM-polarization. The threshold current of the TS-SQW wire laser was only 16 mA ($J_{th} = 816$ A/cm^2) and was twice as high as that of the TS-SQW film laser ($I_{th} = 8$ mA). Because the threshold current density of this TS-SQW film laser was only 421 A/cm^2 and the differential quantum efficiency η_d was 40%, which may be comparable with those of a similar structure laser fabricated by one-step LP-OMVPE growth, the regrowth process employed here was quite successful for regrowth on a planar wafer or an increase in the threshold carrier density was negligible due to larger optical confinement factor of the active layer ($\xi = 1.90\%$).

On the other hand, since the light output characteristics of the TS-SQW wire laser were poorer than those of the TS-SQW film laser, the temperature dependences of the threshold current I_{th} and the differential quantum efficiency η_d of those lasers were measured and compared. As can be seen in Figure 6.3.6, the temperature dependence of these lasers are almost the same; the I_{th} of the TS-SQW wire laser was approximately twice high as that of the TS-SQW film laser within the investigated temperature range, and η_d of the TS-SQW wire laser was degraded to 60–70% of that of the TS-SQW film laser. However, the slope efficiency of the light output below the threshold was almost the same for both samples, which means that the fraction of nonradiative recombination current in the TS-SQW wire laser is almost the same as that in the TS-SQW film laser. Hence the poor η_d property of the TS-SQW wire laser can be attributed to a larger optical loss in the cavity. From a temperature dependence of η_d, the absorption loss coefficient of the TS-SQW film laser was evaluated to be 1–2 cm^{-1} at $T < 230$ K, while that for the TS-SQW wire laser was evaluated to be 7–8 cm^{-1}. This discrepancy

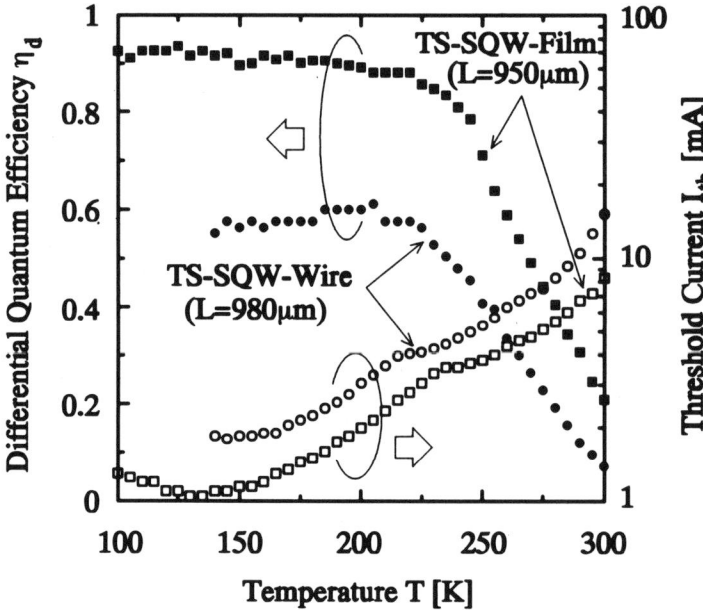

Figure 6.3.6
Temperature dependences of threshold current and differential quantum efficiency of a tensile-strained (TS) GaInAs/GaInAsP/InP quasi quantum wire laser and those of a TS quantum film laser cut from the same wafer

might be caused by carrier-induced absorption mechanisms, such as the free carrier plasma effect and inter-valence band absorption (IVBA), because the threshold carrier density of the TS-SQW wire laser is higher than that of the TS-SQW film laser. An increase in the optical confinement factor of the active region ξ is effective in reducing the threshold carrier density, and hence in improving the differential quantum efficiency.

The spontaneous emission peak wavelength of both the CS-MQW wire and the TS-SQW wire lasers were measured and compared with those of the CS-MQW film and the TS-SQW film lasers as shown in Figure 6.3.7 as a function of the square root of the injection current density, which corresponds to the injected carrier density because the spontaneous emission carrier lifetime in GaInAsP/InP lasers can be regarded as inversely proportional to the injected carrier density. In both types of lasers a blue shift with an increase in the injection current was observed, and tends to saturate at a certain current density. The space-filling factor of quasi quantum wire lasers is one-third to one-half of that of quantum film lasers and the emission peak wavelength should be compared at the same carrier density, i.e. solid circles should be compared with solid squares in Figure 6.3.7. The transition energy shift of the TS-SQW wire laser was approximately 10 meV and was almost half that of the CS-MQW wire laser; this fact indicates that the effective mass of holes in the lateral direction of the CS quantum well is much smaller than that of the TS quantum well, as expected by theoretical prediction [11].

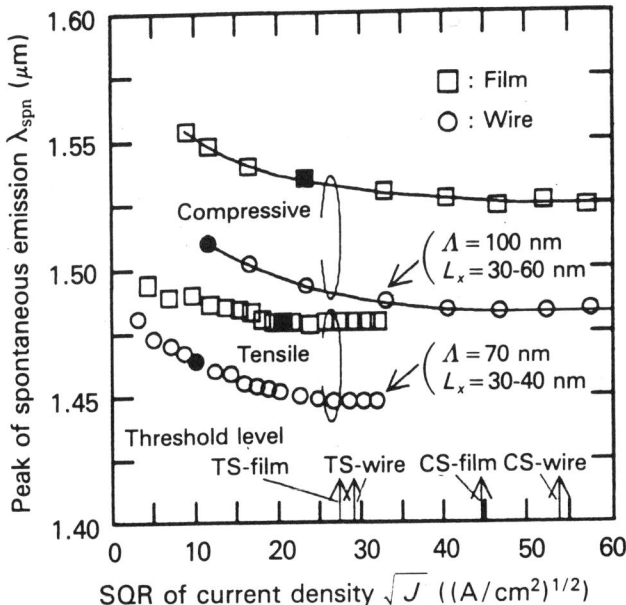

Figure 6.3.7
Spontaneous emission peak wavelength of tensile-strained (TS) GaInAs/GaInAsP/InP quantum well lasers and compressively-strained (CS) quantum well lasers as a function of the square root of the injection current density. Open circles and open squares indicate data for quasi quantum wire lasers and quantum film lasers cut out from the same wafer [12]

4 Conclusion

The preheating process in a pure H_2 atmosphere prior to the regrowth by LP-OMVPE and the use of a p-type InP substrate are very effective in improving the regrowth interface of two-step grown GaInAs/GaInAsP/InP lasers. These new methods enabled the realization of room-temperature CW operation of quasi quantum wire lasers with a moderately low threshold current density. However, it is still much higher than that predicted by theory and the differential quantum efficiency is lower.

In order to overcome these problems, an increase in the optical confinement factor of the active region seems to be the most important issue. The development of fabrication technologies for ultra-fine size structures (10-20 nm size with an accuracy of a few nm), such as a damage-free dry etching process with a high aspect ratio, is strongly required.

References

[1] Y. Arakawa and H. Sakaki, *Appl. Phys. Lett.*, **40** (11), 939-941 (1982).
[2] M. Asada, Y. Miyamoto and Y. Suematsu, *IEEE J. Quantum Electron.*, **QE-22** (9), 1915-1921 (1986).
[3] K.G. Ravikumar, T. Kikugawa, T. Aizawa, S. Arai and Y. Suematsu, *Appl. Phys. Lett.*, **58** (10), 1015-1017 (1991).
[4] K. Shimomura, S. Arai and Y. Suematsu, *IEEE J. Quantum Electron.*, **28** (2), 471-478 (1992).
[5] Y. Miyake and M. Asada, *Jpn J. Appl. Phys.*, **28** (7), 1280-1281 (1989).
[6] M. Tsuchiya, P.M. Petroff and L.A. Coldren, *IEEE Trans. Electron Dev.*, **ED-36** (11), 2612-2613 (1989).
[7] E. Kapon, S. Simhony, R. Bhat and D.M. Hwang, *Appl. Phys. Lett.*, **55** (26), 2715-2717 (1989).
[8] K.Y. Cheng, K.C. Hsieh and J.N. Baillargeon, *Appl. Phy. Lett.*, **60** (23), 2892-2894 (1992).
[9] H. Kamada, R. Nötzel, J. Temmyo, T. Furuya and T. Tamamura, *The 41st Spring Meeting of the Jap. Soc. of Appl. Phys., 28p-S-17, digest no. 3*, 1994, p. 1169.
[10] Y. Miyake, H. Hirayama, K. Kudo, S. Tamura, S. Arai, M. Asada, Y. Miyamoto and Y. Suematsu, *IEEE J. Quantum Electron.*, **29** (6), 2123-2133 (1993).
[11] S. Ueno, Y. Miyake and M/Asada, *Jpn J. Appl. Phys.*, **31** part 1, (2), 286-287 (1992).
[12] K. Kudo, Y. Nagashima, S. Tamura, S. Arai, Y. Huang and Y. Suematsu, *IEEE Photon. Technol. Lett.*, **5** (8), 864-867 (1993).

6.4

Ultrafast Guided Wave Electrooptic LiNbO$_3$ Modulators

Tadasi Sueta and **Masayuki Izutsu**

Abstract

Electrooptic light modulators of LiNbO$_3$ guided wave structure for use in microwave and millimeter-wave regions are reviewed. Design practice is discussed to point out the importance of using band operation instead of the conventional baseband operation to realize ultrafast operation of the modulators in future photonic systems. Possible applications are also discussed for the case when lightwaves and microwaves/millimeter-waves are related.

1 Introduction

There has been significant development in recent years of guided wave light modulators for ultrafast operation [1]. For use in digital transmission systems, for example, modulators up to 10 GHz frequency range have reached the stage of practical use, and certain 20–30 GHz bandwidth devices are now at the final stage of development. In most cases these devices are built as intensity modulators of a Mach–Zehnder interferometric structure by using a Ti indiffused LiNbO$_3$ optical waveguide together with a coplanar transmission line electrode for travelling-wave operation [2].

Although direct modulation of the laser diodes is a simple and important method for controlling lightwaves, devices specific for the modulation/switching of lightwaves by rf, microwave (mw) and millimeter-wave (mmw) signals are substantial for constructing photonic systems to handle these signals by light. The performances of the devices have been improved dramatically by using optical waveguide structures. Light modulation even by mmw signals has already been achieved with a reasonable several hundreds mW drive power.

Here, electrooptic light modulators of LiNbO$_3$ guided wave structure for use in the mw and mmw regions are reviewed. The design practice is discussed to point out the importance of using the band operation instead of the conventional baseband operation to realize ultrafast operation of the modulators in future photonic systems. Possible applications are also discussed for the case when lightwaves and mw/mmw are related.

2 Ultrafast Modulators

Guided-wave light modulators/switches are mainly fabricated to be driven using the electrooptic effect. According to the applied modulating voltage, the refractive index of the guide is changed to give the transmitting lightwave a phase retardation proportional to the voltage, so that the structure of the electrooptic phase modulator is quite straightforward. For light intensity modulation or switching, Mach–Zehnder and directional coupler type structures have largely been used. The acousto-optic effect is also available for constructing wideband light control devices. Using the Bragg diffraction of the guided light by a surface acoustic wave, a light deflector/modulator of several GHz bandwidth has already been demonstrated. Another significant effect for controlling lightwaves is the nonlinear optic effect, which may be especially important for the interaction of multiple lightwaves to generate or control mw/mmw signals. Light modulators in the mw/mmw range are primarily built to be operated through the electrooptic effect, the response of which is so fast that the effect adjoins the nonlinear optic effect by which the index changes with the optical field amplitude. There is therefore no limitation in principle in outlining a device having an ultrafast response time or an operation frequency up to or beyond the sub-millimeter range except in certain absorption bands of the electrooptic material. However, from a practical point of view, the design of the electrical circuit of a device, i.e. the structure of the modulator electrodes and the feeding method of the modulating signals, is especially important to actually realize ultrafast light modulators/switches.

2.1 Travelling-wave mode of operation

A powerful way to elevate the modulating frequency to the mw/mmw region is to use the travelling-wave mode of operation (Figure 6.4.1). The modulating mw/mmw is made to propagate in the same direction as the lightwave. The bandwidth is limited by an accumulating phase difference between the light and modulating waves during their propagation through the modulator, and is inversely proportional to the product of the velocity mismatch between two waves and the interaction length (or the electrode length). The response time of the modulator shows a similar dependence on these parameters [3].

The selection of a modulator electrode structure as part of the wide-band mw/mmw circuits for the modulating signal is another important problem for smooth frequency responses. To avoid unwanted reflections at discontinuities in the circuit, an asymmetric coplanar strip line and a three-electrode coplanar waveguide structure were successfully adopted.

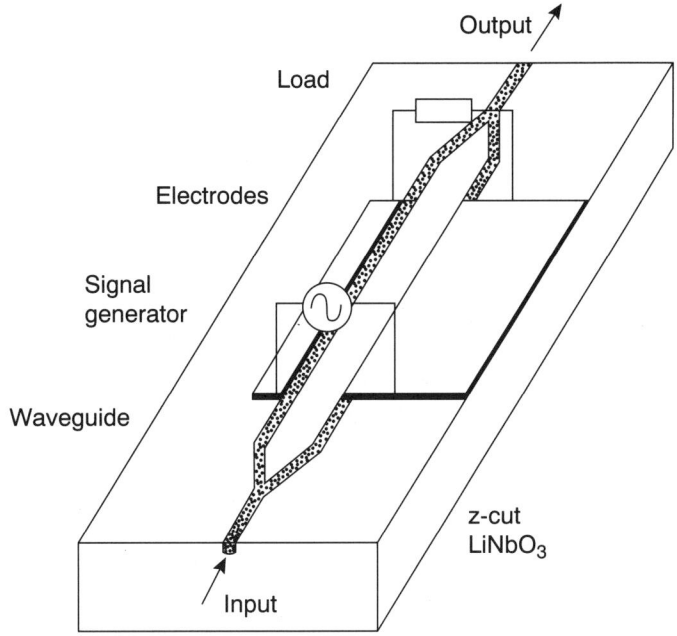

Figure 6.4.1
Schematic of a guided wave light modulator for travelling-wave operation

2.2 Reduction of the velocity mismatch

With a reduced interaction length, faster operation will be achieved, although a higher drive power is needed to decrease the modulation efficiency since the required drive power is inversely proportional to the interaction length squared. The way to achieve efficient and fast, or wide-band, operation at the same time is to reduce the velocity mismatch between the light and modulating waves of the travelling-wave modulator. In the case of $LiNbO_3$ waveguides with a coplanar electrode structure, the lightwave travels twice as fast as the modulating microwave, so there is a need to decrease the velocity of the lightwave and/or to increase the velocity of the modulating wave in order to reduce the mismatch [4].

Several attempts have already been made to realize extremely high modulation efficiency of 10–20 GHz light modulators. Among them the use of extra thick (5–7 μm) electrode is an accepted design practice to decrease the velocity mismatch by increasing the portion of the modulating electric field travelling outside the crystal substrate of a high dielectric constant. A groove between the parallel coplanar electrodes is also applied to reduce the velocity mismatch (Figure 6.4.2).

2.3 Modulation polarity reversal

Another approach to achieving efficient broadband modulation is the reversal of the modulation polarity according to a special sequence by, for example, shifting

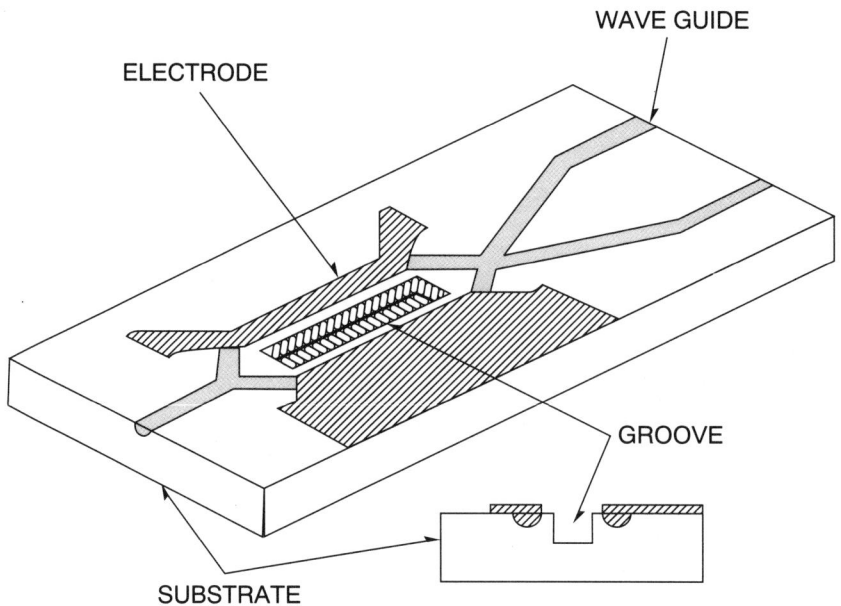

Figure 6.4.2
travelling-wave light modulator with an etched groove to reduce velocity mismatch

the position of electrodes relative to the optical waveguides. The cumulative phase difference between the light and modulating waves while they travel down the device with different velocities is canceled out at a certain covering distance by the inversion of the modulation polarity. Applying a special sequence for the reversal, a bandwidth up to 40 GHz has been reported.

2.4 Band operation

As a different way to achieve light modulation in the higher frequency range, the use of band operation has been proposed. Insofar as conventional lumped or travelling wave structures are employed to build modulators, a wider bandwidth will be needed to modulate lightwaves at higher frequencies since the lower end of their operation band is rooted at zero frequency. To double the bandwidth, the electrode length should be halved, and with a fixed drive power the modulation depth decreases to a quarter of the initial value.

The band modulation scheme reduces the drive power dramatically at the expense of narrowing the bandwidth. The concept of band-limited modulation is believed to play an important role in future photonic systems. The device will be used to mix lightwaves with mw/mmw. Bulk type devices have been built by placing crystals in electrical or optical resonators, while for waveguide types it was initially proposed to build with a DFB structure and a periodic polarity reversal, and was realized by using a standing-wave electrode. We have demonstrated efficient modulation in the millimeter-wave region by introducing novel resonant electrode structures (Figure 6.4.3) [5]. Recently,

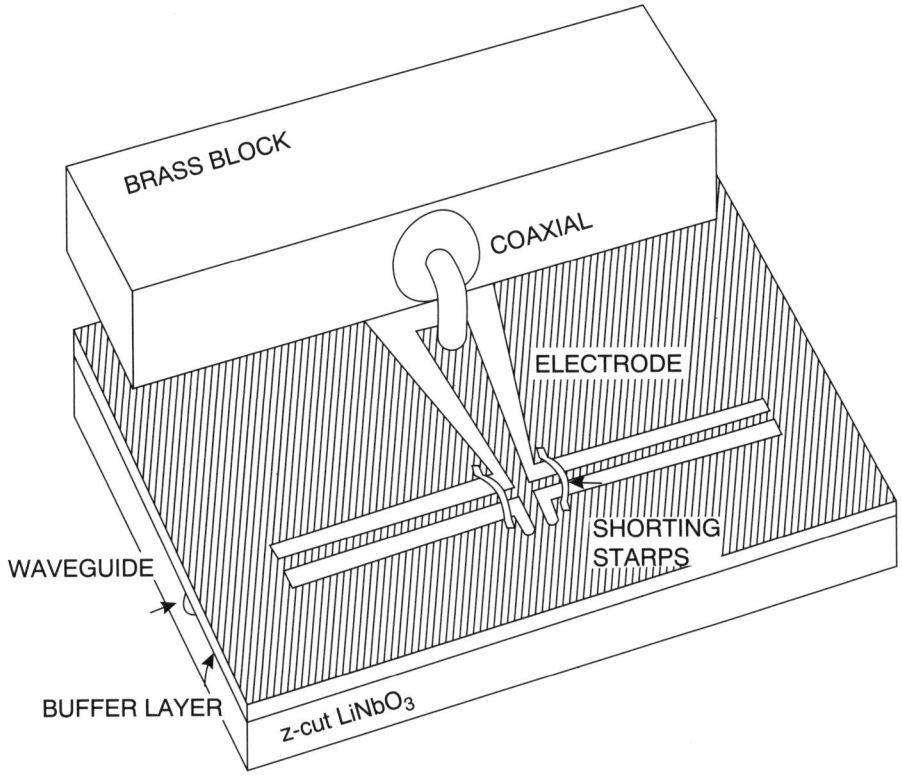

Figure 6.4.3
Guided wave band modulator using a resonant coplanar electrode

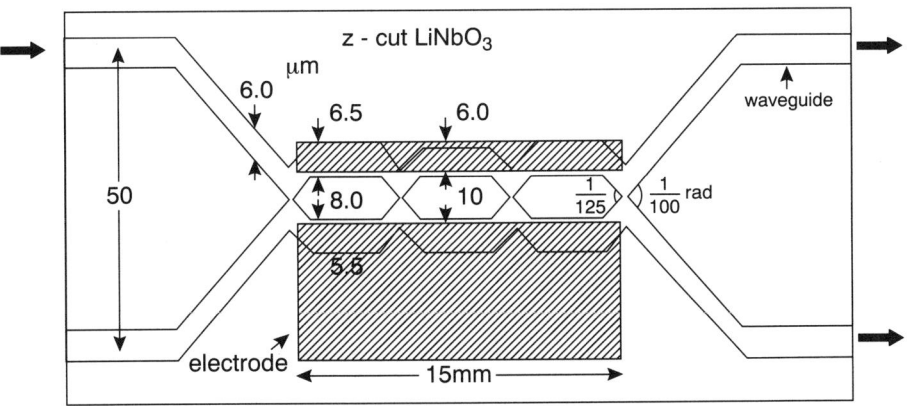

Figure 6.4.4
Guided wave band modulator using a twisted Mach–Zehnder interferometer

it has also been operated by periodic reversal [6] and an electrode of fast wave structure loading inductances periodically on a travelling-wave electrode (Figure 6.4.4).

3 Modulators for Microwave/Millimeter-wave Applications

Recently much attention has been directed toward research fields where lightwaves and mw/mmw are related [7]. Various high-speed optical devices, especially excellent integrated optical components, have become available for practical microwave applications.

3.1 Lightwaves and microwave/millimeter-waves

The frequency, or the photon energy, of lightwaves is some tens of thousands times higher than that of the mw/mmw, and thus to control lightwaves is much more difficult compared with mw/mmw. Lightwaves couple with matter through the quantum electronic mechanism at an energy around the eV range. This might be the basic reason why we have such a nice transmission line as optical fibers in the optical region.

One of the chief advantages in the application of lightwave technology to mw/mmw systems comes mainly from the fact that it is a better transmission medium than coaxial cables. The recent rapid development of optoelectronic elements and devices has granted the privilege of using fiber optics in microwaves. Applications to areas such as signal distribution and delay line systems are included in the promising areas.

Another important feature of using lightwaves to control mw/mmw is the possibility of realizing shorter response times, higher isolation between controlling and controlled signals, and so on. Intensive research activities in these fields have already been started.

In the first stage of these applications direct modulation of semiconductor lasers may be utilized in combination with single-mode optical fibers and photodiode receivers. To realize various complicated functions in handling mw/mmw through light, however, it is desirable to introduce devices to process light signals. Integrated optic technologies will play significant roles, for example, in multiple delay line systems by selecting a fiber cable of desirable delay, or the switching of light signals in signal distribution systems.

3.2 Integrated ultrafast devices

One of the most important assets of guided wave devices is the possibility of combining different devices or elements on a single substrate to realize optical integrated circuits. The integration of ultrafast guided wave components onto a substrate provides compact and stable optical circuits which enable us to construct novel optoelectronic functional devices that are difficult to realize without the integrated optics technologies.

Switch matrices composed of multiple optical switch elements have attracted attention for use by space-division switching in optical communication systems, and much work has been reported. Although each element is rather low-speed switches, lightwave signals modulated by microwaves can be treated. For the integration of

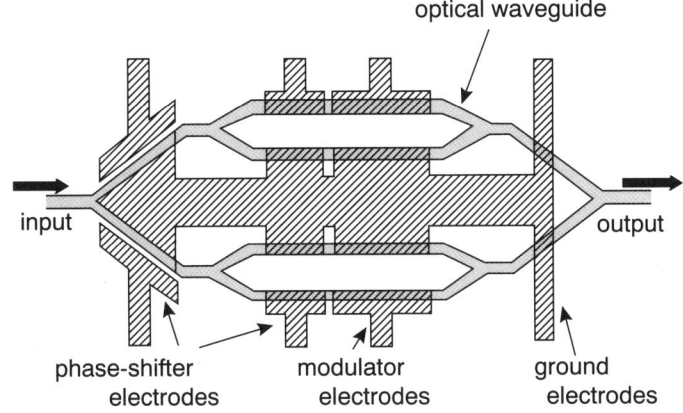

Figure 6.4.5
Integrated optic light SSB modulator/frequency shifter

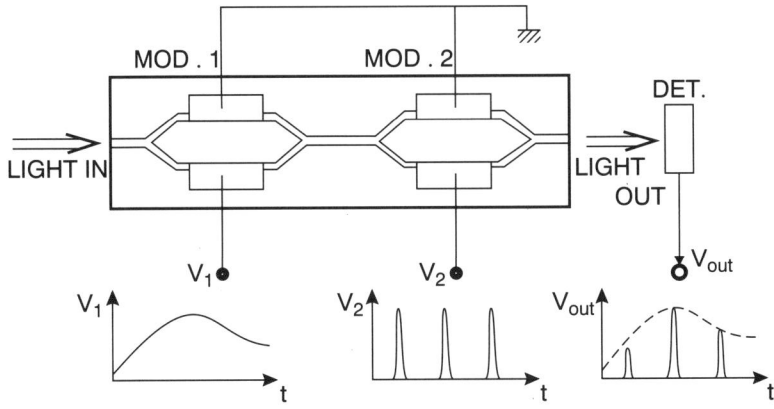

Figure 6.4.6
Integrated tandem light modulators for picosecond signal sampling and multiplication

high-speed electrooptic devices, however, only a few examples have been reported because of difficulties in integration. Examples are an SSB modulator/frequency shifter (Figure 6.4.5) [8], a signal sampler/multiplier (Figure 6.4.6) [9], a time multi/demultiplexer (Figure 6.4.7) [10], and an analog-to-digital converter.

3.3 *Microwave application of integrated-optic devices*

The application of integrated optical devices and components to mw/mmw systems, such as analog signal transmission [11], high-speed electronics and optoelectronics instrumentation for measurement, signal processing, or other various systems, is the inherent area. Recent rapid progress in integrated optics technologies urges us to evolve promising new applications and to evaluate their feasibility, although only a limited

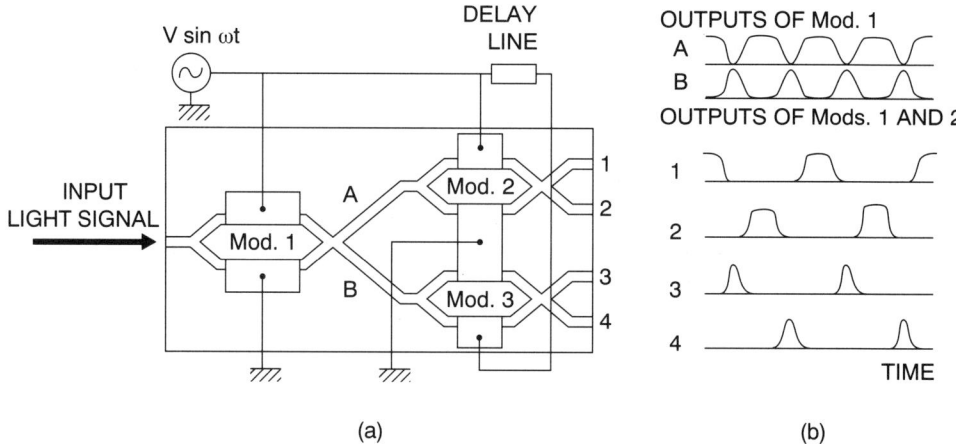

Figure 6.4.7
An integrated 1 × 4 high-speed optical switch for time demultiplexing

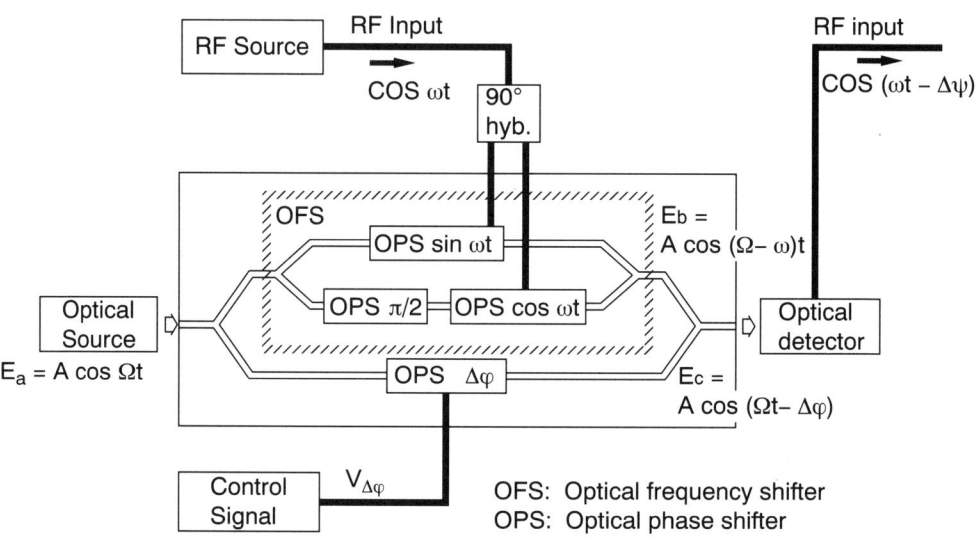

Figure 6.4.8
Microwave phase shifter using the optical waveguide structure

number of examples have been reported as straightforward applications of integrated optics to mw/mmw systems.

Utilization of integrated-optic devices to digital transmission systems is, of course, one of the most promising and leading topics. The other direct application is mw/mmw signal transmission with a fiber-optic link. To deliver mw or mmw signals from one place to distant places, the rf signal is transformed to an optical signal by modulating lightwaves. Then, making use of the low-loss property of the optical fiber, it is delivered to its destination, where the lightwave is demodulated and amplified by an electronic circuit (*e.g.* MMIC) if necessary, to yield the original electrical signal. Endeavors to realize stable, low-noise optical signal generation and detection are also being pursued adopting integrated-optic technologies.

Another example is the application to mw/mmw signal distribution and scanning of array antennas [12]. An optoelectronically controllable microwave phase shifter has been proposed and demonstrated experimentally (Figure 6.4.8). Measurement is another field of interest. For example, the combination of an optical fiber and an ultrafast modulator would be used to measure field distributions in microwave devices by electrooptic sampling, and of an antenna by an integrated-optic microwave field sensor with minimum field disturbance. The precise measurement of distance, or a range finder, is also a promising application of integrated optics.

4 Conclusion

Ultrafast light modulators are one of the key elements in fabricating future photonic systems for information and instrumentation technologies. There have been significant developments to devise novel schemes for the ultrafast operation of electrooptic light modulators, as indicated, while continuing work is needed to materialize ultrafast modulators applicable to practical photonic systems.

A number of possible mw/mmw applications of integrated optics have been proposed and investigated, while a few practical devices have been implemented. With the recent growing interest in this area, we can expect rapid development in the utilization of lightwave technologies in mw/mmw engineering, exploiting the potential of mw/mmw and lightwave interactions.

References

[1] T. Sueta and M. Izutsu, in *Optical Computing in Japan*, S. Ishihara, Ed., Nova Science Publishers, 1990, pp. 383–392.
[2] M. Izutsu and T. Sueta, *Trans. IEICE-C*, **J64-C** (12), 264–271 (1981).
[3] M. Izutsu and T. Sueta, *IEEE J. Quantum Electron.*, **QE-19** (4), 668–674 (1983).
[4] M. Izutsu, H. Haga and T. Sueta, in *Picosecond Electronics and Optoelectronics*, Springer-Verlag, Berlin, 1985, G.A. Mourou, D.M. Bloom and C.H. Lee, Eds., pp. 172–175.
[5] M. Izutsu and T. Sueta, *Trans IEICE*, **J71C** (5), 653–657 (1988) [in Japanese].
[6] Y, Zhou, M. Izutsu and T. Sueta, *IEEE J. Lightwave Technol.*, **9** (6), 750–753 (1991); and also, M. Izutsu, K. Matsumoto and T. Sueta, *Tech. Digest of Fifth Optoelectronic Conference*, 1994, 15B3-3, pp. 346–347.
[7] T. Sueta and M. Izutsu, *IEEE Trans. MTT*, **38** (5), 477–482 (1990).
[8] M. Izutsu, S. Shikama and T. Sueta, *IEEE J. Quantum. Electron.*, **QE-17** (11), 2225–2227 (1981).
[9] M. Izutsu, H. Haga and T. Sueta, *IEEE J. Lightwave Technol.*, **LT-1** (1), 285–289 (1983).
[10] H. Haga, M. Izutsu and T. Sueta, *IEEE J. Lightwave Technol.*, **LT-3** (1), 116–120 (1985).
[11] T. Mizuochi, T. Kitayama, S. Kawanaka, M. Izutsu and T. Sueta, *Trans. OFC '90*, 1990.
[12] Y. Kamiya, W. Cyujou, K. Yasukawa, K. Matsumoto, M. Izutsu and T. Sueta, *Tech. Digest of IEEE AP-S Int. Symp. and URSI Radio Science Meeting*, 40-5, **2**, 1990, pp. 774–777.

7
OPTOELECTRONIC DEVICES FOR PARALLEL PROCESSING AND INTERCONNECTS

In this Chapter we feature several optoelectronic devices mainly employing surface operation and having a two-dimensional (2D) arrayed configuration. These devices will play important roles in parallel optical processing and interconnects. First, the latest research results of surface-emitting lasers and related 2D components are introduced. Section 7.2 deals with semiconductor-based spatial modulators employing an electron-depleting effect. Section 7.3 reviews the research on 3D LSI technology for optical interconnects. Integrated optoelectronic devices for neural computing are discussed in section 7.4. In section 7.5, liquid crystal cells are treated for optical neural computing.

7.1

Surface-emitting Laser Diodes and 2D Arrayed Devices

Kenichi Iga and **Fumio Koyama**

Abstract

In recent years, two-dimensional optical devices and sub-systems based on vertical cavity surface-emitting lasers and planar microlens arrays have been developed for future parallel lightwave systems. The research field of surface-emitting lasers is growing fast and improvements in device processes are remarkable, *e.g.* the realization of sub-mA threshold surface-emitting lasers. The operating wavelength region of surface-emitting lasers is now from visible to infrared. In addition, parallel micro-optics has become important to avoid complexities in optical coupling between various optical components.

In this section we review the progress of surface-emitting semiconductor lasers including some new functions as well as related integrations technology. We also present a novel concept of self-aligning optics using planar microlens arrays.

1 Introduction

Massively integrated parallel optical devices are becoming important for use in future parallel optical fiber communication systems, optical interconnects, optical image processing, parallel optical recording, and so on. For this purpose, vertical cavity surface-emitting lasers (VCSELs) [1] as well as planar microlens arrays [2] have been developed. The vertical cavity is formed by the surfaces of epitaxial layers, and the light output is taken vertically from one of the mirror surfaces. The surface-emitting laser structure may provide a number of advantages: (i) a huge number of laser devices can be fabricated by fully monolithic processes; (ii) a densely packed two-dimensional (2D) laser array can be formed; (iii) an ultra-low threshold operation is expected from the small cavity volume; (iv) the initial probe test can be performed before separating the devices into discrete chips; (v) dynamic single-mode operation is possible; (vi) vertical stacks of multi-thin-film functional optical devices can be integrated intact into an VCSEL resonator; and (vii) it is easy to couple the output to optical fibers.

The first lasing operation of a GaInAsP/InP SE laser device was achieved in 1979 under pulsed conditions at 77 K. The authors' group has spent much effort developing VCSELs with GaInAsP/InP, GaAlAs/GaAs, and GaInAs/GaAs systems. The first room-temperature pulsed operation, low-threshold operations using buried microcavity structures, and the first room-temperature cw operation in GaAlAs/GaAs VCSELs were realized in 1983, in 1987, and in 1988, respectively [1,3]. After that, much attention has been paid to VCSELs, and AT&T, Bellcore, UCSB, and many other research groups started the vertical-cavity SE laser research [4,5]. Quantum well structures and high reflectivity semiconductor multilayer reflectors have been successfully used in GaAs/GaAlAs and GaInAs/GaAs systems. In recent years, the research field of VCSELs has been growing fast. Improvements in device processes have been performed, resulting in sub-mA threshold VCSELs [4,5] in this wavelength region. The operating wavelength region of surface-emitting lasers is now from visible to infrared [1–6]. Although the laser performances of long wavelength (1.3 or 1.55 μm) VCSELs have been limited due to noticeable nonradiative recombination and large cavity losses, there is a strong demand for high-performance long wavelength VCSELs, which are applicable to ultra-parallel optical fiber communication networks.

A distributed index planar microlens array, which is made with a glass substrate using the selective ion exchange technique, has been developed since 1980. A 10 × 10 cm squared array samples are now commercially available. We have developed novel applications of microlens array for arrayed device coupling and for parallel optical image processing.

In this section we describe recent progress in surface-emitting semiconductor lasers, including some new functions as well as integrations. In addition, the novel concept of a self-aligning coupling scheme using planar microlens arrays is presented.

2 Surface-Emitting Lasers

2.1 Performances of surface-emitting lasers

For long wavelength VCSELs, the development of high reflectivity mirrors is one of the crucial issues. Also, thermal problems for cw operations are very important, especially in this wavelength region. Another key issue is to realize an effective current injection scheme. We have developed a MgO/Si mirror with good thermal conductivity and buried heterostructures with a tight current confinement. We achieved the first room-temperature cw operation of 1.3 μm VCSELs, as shown in Figures 7.1.1 and 7.1.2 [7]. The threshold current was as low as 22 mA. We found a circular spot profile as well as a narrow far-field divergence angle (4.2°), which would enable easy coupling to single-mode fibers. High-temperature operation and output power increase are the next targets, which will be achieved by a further reduction in cavity losses and the optimization of mirror reflectivities.

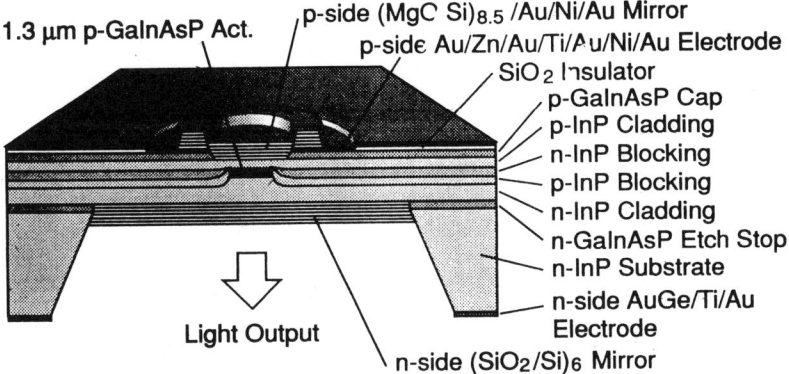

Figure 7.1.1
1.3 μm GaInAsP/InP planar circular buried heterostructure surface-emitting laser

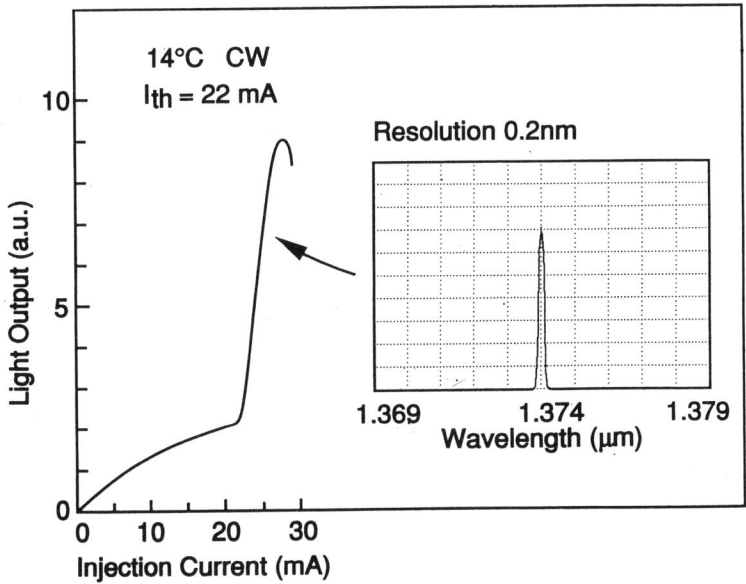

Figure 7.1.2
L/I characteristic of 1.3 μm VCSEL under room-temperature cw operation

A hybrid mirror technology has been developed to realize high reflectivity mirrors in long wavelength regions. One is to use a semiconductor/dielectric reflector, which is demonstrated by chemical beam epitaxy (CBE). Submilliamp low-threshold operations at 77 K were demonstrated in 1.55 μm CBE grown VCSELs with hybrid mirrors, as shown in Figure 7.1.3 [6]. The other is epitaxial bonding of a quaternary/GaAs–AlAs mirror. Recently, an epitaxially bonded mirror made of GaAs/AlAs was introduced into surface-emitting lasers operating at 1.3 μm providing 9 mA of room-temperature pulsed threshold.

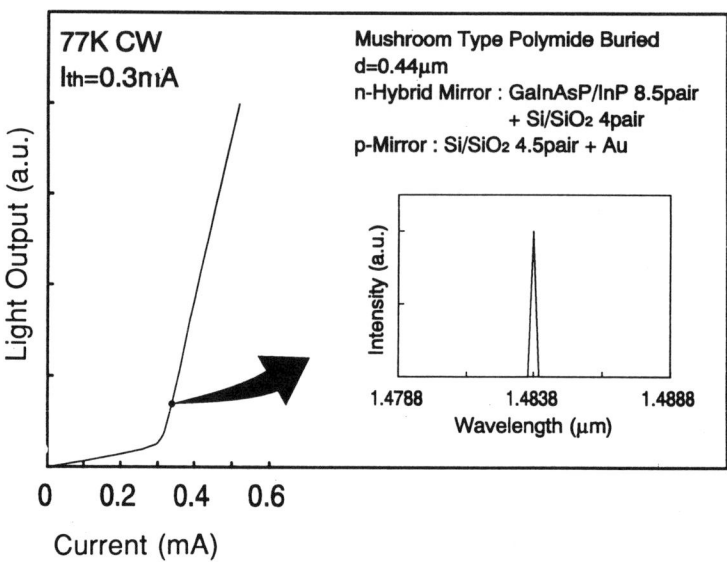

Figure 7.1.3
L/I characteristic of 1.55 μm CBE grown VCSEL with hybrid mirror at 77 K

Table 7.1.1 summarizes the performances of some surface-emitting lasers operating in the wavelength region from visible to infrared. The lowest threshold of 0.98 μm InGaAs/GaAs SE lasers has been reduced to 190 μA with the help of the low surface recombination velocity of the InGaAs system and the high reflectivity of GaAs/AlAs reflectors [5]. The reduction in the surface recombination by using surface passivation as well as the regrowth technique will be a key issue for further threshold reduction. Also, the room-temperature operation has been achieved in 630 nm wavelength regions with a low threshold. Blue and green SE lasers using ZnSe or GaN systems are also the next challenging subjects.

2.2 Polarization control and wavelength tuning

Surface-emitting lasers with symmetric circular resonators have two degenerated polarization states. In order to achieve very low noise operations of SE lasers, polarization switching or fluctuations must be avoided. So far, several polarization control methods using anisotropic stress or strained quantum wells on off-angled substrates have been proposed. However, polarization control has not yet been perfectly achieved.

We have proposed a novel polarization control method [8] as shown in Figure 7.1.4. The structure consists of a semiconductor multilayer reflector and a laminated metal/dielectric polarizer. The polarizer has a metallic nature and a dielectric nature for a parallel electric field $E_\parallel$ and perpendicular electric field $E_\perp$, respectively. Therefore, the phase difference in the reflected light from the DBR and the polarizer provides a large difference in reflectivities between the two polarization states. The metal/dielectric periodic structures can be easily formed, for example, by the holographic exposure

Table 7.1.1
Some performances of vertical cavity surface-emitting lasers in various wavelength regions

	GaAlInP	GaAlAs	InGaAs	GaInAsP
I_{th}	1.25 mA (Sandia)	1.5 mA (AT&T)	0.8 mA (AT&T) 0.7 mA (UCSB) 0.65 mA (Ulm) 0.19 mA (300 K, P) (NEC)	0.3 mA (77 K, CW) (Tokyo IT) 150 mA (300 K, P) (Furukawa) 9 mA (300 K, P) (UCSB) 22 mA (14 °C, CW) (Tokyo IT)
J_{th}		1.4 kA/cm^2 (AT&T)	600 A/cm^2 (UCSB) 450 A/cm^2 (NEC)	9.5 kA/cm^2 (300 K, P) (Tokyo IT) 19 kA/cm^2 (14 °C, CW) (Tokyo IT)
η_d		78% (AT&T)	33% (CW) (UCSB) 50% (P) (UCSB) 28% (P) (AT&T)	12% (77 K, QCW) (AT&T) 28% (77 K, CW) (Tokyo IT) 0.6% (300 K, P) (UCSB)
$\Delta\nu$		50 MHz (Tokyo IT)	85 MHz (UCSB)	<1 Å (Tokyo IT)
$\Delta\nu \cdot P$		89 MHz·mW (Tokyo IT)	5 MHz·mW (UCSB)	
RIN	*	<−140 dB/Hz (Tokyo IT)		*
P_{out} (CW)	0.3 mW (Sandia)	11 mW (AT&T)	40 mW (UCSB)	0.01 mW (14 °C, CW) (Tokyo IT)
P_{out} (Pulse)		120 mW (TRW) 1 W (AT&T)	500 mW (UCSB) 127 mW (NEC)	0.13 mW (20 °C, P) (Tokyo IT) 0.1 mW (300 K, P) (UCSB)
f_m	*	300 ps pulse (Tokyo IT)	8 GHz (AT&T) 6 ps (UCSB)	

Notes:
CW: Continuous wave   P: Pulsed
QCW: Quasi-CW   *: Not measured
Tokyo IT: Tokyo Institute of Technology   UCSB: Univ. California, Santa Barbara
Furukawa: Furukawa Electric   Ulm: Univ. Ulm

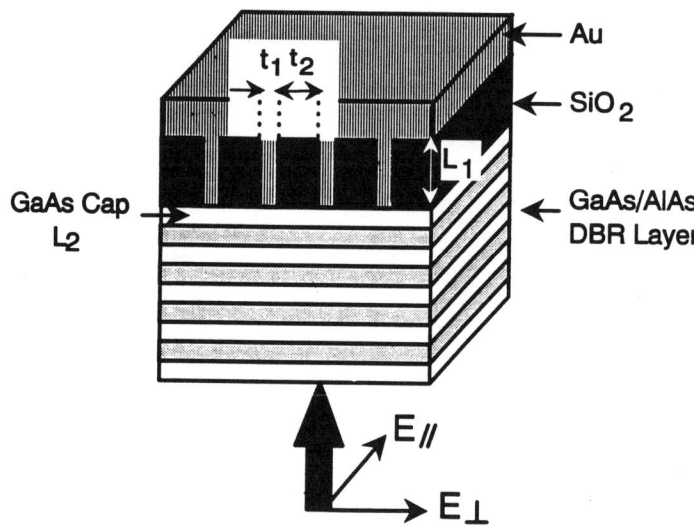

Figure 7.1.4
Polarization control structure by a metal/dielectric polarizer

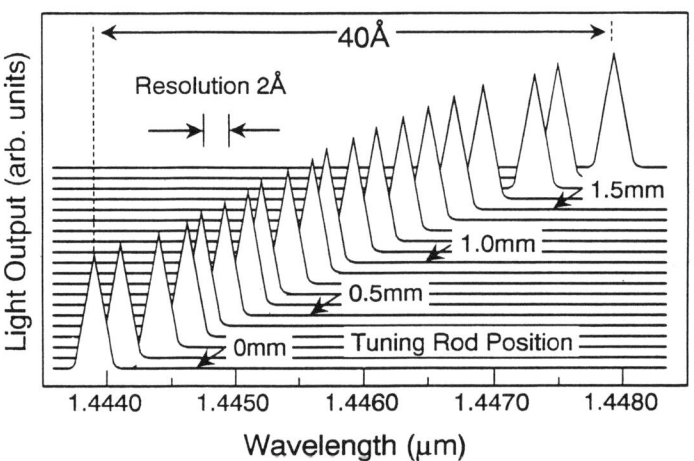

Figure 7.1.5
Wavelength tuning using an external reflector

technique, which has been well developed for DFB lasers. We can estimate a 9% reflectivity difference at the cavity resonant wavelength, while the reflectivity of $E_\perp$ remains at more than 99.9%. The corresponding threshold gain difference is more than 30 000 cm^{-1}, which is sufficiently large for stable polarization control. We fabricated a Au(400 nm)/SiO$_2$(1 μm) polarizer terminating a 10 pair GaAs/AlAs DBR. A reflectivity difference of ~5% was obtained at a Bragg wavelength of 1 μm. The observed large reflectivity difference would enable us to achieve reliable polarization control of SE lasers.

Various functions, such as frequency tuning and filtering, can be integrated into surface-emitting lasers by stacking. We can expect wide continuous wavelength tuning by using a vertical microcavity scheme. We have demonstrated 40 Å continuous tuning by the use of an external reflector for 1.55 μm VCSELs grown by CBE as shown in Figure 7.1.5.

2.3 Unique features in microcavity structures

Spontaneous emission control using a microcavity has been attracting much interest to realize thresholdless lasers or high efficient LEDs. The spontaneous emission factor has been estimated by performing a three-dimensional mode density analysis as shown in Figure 7.1.6 [9]. If we are able to realize a submicron scaled microcavity with a quantum wire structure, the spontaneous emission would be almost coupled to only one cavity mode, resulting in no district threshold devices.

Another interesting topic is photon recycling in microcavities. A part of the wasted spontaneous emission can be recycled by using a nearly closed cavity, which is realizable by covering the sidewall with metal reflectors. Threshold reduction can be expected by the photon recycling effect, which was experimentally demonstrated by Numai *et al.* [10]. The microfabrication of semiconductors with low induced damages and reduction of surface nonradiative recombination are crucial issues.

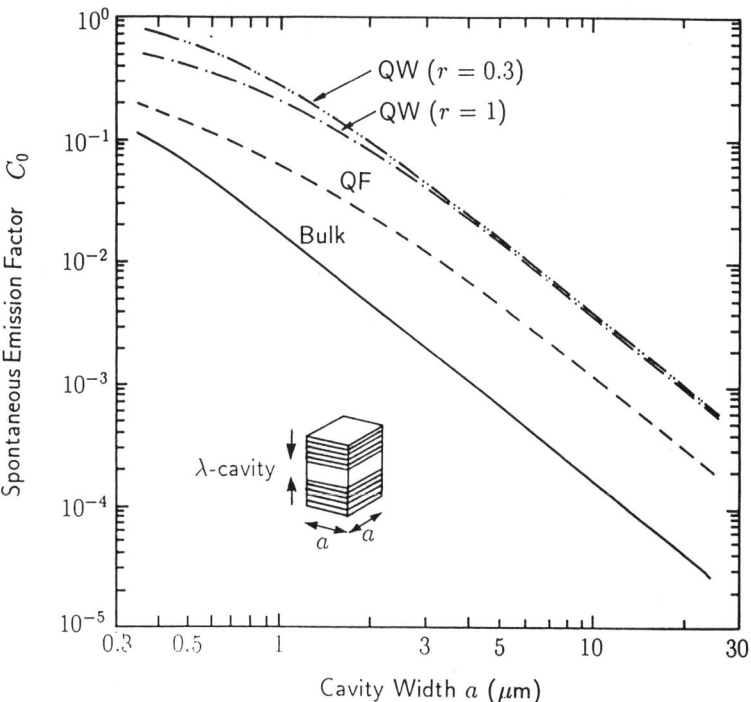

Figure 7.1.6
Spontaneous emission factor of microcavity lasers

3 Planar Microlens Arrays and Self-aligned Optics

3.1 Self-aligning schemes

If we can realize 2D arrayed devices that are available for actual use, we can expect to align a tremendous number of optical components simultaneously, as in the parallel multiplexing lightwave systems shown in Figure 7.1.7. We have developed a novel self-aligning scheme which uses put-in microconnectors as shown in Figure 7.1.8 [11]. It consists of many microconnectors utilizing the coupling between the optical plugs on the microlens substrate and the fiber jacks made on the core of the fibers. The light coupling characteristic of fabricated devices has been examined. A coupling loss of 6 dB has been achieved without any precise optical alignment. The excess loss and deviation will be reduced by the careful choice of a polyimide having a small absorption

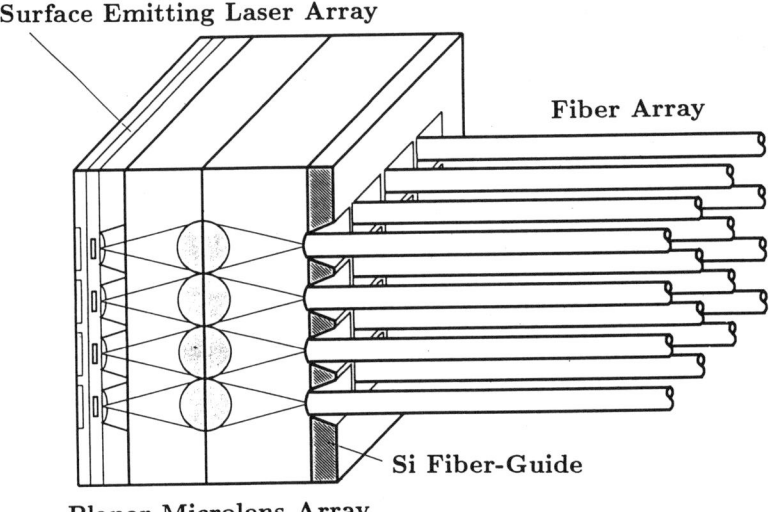

Figure 7.1.7
Parallel lightwave subsystem using a surface-emitting laser array and a planar microlens array

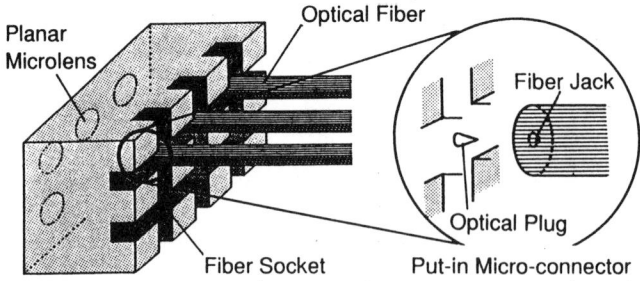

Figure 7.1.8
Put-in self-aligning scheme

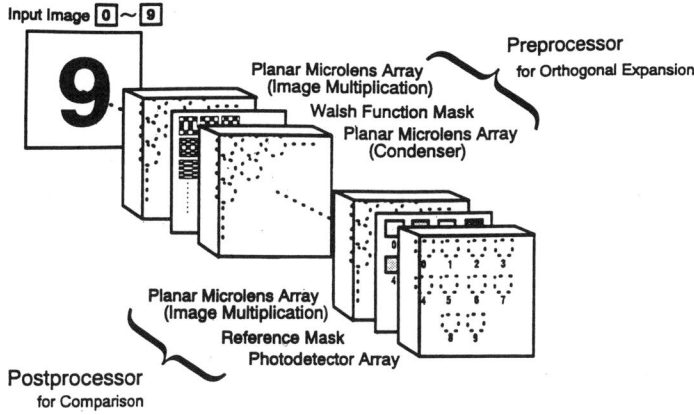

Figure 7.1.9
Schematic configuration of proposed optical processor

and refractive index close to that of fibers and by a reduction in the angle errors in the inserted fibers. The proposed concept will be a key technology for low-cost optical couplers for advanced optical fiber communication systems handling a huge number of fiber and laser arrays.

3.2 Parallel optical processors using planar microlens arrays

Optical parallel processing is attractive for processing large amounts of image information because of its high speed and essential parallelism of light, if it is realized. We have proposed and studied an optical parallel processor which is composed of two stacked sections, a pre-processor and a post-processor, as shown in Figure 7.1.9 [12]. The multi-imaging characteristics of the planar microlens array has been fully used for this purpose. We have performed a computer simulation as well as optical experiments separately both for the pre-processor and the post-processor. The experimental results almost correspond to the computer simulation, showing the possibility of ultra-parallel optical image processors based on planar microlens arrays.

4 Future Prospects

Optical interconnects between computers may be the most interesting field related to VCSELs. For this purpose, GaInAs–GaAs or GaAs–GaAlAs VCSELs will be in the production stage very soon. A further reduction in the threshold and operating voltages is needed to realize large-scale 2D VCSEL arrays. The advance of micro-fabrication and fine crystal growth technologies may accelerate further development of VCSELs in various wavelength regions, which may open up new applications of VCSELs in the next era optoelectronics. The arrayed micro-optics technology would be very helpful for advanced ultra-parallel optical systems.

References

[1] K. Iga, F. Koyama and S. Kinoshita, *IEEE J. Quantum Electron.*, **QE-24**, 1845–1855, (1988).
[2] S. Misawa, M. Oikawa and K. Iga, *Appl. Opt.*, **23**, 1884–1786, (1984).
[3] F. Koyama, S. Kinoshita and K. Iga, *IEICE*, E71, 1089–1090, (1988).
[4] J.L. Jewell, A. Sherer, S.L. McCall, Y.H. Lee, S.J. Walker, J.P. Harbison and L.T. Florez, *Electron. Lett.*, **25**, 1123–1124, (1989).
[5] T. Numai, T. Kawakami, T. Yoshikawa and M. Sugimoto, *Jpn J. Appl. Phys.*, **32**, L1533–L1534, (1993).
[6] T. Uchida, T. Miyamoto, N. Yokouchi, Y. Inaba, F. Koyama and K. Iga, *IEEE J. Quantum Electron.*, **29**, 1975–1980, (1993).
[7] T. Baba, Y. Yogo, K. Suzuki, F. Koyama and K. Iga, *IEICE Trans. Electron.*, **E76-C**, 1423–1424 (1993).
[8] T. Mukaihara, F. Koyama and K. Iga, *Jpn. J. Appl. Phys.*, **2B**, L227–L229 (1994).
[9] T. Baba, T. Hamano, F. Koyama and K. Iga, *IEEE J. Quantum Electron.*, **28**, 1310–1319 (1992).
[10] T. Numai, M. Sugimoto, I. Ogura, H. Kosaka and K. Kasahara, *Jpn. J. Appl. Phys.*, **30**, L602–L603 (1991).
[11] A. Sasaki, T. Baba and K. Iga, *Photon. Technol. Lett.*, **4**, 908–911 (1992).
[12] K. Murashige, A. Akiba, T. Baba and K. Iga, *Jpn. J. Appl. Phys.*, **31**, 1666–1671 (1992).

7.2

Liquid Crystal Cells for Optical Neural Computing

Shunsuke Kobayashi, Yasufumi Iimura, Koji Maeda, Mayuki Hashi and **Hironori Kikkawa**

Abstract

The operating principles, structures, and performance of liquid crystal (LC) electrooptical (EO) devices, which are actually being utilized or are potentially to be utilized for constructing various types of optical logic calculations and optical neural computing (ONC) systems, are reviewed. Furthermore, as examples of demonstrations a newly designed reflective spatial light modulator and a feedback type ONC using a ferroelectric liquid crystal cells as neurons and a twisted nematic cell as a weight matrix, are described.

1 Introduction

Liquid crystal media are useful for fabricating a spatial light modulator, neurons, and a weight matrix for optical data processing or optical neural computing (ONC) owing to their easiness in fabricating a flat panel matrix cell with the features of low voltage and low power operation. These properties promise a high potential in parallel data processing, even though their response speed is slow compared with photorefractive crystals.

This section first introduces the operating modes of liquid crystal (LC) electrooptic (EO) devices and their characteristics; then discusses how each operation principle is useful for fabricating a particular component for optical logic or ONC; and finally describes demonstrations of ONC, which has been developed by the authors' group, using a ferroelectric LC cell for neurons and a twisted nematic LC matrix for a weight matrix. Also, a new reflective type spatial light modulator is demonstrated.

2 Modes of Liquid Crystal Electrooptic Devices, their Principles, and their Characteristics

2.1 General introduction

Except for a thermally addressed smectic A device, all other LC devices are operated in forms of EO effects. Table 7.2.1 summarizes the operating modes of LC devices that are known through existing commercially available products, experimental demonstrations, and the literature [1-3]. The addressing of these devices is done mostly electrically and optical addressing is also possible in term of photoelectric devices.

All the types of LC devices listed in Table 7.2.1 act as optical shutters. However, the principles or phenomenon are basically divided into the following five major categories: (1) light scattering; (2) birefringence; (3) rotatory power; (4) selective reflectance; and (5) optical absorption by dichroic dyes dissolved in an LC medium. By taking advantage of these effects, almost all types of LC devices can actually or potentially to be utilized as components for optical data processing or ONC.

Among the operating modes shown in Table 7.2.1, we concentrate mainly on nematic tunable birefringence (NTB), twisted nematic (TN), surface stabilized ferroelectric liquid crystal (SSFLC), and guest host (GH) devices due to their popularity and usefulness in optical logic circuits and ONC.

2.2 Operating principles and characteristics of LC-EO devices

Tunable birefringence (TB) mode using nontwisted nematic liquid crystal [4]

This mode is also called the electrically controllable birefringence (ECB) mode. The nontwisted (parallel) LC molecular conformations and their variations by the application of electric fields are shown in Figures 7.2.1(a)-7.2.1(d). A pretilted homogeneous (planar) LC conformation changes to a vertical conformation (Figures 7.2.1(a), 7.2.1(b)) in an NLC having a positive dielectric anisotropy ($\varepsilon_\parallel - \varepsilon_\perp = \Delta\varepsilon > 0$). The opposite change, i.e. from vertical to planar (Figures 7.2.1(c), 7.2.1(d)), takes place in an NLC with a negative $\Delta\varepsilon$. The necessary voltage to cause these transitions, called the Freedericksz transition, is 1-2 V for ordinary NLC materials.

The deformation of NLC conformations shown in Figure 7.2.1 can be detected optically. The set-up is shown in Figure 7.2.2. A light beam polarized by a polarizer P will be split into ordinary and extraordinary rays, which can be to interfere at after passing through an analyzer. The difference in the phase between the two rays (polarized along the x and y directions in Figure 7.2.1) is

$$\Delta R = \frac{2\pi}{\lambda} \int_0^d [n(z) - n_o] dz \quad (1)$$

where $n(z)$ is the local refractive index for light polarized along the y direction and $n(z)$ is given by

$$n(z) = n_e n_o (n_e^2 \sin^2\theta + n_o \cos^2\theta)^{-1/2} \quad (2)$$

Liquid Crystal Cells for Optical Neural Computing

Table 7.2.1
Operation modes of LC Devices

LDs	Modes	Abbr.	Effects	Principles	Reflective	Transmissive	Projection	Color	MPX	Memory	Number of polarizers
	Tunable Birefringence	TB	E	Birefringence	Y	Y	Y	Y	Y excellent	N	2
		STN	E	Birefringence	Y	Y	Y	Y	Y	N	2
	Twisted Nematic	TN	E	Birefringence	Y	Y	Y excellent	Y	Y	N	2
N(N*)	Guest Host	GH	E	Dichroism	Y	Y	Y	Y	Y weak	N	1 or 0
	Dynamic Scattering	DS	E	Light Scattering	Y	Y	Y	weak	Y	N	0
	Polymer Dispersed	PD	E	Light Scattering	Y	Y excellent	Y	Y	N	N	0
Ch	Optical Bragg Refl	BR	E or T	Selective refl	Y excellent	N	N	Y	N	Y N	0
	Phase change (unwinding)	PC	E	Light Scattering	Y	Y	Y excellent	Y	Y	Y	0
SA	Thermally Addressed	TA	T + E	Light Scattering	Y	Y	Y excellent	Y	Y N	Y excellent	0
	Electroclinic	EC	E	Birefringence	Y	Y	Y	Y	N/?	N	2
	Tunable Birefringence	TB	E	Birefringence	Y	Y	Y	Y	Y	Y excellent	2
SC*	Surface Stabilized	SSFLC	E	Birefringence	Y	Y	Y	Y	Y	Y excellent	2
	Guest Host	GH	E	Dichoism	Y excellent	Y	Y	Y	Y	Y	1 or 0
	Unwinding	UW	E	Light Scattering	Y	Y	Y	Y	Y	N	0

Legend: E for electric field; I for electric current; T for thermal effect; Y for Yes; N for No; MPX for Multiplexability.

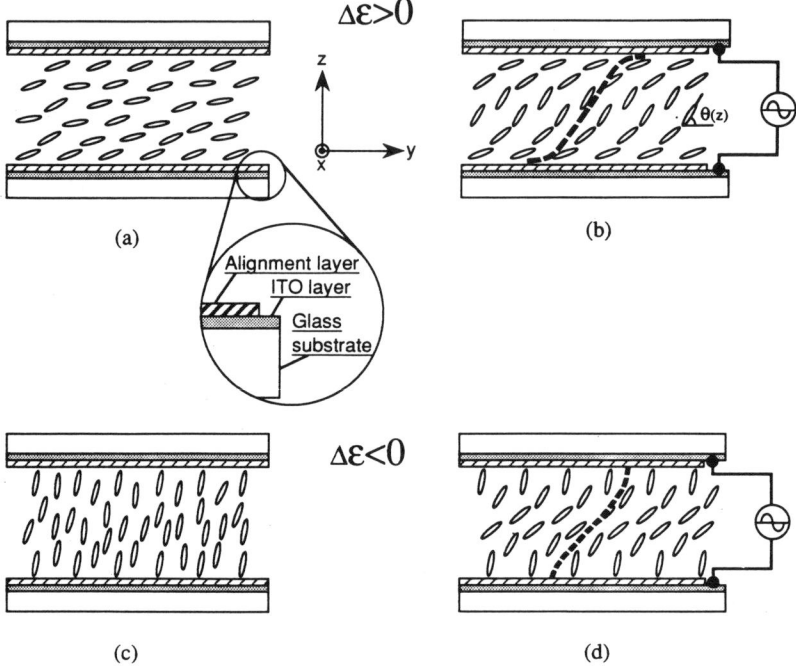

Figure 7.2.1
Nontwisted parallel nematic liquid crystal conformations and their variations by the application of electric fields: (a) a pretilted homogeneous (or planar) alignment by an antiparallel surface orientation; (b) a spatial variation of the tilt angle $\theta(z)$ of NLC molecules (having a positive dielectric anisotropy ($\Delta\varepsilon > 0$)) by the application of an electric field; (c) a pretilted homeotropic (or vertical alignment); and (d) distortion of NLC conformation with a negative $\Delta\varepsilon$. Surface alignment of NLCs can be done by controlling the capability of the surface alignment layers [1–3]

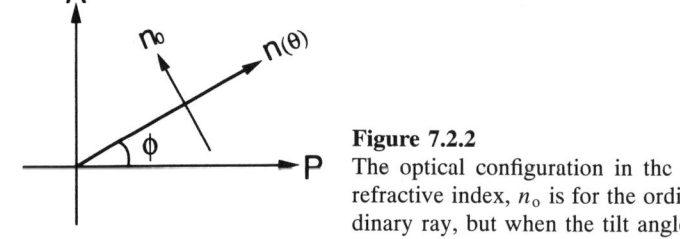

Figure 7.2.2
The optical configuration in the tunable birefringence mode. The refractive index, n_o is for the ordinary ray, and n is for the extraordinary ray, but when the tilt angle θ is zero, then $n = n_e$

where n_e, n_o, and θ stand for the refractive index of the ordinary and the extraordinary rays, respectively, and the local tilt angle of the LC molecule (actually the tilt angle of the director vector).

The intensity and color formation of the transmitted light is expressed by [4]

$$I = I_o \sin^2(2\phi) \sin^2\left(\frac{\Delta R}{2}\right) \quad (3)$$

The best conditions for giving the maximum intensity of the transmitted light is obtained by taking $\phi = 45°$, and equation (1) tells us that I oscillates as

$\sin^2\{\pi\Delta n(E)d/\lambda\}$ as $\Delta n(E)$ changes with the field E. The LC cells shown in Figure 7.2.1 play the role of a phase retarder. In a completely planar cell $\Delta R = 2\pi\Delta nd/\lambda$, where $\Delta n = n_e - n_o$ is the birefringence. In particular, in the case where the range of electrically controllable ΔR is 180°, then the medium acts to rotate the linearly polarized light by 90°. This mode is called the π cell [5].

Twisted nematic (TN) EO device

Figures 7.2.3(a) and 7.2.3(b) illustrate the operating principles of a TN–LC device [6] which is currently the major LC display device.

A defect-free TN device can be fabricated by choosing an appropriate pretilt angle; by twisting below 90°, say 87°; and by doping with a chiral agent [1].

As is shown in Figure 7.2.4, by choosing a crossed or a parallel configuration of polarizers, one obtains two possible modes, called normally white or normally black, respectively.

A high information content TN-LC device extending from 5×10^5 to 1×10^6 pixels (including RGB colors) can be fabricated as an active matrix (AM) driven TN-LCD in which each subpixel is installed with a switching element such as a thin film transistor or diode. A commercially available AM-LCD screen of a TV set is also useful as a unit of optical data processing [7]. It is also possible to fabricate a direct multiplexed TN-LCD whose information content is of the order of 10^4 pixels or less. A multicolor TN-LCD is also possible by using selective polarizers [1].

A supertwisted nematic (STN) device, which is similar to TN-LCD but features a larger twisting angle reaching 180–270°, is capable of displaying the information

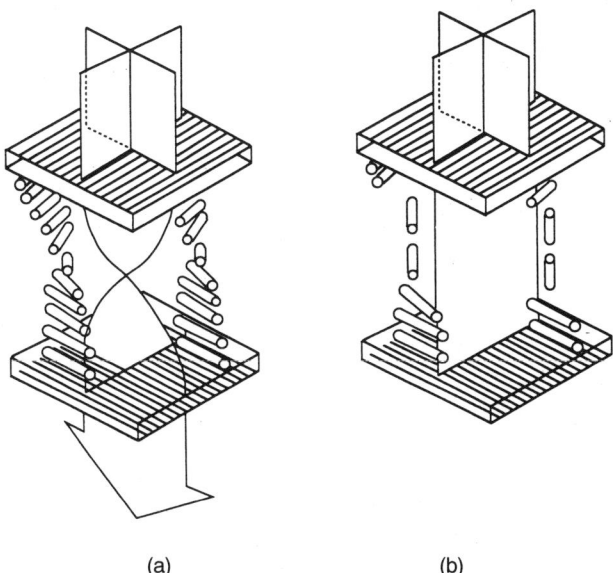

Figure 7.2.3
An explanation for the operation principle of TN-LCD. (a) off ($V = 0$); (b) on ($V > V_{\text{threshold}}$)

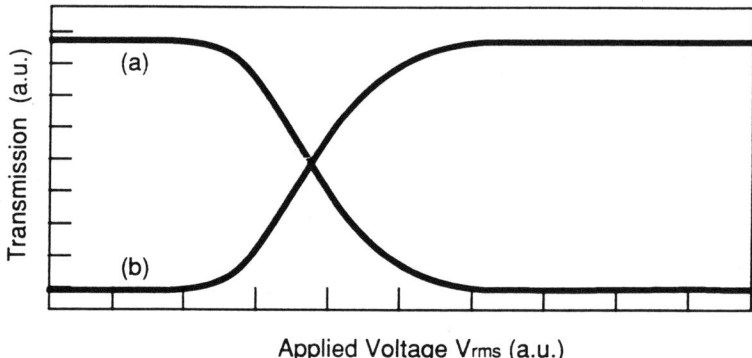

Figure 7.2.4
The EO performance of TN-LCD; (a) normally white (open) (crossed polarizers) and (b) normally black (closed) (parallel polarizers)

content of the order of 5×10^5 pixels by direct multiplexing [2,8]. Both TN and STN EO devices are able to be used as a weight matrix, a device for boolian algebra, and neurons. There are some modes that combine the TN and TB modes [9]. An example of this approach will be introduced in section 3 on a spatial light modulator.

Ferroelectric liquid crystal EO device

Figure 7.2.5 is an example of molecules that show a chiral smectic (SC*) phase; the molecule has a chiral moiety that prohibits free rotation around the long axis when the molecules form a SC* phase and the molecule bears a dipole moment(s). As is shown on Figure 7.2.6, this kind of molecule forms a layered and helicoid structure (each molecule exists on a cone having an angle θ) [10,11].

There are three kinds of EO effects in a FLC device: one is light scattering during the course of unwinding of the short pitch SC* [12]; the other two are electrically

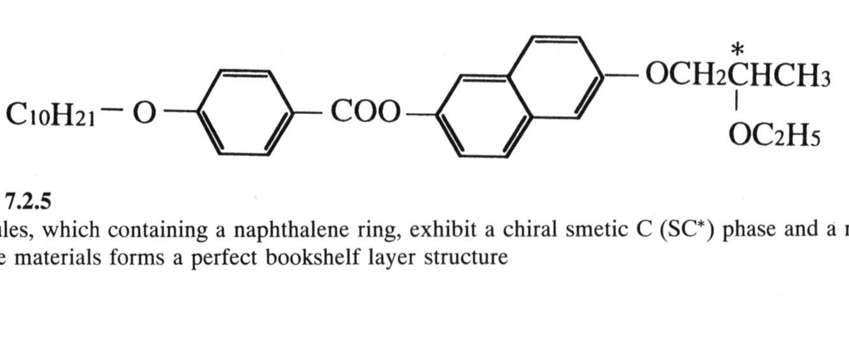

Figure 7.2.5
Molecules, which containing a naphthalene ring, exhibit a chiral smetic C (SC*) phase and a mixture of these materials forms a perfect bookshelf layer structure

Liquid Crystal Cells for Optical Neural Computing

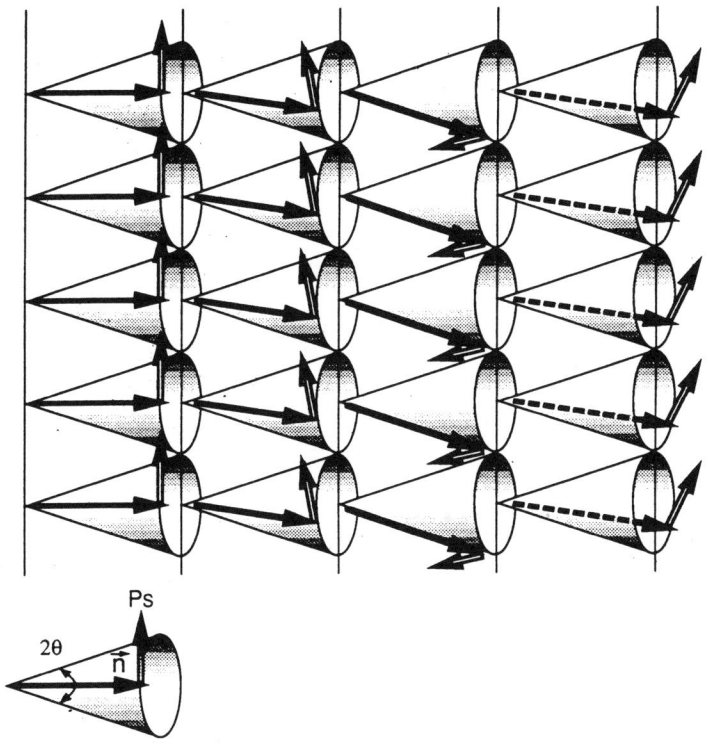

Figure 7.2.6
Helix molecular conformation of SC*. The arrows show the spontaneous polarization, P_s, and the director, *n*

controlled birefringence effects, one of which occurs during unwinding [13] and the other is surface stabilized (SS) FLC that is an unwound and planar structure [13]. The unwinding occurs by inserting FLC molecules into a narrow space of 1–2 μm thick of a sandwich cell. Figures 7.2.7(a) and 7.2.7(b) show the top and side views of an SSFLC cell [10,11,14].

By applying a bipolar voltage, V^+ and V^-, to an SSFLC cell (Figure 7.2.7) the molecules change between two stable states. This transition gives rise to a switch in the direction of the dipoles from up to down, and vice versa. The origin of the name ferroelectricity is the switching of an ensemble of dipole moments belonging to molecules. The movement of molecules between two stable states separated by the cone or the tilt angle 2θ can be visualized by inserting an FLC medium between two polarizers, as illustrated in Figure 7.2.2. The optical transmission is given by equation (1), where the angle ϕ changes by the application of a bipolar waveform while keeping Δn constant.

The SSFLC EO device features are (1) bistability (memory effect) and (2) a fast response. However, to fabricate a defect-free SSFLC device exhibiting perfect bistability and good contrast ratio is not easy. To realize these effects, it is necessary to adopt the following technologies and materials: (1) ultra-thin LC molecular alignment layers, such as polyimide Langmuir Blodgett films or electrically conductive films such as polypyrrole films [10]; (2) FLC material containing naphthalene rings

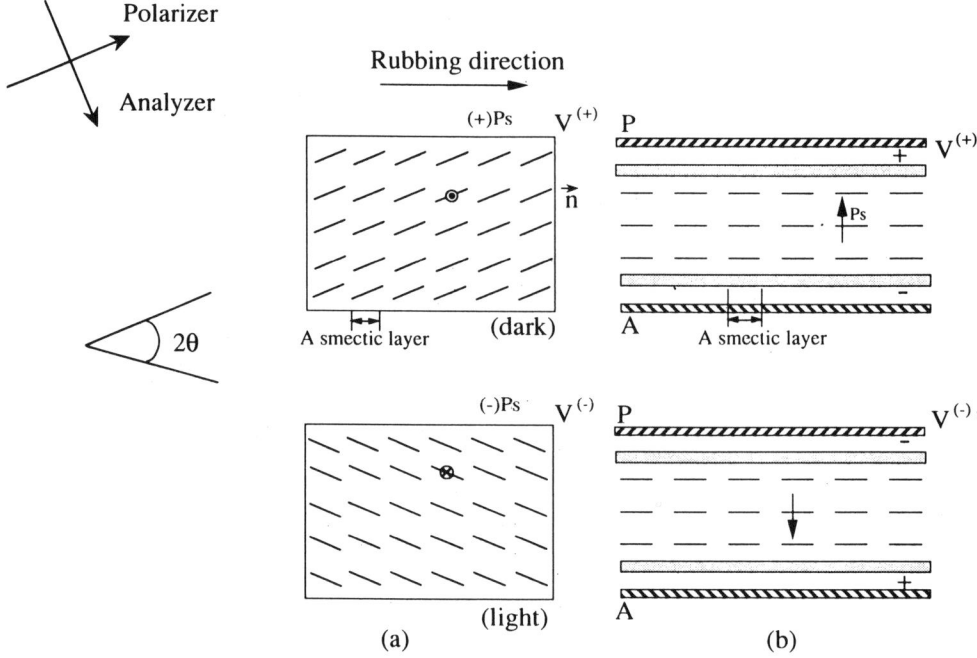

Figure 7.2.7
(a) Top views of the surface stabilized FLCD; the operation of the tunable birefringence mode is shown; (b) side views which are called the bookshelf configuration

[15]; (3) polyimide alignment layers which are capable of generating a high angle and choose a special texture [16,17]. Otherwise the so-called zig-zag defects appear due to the chevron structure of smectic LC layers. An example of good bistability is shown in Figure 7.2.8(a) compared with an imperfect case, Figure 7.2.8(b). For the birefringence mode, in the case where the cone angle 2θ (Figure 7.2.6) equals ϕ (equation (1)) and $2\phi = \pi/2$, the optimum cone angle θ that gives rise to the best contrast ratio is $\theta = 22.5°$. However, the value of θ depends on the material and temperature.

The SSFLC device is useful for a spatial light modulator [18,19] or a spatial retarder that is utilized for real-time holography applications [20]. One of the most advanced spatial light modulators is a 256×256 LC on silicon tip SLM [18,19]. Gray scale operation of the SSFLC is also possible, but the texture is a multidomain type [10].

An antiferroelectric LC, which exhibits tristable states, is another possible medium for optical data processing [21].

Guest–host (GH) LC–EO device

Dichroic dyes dissolved in an LC host medium rotate according to the rotation of the host LC molecules due to the application of an electric field. One of the most useful forms is the so-called White–Taylor type that uses a chiral NLC as the host medium [1–3,22].

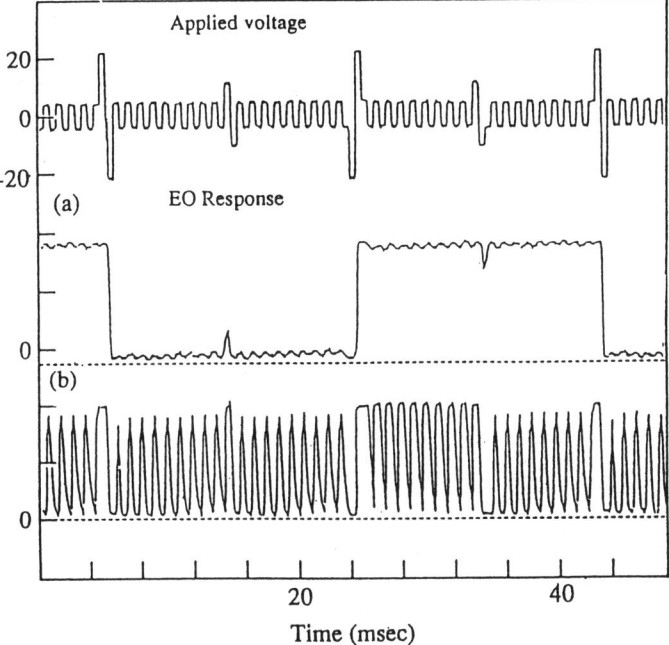

Figure 7.2.8
The difference in dynamic EO performances of FLCDs containing naphthalene compounds in a FLC mixture (a) adopting the polypyrrole film for molecular orientation and (b) using conventional spin-coated PI films for the molecular orientation layers

The LCD modes that do not need polarizers or that need to use one polarizer, are more optically transparent compared with LCDs that need two polarizers (refer to Table 7.2.1). For this reason the polarizer free GH-LCD is useful as a reflective type. This characteristic is also useful for ONC application.

3 Spatial Light Modulators Using Liquid Crystals

A spatial light modulator (SLM) is an light signal converter capable of converting an incoherent light into a coherent light, one wavelength to another one, and of performing intensity amplification.

The conventional structure of an SLM consists of a photoconductor (PC) layer, a reflecting layer, and an LC layer. The relationships for the impedances in these layers to obtain a good performance as an SLM are as follows:

$$Z_{LC}(\text{dark}) \ll Z_{PC}(\text{dark}) \quad \text{and} \quad Z_{LC}(\text{light}) \geq Z_{PC}(\text{light}),$$

where dark or light corresponds to 'off' and 'on' of the writting light signal. Conventionally, the PC layer is made of amorphous (a)Si prepared by CVD due to its popularity; the PC layer prepared in this way has a good responsivity as a PC but the necessary thickness of the layer is about 30 μm, which degrades the resolution.

We have developed a new SLM consisting of a-Si:H deposited by sputtering, an Al mirror as a reflector, and a degree TN-LC layer, as shown in Figure 7.2.9; the a-Si:H prepared by this method has a large resistivity in the dark, even though the relatively is small. To fulfil the aforementioned relations for impedances, we chose the area ratio between the PC layer and the Al layer to be 1/25. The performance obtained is shown in Figure 7.2.10 [23]. The TN-LC used in this system can be replaced by an SSFLC cell.

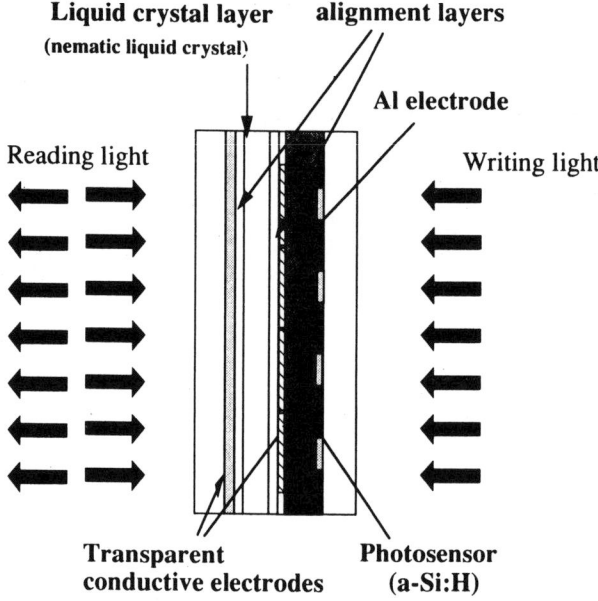

Figure 7.2.9
A schematic cross-sectional view of SLM consisting of an amorphous Si photoconductive layer deposited by sputtering, a two-dimensional array of Al mirror pads, and ITO stripes

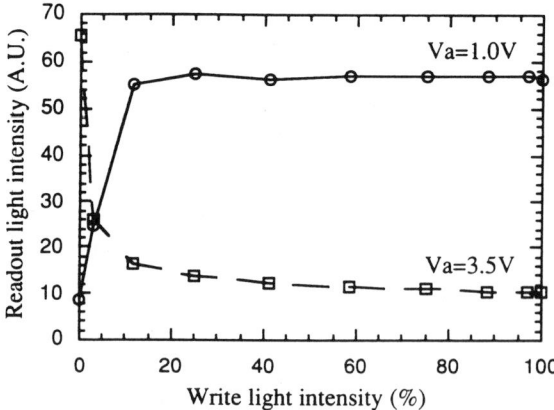

Figure 7.2.10
Relation between readout light intensity and write light intensity. V_a is the applied voltage. The maximum write light intensity is 42.3 mW/cm^2

4 An Example of Optical Neural Computing Using Liquid Crystal Cells

We have developed a defect-free SSFLC cell that exhibits perfect bistability and grayscale capability [10]. Furthermore, the authors' group has experience in fabricating defect-free TN-LCDs [1–3]. Taking advantage of this experience, we have constructed an ONC system, as shown in Figure 7.2.11 [24–27]. The system is a feedback type as proposed by Hopfield [28]. The neurons are nine stripes of SSFLC, and the weight matrix is a 9×9 TN matrix. The TN matrix and the FLC cell are inserted between the crossed polarizers. This system plays the role of an 'exclusive OR' operation accompanying a gray scale, as shown in Figure 7.2.12.

In this case the optical steady state of each neuron is binary, i.e. dark (black) or light (white). A state of a set of neurons can be made according to a learning rule by setting up the corresponding state of the weight matrix; expressed by a black and white checker board, where each pixel is divided into two parts.

The operation sequence of the system shown in Figure 7.2.11 is as follows: (1) as an example, the state of the weight matrix is selected electrically, in this way corresponding patterns are selected as shown in Figure 7.2.13(a) (each learned pattern has a complementary one, but this degeneracy can be eliminated by offseting the output voltages of the photosensor [26]); (2) then, as an input, a pattern is displayed on the neuron FLC cell. An example is the pattern of Figure 7.2.13(b); (3) in the next turn the switch to the right and the FLC is illuminated from the right; (4) the light beam through the analyzer is collected by a cylindrical lens and detected by a photosensor array, the outputvoltages of these photosensors are fed back to the neuron FLC cell; finally (5) as a result, a pattern is selected from the memorized patterns by finding one the same as, or very similar to, the input pattern. The resulting pattern, which has the shortest Hamming distance, is shown in Figure 7.2.13(c).

This computation is performed for about 1 ms. The result is shown to agree with the analytical calculation of the dynamics of the system [25–27].

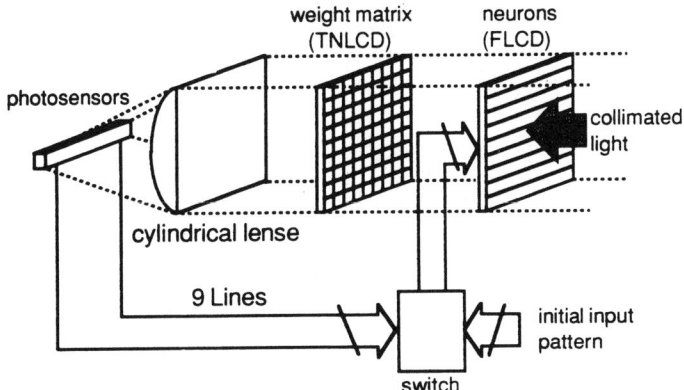

Figure 7.2.11
A feedback type optoelectrical neural computing system consisting of neurons made of an FLC cell with nine stripes, a weight matrix of TN cell (9×9 matrix) and opto and electrical circuits

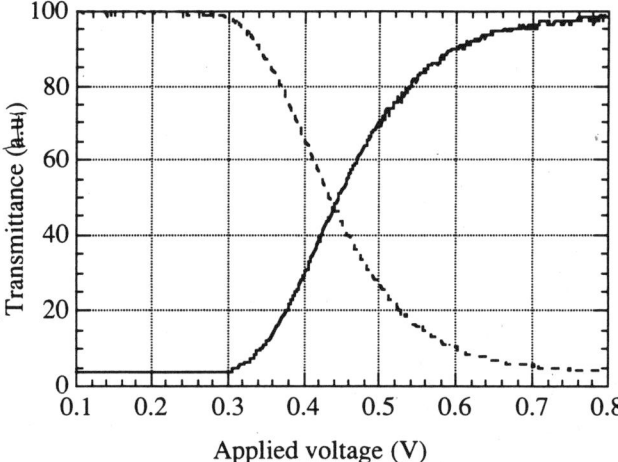

Figure 7.2.12
A four-quadrant multiplication implemented by an optical system of double layered LCDs comprising an FLCD (with polypyrrole alignment layers) and a TNLCD. The ordinate is the optical transmission through the system and the abscissa represents the rms voltages applied to the TN matrix, while the states of the FLC cell are selected in advance and are kept using memory capability

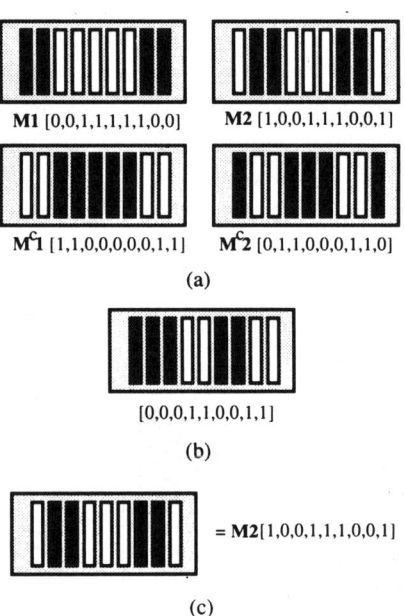

Figure 7.2.13
An experimental result of the computation. (a) patterns memorized in the TN-LCD weight matrix; (b) a pattern initally inputted to the FLC meurons; and (c) the finally recollected pattern after the feedback process

5 Concluding Remarks

Liquid crystal (LC) media are useful for fabricating various kinds of devices for optical data processing such as spatial light modulators or optical neural computing systems.

One of the important technical problems in LC-EO devices for optical data processing is to increase resolutions or pixels upto 1 M pixels or more. In the future, the function of information displays will be merged with optical neural computing by developing materials, operational modes and fabrication processes.

References

[1] K. Okano and S. Kobayashi, eds., *Liquid Crystals, Fundamentals and Applications — Fundamentals and Applications*, Baifukan, Tokyo, 1985 [in Japanese].
[2] S. Kobayashi, *Advances in Direct Multiplexed LCDs*, SID Seminar Notes, **2**, 1990, F-3.
[3] B. Bahadur, ed., *Liquid Crystals — Applications and Uses*, **1-3**, World Scientific, 1990.
[4] S. Kobayashi and A. Mochizuki, in Reference [3], **3**, Chapter 19.
[5] P. Boss, T. Buzak and R. Vatne, *Proc. Eurodisplay*, **84**, 7-10 (1984).
[6] M. Schadt and W. Helfrich, *Appl. Phys. Lett.*, **18**, 127-128 (1971).
[7] N. Hashimoto and S. Morokawa, *J. Electron. Imaging*, **2**, 93 (1993).
[8] T. Scheffer and J. Nehring, in Reference [3], **1**, Chapter 10.
[9] W.P. Bleha, L.T. Lipton, E. Wiener-Aunear, J. Grinberg, P.G. Reif, D. Casasent, H.B. Brown and B.V. Markevitch, *Opt. Eng.*, **17**, 371-384 (1978).
[10] S. Kobayashi, Kogaku-Optics- (in Japanese) 19, 410 (1990); H. Maeda, M. Yoshida, B.Y. Zhang, M. Kimura and S. Kobayashi, *Proc. SID*, **32**, 409-412 (1991).
[11] J. Dijon, in Reference [3], **2**, Chapter 13.
[12] K. Yoshino and M. Ozaki, *Jpn J. Appl. Phys.*, **23**, L385 (1984).
[13] J. Fuenschilling and M. Schadt, *J. Appl. Phys.*, **66**, 3877-3882 (1989); T. Tanaka, K. Sakamoto, K. Tada and J. Ogura, *Digest of Technical Papers*, SID 94 Int'l. Symp., **25**, 1994, 430-433.
[14] N.A. Clark and S. Lagerwall, *Appl. Phys. Lett.*, **36**, 899-901 (1980).
[15] A. Mochizuki and S. Kobayashi, *MCLC* **243**, 77-90 (1994).
[16] M. Koden, T. Shinomiya, N. Itoh, T. Kutate, T. Taniguchi, K. Awane and T. Wada, *Jpn. J. Appl. Phys.*, **30**, L1823 (1991).
[17] J. Kanbe, H. Inoue, A. Mizutome, Y. Hanyu, K. Katagiri and S. Yoshihara, *Ferroelectrics*, **114**, 3-26 (1991).
[18] N.A. Clark and K.M. Johnson, in Reference [3], **3**, Chapter 17.
[19] *Appl.Optics*, a spatial issue on SLM, **33** (14) (1994).
[20] S. Fukushima, T. Kurokawa, S. Matsuo and H. Kozawaguchi, *Opt. Lett.*, **15**, 285 (1990); S. Yamamoto, "FLC devices for SLM" Ohyobutsuri (in japanese), (1994).
[21] A.D.L. Chadani, T. Hagiwara, Y. Suzuki, Y. Duchi, H. Takezoe and A. Fukuda, *Jpn. J. Phys.*, **27**, L729 (1988).
[22] B. Bahadur, in Reference [3], **3**, Chapter 11.
[23] M. Hashi, Master's Thesis, The Graduate School of Technology, Tokyo Univ. of Agri. and Tech. (1995); to be published.
[24] C.M. Gomes, H. Sekine, T. Yamazaki, A. Nakagawa and S. Kobayashi, *Digest of Technical Papers, SID Int'l. Symp.*, **23**, (1992), 481-484.
[25] T. Yamazaki, K. Maeda, M. Hashi, Y. Iimura and S. Kobayashi, *Optoelectronics-Devices and Technology*, **8**, 53-59 (1900); and to be published in *Applied Optics*.

[26] K. Maeda, Master's Thesis, The Graduate School of Technology, Tokyo Univ. of Agri. & Tech. (1995); to be published.
[27] S. Kobayashi, Y. Iimura, M. Hashi, K. Maeda and H. Kikkawa, *Int'l Symp. on Ultrafast and Ultra-parallel Optoelectronics*, Chiba, 12 July 1994, 3-2, 1994, 147–150.
[28] J.J. Hopfield, *Proc. Nat'l. Acad. Sci. USA*, **79**, 2254–2258 (1982).

7.3

Realization of High-speed 3D LSIs by Optical Interconnections

Mitsumasa Koyanagi

Abstract

The optical waveguide, micro-lens and micro-mirror for the interchip and intrachip optical interconnection have been constructed using LSI technology and their basic characteristics have been evaluated. A new micro-bonding technique which integrates the photonic devices onto the silicon chip and a low temperature wafer bonding technique which produces 3D structures are described. In addition, a new ORAM-bus memory with the parallel data transfer function using the optical interconnections has been introduced and its basic operation evaluated using a computer simulation method.

1 Introduction

To improve computer system performance it is important to increase the data-transfer speed between the processor and the memory. The clock frequency of the processor chip has rapidly increased in recent years owing to dramatic progress in LSI technology. Several microprocessor chips with a clock frequency above 100 MHz have already been reported. On the other hand, the access time of the main memory has not been reduced much. As a result, the difference in the operating speed between the microprocessor and the main memory has been extended. Therefore, in order to overcome such problems as the difference in increased operating speed, we have proposed introducing an optical interconnection into the main memory [1-5]. We call this new memory an optical RAM-bus (ORAM-bus) memory. Here, we first describe the method of introducing the optical interconnection into memory LSIs, and then describe the basic concept of the ORAM-bus memory and a new parallel processing system with an ORAM-bus memory.

2 Formation of an Optical Interconnection on LSIs and a MCM

The optical interconnection is very attractive because it can transfer a huge amount of data at very high speed. However, it is not clear how we can reduce the interconnection distance without losing the advantages of the optical interconnection over the electrical interconnection. Furthermore, in the case of replacing the intra-chip or inter-chip electrical interconnection with a relatively short interconnection distance by the optical interconnection, we cannot simply say that the optical interconnection has advantages over the electrical interconnection because not only the data-transfer speed but also the power dissipation, the compatibility with LSI, and the cost should be taken into account. A significant improvement in the data-transfer speed and the power dissipation is required in order to employ the optical interconnection as the intra-chip or inter-chip optical interconnection. When only the data-transfer speed is taken into consideration, the minimum optical interconnection distance with the advantage over

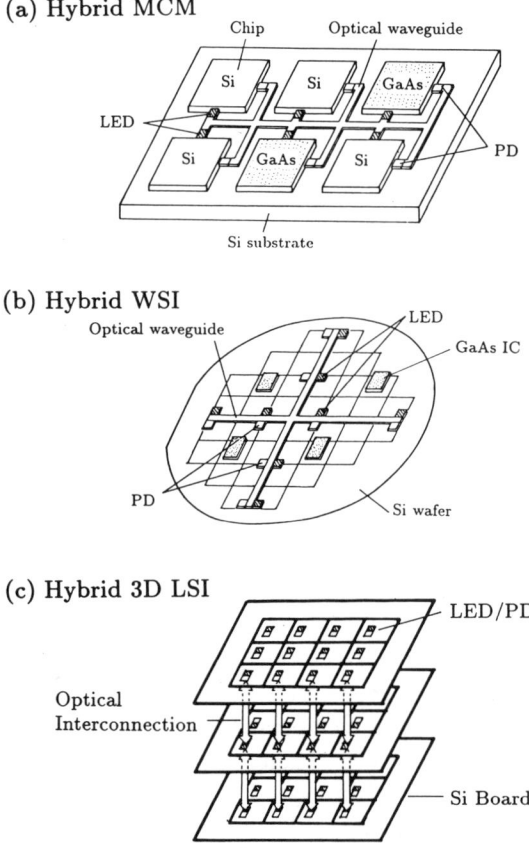

Figure 7.3.1
MCM, WSI and 3D LSI with optical interconnections. (a) hybrid MCM; (b) hybrid WSI; (c) hybrid 3D LSI

the electrical interconnection will be around 0.5 cm even if the conversion time of the photonic devices is included. However, it will be more than 1 cm if the compatibility with LSI and the cost are taken into consideration. Therefore, at present it is recommended employing the optical interconnection as the inter-chip interconnection of a MCM (multi-chip module), WSI (wafer scale integration), or three-dimensional (3D) LSIs, as shown in Figure 7.3.1, rather than the intra-chip interconnection inside LSI chips. However, in the future it will become possible to use the optical interconnection inside LSIs as well because the chip size will be increased to 3 cm × 3 cm or more. The optical interconnection has many advantages over the electrical interconnection, *e.g.* (1) there is no mutual interference between the signals; (2) there is no need for a ground; (3) there is no impedance mismatch; (4) there is a large fan-out; and (5) there is the capability of crossing within the identical plane. Therefore, if these advantages can be applied to practical use, the optical interconnection will become the key technology for future interconnections.

In this section we describe how such an optical interconnection can be formed on the LSI chip or on the silicon board of a MCM. The method of forming a vertical optical interconnection for 3D LSIs is also described. The basic technologies needed to form the optical waveguide, the micromirror and the microlens, and to integrate the photonic devices onto silicon chips have to be developed in order to fabricate LSIs or a MCM with optical interconnections. We have developed these technologies utilizing the LSI fabrication technology as described below;

2.1 *Formation of the optical waveguide*

In order to form the optical waveguide with excellent optical transfer characteristics, two kinds of materials with a large refractive index difference should be used for the core layer and the cladding layer. When such a waveguide is fabricated on a LSI chip using LSI technology, these materials should be chosen so as to be compatible with the LSI fabrication sequence. Before the metallization process, the silicon nitride (refractive index: around 2.0) and the silicon oxide (refractive index: around 1.45) with excellent film qualities can be used for the core and the cladding layers, respectively. These materials are very popular in LSI technology. After the metallization process, these materials cannot be used because of their higher formation temperature than the Al melting temperature. Then, after metallization, organic materials such as polymers are preferable owing to their low formation temperature after metallization. An SEM cross-section of the waveguide with a $SiO_2/Si_3N_4/SiO_2$ structure formed on a LSI chip is shown in Figure 7.3.2. As is obvious in the figure, the waveguide with a narrow width can be easily fabricated by using LSI fabrication technology. However, it was observed that the optical transfer characteristics are rapidly degraded as the waveguide width is decreased to less than 10 μm. This is because the small surface roughness on the sidewall of the core layer influences the optical transfer characteristics [6]. The core pattern is formed by RIE (reactive ion etching). RIE is inclined to cause such surface roughness because the roughness of the photoresist pattern edge is transferred to the etched material as it is. However, the propagation loss decreases if the surface roughness on the core sidewall is reduced by isotropic wet etching after forming the

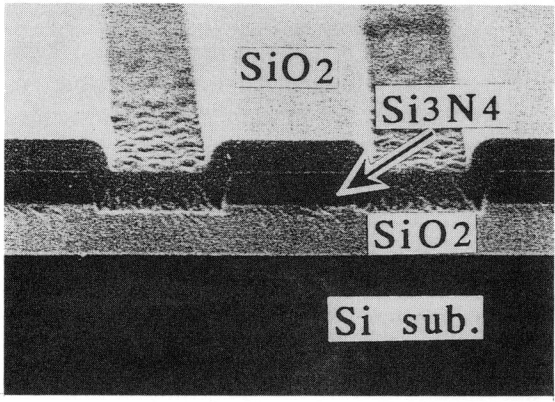

Figure 7.3.2
SEM cross-section of the optical waveguide

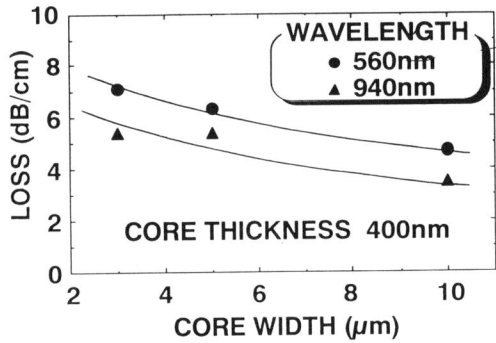

Figure 7.3.3
Propagation loss of the optical waveguide as a function of core pattern width

core pattern by RIE. The propagation loss characteristics measured in the fabricated waveguide are shown in Figure 7.3.3, where the propagation loss is plotted versus the core pattern width. The propagation loss increases as the core pattern width is decreased. Accordingly, the minimum core pattern width should be around 10 μm at present. The propagation loss was 1.5 dB/cm when the fabrication condition was optimized for the waveguide with a width of 10 μm. Such propagation loss is acceptable for intra-chip or inter-chip interconnections.

2.2 Change of propagation direction by micromirror

To use the optical waveguide as an interconnection inside LSI chip it is required that the propagation direction can be changed in a small area. To do this, it is useful to utilize the

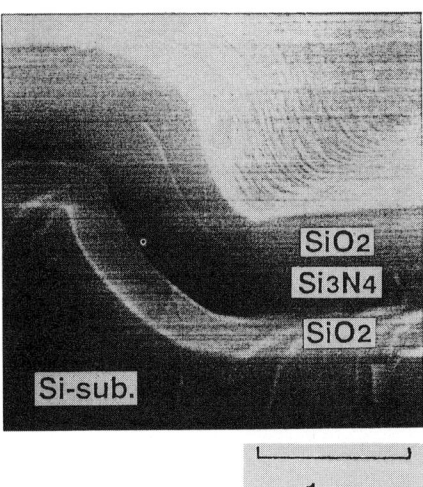

Figure 7.3.4
SEM cross-section of the optical waveguide formed on the tapered sidewall surface which acts as a micromirror

reflection of light. For example, in the case of changing the propagation direction by 90° within the identical plane, we can use an L-shaped optical waveguide after cutting its bending edge 45° and then forming the reflecting film on the cutting edge. Furthermore, in the case of changing the direction from horizontal to vertical or vice versa, we can utilize the tapered slanting sidewall surface formed by etching as the reflecting surface. An SEM cross-section of the optical waveguide formed on the tapered sidewall surface is shown in Figure 7.3.4, where the tapered surface is formed on the silicon substrate by using isotropic wet etching [6]. The propagation direction of the signal light can be changed from the vertical direction to the horizontal direction by utilizing the tapered sidewall surface as a micromirror. The signal propagation characteristics of the optical waveguide with such a micromirror are shown in Figure 7.3.5. It is obvious from the figure that, by bending the light with the micromirror, the signal loss can be ignored because the slope of the output intensity versus the waveguide length is almost identical to that for the straight waveguide without bending.

2.3 Fabrication of a microlens

It is important to use materials with a large refractive index in order to increase the light-collecting efficiency of a lens. Of the LSI materials, the poly-crystalline silicon (poly-Si) can satisfy such a requirement. Therefore we fabricated a microlens using poly-Si based on LSI fabrication technology. First, a photoresist lens is formed by baking the photoresist pattern, and then the photoresist lens pattern is transferred to poly-Si by RIE. Microlenses with a diameter of $1-10$ μm can be easily fabricated by

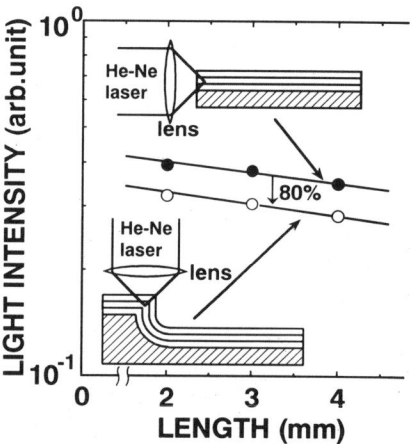

Figure 7.3.5
Signal propagation characteristics of two kinds of optical waveguides

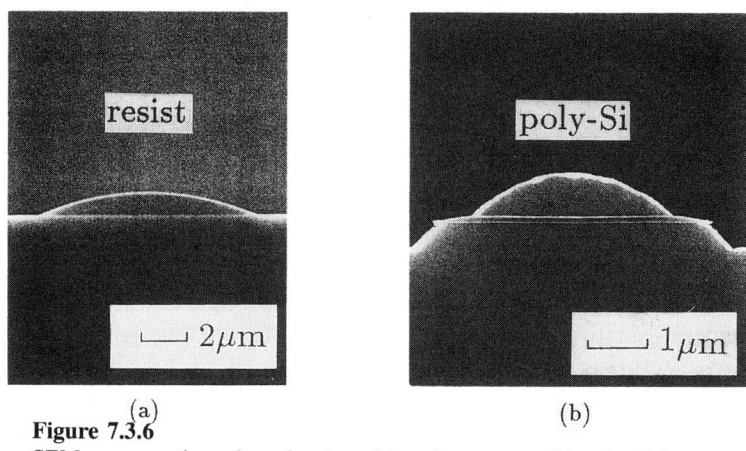

Figure 7.3.6
SEM cross-section of a microlens (a) resist pattern; (b) poly-Si lens

using such a method. Excellent light-collecting characteristics are obtained in these microlenses. An SEM cross-section of a microlens is shown in Figure 7.3.6.

2.4 Integration of photonic devices on a Si chip

A monolithic method using the hetero epitaxial growth technique (GaAs-on-Si) and a hybrid method using the flip-chip bonding technique have been proposed for integrating compound semiconductor photonic devices on Si chip. The hetero epitaxial growth technique is suitable for forming small photonic devices on a Si chip. However, it is not easy to obtain the excellent device characteristics using this technique because at present the crystal quality of the compound semiconductor formed on Si is not good. On the other hand, in the hybrid method, it is difficult to form small photonic devices on Si. However, we can integrate photonic devices with excellent characteristics if a relatively

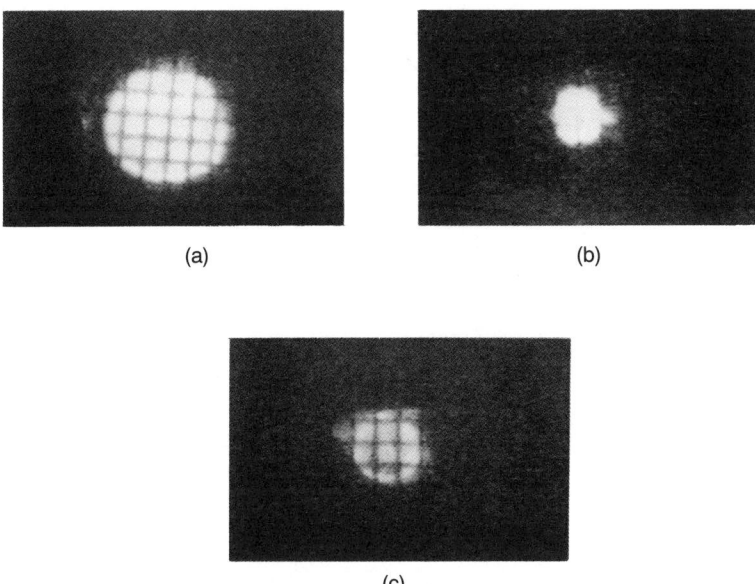

Figure 7.3.7
Photon emission pattern from a LED mounted on a Si chip ($I_{LED} = 0.1$ mA). (a) $\phi = 240$ μm; (b) $\phi = 150$ μm; (c) $\phi = 100$ μm

large size is allowed. Thus, we tried to integrate photonic devices on a Si chip using the micro-bonding technique, which is a kind of hybrid method. Figure 7.3.7 shows the photon emission from LEDs formed on a Si chip using the micro-bonding technique. As is obvious in the figure, a uniform emission pattern is observed. Thus, we can easily integrate photonic devices by using the micro-bonding technique when the device size is as large as 100 μm. We will develop a new method combining the micro-bonding technique with the epitaxial lift-off technique for the integration of photonic devices with the size smaller than 100 μm. Eventually, we will use the monolithic method.

2.5 Three-dimensional integration

Three-dimensional (3D) integration becomes very important in the case of forming a free-space optical interconnection in the vertical direction, as shown in Figure 7.3.1(c). Therefore we have developed a new 3D integration technique based on the wafer bonding method [7]. In this new method a Si wafer with devices is glued to the quartz glass using a wax, and then is thinned to 5 μm by chemical–mechanical polishing (CMP) from the reverse side as shown in Figure 7.3.8. In this case the trench filled with the oxide is formed on the Si surface before polishing. The depth of the trench is 5 μm. Therefore, the bottom of the trench is exposed when the Si is thinned to 5 μm by polishing from the reverse side. The oxide is formed at the bottom of the trench and this oxide acts as a stopper for chemical–mechanical polishing. Thus, we can precisely

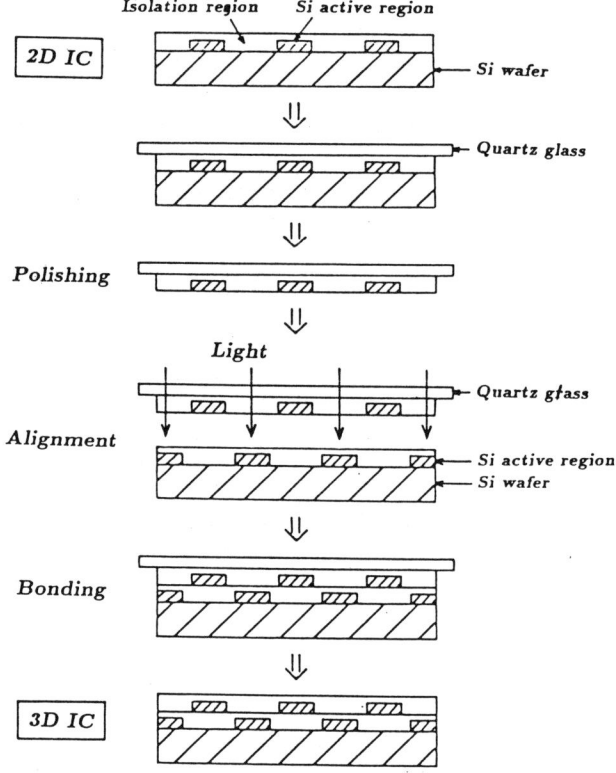

Figure 7.3.8
Fabrication process sequence for a 3D-LSI

control the Si wafer thickness to within (5 ± 0.5) μm. The quartz glass glued with such thinned Si wafer can be used as a mask in a contact printing exposure system. Therefore, this 'mask' can be aligned to another Si wafer with devices using infrared light and glued to it using an UV sensitive adhesive agent. The quartz glass is removed from the Si wafer after gluing. We can achieve a 3D LSI with a multi-layer structure by repeating this sequence. We have developed a new 3D wafer aligner, as shown in Figure 7.3.9, to fabricate such a 3D LSI [8]. More than 10 thinned Si wafers can be aligned with an accuracy of 1 μm by using this new aligner. In order to form a vertical optical interconnection in such an LSI with 3D structure, the photonic devices have to be formed on a Si wafer before bonding the wafer, as shown in Figure 7.3.10. In this case the LED or LD should be placed face to face with the photo-detector in every second layer of the 3D LSI.

3 ORAM-bus Memory

In this section we describe the basic concept of the ORAM-bus memory. Furthermore, the basic operation of the ORAM-bus memory is examined by computer simulation.

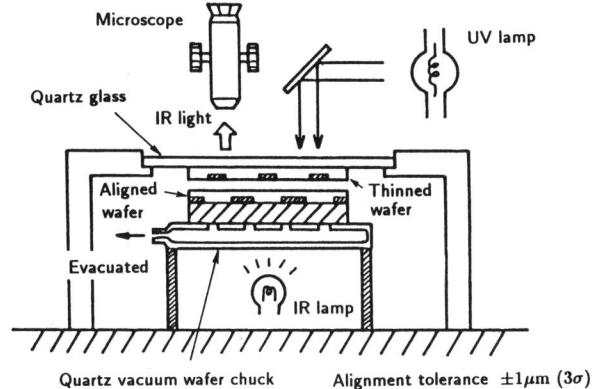
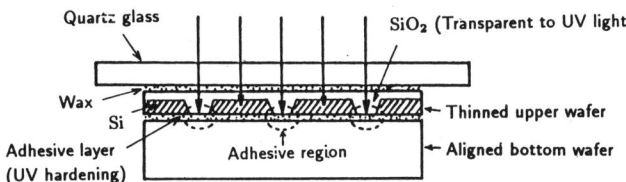

Figure 7.3.9
3D wafer aligner

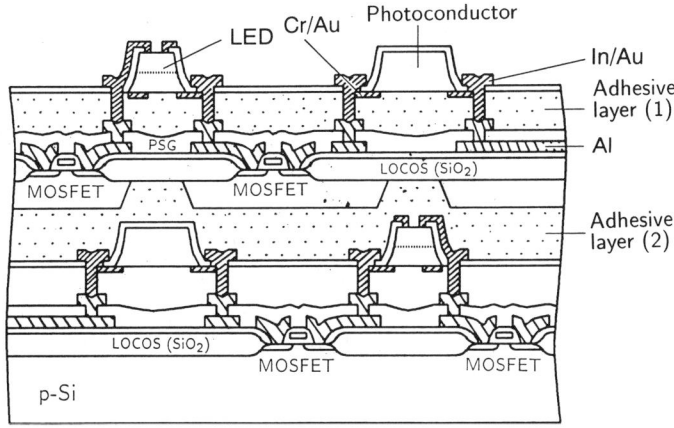

Figure 7.3.10
Cross-sectional structure of a 3D-LSI with an optical interconnection

3.1 Concept of the ORAM-bus memory

The basic concept of the ORAM-bus memory is similar to that of RAM-bus DRAM (dynamic random access memory) where CMOS latch-type sense amplifiers are also used as built-in cache memories [9]. However, ORAM-bus memory consists of a 3D structure with several memory layers. These memory layers are connected to each other by optical interconnections. Therefore, in ORAM-bus memory the data are directly transferred from the built-in cache memories of a memory layer to those of other memory layers through the optical interconnections without using the electrical data bus outside the chip. The configuration of RAM-bus memory is shown in Figure 7.3.11. As is clear in the figure, several thinned memory chips (memory layers) are stacked on top of each other and are connected by a number of vertical optical interconnections in the ORAM-bus memory in order to extend the data band width and to increase the data-transfer speed. Data transfer using the vertical optical interconnection is performed through the optical coupling sense amplifiers, as shown in Figure 7.3.12, which also act as cache memories. The optical coupling operation between the upper and lower memory layers forming the vertical optical interconnection is carried out through an optical coupling flip-flop which is included in the optical coupling sense amplifier circuit, as shown in Figure 7.3.13. The optical coupling flip-flop (OC-FF) circuit consists of a data-storing portion and a data-transfer portion. The flip-flop with two high resistive loads is used in the data-storing portion. The resistive load in the flip-flop also acts as a photoconductor. Two LEDs are included in the data-transfer portion. The data are optically transferred by the LEDs to the upper and lower memory layers. When the transferred light signal impinges on to either of the two photoconductors in the upper or lower memory layer, the node voltage of the data-storing flip-flop is inverted. Thus, the transferred data are directly written into the data-storing flip-flop. The data

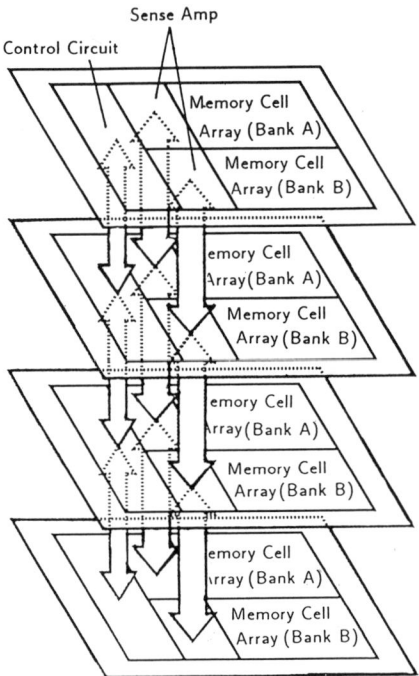

Figure 7.3.11
Configuration of the ORAM-bus memory

Realization of High-speed 3D LSIs by Optical Interconnections

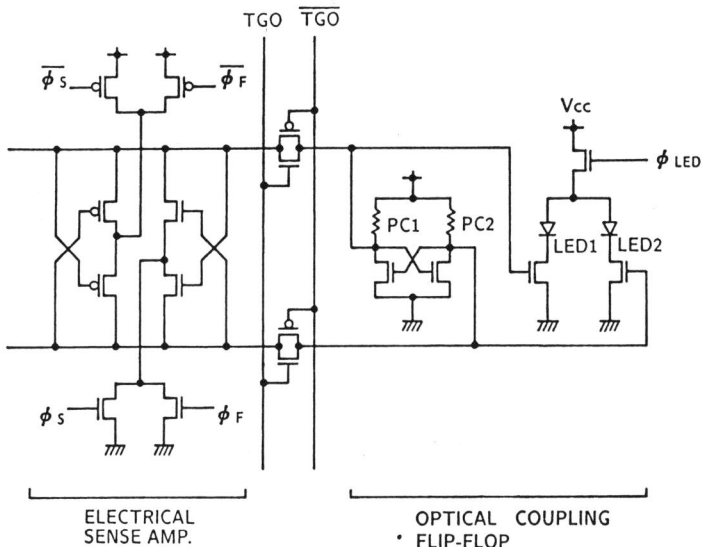

Figure 7.3.12
Optical coupling sense amplifier

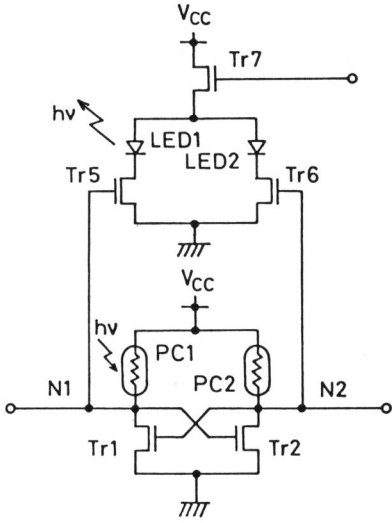

Figure 7.3.13
Optical coupling flip-flop

are transferred to the vertical direction by repeating this sequence. Therefore, the data transfer is sequential in the OC-FF. However, the optical positive feedback caused between the two OC-FFs in the upper and lower memory layers enhances the data transfer speed. In addition, many optical interconnections are easily formed in the vertical direction to extend the data bandwidth.

3.2 Design of ORAM-bus memory

The ORAM-bus memory with a configuration of 256 kbit × 4 layers was designed on the basis of the 2 μm CMOS design rule. The circuit block diagram for one memory layer is shown in Figure 7.3.14. The electrical data writing and reading operations inside one memory layer of the ORAM-bus memory are controlled by the address signals (A0–A9) and the control signals (RAS, CAS, WE), as well as DRAM. It is different from DRAM in that the extra circuits such as the optical coupling sense amplifiers, the optical row address registers, the optical block address buffers, the optical block decoders, and the optical transfer control circuits are necessary in the ORAM-bus memory in order to transfer the data through the optical interconnections. These circuits are controlled by the optical transfer enable signal (TE) and the optical transfer masking signal (TM). The optical data-transfer operation is carried out through the optical coupling sense amplifiers. Many memory cells connected to a pair of bit lines share one optical coupling sense amplifier as the optical data-transfer circuit. The optical coupling sense amplifier consists of an OC-FF and an electrical sense amplifier. The optical coupling operation between the upper and lower memory layers to transfer the data through the optical interconnection is performed by an OC-FF. The electrical sense amplifier consists of a CMOS flip-flop circuit. Therefore, the electrical sense amplifier can be used as a memory cell for the cache memory. The electrical sense amplifier is also used to amplify the small signal read-out from the memory cell to the bit line and to restore the data into the memory cell in the reading and refreshing

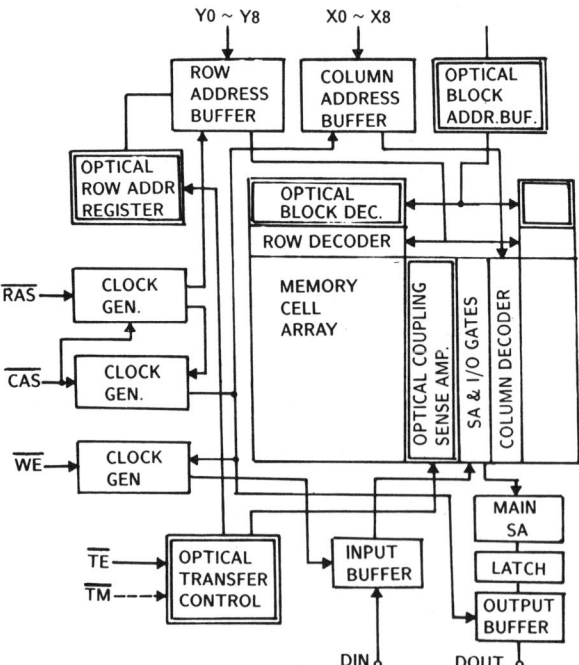

Figure 7.3.14
Circuit block diagram of the ORAM-bus memory

operations. Furthermore, the electrical sense amplifier has the role of restoring the signal transferred from the upper or lower memory layer through the OC-FF into the memory cell after amplifying it. The optical block address buffer and the optical block decoder are used to select the memory cell array block with the data transferred. It is possible in the optical data-transfer operation of the ORAM-bus memory to simultaneously transfer a block of data which are stored in many memory cells connected to one word line (one row) by activating the corresponding optical coupling sense amplifiers. However, a number of LEDs should be simultaneously turned on and consequently the power consumption is increased to simultaneously transfer much data through the optical interconnections. Therefore, the size of the data which can be simultaneously transferred is limited due to the power consumption. Here, the chip was designed so that 1 kbit of data are simultaneously transferred. However, it was shown by computer simulation that a very high data-transfer speed of 64 Gbits/s can be achieved. The extra power penalty is only 130 mW. The address signals are also transferred through the optical interconnections in the ORAM-bus memory. The optical address register is used to transfer these address signals. The timing of the optical data-transfer operation is controlled by the optical transfer control circuit.

3.3 Fabrication of the ORAM-bus memory test chip

As mentioned above, we have succeeded in developing a wafer-bonding technique for 3D LSIs. However, we have not succeeded yet in stacking several LSI wafers with photonic devices to make 3D LSIs with optical interconnections. This is because very thin photonic devices have not been obtained. It is necessary to reduce the thickness of the photonic devices to 1 μm in order to stack several wafers with photonic devices. Therefore, we tried putting a thick LED on the top surface of such a 3D LSI test chip by using the micro-bonding technique in order to confirm the basic memory operation of

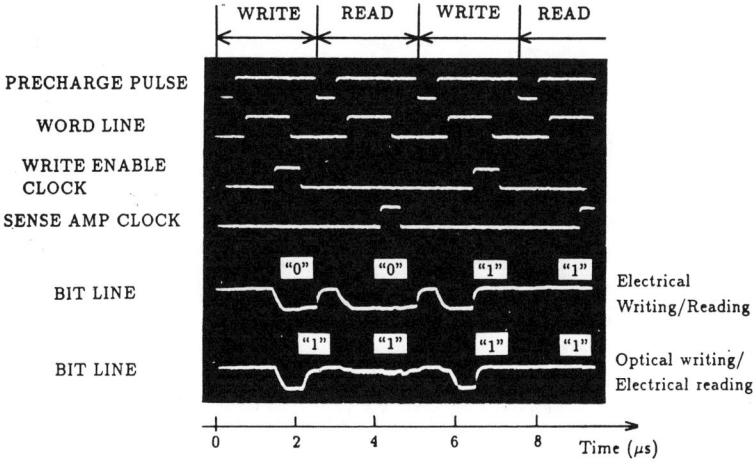

Figure 7.3.15
Waveforms measured in the ORAM-bus memory test chip

the ORAM-bus memory, although the data-transfer operation through several memory layers in the vertical direction cannot be evaluated in this test chip. The test chips were fabricated using 2 μm CMOS technology. The waveforms measured in this test chip are shown in Figure 7.3.15. Both the input signal to the memory cell and the output signal from the memory cell appear on one bit line. Two basic operations of electrical writing/electrical reading and optical writing/electrical reading are evaluated in the figure. In the electrical writing/electrical reading operation of Figure 7.3.15, the data '0' is written in the first cycle and the data '1' in the second cycle. It is clear from the figure that the written data are correctly read-out to the bit line in the electrical reading operation. Meanwhile, in the optical writing/electrical reading operation, the data '0' is electrically written first and then the data '1' is optically written in the first cycle. As is clear in the figure, the optically written data '1' is correctly read-out in the reading operation. Thus, we could confirm the basic operation of ORAM-bus memory using a test chip.

4 *Parallel Processor System with the ORAM-bus Memory*

A newly proposed parallel processor system with the ORAM-bus memories is shown in Figure 7.3.16, where each memory layer of the ORAM-bus memory is connected to the respective CPU. In this system, a large block of data can be simultaneously transferred at very high speed through the optical interconnections in the vertical direction, while the conventional memory operations are carried out in the horizontal planes. Memory cells with identical addresses in all memory layers have the same data after the optical

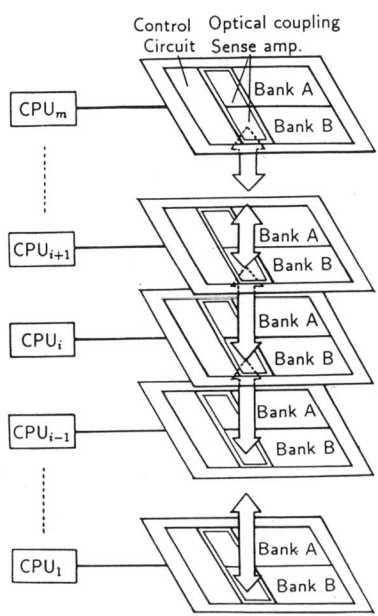

Figure 7.3.16
Parallel processor system with the ORAM-bus memory

data transfer. The stored data in the cache memories can be simultaneously read by many CPUs without conflict. Therefore, the ORAM-bus memory acts as a real shared memory in this system. Thus, the real shared memory connection-type parallel processor system with a very high performance can be achieved using the ORAM-bus memory.

5 Summary

The basic technologies needed to form optical interconnections have been developed using LSI technology. In addition, a new ORAM-bus memory with a parallel data-transfer function using optical interconnections was proposed and its basic operation was evaluated by using a computer simulation method. As a result, it was revealed that a very high data transfer speed of 64 Gbits/s can be achieved with only a small penalty in the extra power consumption of 130 mW in the 256 kbits × 4 layers ORAM-bus memory although the chip was designed using the 2 μm CMOS design rule. Futhermore, the test chip of the ORAM-bus memory was fabricated and the basic operation of ORAM-bus memory was confirmed. A new real shared memory connection-type parallel processor system was also proposed using the ORAM-bus memory.

References

[1] M. Koyanagi, *Proc. SPIE Int. Symp. on Advances in Interconnection and Packaging*, 1990, pp. 109-116.
[2] M. Koyanagi, *Proc. VLSI 91* (IFIP wg 10.5), 1991, pp. key3.1-3.10.
[3] H. Takata, H. Mori, J. Iba and M. Koyanagi, *Jpn J. Appl. Phys.*, **28**, L2305-L2308 (1989).
[4] M. Koyanagi, T. Nakano, T. Etoh and R. Aibara, *Proc. SPIE OE/LASE '93-Optoelectronic Packaging and Interconnects*, **1849**, 1993, pp. 233-241.
[5] K. Miyake, T. Tanaka, T. Etoh, M. Tsuno, S. Yokoyama and M. Koyanagi, *Extended Abstracts of the 1993 Int. Conf. on Solid State Devices*, 1993, pp. 673-675.
[6] T. Nagata, T. Tanaka, K. Miyake, H. Kurotaki, S. Yokoyama and M. Koyanagi, *Extended Abstracts of the 1993 Int. Conf. on Solid State Devices*, 1993, pp. 1047-1049.
[7] M. Koyanagi, *Extended Abstracts of 8th Symp. on Future Electron Devices*, 1989, pp. 50-60.
[8] H. Takata, T. Nakano, S. Yokoyama, S. Horiuchi, H. Itani, H. Tsukamoto and M. Koyanagi, *Proc. Int. Semiconductor Device Research Symp.*, 1991, pp. 327-330.
[9] N. Kushiyama, S. Ohshima, D. Stark, K. Sakurai, S. Takase, T. Furuyama, R. Barth, J. Dillon, J. Gasbarro, M. Griffin, M. Horowitz, V. Lee and W. Leung, *Dig. Tech. Papers of 1992 Symp. on VLSI Circuits*, 1992, pp. 66-67.

7.4

Semiconductor-based, Two-dimensional Spatial Light Modulator

Minoru Yamada and **Yuji Kuwamura**

Abstract

Establishment of optical computing systems is expected for further development of optoelectronics. Spatial light modulators or optical modulator arrays are going to be key devices in these systems.

The operating mechanisms of various types of semiconductor light (optical) modulators for the optical parallel processing are reviewed here. Also the author's work on the semiconductor light modulators, which utilize the electron–depleting effect in semiconductors such as AlGaAs/GaAs, is reported.

1 Introduction

The establishment of optical computing systems is expected in the further development of optoelectronics. An imaginative scheme for a digital optical parallel computer is illustrated in Figure 7.4.1, where data are stored in spatial light modulators (or optical modulator arrays) and the light source (or optical emitter) arrays give an operation or calculation method. The results are obtained through detector arrays. An important subject to realize this computing system is to develop spatial modulators made of semiconductor materials.

We review here the varieties and operating mechanisms of semiconductor light (optical) modulators for optical parallel processing and report our current work for this purpose [1,2].

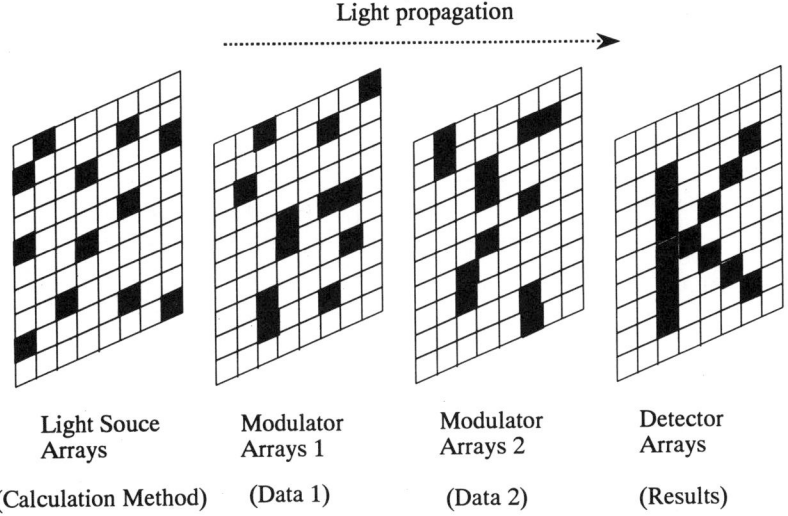

Figure 7.4.1
An imaginative scheme for a digital optical parallel computer

2 Varieties of Semiconductor Light Modulators

2.1 Configuration of the spatial modulators

Spatial light modulators are classified into two types on the basis of configuration of the input light and the output light, as shown in Figure 7.4.2. One is a transmission type and the other is a reflection type.

The modulators are also classified on the basis of the type of material parameters used to control the optical light, as illustrated in Figure 7.4.3. The first one is due to

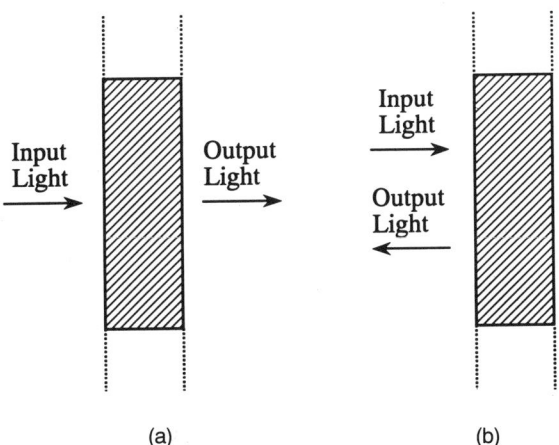

Figure 7.4.2
Configurations of spatial light modulators. (a) Transmission type; (b) reflection type

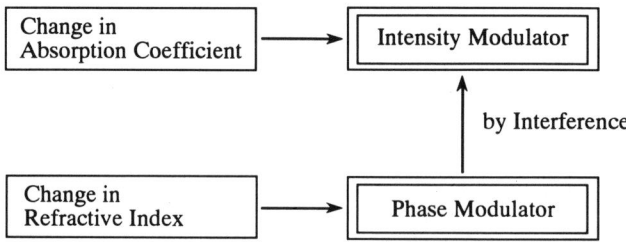

Figure 7.4.3
Classification of a light modulator based on operating mechanisms

a change in the absorption coefficient which works like an intensity modulator. The second one is due to a change in the refractive index. A phase modulator is made up by this mechanism directly. We can also compose another intensity modulator with a change in the refractive index by interfering with the input light.

2.2 Modulation mechanisms

The possible mechanisms for changing the absorption coefficient and/or the refractive index are listed in Figure 7.4.4, where one group is based on the effects of the applied electric field and the other one is based on changes in the carrier (the electron or the hole) number. We may expect for all mechanisms both changes in the absorption coefficient and the refractive index because these are combined by the Kramer–Kronig relation.

The electrooptic effect in the infrared region is caused by stable electrons such as the valence band electrons or the bounded electrons, and is weaker than the other effects which are remarkable at the photon energy around the bandgap.

The Franz–Keldysh effect in a bulk semiconductor is illustrated in Figure 7.4.5, where waves of the electron and the hole penetrate into the bandgap with application of the field as in Figure 7.4.5(b) resulting in effective reduction of the bandgap.

The quantum confined Stark effect (QCSE) is basically the same as the Franz–Keldysh effect, but is explained by a change in the potential well due to the applied field in the quantum well structure [3]. A strong effect is observed at the exciton's energy levels. This effect may be strongest in semiconductor so far observed

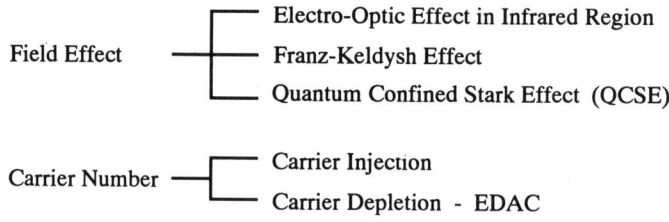

Figure 7.4.4
Possible mechanisms in a semiconductor crystal to be a light modulator

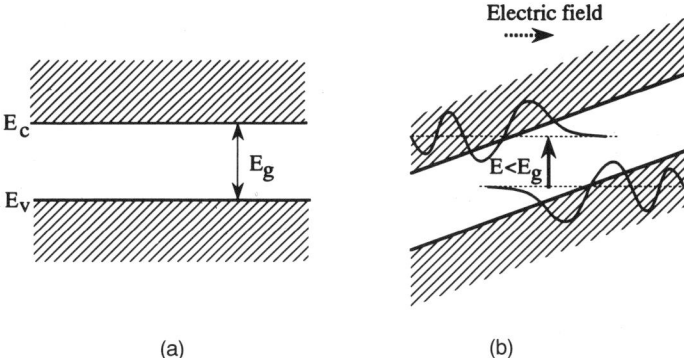

Figure 7.4.5
Mechanism of the Franz–Keldysh effect in a bulk semiconductor. (a) No electric field; (b) under an applied electric field

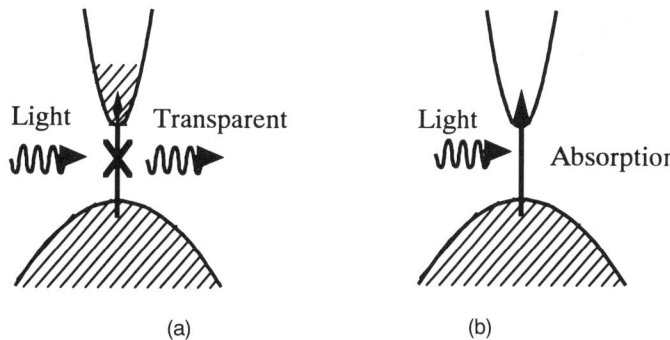

Figure 7.4.6
Modulation mechanism by changing the electron numbers in the conduction bands. (a) Sufficient electrons in conduction band; (b) no electrons in conduction band

but the effective range of wavelength for the operation is restricted within several nm around at the exciton's level. Precise stabilization of the temperature of the device is required to adjust the energy level to the incident light if we use the change in the exciton's level.

The absorption coefficient and the refractive index are also varied by changing the carrier number in a direct transition type semiconductor, such as GaAs. In an intrinsic semiconductor without any impurities, the electron is absent in the conduction band, as shown in Figure 7.4.6(b), then the incident light is absorbed by the electron transition from the valence band to the conduction band. If sufficient electrons are injected into the conduction band, the transition from the valence band is weakened because the energy states in the conduction band were already occupied by the other electrons. This idea is identical to the way to get an optical gain in a semiconductor laser. So the modulator actually gives spontaneous emission, which is counted to be noise generated in the modulator, by electron injection.

An alternative method to change the carrier number is to use doped materials. The electron transition from the valence band to the conduction band is suppressed due to a large number of electrons in the conduction band in a n-type semiconductor, as in Figure 7.4.6(a), but is enhanced by depleting the electrons from the conduction band, as in Figure 7.4.6(b). This condition is achieved in the depletion region around the p–n junction. The same effect is also expected in the p-side depletion region.

The authors fabricated modulators on the basis of the last mechanism, and named them electron depleting absorption control (EDAC) modulators [1,2].

3 Basic Structure of the EDAC Modulator

The basic structure of our EDAC modulator is shown in Figure 7.4.7. The absorption is controlled at the p^+–n^- junction planes. Since the variable range of the depletion region is less than several tens of nanometers, we adopt multiple junction planes to get sufficient width of the depletion regions. The operating voltage is applied through the p^+ and n^+ regions, which are located not only in the multi-layer structure but also at the side corners in the transverse directions forming the comb-shaped structure.

The devices were fabricated with the liquid-phase epitaxy.

4 Operating Characteristics of the EDAC Modulator

The extinction (or on–off) ratio is defined to examine the quality of the intensity modulator by $(I_{max} - I_{min})/I_{max}$ in the linear scale, or by $10 \log_{10}(I_{max}/I_{min})$ in dB, where I_{max} is the maximum intensity and I_{min} is the minimum intensity of the transmitted or reflected light under variation of the drive voltage keeping the intensity of the input light constant.

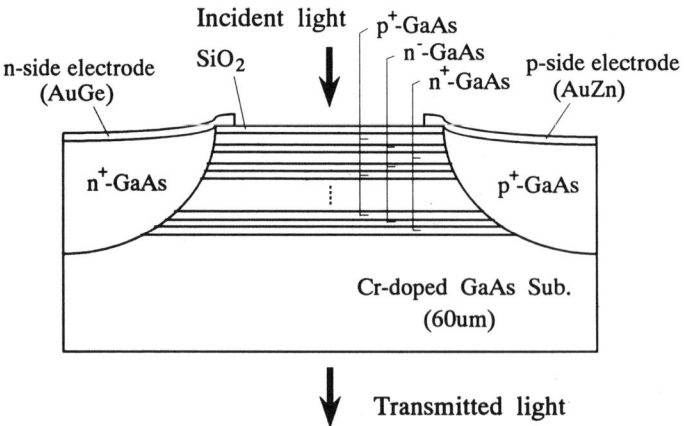

Figure 7.4.7
Basic structure of the EDAC (electron depleting absorption control) modulator

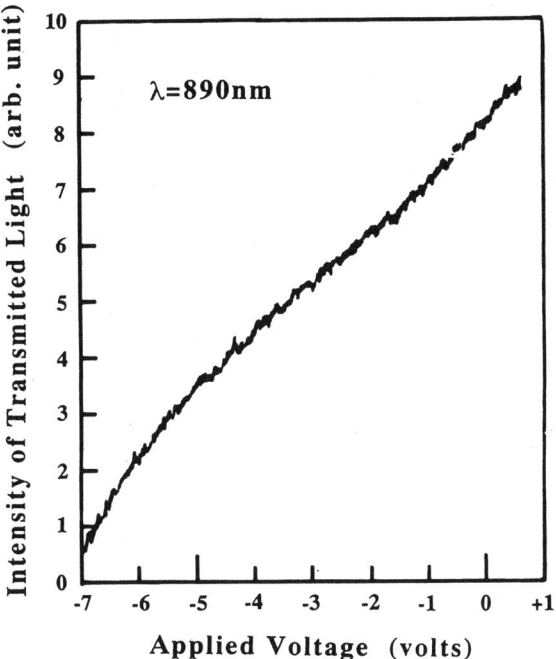

Figure 7.4.8
Variation of the transmitted light with an applied voltage in the EDAC modulator

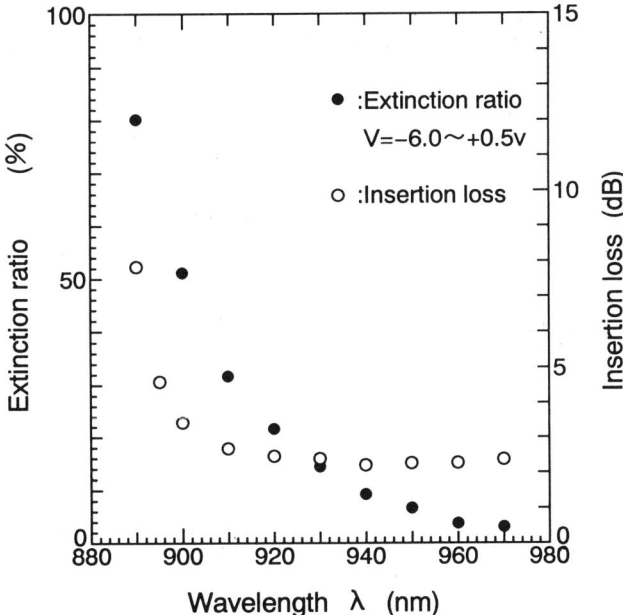

Figure 7.4.9
Wavelength dispersion of the extinction ratio in the EDAC modulator

Variations of the transmitted light with an applied voltage is shown in Figure 7.4.8 for a device with 16 sets of p$^+$-n$^-$-n$^+$ junctions. More than 93% (11 dB) of the extinction ratio is obtained with a voltage variation from +0.5 V to −7 V. The change in absorption coefficient in GaAs material was estimated to be $\Delta\alpha \approx 5000$ cm^{-1}.

The wavelength dispersion of the extinction ratio is given in Figure 7.4.9 together with insertion loss data. The wavelength width giving a sufficient extinction ratio is around 20 nm. The insertion loss of this device seems too large. However, the main part of the loss comes from absorption in the GaAs substrate. The loss will be reduced by opening a window in the substrate after fabrication, or by changing the substrate to a material that has a wider bandgap than GaAs.

5 Comparison of Operating Characteristics of the EDAC Modulator with Other Semiconductor Light Modulators

The data obtained are compared with those of other types of modulators in Figure 7.4.10. A higher extinction ratio with a lower applied voltage is expected for wider applications of the light modulator. In the figure, open circles indicate reflection types, while solid circles inside transmission types. The modulation efficiency in the reflection-type modulator was enhanced by forming a Fabry–Perot resonator, making the front and the back surfaces reflecting mirrors and changing the resonance condition with

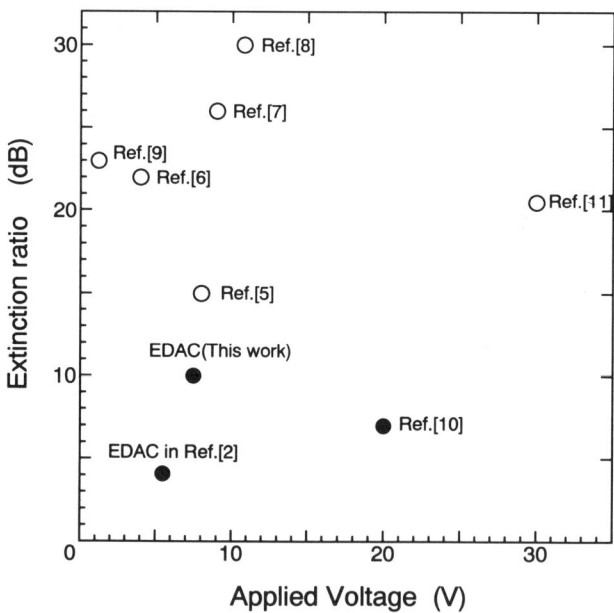

Figure 7.4.10
Comparison of the extinction ratios for several types of modulators reported so far

variations in the absorption coefficient and/or the refractive index [4]. Almost all data of the reflection type are increased with the help of the resonance effect [5-9,11]. Naturally, the operating ranges of the wavelength will become narrower by enhancing the resonance effect.

The resonance effect is useful in the reflection type but is not effective in the transmission type. The extinction ratio in the transmission type directly comes from a change in the absorption coefficient. The obtained extinction ratio of EDAC modulator was the highest as the transmission type modulator. All modulators except EDAC in Figure 7.4.10 are based on the QCSE in the quantum well structure. The operating range of the wavelength in the EDAC is wider than that in the QCSE in principle. So we can say the EDAC mechanism is also a promising candidate in establishing spatial light modulators.

6 Integration of Light Modulators

Our EDAC devices were integrated in $11 \times 9 = 99$ two-dimensional arrays, as shown in Figure 7.4.11.

The EDAC structure can also work as a light detector since it is, or could be, a light-emitting diode by current injection. So, integration of many EDAC devices will give new functional operations in the future.

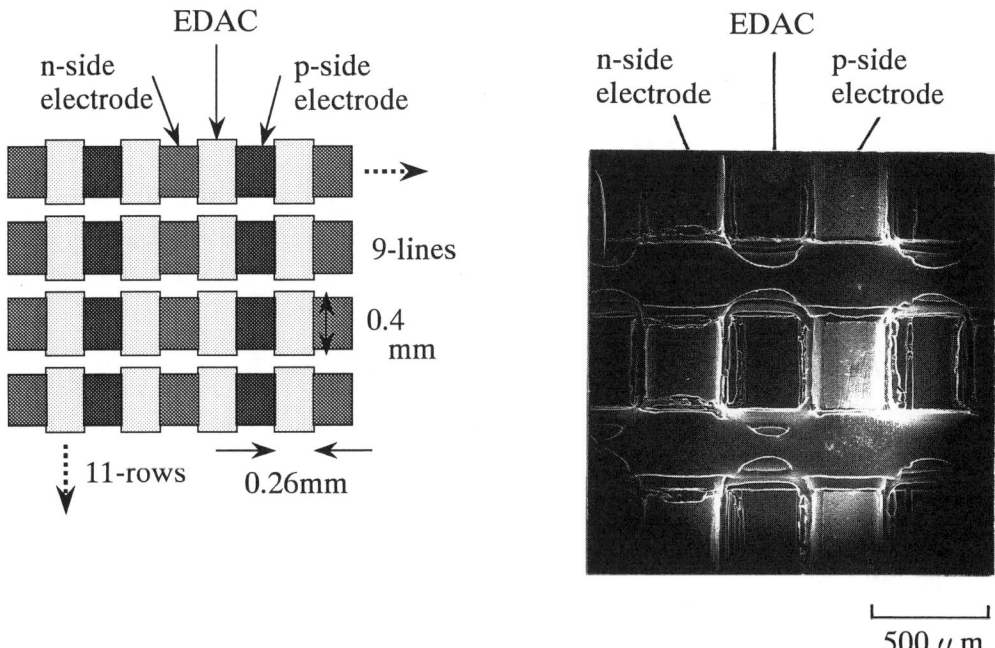

Figure 7.4.11
Two-dimensional integration of EDAC modulators

7 Detailed Discussion of the Absorption Mechanism in the EDAC Modulator

We set out to fabricate our EDAC modulator on the basis of an idea to change the carrier numbers in the bands and change the absorption, like the mechanism illustrated in Figure 7.4.6. However, we found that two other mechanisms also work to enhance modulation in the EDAC.

The second mechanism working in the EDAC is the Franz–Keldysh effect, as already introduced in Figure 7.4.5. Since we doped the donor and the acceptor impurities in the n and p regions, respectively, a strong electric field is induced by ionization of the impurities in the depletion layers.

The third mechanism is the screening effect of the impurity atoms. The energy levels of the impurity atoms are determined with a profile of the Coulomb potential, which can be shielded or screened by the existing carriers. We call this effect the screening effect. The impurity levels then rise and approach the band edge when the carrier number is large, but shift down to the longer wavelength side when the carriers are depleted due to changes in the potential profile. That is to say, the optical absorption edge by electron transition shifts to longer wavelength side by carrier depleting because the screening effect has less affect on the impurity atoms.

We may improve the operating characteristics of our device by more detailed investigation of these mechanisms.

8 Conclusions

We have reviewed the possible mechanisms for a spatial light modulator on the basis of a semiconductor crystal and reported our work on the EDAC (electron depleting absorption control) modulator. We showed that utilization of the impurity atoms in a semiconductor is also effective in controlling the intensity of light. We believe that an optical computation system will be realized by utilizing semiconductor-based spatial light modulators in the near future.

References

[1] M. Yamada, K. Noda, Y. Kuwamura, H. Nakanishi and K. Imai, *Trans.IEICE*, **E75-C** (9), 1063–1070 (1992).
[2] Y. Kuwamura, M. Yamada and M. Suzumi, *Jpn J. Appl. Phys.*, **32**, Pt. 1, (1B), 578–582 (1993).
[3] D.A.B. Miller, Optical and Quantum Electronics, **22**, Chapman and Hall Ltd, 1990, pp. S61–S98.
[4] R.H. Yan, R.J. Simes and L.A. Coldren, *IEEE J. Quantum Electron.*, **27** (7), 1922–1931 (1991).
[5] R.H. Yan, R.J. Simes and L.A. Coldren, *IEEE J. Quantum Electron.*, **25** (11), 2272–2280 (1989).
[6] K.K. Law, M. Whitehead, J.L. Merz and L.A. Coldren, *Electron. Lett.*, **27** (20), 1863–1865 (1991).

[7] M. Whitehead, A. Rivers, G. Parry, J.S. Roberts and C. Button, *Electron. Lett.*, **25** (15), 984–985 (1989).
[8] D.S. Gerber, R. Droopad and G.N. Maracas, *IEEE Photon. Technol. Lett.*, **5** (1), 55–57 (1993).
[9] M.R. Stead, R.P. Leavitt and G.J. Simonis, *CLEO'94 Technical Digest*, **8**, 59–60 (1994).
[10] T.Y. Hsu, W.Y. Wu and U. Efron, *Electron. Lett.*, **24** (10), 603–605 (1988).
[11] C. Amano, S. Matsuo and T. Kurokawa, *IEEE Photon. Technol. Lett.*, **3** (8), 736–738 (1991).

7.5

Integrated Optoelectronic Neuro Devices

Hiroo Yonezu

Abstract

Fundamental optoelectronic adaptive devices and circuits have been developed, which are essential for synaptic connection in neural networks. They were constructed using CMOS and EEPROM processes. They realized three essential functions for synaptic connection, which were the long term memory, a fixed total amount of synaptic weights and lateral inhibition. The self-organization was confirmed in primitive networks for supervised and unsupervised learning in experiments and computer simulations.

New technologies have also been developed for producing optical devices on Si substrates for through-wafer interconnection.

1 Introduction

Neuro-computing uses massively parallel information processing with self-organization [1]. Hardware implementation is required because currently computers have in principle series information processing. Research has been done in three fields: optics, optoelectronics (OEIC), and electronics (LSI) [2]. In a combination of the present microprocessor chip and a neuro-chip, the LSI technology is preferred. The LSI technology is advantageous to large-scale arrays of neurons and synaptic connections in particular. However, the large number of interconnections among neurons have prevented the realization of a large-scale neural network. Thus, parallel optical interconnections should be involved in the neuro-chip to solve the problem [3].

We have studied fundamental optoelectronic devices and technologies for synaptic connections and interconnections, which could be applicable to large-scale OEICs in the future.

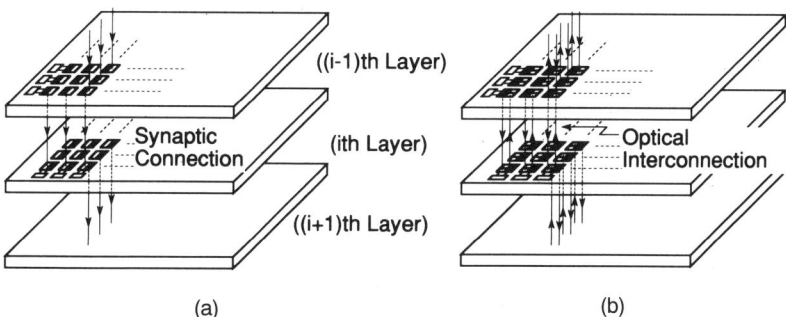

Figure 7.5.1
Architectures of multi-chip OEICs with optical interconnections. (a) Feedforward; (b) Feedback

2 Architectures of OEIC Chips

Neural networks are generally formed in multi-layers, which consist of input, intermediate, and output layers. Neurons are connected through a lot of synapses in each layer and between neighboring layers. An ideal chip architecture is shown in Figure 7.5.1, where presynapses and postsynapses are arrayed in Si chips [4]. The synapses are connected electrically in each chip and optically between chips. Thus, the most fundamental devices and circuits are ones for synaptic connection and optical interconnection. The power consumption should be low for large-scale integration. Other devices and circuits are formed with conventional LSI technology.

3 Synaptic Devices and Circuits

The following three functions are essential in synaptic connections for self-organization: (1) a long-term memory of synaptic weights; (2) a fixed total amount of synaptic weights; and (3) lateral inhibition. Function (1) is the fundamental requirement for learning. Function (2) is generally required to avoid divergence of the synaptic weights. Function (3) is particularly required for spacific self-organization in biological networks such as topological feature mapping.

3.1 Synaptic connection circuit

The synaptic connection circuit was constructed by a differential amplifier operating in the sub-threshold region [5], as shown in Figure 7.5.2. When the optical output from the jth neuron is received by the photodiode PD, the photocurrent $I_{\mathrm{ph}j}$ is divided into $I_{\mathrm{d}1}$ and $I_{\mathrm{d}2}$. The difference of both currents, $I_{\mathrm{d}1} - I_{\mathrm{d}2}$, is given by

$$I_{ij} = I_{\mathrm{d}1} - I_{\mathrm{d}2} = I_{\mathrm{ph}j} \cdot \tanh\left\{\frac{\kappa}{2} \cdot (V_{\mathrm{TH}2} - V_{\mathrm{TH}1})\right\} \propto (V_{\mathrm{TH}2} - V_{\mathrm{TH}1}) \cdot I_{\mathrm{ph}j}$$

depending on the threshold voltage V_{TH} of the two transistors T1 and T2 called the optical adaptive device (OAD) [6], where κ is constant. When $I_{\mathrm{d}1} - I_{\mathrm{d}2}$ is defined as an

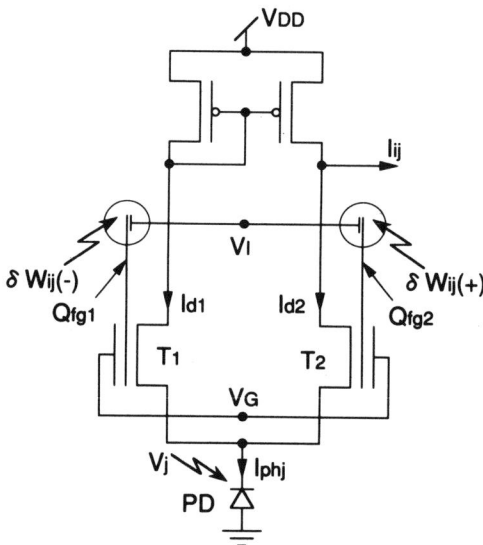

Figure 7.5.2
Synaptic connection circuit

input I_{ij} into the ith neuron through the ijth synaptic connection, the synaptic weight W_{ij} is given by the difference of charge quantities $\Delta Q (= Q_{FG2} - Q_{FG1})$ in the floating gate between both OADs.

3.2 Optical adaptive device and long-term memory of synaptic weights

The synaptic function (1) has been realized with the OADs shown in Figure 7.5.3. The charge quantity is controlled according to the correction signals (δW_{ij}) of the optical

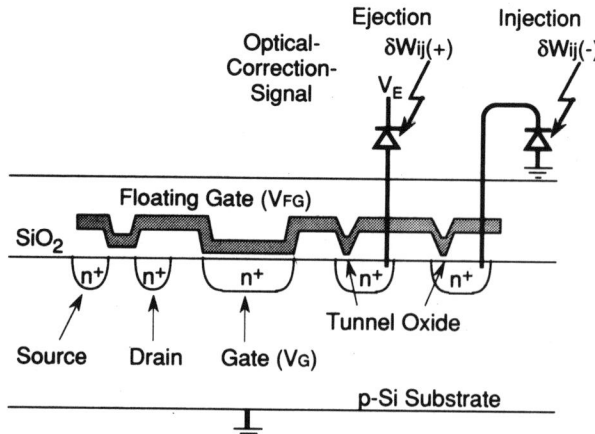

Figure 7.5.3
Cross-sectional view of the optical adaptive device

Integrated Optoelectronic Neuro Devices

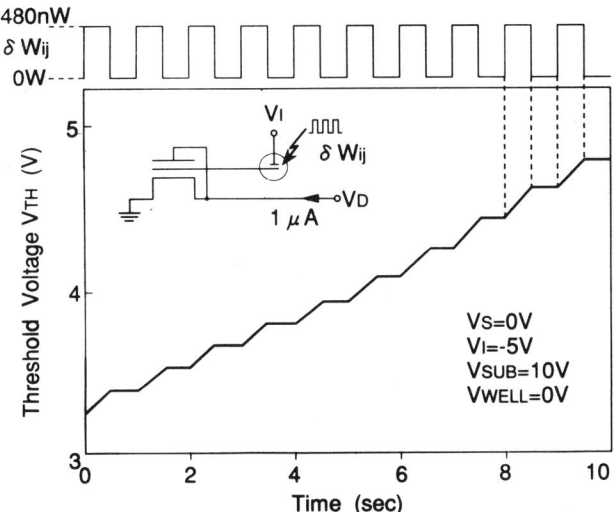

Figure 7.5.4
Variation of the threshold voltage of the optical adaptive device when pulsed optical correction signals are applied

pulses by injecting electrons into the floating gate or by ejecting electrons from the gate through a thin SiO_2 film.

The OAD and synaptic connection circuit were fabricated with our CMOS and EEPROM processes. The threshold voltage V_{TH} of the OAD was varied during the period when the optical pulse δW_{ij} was applied, as shown in Figure 7.5.4. When no optical pulse was applied, the threshold voltage stayed at the previous value. Thus, the synaptic weights were varied in a nonvolatile manner by the optical PDM or PWM signals.

A similar adaptive device was also developed whose charge quantity in the floating gate was controlled by electrical correction signals [6]. The synaptic weights were also varied in a nonvolatile manner.

3.3 Fixed total amount of synaptic weights

The synaptic function (2) was realized in a synaptic circuit using OADs (T1–T4), as shown in Figure 7.5.5(a), where four synapses were included. Each synaptic weight was determined by the voltage drop across the resistance instead of the difference in the threshold voltage of the OADs. Thus, the OADs in Figure 7.5.2 were replaced by conventional nMOS transistors. The current I_j flowing through each resistance is controlled by the OAD, while the total sum of the currents is fixed to a constant current, I_s.

The circuit in Figure 7.5.5(a) was fabricated with our CMOS process. When the synaptic weight W_{i1} is increased by applying the optical pulses δW_1 to OAD T1, the others, W_{i2}, W_{i3}, and W_{i4} are decreased, as shown in Figure 7.5.5(b). The total sum $\Sigma_j W_{ij}$ was kept nearly constant. The varied synaptic weights were kept in a nonvolatile manner.

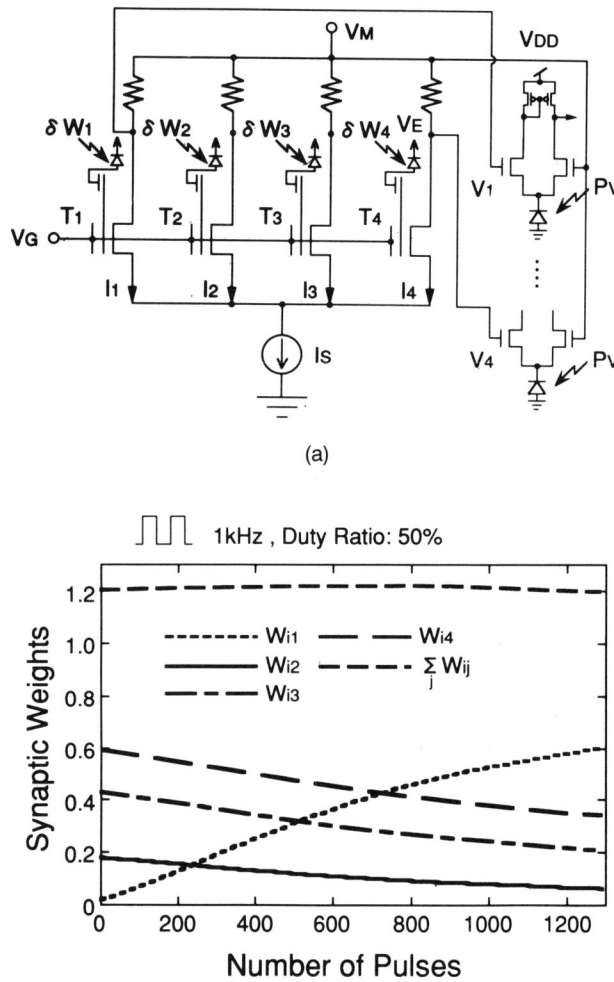

Figure 7.5.5
Circuit for a fixed synaptic weight amount. (a) Circuit; (b) Experimental variation of synaptic weights

3.4 Lateral inhibition

A lateral excitation connection (LEC) is the main part of lateral inhibition. Thus, a LEC circuit was realized which followed a winner-take-all (WTA) circuit [7], as shown in Figure 7.5.6(a). The gate voltage (V_{tg}) of the pMOS transfer gates controls the lateral connection strength. The WTA circuit is connected to the input currents $I_{in1}, I_{in2}, \ldots, I_{ini}, \ldots,$ and I_{inn} through the current mirrors. The output current I_{outi} becomes nearly I_{WTA} from a current source when the input current I_{ini} in the ith node is the largest and V_{tg} is far below the threshold voltage.

Integrated Optoelectronic Neuro Devices

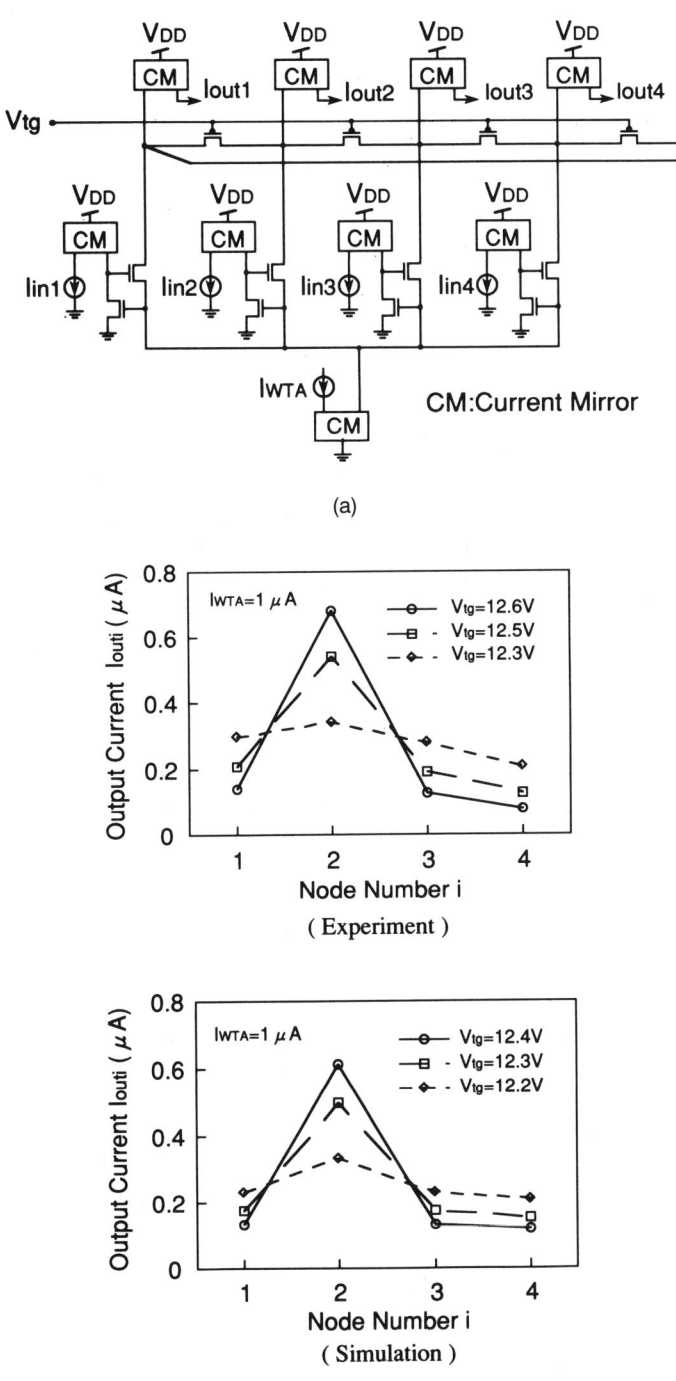

Figure 7.5.6
Lateral excitation connection (LEC) circuit followed by a WTA circuit. (a) Circuit; (b) Experimental and simulation results

A circuit with four nodes was fabricated with our CMOS process. The results are shown in Figure 7.5.6(b), where the second node has the largest input current. As the lateral connection strength was increased by decreasing the gate voltage of the pMOS transfer gates, the output current spread over the neighboring nodes. The experimental results are in good agreement with simulation results.

4 Technologies for Optical Interconnection Devices

Optical interconnection devices are composed of a light emitter and a detector, as seen in Figure 7.5.3. The wavelength of the optical beam should be longer than 1.1 μm in order to reduce the optical absorption loss in the Si substrate. Thus, the optical devices should be formed on the Si substrate. However, GaAs epitaxial layers grown on a Si substrate contain very many threading dislocations. In addition, the threading dislocations are also formed in high density in InGaAs epitaxial layers grown on GaAs substrates, which can emit light with a wavelength longer than 1.1 μm. Such generation of threading dislocations is caused by a large lattice mismatch of about 4%. A large number of threading dislocations lower the quantum efficiency and reduce the operating life. Thus, we have developed fundamental technologies to reduce the dislocation density in these epitaxial layers.

4.1 GaAs-on-Si

Two methods were tried to reduce the number of threading dislocations in GaAs epitaxial layers grown on Si substrates. The first one was high-temperature annealing, and the second was epitaxial growth with strained short-period superlattices (SSPS).

In high-temperature annealing 2 μm thick GaAs layers grown on Si substrates were annealed in As_4 ambient at 1050 °C for 1 h. As a result, the number of threading dislocations was remarkably reduced, as reported in Reference [8].

In the growth with SSPS, a GaP layer was initially grown on Si substrates. Then, $(GaP)_m(GaAs)_n$ SSPSs were grown following the growth of the GaAs epitaxial layer. The GaP layer suppressed the generation of defects because the lattice constant is close to that of Si. The large lattice mismatch was accommodated in the SSPSs with (m, n) of (3, 1), (1, 1), and (1,3). As a result, the number of threading dislocations was markedly reduced, as shown in Figure 7.5.7, although misfit dislocations were introduced in high density at each hetero-interface [9].

4.2 $(GaAs)_m(InAs)_1$ SSPS/GaAs quantum wells

It was found in the growth of $(GaAs)_m(InAs)_1$ SSPSs on GaAs substrates that the critical thickness of SSPSs is increased about ten times, as large as that of $In_xGa_{1-x}As$ alloys. Thus, the width of the dislocation-free well was increased in the SSPS/GaAs quantum wells. As a result, the wavelength range of optical emission was increased, as shown

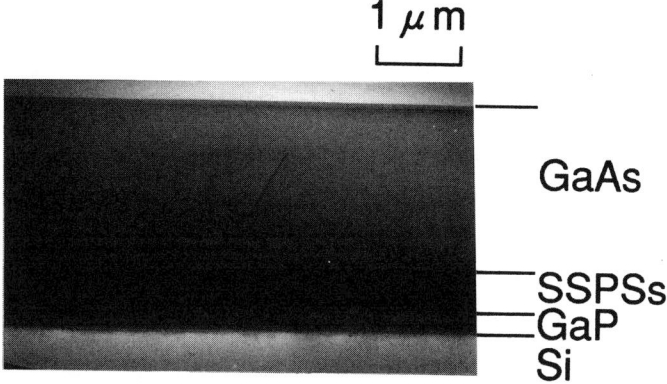

Figure 7.5.7
An XTEM image of the GaAs/(GaP)$_m$(GaAs)$_n$ SSPSs/GaP/Si

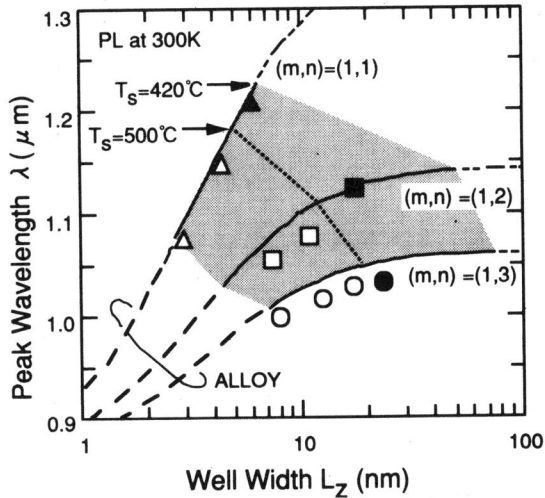

Figure 7.5.8
Peak wavelength of photoluminescence in (InAs)$_m$(GaAs)$_n$ SSPS/GaAs quantum wells

in Figure 7.5.8 [10]. The photoluminescence (PL) peak wavelength λ extended over 1.1 μm.

A quantum well structure was formed on the GaAs layer grown on a Si substrate after high-temperature annealing, as shown in Figure 7.5.9. The (GaAs)$_1$(InAs)$_1$ SSPS well was sandwiched between Si-doped n-GaAs and Be-doped p-GaAs barriers. The peak PL wavelength was 1.14 μm.

These fundamental technologies could support the realization of optical devices for through-wafer interconnection. The most preferred light emitter is a vertical cavity surface emission laser (VCSEL) with ultimately a low threshold current [11].

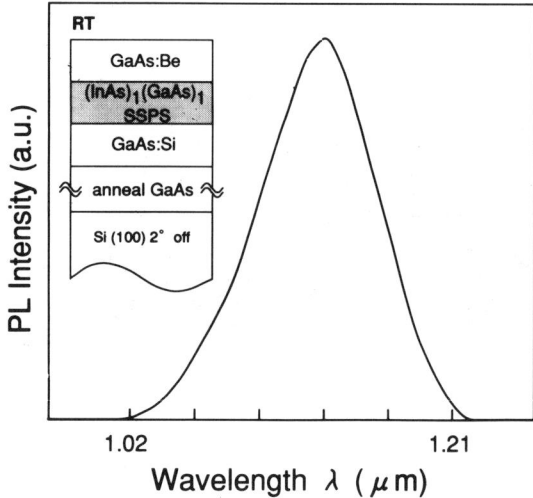

Figure 7.5.9
Photoluminescence spectrum of $(InAs)_1(GaAs)_1$ SSPS/GaAs quantum well grown on a Si substrate

5 Self-organization

Self-organization can be realized in a learning network constructed with the above-mentioned synaptic devices and circuits.

A primitive two-layers-network with two neurons was constructed for supervised learning with the back propagation (BP) algorithm. The experimental performance of self-organization was in good agreement with the simulated one. The simulated results showed successively performed self-organization in three-layer networks, as well.

Figure 7.5.10(a) is a primitive network with four neurons for unsupervised learning with the competitive learning algorithm. The learning properties were evaluated experimentally and compared with computer simulations based on the device characteristics of Figure 7.5.4 and the analog circuits of Figure 7.5.5 and others [12].

Different four bits digital patterns 1–4 were applied as input patterns. The different output neurons responded to different input patterns, as shown in Figure 7.5.10(b). Thus, it appeared that the input patterns were successfully classified by the output neurons. The performance was in good agreement with simulation results.

When the LEC circuit was applied to a similar primitive network, the ordering of the winners appeared according to the varied synaptic weights. The results could lead to the realization of topological mapping.

6 Conclusion

Fundamental devices and circuits were developed which were essential for self-organization in neural networks. They were fabricated with CMOS and EEPROM processes. Their performances were evaluated. Self-organization was confirmed in

Integrated Optoelectronic Neuro Devices

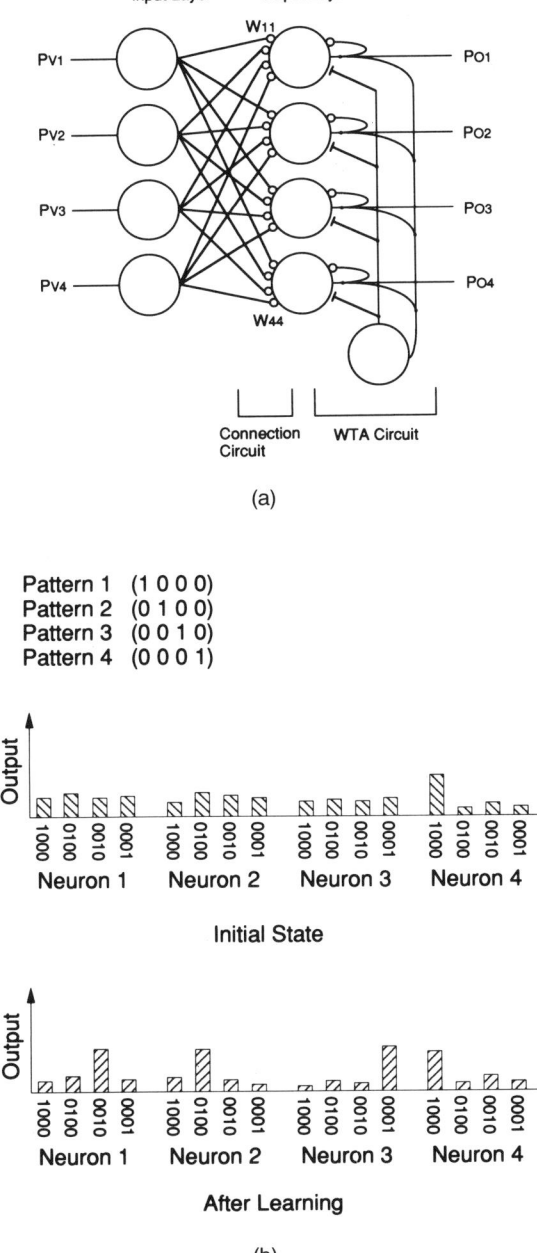

Figure 7.5.10
Unsupervised competitive learning network. (a) Network; (b) Learned results for four learning patterns

experiments or computer simulations in primitive networks for supervised and unsupervised learning. New technologies were developed for realizing the optical devices for through-wafer interconnection.

However, the scale of integration was limited by complicated wiring for feedback in learning and by a relatively large number of transistors in the neuron circuits. Thus, for large-scale integration in the future it is required to decrease the number of interconnections by applying a local Hebbian rule and to condense the fundamental functions into devices and circuits.

References

[1] D.E. Rumelhart, J.L. McClelland, and the PDP Research Group, *Parallel Distributed Processing*, MIT Press, Cambridge, MA, 1986.
[2] C. Mead, *Analog VLSI and Neural Systems*, Addison-Wesley, 1989.
[3] S. Usui and H. Yonezu, *Oyo Buturi*, **57**, 668 (1988) [in Japanese].
[4] H. Yonezu, *J. IEICE*, **75**, 350 (1992) [in Japanese].
[5] H. Yonezu, T. Himeno, K. Kanamori, K. Pak and Y. Takano, *Jpn J. Appl. Phys.*, **29**, L1314 (1990).
[6] K. Tsuji, H. Yonezu, K. Shibao, K. Hosono and K. Pak, *Electron. Lett.*, **29**, 1774 (1993).
[7] J.A. Starzyk and X. Fang, *Electron. Lett.*, **29**, 908 (1993).
[8] Y. Takagi, H. Yonezu, Y. Hachiya and K. Pak, *Jpn J. Appl. Phys.*, **33**, 3368 (1994).
[9] Y. Takagi, H. Yonezu, T. Kawai, K. Hayashida, K. Samonji, N. Ohshima and K. Pak, *J. Cryst, Growth*, to be published (1995).
[10] D. Saitoh, H. Yonezu, T. Kawai, M. Yokozeki and K. Pak, *Jpn J. Appl. Phys.*, **33**, L1205 (1994).
[11] F.M. Matinaga, A. Karlsson, S. Machida, Y. Yamamoto, T. Suzuki, Y. Kadota and M. Ikeda, *Appl. Phys. Lett.*, **62**, 443 (1993).
[12] K. Tsuji, H. Yonezu, K. Hosono, K. Shibao, N. Ohshima and K. Pak, *Jpn. J. Appl. Phys.*, **34**, 1056 (1995).

8
ULTRA-HIGH CAPACITY OPTICAL COMMUNICATIONS

In Chapter 8 we discuss the transmission technology for ultra-high capacity handling. In section 8.1 the ultimate bandwidth of optical fiber transmission will be considered. Next, section 8.2 reports detailed studies that have been made for polarization-maintaining fibers. In section 8.3 ultrafast transmission and transmission are discussed on the basis of optical fiber nonlinearity. In section 8.4 the compensation of dispersion and the nonlinearity of optical fibers is introduced for the purpose of expanding the transmission bandwidth. In section 8.5 a very precise analysis is presented for optical soliton transmission. Section 8.6 considers the sub-carrier multiplexing scheme.

8.1

Ultrafast Transmission and Processing Using Optical Fiber Nonlinearity

Yoichi Fujii

Abstract

In this section the development of solitonics, or the new members of the soliton family are briefly summarized, focussing on recent results of the fiber-optic logic device and a new concept of a 'coupled soliton'.

1 Introduction, Family of Solitons, Solitonics

Soliton technology is a fundamental element in the field of optical technology. Therefore the introduction of a new concept 'solitonics' — is introduced to describe soliton technology.

Various members of the soliton family are described. A variety of solitonic applications, such as the dynamic soliton, the 'supersoliton', the logical soliton, the switching property of fiber-optic solitons, and a new concept of the coupled soliton, are discussed.

2 Fundamental Solitons, Nonlinear Schrödinger Equation

The nonlinear interaction of the electromagnetic wave through the Kerr effect of the waveguiding material is expressed as

$$n = n_0 + n_2^2 I \qquad (1)$$

This nonlinearity, coupled with the dispersion of the waveguide, is generally expressed by the nonlinear wave equation. This equation can be simplified if the

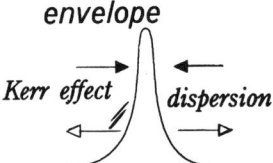

Figure 8.1.1
The fundamental soliton

pulse duration is long enough to the frequency. This is derived from a slowly-varying approximation of the nonlinear wave equation. This equation is termed the nonlinear Schrödinger wave equation. In this formalism, the envelope Ψ of the wave follows the equation with the first-order derivative of the propagation direction.

The soliton is an eigensolution of this nonlinear Schrödinger wave equation, i.e.

$$\left(i\frac{d}{dz} + \tfrac{1}{2}\frac{d^2}{dt^2} + |\Psi|^2 \right) \Psi = 0 \qquad (2)$$

where the dispersion and the nonlinear terms are included implicitly in this normalized equation. The eigen-waveform of this equation is termed the fundamental soliton, i.e.

$$\Psi(t) = \Psi_0 \operatorname{sech}(t) \qquad (3)$$

This waveform is realized at the balance of the dispersion and the nonlinear self-focusing effect as shown in Figure 8.1.1. This condition is not stable. So there exists a condition, called the soliton condition, which limits the relation between the amplitude and the pulse width. The higher the peak power, the narrower the pulse width.

The soliton propagates without changing its waveform on the basis of the balance between the dispersion and the nonlinear effect.

The soliton is considered to be very promising because its waveform is unchanged while it propagates. However, the waveform of the soliton cannot be kept constant due to the attenuation of the fiber. The next problem is the interaction of the solitons in the pulse train which carries the information.

The third problem is that the nonlinear Schrödinger equation is an approximation which cannot be used for femtosecond pulses.

These problems are solved in the following research works.

3 Dynamic Soliton, Pre-emphasis

The transmission length can be optimized by introducing slightly higher-order solitons to compensate for the pulse broadening due to the attenuation in the fiber. A simple numerical calculation shows that a soliton with order of $N = 1.3–1.4$ may be the optimum for long transmission distances. This result is widely used in soliton transmission systems.

4 WDM Solitons, Kidnapping

Solitons with different wavelengths can be transmitted in one single-mode optical fiber. This scheme, WDM solitons, can extend the transmission capacity of the fiber and also can be applied to the wavelength switching system.

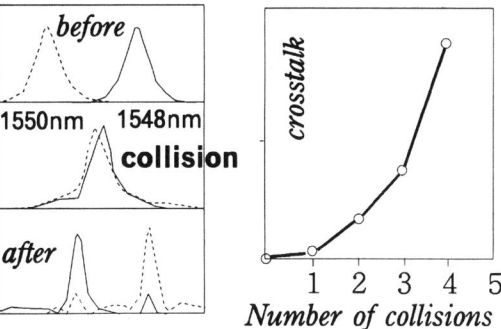

Figure 8.1.2
The WDM soliton

Solitons with different wavelengths have the different group velocities due to the dispersion of the fiber. Thus these solitons may suffer collisions between the transmitting solitons even with the same group velocity. In such collisions the solitons are interacting with each other and, as a result, one soliton may remove a part of the other soliton. Such an interaction, or kidnapping of the soliton pulse, may cause cross-talk between the transmission channels, as shown in Figure 8.1.2.

In the case of ideal fundamental solitons, these solitons can be transmitted without any change in the waveform, since such a collision does not affect the soliton waveform. In the case of higher-order solitons, especially those applied to dynamic soliton transmission, the collision may change the waveform and cause kidnapping of the soliton.

The effect of soliton kidnapping is calculated numerically and the effect of crosstalk between communication channels is estimated. From this estimation the maximum number of WDM soliton channels, and thus the maximum channel capacity, are calculated.

Crosstalk increases exponentially with the number of collisions, the maximum number of channels may be limited to 5 for conventional transmission lengths.

5 Supersoliton, Generalized Nonlinear Wave Equation

The nonlinear Schrödinger equation, conventionally used in the analysis of solitons, is essentially an approximation of the slowly-varying envelope of the carrier sinusoidal wave. In this approximation the dispersion is assumed to be linearized to the frequency (wavelength). Thus only a small dispersion of the narrow bandwidth in the fiber can be considered.

This approximation cannot hold for pulses (faster than 150 fs as is shown by one of the Authors) considering a third-order dispersion and Raman effect. For shorter pulses, the approximation of the dispersion by a power series of the wavelength becomes inhibiting because the calculations are cumbersome and the coefficients are meaningless.

An improvement is proposed in which a simple term is introduced directly into the wave equation (of second-order, not the nonlinear Schrödinger equation)

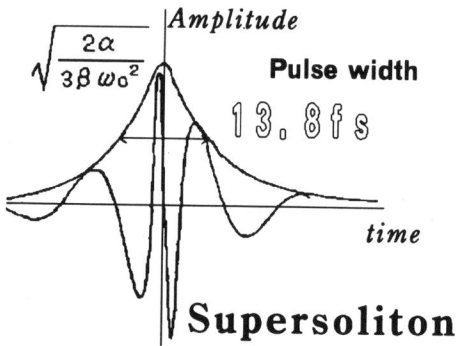

Figure 8.1.3
The supersoliton

$$\left(\square + \alpha + \beta \frac{d^2}{dt^2}|\Psi|^2\right)\Psi = 0 \tag{4}$$

This term is very simple but can give a good approximation of the existing dispersion of the optical fiber for extremely wide bandwidths except near the absorbing region.

This generalized nonlinear wave equation is expressed by a generalized dispersion coefficient and the conventional nonlinear term, as follows:

This generalized wave equation is solved analytically, the eigensolution of which is expressed by a very narrow (about 14 fs) pulse with a fixed amplitude. These two values are independently fixed since the wave equation is of second order, as shown in Figure 8.1.3. This eigensolution is termed as a 'supersoliton'.

6 Switching Soliton, Nonlinear Fiber-optic Coupler

We assume a fiber-optic coupler, in which the coupling is linear but the transmission is nonlinear.

A pulse injected into one end of the core of the coupler is nonlinearly distorted so that the power transmission coefficients change by the input power. Thus the pulse distortion can be improved by introducing the solitonic property.

In a fiber-optic switch the input pulse power is below the power level called the switching threshold. The soliton pulses are periodically coupled between the two fibers. However, if the input pulse power exceeds the switching threshold, switching of the soliton power is realized. Thus switching of the soliton pulse is achieved by its own power.

A schematic of the asymmetrical dual core nonlinear directional coupler is shown in Figure 8.1.4.

6.1 Asymmetric fiber coupler

In this section this principle is applied to a nonlinear fiber coupler with different core diameters. The asymmetry of the cores makes the coupling of the solitons between two

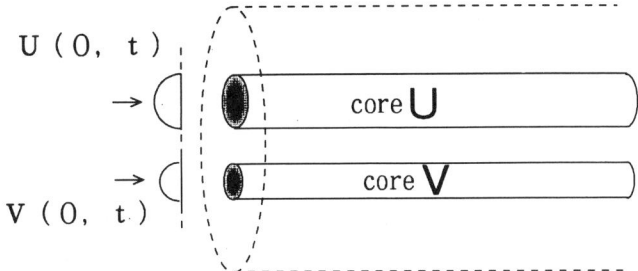

Figure 8.1.4
Schematic of the asymmetrical dual core nonlinear directional coupler

cores asymmetric. Thus, under certain conditions, it is possible to realize a sort of logic operation by using the asymmetric coupler. The condition and the optimum parameters for this logic circuit are discussed and its possible application is considered.

6.2 Basic equations of the nonlinear fiber coupler

The refractive index is assumed to be dependent on the optical intensity, as shown in equation (1). The nonlinearity in each core is defined by

$$K_u|u|^2 u \text{ and } K_v|v|^2 v \qquad (5)$$

The Kerr coefficient is dependent on the core area and is defined by the average of the square of the field amplitude Ψ over the cross-section A of the fiber, as is used in Reference [4]:

$$K = \frac{n_2 \beta \int_A \Psi^4 dA}{\left(\int_A \Psi^2 dA^2\right)^2} \qquad (6)$$

This equation is valid for u and v fibers and $\beta = 2\pi/\lambda$.

Because the core areas are different, each power density of the light is different. Hence, the nonlinear coefficients for each mode become different. An asymmetric coefficient m is defined by the ratio of the nonlinear coefficients as

$$m = \frac{K_v}{K_u} \qquad (7)$$

Since the nonlinear coefficient is inversely proportional to the core area, the asymmetric coefficient is inversely proportional to the ratio of the core area.

The propagation of pulses in an asymmetric fiber coupler can be described by two linearly coupled nonlinear Schrödinger equations. After all quantities are transformed to the soliton units, the nonlinear Schrödinger equations reduce to the following equations:

$$\frac{i\,du}{dz} + \frac{1}{2}\frac{d^2u}{dt^2} + |u|^2 u + \kappa v = 0$$
$$\frac{i\,dv}{dz} + \frac{1}{2}\frac{d^2v}{dt^2} + m|v|^2 v + \kappa u = 0 \qquad (8)$$

where u and v are the mode amplitudes of the cores U and core V, respectively. z, t, and κ are the normalized length, the normalized retard time, and the normalized coupling coefficient, respectively. We neglect the higher-order dispersion, Raman effect, and the shock term because the pulse width is assumed to be above a picosecond.

The difference in the velocity and the dispersion between the two cores is neglected because the asymmetric coefficient m is considered to range from 1.0 to 2.0. The coupling coefficient κ is assumed to be 0.250 and 0.333.

In all cases the length of the coupler is the half beat length $(= \pi/2\kappa)$. The initial conditions used in each case are

$$U(0, t) = U(0)\,\text{sech}(t)$$
$$V(0, t) = V(0)\,\text{sech}(t)\exp(j\theta) \qquad (9)$$

where $U(0), V(0) = 0$ or 1.

The coupled nonlinear Schrödinger equation was solved by the finite difference method with periodic boundary conditions.

6.3 Logic switching characteristics of a nonlinear fiber coupler

The coupled nonlinear Schrödinger equation is numerically solved for the noncoupled soliton inputs. The inputs are the solitons for each core '1' or none '0'.

6.4 Input $(U,V) = (1,0)$ case

The transmission is calculated as a function of m. In this case, the fundamental soliton is injected into the wider core U. Because the half-beat length coupler is considered, most of the pulse energy injected into core U emerges from core V. The coupling coefficient κ does not affect the transmission much. Asymmetry can improve the transmission a little.

6.5 Input $(U,V) = (0,1)$ case

The transmission is calculated as a function of m. In this case the fundamental soliton is injected into the narrower core V. At a certain m (this value is defined as m_0), most input energy emerges from the same core V as the input core. This is essentially unlike the symmetric case. The value of m_0 is 1.51 and 1.77; hence the ratios of the core diameters of core V to core U are 0.814 and 0.752 when the coupling coefficients are 0.250 and 0.333, respectively. A low coupling coefficient appears when m_0 is low.

This can be explained as follows: when the core area is small, the optical power density and hence the nonlinearity is large, even when the same optical power is injected.

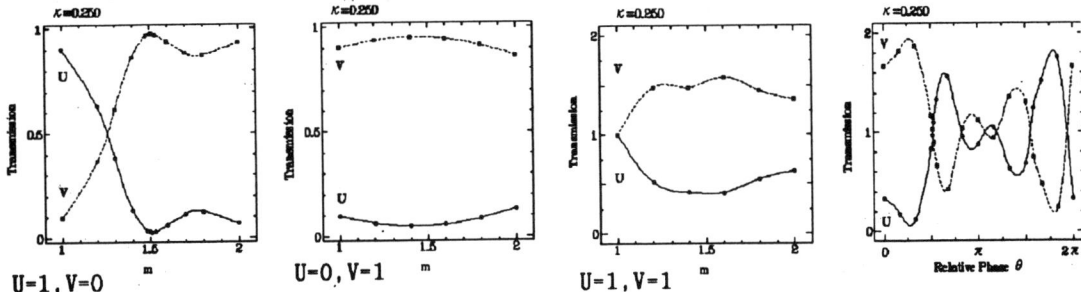

Figure 8.1.5
The transmission of an asymmetrical nonlinear fiber coupler as a function of the asymmetric coefficient m, for the input (0,1), (1,0), and (1,1), $\kappa = 0.250$, and the transmission of an asymmetrical nonlinear fiber coupler as a function of the relative phase θ. Here the input is (1,1), $\kappa = 0.250$, and $m_0 = 1.51$

The phase matching condition is broken by the large nonlinearity. So, the two cores are coupled, but pulse energy swapping does not appear.

6.6 Input (U,V) = (1,1) case (with the same phase)

In this case, fundamental solitons with same phase are injected into both cores. More energy emerges from the narrower core V because its nonlinearity is larger. The coupling coefficient κ does not affect the transmission much.

At the value of m_0, the transmitted powers of the core U are 0.470 and 0.312 when the coupling coefficients are 0.250 and 0.333, respectively.

6.7 Input (U,V) = (1,1) case (with relative phase difference)

At the value of m_0, the transmitted power of the core U is not the same as the input power. So, the relative phase difference between both cores is taken into account.

Figure 8.1.5 shows the transmission of an asymmetrical nonlinear fiber coupler as a function of the asymmetric coefficient m, for the input (0,1), (1,0) and (1,1), $\kappa = 0.250$.

The transmission is also calculated as a function of the relative phase θ. It shows that transmission is greatly dependent on the relative phase and thus there are several phases at which transmission of both the cores U and core V is 1.00. Figure 8.1.5 also shows the transmission of an asymmetrical nonlinear fiber coupler as a function of the relative phase θ. Here input is (1,1), $\kappa = 0.250$, and $m_0 = 1.51$.

7 Logic Soliton, Solitonic Device Application

7.1 All-optical and/or logic circuit

From these results an all-fiber-optic logic circuit can be realized. The logical '0' corresponds to a signal of zero power and the logical '1' corresponds to a signal of the

Table 8.1.1
Truth table of core U and core V
$\kappa = 0.250, m = 1.51, \theta = 0.517\pi, L = 6.28$

$V(0)/U(0)$	0	1	0	1
0	0.000	0.0626	0.000	0.937
1	0.0278	1.00	0.972	0.998

core $U \to$ AND; core $V \to$ OR

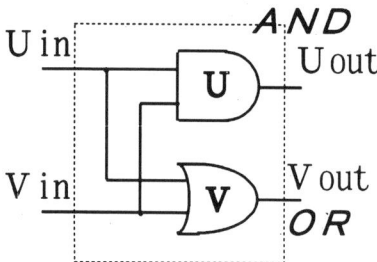

Figure 8.1.6
A logical circuit with an AND output from the core U and an OR output from the core V

power of the fundamental soliton. The output of the symmetric half-beat length coupler is (1,0) when the input is (0,1). On the other hand, in the case of asymmetric cores, the output can be (0,1). In the asymmetric case the output power is asymmetric when both signals have the same phase. Thus, a specific relative phase difference between two solitons can make the output symmetric.

The parameters were optimized such that the coupling coefficient κ is 0.250 and 0.333, the coupling length is the half beat length ($= \pi/\kappa$), the asymmetric coefficient is m_0, and the relative phase difference is the value at which transmission of both cores is 1.00 when the input is (1,1).

A logical circuit can be obtained by optimizing the parameters such that the two outputs are an AND output from the core U and an OR output from the core V. The logic function obtained is shown in Table 8.1.1 and a schematic diagram is shown in Figure 8.1.6.

7.2 Application of a soliton logic circuit

Assuming that $\lambda = 1.55 \ \mu$m, $\tau_0 = \tau_{\text{FWHM}}/1.763$, $D = -14$ (ps/nm)/km, $A_{\text{eff}} = 6.5 \times 10^{-7}$ cm^2, and $n_2 = 3.18 \times 10^{-16}$ cm^2/W, the coupler length and the fundamental soliton power are shown in Table 8.1.2.

For the conventional parameters of the silica fiber, the coupling length for the realizable soliton is much longer for the integrated circuit. So this switching device can be useful for an all-optical route switch for future optical soliton communications. This logic circuit can be useful for transmission lines combined with logic processing, such as for multipoint sensors.

Table 8.1.2
The coupler length and the optical power

κ	Pulse width	Coupler length	Optical power
0.250	10 ps	11.4 km	279 W
	1 ps	114 m	27.9 W
	100 fs	1.14 m	2.79 kW
0.333	10 ps	8.58 km	372 mW
	1 ps	85.8 m	37.2 W
	100 fs	85.8 cm	3.72 kW

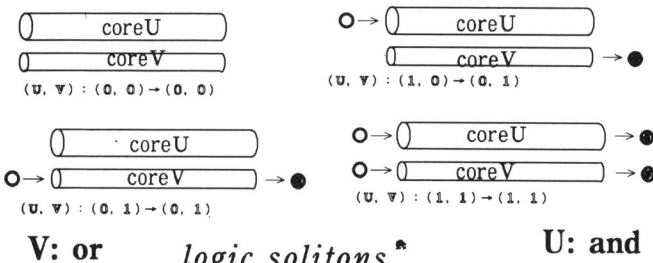

Figure 8.1.7
Schematic diagram of a fiber-optic soliton logic circuit

For an optical waveguide with higher nonlinearity, the coupled soliton will be useful and the logic circuit can be integrated.

Figure 8.1.7 shows a schematic diagram of the fiber-optic soliton logic circuit.

8 Coupled Soliton: Unchanged for a Coupled Waveguide

Generally, this kind of nonlinear fiber coupler can transmit a so-called 'coupled soliton'. This new concept of a coupled soliton and its condition, and its application are discussed.

The coupled nonlinear Schrödinger equation for the nonlinear fiber-optic coupler gives some special cases in which the soliton waveforms are transmitted unchanged or periodically changed. For such cases the new concept of a 'coupled soliton' is introduced. In the symmetrical dual-core fiber coupler fundamental coupled solitons are obtained for in-phase and out-of-phase input solitons.
But in the asymmetrical dual-core fiber coupler, they appear only in the out-of-phase soliton input.

For a numerical example, for a coupling coefficient of $\kappa = 1.00$ and an asymmetric coefficient $m = 1.50$, the fundamental coupled solitons propagate as

$$U(0, t) = 1.00 \text{ sech}(t)$$
$$V(0, t) = 0.958 \text{ sech}(t) \exp(i\pi)$$
(10)

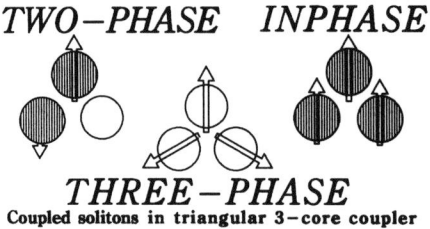

Figure 8.1.8
Coupled solitons in a three-core fiber

The coupled soliton propagates without pulse distortion and relative core-power distribution. Because the core diameters are different, the soliton powers are different.

In a three-core fiber coupler, three types of coupled solitons can be obtained, as shown in Figure 8.1.8.

9 New Solitons

There are many other possibilities. For example, solitonic sensors, two-fiber diversity transmission, and phase conjugation may be promising.

10 Conclusion

In this section various members of the soliton family and the new technology of solitonics are discussed. The new concept of a coupled soliton is proposed. An all-fiber-optic AND/OR logic circuit using an asymmetrical nonlinear directional coupler is demonstrated. These solitonic devices are the key to the future of optical communication and optical computing.

References

[1] S.M. Jensen, *IEEE J. Quantum Electron.*, **QE-18**, 1580 (1982).
[2] S. Trillo, S. Wabnitz, E.M. Wright and G.I. Stegeman, *Opt. Lett.*, **13**, 672 (1988).
[3] For example: M. Romagnoli, S. Trillo and S. Wabnitz, *Opt. Quantum Electron.*, **24**, S1237 (1992).
[4] D.R. Rowland, J. Lightwave. Technol., no. 1, **9**, 1074 (1991).

8.2

High Precision Simulation of Soliton Problems

Kazuo Horiuchi, Shin'ichi Oishi, Ryogo Hirota and **Satoshi Tsujimoto**

Abstract

The fundamental problems of analyzing the characteristics of optical soliton transmission are discussed in this section. First, for the nonlinear Schrödinger equation model, it is briefly reviewed that this model equation can be solved exactly. Using this, a new discretization method for the nonlinear Schrödinger equation is presented which preserves the integrability of the equation. The method will be useful for developing new kinds of soliton transmission systems. Then, a computer-assisted analyzing method of optical soliton systems is presented. In this method, self-validating numerics is utilized. Self-validating numerics is based on fixed point theorems, and by this method an approximate solution is obtained together with its confidence interval. This method is implemented as an automatic numerical validation system, which will be useful for analyzing optical soliton systems.

1 Introduction

We have studied the various properties of solitons using an exact analytical method called the bilinearization method. Moreover, we also have studied numerical methods for the high precision simulation of soliton problems. First, a new discretization method for the nonlinear Schrödinger equation is presented, which preserves the integrability of the equation. The method will be useful for developing new kinds of soliton transmission systems. Then, a computer-assisted analyzing method of optical soliton systems is presented. In this method, self-validating numerics is utilized. Self-validating numerics is based on fixed point theorems, and by this method an approximate solution is obtained together with its confidence interval. This method is implemented as an automatic numerical validation system, which will be useful for analyzing optical soliton systems.

2 Exact Analyzing Method for Optical Soliton Systems

Optical soliton transmission in single mode fibers is modeled by the following nonlinear Schrödinger equation

$$i\frac{\partial}{\partial t}\psi + \frac{\partial^2}{\partial x^2}\psi + 2|\psi|^2\psi = 0. \tag{1}$$

The nonlinear Schrödinger equation is transformed into the following bilinear equation

$$D_x^2 F \cdot F = 2GG^* \tag{2}$$

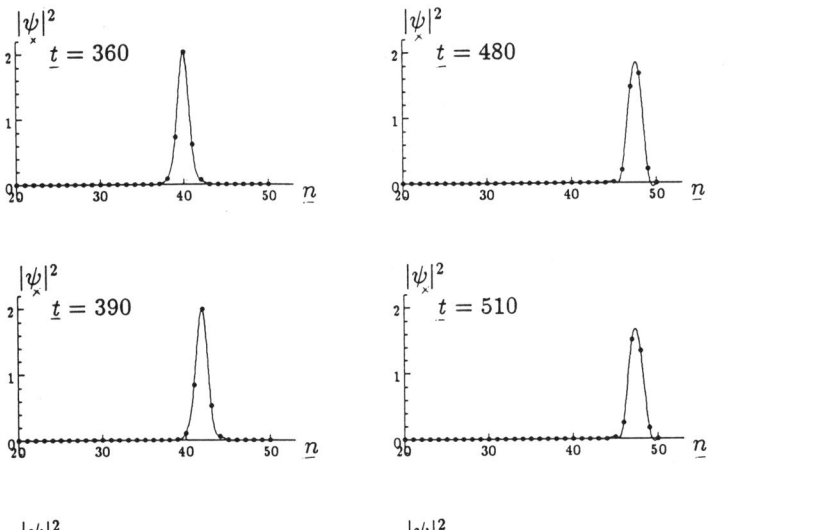

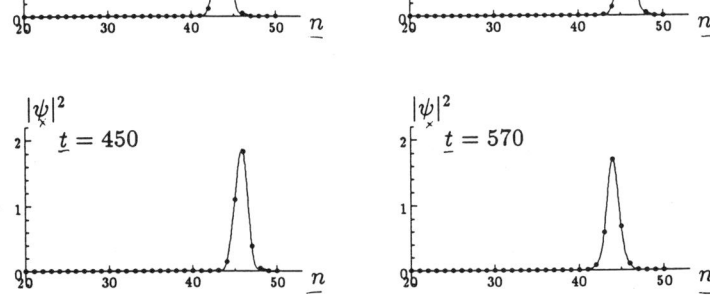

Figure 8.2.1
Simulation of the collision of solitons at a fixed right end using equation (5)

$$(iD_t + D_x^2)G \cdot F = 0 \tag{3}$$

through a dependent variable transformation

$$\psi = \frac{G}{F}, \quad F = \text{real}. \tag{4}$$

Using this bilnear form, Oishi [1] derived a generalized soliton solution of equation (1). It is shown that using this solution, its initial value problem can be solved.

Descretizing a bilinear form of equation (1), we have the following partial difference equation [2]:

$$\begin{aligned} i\delta^{-1}(\psi_n^{t+1} - \psi_n^t) - (\psi_n^{t+1} + \psi_n^t) + (\psi_{n-1}^{t+1} + \psi_{n+1}^t)(1 + |\psi_n^t|^2)\Gamma_n^t = 0, \\ \Gamma_{n+1}^t = \frac{1 + |\psi_n^t|^2}{1 + |\psi_n^{t+1}|^2}\Gamma_n^t. \end{aligned} \tag{5}$$

This equation preserves the integrability of equation (1). This discrete nonlinear Schrödinger equation, or DNLS for short, is shown to be stable so that it is suited for long time numerical integration. Using this scheme, we have studied the following problems:

(1) The multi-body collision of solitons.
(2) The response to excitation.
(3) The collision of solitons at a fixed end point.

Applying this discretization method to a certain Volterra equation, new soliton phenomena have been found. For example, we will report brother solitons, which is the exact solution representing two solitons run at the same speed.

3 Validation of the Simulation Results

In this section we treat the case when the original system is not integrable. In this case we must solve the equation numerically. The problem is to estimate the validity of such a numerical simulation. We assume that we have obtained a certain numerical solution. We present a computer-assisted method of proving the existence of an exact solution around this approximate solution.

Let us first consider the nonlinear boundary value problems of a system of first-order real differential equations

$$\begin{aligned} \frac{dx}{dt} &= f(x, t), \quad t \in I = [-1, 1], \\ g(x) &= 0, \end{aligned} \tag{6}$$

where x is an n-dimensional vector, and $f(x, t)$ an n-dimensional vector-valued function, and g an n-dimensional vector-valued functional. Such nonlinear boundary value problems are hard to solve analytically. Thus, in the present study it may be solved by some numerical method to obtain an approximate solution. However, the numerical solution obtained does not necessarily guarantee the existence of an exact solution of the problem. Therefore, it is important to give a sufficient condition under which the problem has an exact solution in a domain containing an approximate solution and to find a sharp error bound. For the purpose we reformulate the problem as an operator equation. Then, we consider a Newton operator associated with this operator equation. We then check whether or not this Newton operator becomes a contraction mapping from a small ball centered at a given approximate solution into itself.

In our argument we will show that an infinite-dimensional extension of the Krawczyk operator can be defined associated with the Newton-like operator. Then, using Caprani-Madsen-Rall's theory of integration of interval function [4] we will show that the range of the Newton-like operator can be evaluated numerically.

Next let us consider the following nonlinear operator equations

$$Lu + Nu = 0. \tag{7}$$

Here L is a closed linear operator from a Banach space X to another Banach space Y and N a nonlinear operator from X to Y. This type of equation occurs in a variety of situations in both pure and applied sciences. Equation (7) is sometimes called a coincidence equation because one wants to find a point u for which the images under L and $-N$ coincide. We also have developed a method for numerical verification and inclusion of solutions for these equations [5]. That is, in association with a certain approximate solution of equation (7), we present an algorithm which may answer the question as to whether there exists an exact solution u^* in some neighborhood of $\tilde{u}$, and in the affirmative case may give a bound for $u^* - \tilde{u}$. If an error bound for $u^* - \tilde{u}$ can be obtained, we shall say that an inclusion of a solution u^* is obtained. This theory can be seen as an extension of the Galerkin-Urabe method [6].

References

[1] S. Oishi, *Memoirs of the School of Science and Engineering, Waseda University*, **46**, 191-225 (1982).
[2] R. Hirota, S. Tsujimoto and T. Imai, in *Proceeding of 'Future Directions of Nonlinear Dynamics in Physics and Biological Systems'*, P.L. Christiansen, J.C. Eilbeck and R.D. Parmentier, Eds., Plenum, New York, to appear.
[3] Shin'ichi Oishi, submitted to *Journal of Computational and Applied Mathematics*.
[4] O. Caprani, K. Madsen and L.B. Rall, *SIAM J. Math. Anal.*, **12**, 321-341 (1981).
[5] S. Oishi, *Technical Report No. 93-11*, Advanced Research Center for Science and Engineering, Waseda University.
[6] M. Urabe, *Arch. Rational Mech. Anal.*, **20**, 120-152 (1965).

8.3

Dispersive and Nonlinear Degradation Compensation in Fiber-optic Systems

Kazuo Kikuchi and **Chaloemphon Lorattanasane**

Abstract

In long-distance optical communication systems using in-line optical amplifiers, the waveform distortion is induced by both the Kerr effect and the group-velocity dispersion of optical fibers. In this section we show that the optical phase conjugator (OPC) placed at midpoint of the transmission line can compensate for the Kerr effect as well as the group-velocity dispersion, thus preventing the waveform distortion. We also discuss the advantages of the OPC system when it is introduced into the WDM systems and coherent optical communication systems.

1 Introduction

The introduction of the erbium-doped fiber amplifier (EDFA) has opened the possibility of constructing optical fiber communication systems spanning very long distances without the need for the electronic regeneration of optical pulses. The fiber loss is periodically compensated by an in-line optical amplifier gain $G = \exp(\alpha \ell)$, where α denotes the attenuation constant of the fiber and ℓ the amplifier spacing. Since the signal power in such systems is maintained at a high level along the entire length of the fiber, the dependence of the fiber refractive index on the optical power (Kerr effect) can no longer be neglected.

When the signal power is weak and the Kerr effect is not involved, the waveform distortion is induced only by the group-velocity dispersion. Such a waveform distortion can be prevented by using the zero-dispersion fiber or compensated by the equalizer fiber [1]. However, the presence of the Kerr effect makes these methods less effective.

The Kerr effect leads to the self-phase modulation of optical pulses, which in turn interplays with the group-velocity dispersion of the fiber causing the nonlinear waveform distortion. In the anomalous-dispersion region of fibers, such interplay causes serious waveform distortion, known as modulation instability [2], which cannot be compensated for by equalizer fibers. Even if we use the zero-dispersion fiber for transmission, part of the modulation sideband falls in the anomalous-dispersion region, also causing waveform distortion [3].

It should be noted that the interplay of the Kerr effect with the group-velocity dispersion is diminished by increasing the positive dispersion of the fiber. Therefore, the normal-dispersion region is favorable for reducing the nonlinear effect [4]. In such a case, equalizer fibers periodically placed at rather short intervals can compensate for the group-velocity dispersion and prevent the waveform distortion to some extent. Recently, the possibility of this technique has been confirmed by using recirculating loop experiments [5]. However, it should be noted that the compensation is carried out only for the group-velocity dispersion.

The soliton transmission system is one of the candidates to cope with the Kerr effect and the group-velocity dispersion [2]. In such a system the soliton pulse having a definite waveform and peak power, which are determined by the group-velocity dispersion, can be transmitted without distortion, because the Kerr effect is balanced with the group-velocity dispersion along the entire length of the transmission link. However, once such a subtle balance is destroyed by system instability, waveform degradation becomes serious.

As an alternative approach, we propose that the optical phase conjugator (OPC) placed at the midpoint of the in-line optical amplifier chain can compensate for both the group-velocity dispersion and the Kerr effect. When the normal-dispersion region is used for transmission and the repeater spacing is properly chosen, the distortion of the non-return-to-zero (NRZ) transmitted waveform is greatly improved. The requirement for system parameters is found to be less stringent than the soliton system.

2 Theoretical Treatment

It has already been proposed that when an OPC is placed at the midpoint of the dispersive fiber link, the waveform distortion in the preceding half of the fiber will be undone during propagation through the succeeding half of the fiber, whose dispersion characteristics are the same as those of the preceding half [6]. In a practical system, the OPC can be implemented by using four-wave mixing in a single-mode fiber [7] or a semiconductor laser amplifier [8]. Recently, this compensation technique has been demonstrated experimentally in short-distance transmission systems [9, 10]. However, we discuss below the case of long-distance systems in which the Kerr effect must be taken into account. We show how not only the group-velocity dispersion but also the Kerr effect can be compensated for by placing the OPC at the midpoint of the optical amplifier chain.

The propagation of signals in a nonlinear dispersive fiber is governed by the nonlinear Schrödinger (NLS) equation

Dispersive and Nonlinear Degradation Compensation in Fiber-optic Systems

$$\frac{\partial A}{\partial z} = -\frac{\alpha}{2}A - \frac{i}{2}\beta_2\frac{\partial^2 A}{\partial T^2} + \frac{1}{6}\beta_3\frac{\partial^3 A}{\partial T^3} + i\gamma|A|^2 A \tag{1}$$

where A denotes the slowly-varying pulse amplitude, z the propagation distance, α the fiber attenuation coefficient, β_2 and β_3 the first- and second-order dispersion coefficients, respectively, γ the nonlinearity coefficient, and T the normalized time in a frame moving at the group velocity.

Since this equation is generally valid, its complex conjugate must also be valid. We obtain

$$-\frac{\partial A^*}{\partial z} = +\frac{\alpha}{2}A^* - \frac{i}{2}\beta_2\frac{\partial^2 A^*}{\partial T^2} - \frac{1}{6}\beta_3\frac{\partial^3 A^*}{\partial T^3} + i\gamma|A|^2 A^* \tag{2}$$

where an asterisk denotes the complex conjugate. Let us first assume that the second-order dispersion β_3 is zero and that the sign of α is reversed. In this case, equation (2) describes a wave propagating in the negative z-direction, which is the complex conjugate of the forward-propagating wave at every place in the fiber, especially at the fiber input end.

From this consideration we find that if at the midpoint of the transmission link we somehow generate a complex conjugate of the distorted complex pulse amplitude and make it travel backwards, then we obtain exactly the (complex conjugate of the) undistorted input pulse. Both fiber nonlinearity and dispersion have therefore been compensated for.

Instead of backward propagation, we carry out a virtual reversal of the direction of propagation. When the conjugated pulse is propagated forward, but the power distribution is inverted with respect to the system midpoint, perfect compensation can be achieved as in the actual reversal of the direction of propagation. Such a power distribution, where the propagation direction is virtually reversed, is shown in Figure 8.3.1(a). On the other hand, the actual power distribution in an optical amplifier chain is given by Figure 8.3.1(b). The difference between the two distributions will make the distortion compensation less effective.

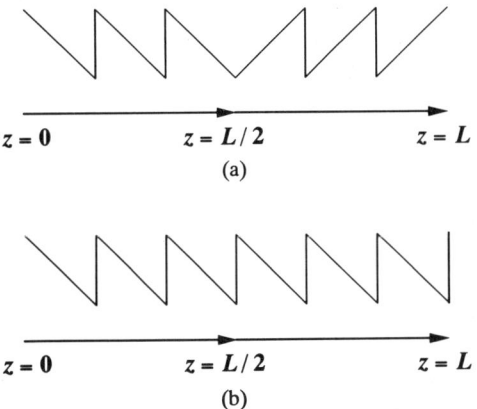

Figure 8.3.1
Power distribution in an optical amplifier chain. (a) Virtual propagation reversal; (b) actual power distribution

In the next section we carry out extensive computer simulations and demonstrate the feasibility of this approach in compensating for the nonlinear waveform distortion even in long-distance systems with asymmetrical power distribution.

In actual systems the transmission fiber has nonzero second-order dispersion ($\beta_3 \neq 0$). It can be seen from the NLS equation that the waveform distortion resulting from the second-order dispersion cannot be compensated for by the OPC. As a result, the waveform distortion determined by the second-order dispersion still remains even in an ideal midpoint OPC system.

3 Computer Simulation

Figure 8.3.2 shows a block diagram of the proposed system. An OPC is placed at the midpoint of the amplifier chain.

The efficiency of the waveform-distortion compensation is evaluated by using the eye penalty defined as the ratio of the eye opening at the transmitter to the eye opening at the receiver. The pulse train at the transmitter consists of 16 bits in NRZ intensity-modulation format. The data transmission rate is 10 Gb/s. The NLS equation describing the pulse propagation in optical fibers is solved by using the split-step Fourier method [2].

A fiber loss of $\alpha = 0.2$ dB/km is compensated for by the in-line optical amplifier gain $G = \exp(\alpha \ell)$ in each section. The fiber nonlinearity has the value $\gamma = 2\pi n^{(2)}/\lambda A_{\text{eff}} = 2.6$ W/km, where $n^{(2)}$ denotes the nonlinear refractive index, and A_{eff} the effective cross-section of the core. The complex amplitude of the optical pulses at the midpoint of the system is conjugated by the OPC. At the receiver, the photo-current is filtered by a 5-GHz-cutoff lowpass filter before measuring the eye opening.

Figure 8.3.3 shows the eye penalty as a function of the signal peak power P_0 for the amplifier spacing $\ell = 100$ km, 50 km, 40 km, and 25 km. The fiber first-order dispersion is assumed to have the value $D = -(2\pi c/\lambda^2)\beta_2 = -4$ ps km^{-1} nm^{-1} (normal dispersion), where λ denotes the carrier wavelength and β_2 the group-velocity dispersion parameter. The total transmission distance, L is 6000 km.

It can be seen that in the low signal power region ($P_0 < 5$ dBm), the OPC can effectively compensate for the waveform distortion, and the eye penalty is kept below 0.5 dB.

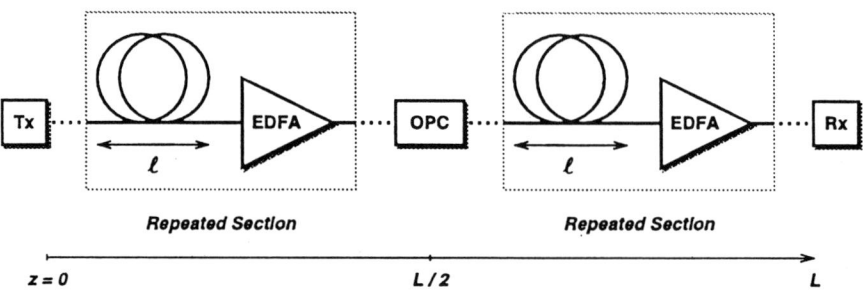

Figure 8.3.2
Long-distance lightwave system using a midpoint optical phase conjugator (OPC)

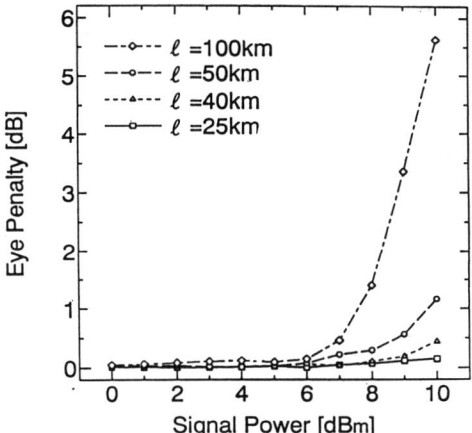

Figure 8.3.3
Eye penalty as a function of the signal peak power P_0 for various values of the amplifier spacing ℓ. The total transmission distance is $L = 6000$ km

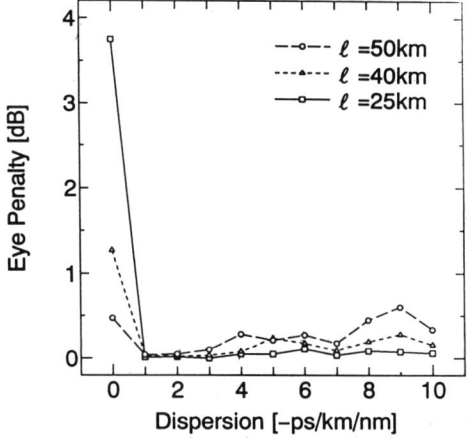

Figure 8.3.4
Eye penalty as a function of the fiber dispersion for various values of the amplifier spacing ℓ. The signal peak power is 10 dBm, and the transmission distance L is 2000 km

When the signal power is increased above 5 dBm, the eye penalty increases rapidly for long amplifier-spacing systems. Note that, for short (25 km) amplifier-spacing systems, the eye penalty is unchanged even though the signal power is increased up to 10 dBm.

The change in the eye penalty cannot be observed when we include the second-order dispersion β_3 in the calculation. This is because the effect of the second-order dispersion can be neglected at this relatively low date rate of 10 Gb/s.

Figure 8.3.4 shows the eye penalty as a function of fiber dispersion for the amplifier spacing $\ell = 25$ km, 40 km, and 50 km. The total transmission distance is 2000 km. The second-order dispersion parameter D_2 is assumed to be 0.06 ps km^{-1} nm^{-2}. The input signal peak power is set to 10 dBm. It can be seen that when D is varied from

-1 ps/km^{-1} nm^{-1} to -10 ps km^{-1}nm^{-1} (normal dispersion), the eye penalty gradually increases. On the other hand, when the fiber dispersion approaches zero, the eye penalty increases rapidly. This effect is caused by modulation instability.

4 Discussion

Section 2 showed that the perfect compensation for the nonlinear distortion due to the Kerr effect can be achieved only if the optical power distribution in the succeeding half is symmetrical to that in the preceding half. This condition is satisfied only in the ideal lossless transmission line, where the distributed optical amplifier gain compensates for the fiber loss at any position along the fiber, and the power distribution becomes constant along the fiber.

However, Figure 8.3.3 shows that the midpoint OPC is still effective even when the power distribution is asymmetric. The success of the midpoint OPC in compensating for the pulse waveform distortion even in the presence of the asymmetrical optical power distribution is attributed to the following reasons.

First, the normal-dispersion region is used for transmission. It should be noted that the interplay of the Kerr effect with the group-velocity dispersion is diminished in the normal-dispersion region [4]. Figure 8.3.4 shows that the compensation mechanism of the OPC system works well as long as the whole signal spectrum stays within the normal-dispersion region. Note that the dependence of the eye penalty on the normal dispersion is very small.

Second, when the signal power is low and the amplifier spacing is short, the effect of fiber nonlinearity is mainly dependent on the path-average power of the signal, $\bar{P} = P_0(1 - \exp(-\alpha\ell))/\alpha\ell$, and is not critically dependent on its detailed power distribution. Thus, the overall system can be considered as a quasi-lossless transmission line with a constant signal power path-averaged over the amplifier spacing, and the OPC can compensate for the waveform distortion to a great extent.

When we increase the signal power, the amplifier spacing, and the total transmission distance, the effect of the actual power distribution can no longer be ignored, and the compensation is less satisfactory as evidenced by the increase in the eye penalty shown in Figure 8.3.3. However, we find that the requirements for system parameters such as the dispersion, the signal power, and the repeater spacing are not so stringent.

We discuss here two other advantages of the system using OPC. First, in wavelength division multiplexing (WDM) systems, the four-wave mixing (FWM) phenomenon in optical fibers seriously degrades system performance [11]. We find that the crosstalk-induced waveform distortion in each channel caused by the FWM phenomenon can be compensated simultaneously by the OPC, provided that a sufficiently wide bandwidth OPC is employed.

Second, in coherent optical communication systems using in-line optical amplifiers, the system performance is strongly affected by the excess phase noise caused by the Kerr effect [12]. The intensity noise generated from each amplifier is converted into phase noise by the Kerr effect while traveling the fiber, as shown in Figure 8.3.5(a). The resulting expression for the accumulated phase-noise variance at the receiver is

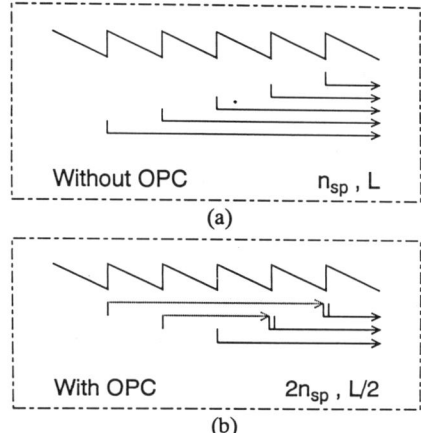

Figure 8.3.5
Mechanism of excess phase noise suppression by a midpoint optical phase conjugator (OPC). (a) Enhancement and accumulation of phase noise from each amplifier in the system without an OPC; (b) phase noise suppression in the system with a midpoint OPC

proportional to $n_{sp}L^3$, where n_{sp} denotes the spontaneous emission factor of optical amplifiers.

In the system using a midpoint OPC, however, the conversion of ASE into phase noise through the Kerr effect is also compensated by the OPC. Therefore, at the $(N - k + 1)$th amplifier stage, the ASE noise level is the same as that generated at the kth amplifier, as illustrated in Figure 8.3.5(b). We easily find that this OPC system is equivalent to a system with a spontaneous emission factor of $2n_{sp}$ and a transmission distance of $L/2$. The accumulated phase-noise variance in this case is therefore approximately proportional to $(2n_{sp})(L/2)^3 = n_{sp}L^3/4$, resulting in 6 dB phase noise suppression.

5 Conclusion

We discussed a long-distance lightwave system where an optical phase conjugator is placed at the midpoint of the transmission system. When the amplifier spacing, the signal power, and the dispersion of transmission fibers are properly chosen, almost distortion-free transmission becomes possible. The remaining waveform distortion is determined by second-order dispersion. This system is also effective in preventing crosstalk in WDM systems, and in suppressing the excess phase noise in coherent optical communication systems.

References

[1] C. Lin, H. Kogelnik and L.G. Cohen, *Opt. Lett.*, **5**, 476–478 (1980).
[2] G.P. Agrawal, *Nonlinear Fiber Optics*, Academic Press, New York, 1989.

[3] D. Marcuse, *J. Lightwave Technol.*, **9**, 356–361 (1991).
[4] K. Kikuchi, *IEEE Photon. Technol. Lett.*, **5**, 221–223 (1993).
[5] N. Henmi, Y. Aoki, T. Ogata, T. Saito and S. Nakaya, to be published in *J. Lightwave Technol.*
[6] A. Yariv, D. Fekete and D. M. Pepper, *Opt. Lett.*, **4**, 52–54 (1979).
[7] K. Inoue and H. Toba, *IEEE Photon. Technol. Lett.*, **4**, 69–72 (1992).
[8] S. Murata, A. Tomiya, J. Shimizu and A. Suzuki, *IEEE Photon. Technol. Lett.*, **3**, 1021–1023 (1991).
[9] S. Watanabe, T. Naito and T. Chikama, *IEEE Photon. Technol. Lett.*, **5**, 92–95 (1993).
[10] M.C. Tatham, G. Sherlock and L.D. Westbrook, *Electron. Lett.*, **29**, 1851–1852 (1993).
[11] K. Inoue and H. Toba, *IEEE Photon. Technol. Lett.*, **3**, 77–79 (1991).
[12] J.P. Gordon and L.F. Mollenauer, *Opt. Lett.*, **15**, 1351–1353 (1990).

8.4

Ultra-multiplexing and Demultiplexing Scheme of Subcarriers

Masamitsu Nakajima and **Young-Kyu Choi**

Abstract

Research has been carried out to attain high-speed optical modulation and demodulation. Yet, an extensive bandwidth remains unused, as the frequency of a lightwave is very high. So far as electronic circuits are employed, there must be some limitation in frequency. To overcome this limitation, a method is presented to modulate a lightwave at a frequency higher than the workable frequency of optical modulators by the use of a proper combination of modulators, and a way to demodulate such signals while using conventional photo-detectors. Also shown is an example of a subcarrier multiplexed scheme to enhance the bandwidth of an optical communication system.

1 Introduction

With a view to attaining high-speed optical modulation and demodulation, research has been done at several tens of gigahertz, for example [1]. Since the frequency of a lightwave is very high, an extensive bandwidth still remains unused. In so far as electronic circuits are employed for the modulation and demodulation of a light signal, there must be some limitation in frequency. We present a way to overcome this limitation by making proper use of the characteristics of optical modulators and demodulators themselves.

The underlying ideas are as follows. We reported previously an optical modulator having a square modulation characteristic [2]. According to the trigonometric formulae, this means that the frequency of modulation is doubled. The kind of operation can be extended to an arbitrary number of frequency multiplications. This fact may suggest that the lightwave could be modulated at a frequency higher than that of the modulation signal.

There are also measures to detect such a high-frequency modulated light signal using conventional photo-detectors. One of the solutions is to excite a photo-detector at a local frequency, so that a signal converted at a lower frequency can be detected. Another more efficient means is to use an optical modulator excited at a local frequency in front of the photo-detector.

If the baseband signal is directly applied to the scheme explained above for modulation, the signal will be distorted because of the nonlinearities involved in this scheme. This scheme therefore should employ a subcarrier multiplexed system, as is common in recent optical communication system.

2 High-frequency Modulation of a Lightwave

2.1 Square modulation

When a sinusoidal signal $\cos \omega t$ is applied to an optical modulator having the square characteristic mentioned above, it produces a light output signal proportional to $\cos^2(\omega t + \phi) = \{1 + \cos 2(\omega t + \phi)\}/2$. The lightwave is thus modulated at twice the input frequency of the electrical signal.

A similar modulation can be achieved by the use of a Mach–Zehnder interferometric modulator, which is very common in optical communication.

This modulator has the characteristic that the output optical power P_o is a function of the applied voltage v : $P_o = P_i \cos^2(\pi v / V_\pi)$, where V_π is the half-wave voltage. When a modulation signal $v = V \cos \omega t$ is impressed without DC bias, we have

$$\frac{P_o}{P_i} = \frac{1}{2}\left\{1 + J_0\left(\frac{\pi V}{V_\pi}\right)\right\} + \sum_{n=1}^{\infty} J_{2n}\left(\frac{\pi V}{V_\pi}\right) \cos 2n\omega t$$

The amplitude of the modulated signal at double the fundamental frequency is proportional to $J_2(\pi V / V_\pi)$.

2.2 Cascaded modulation

It is also possible to obtain a square modulation characteristic by the use of linear optical modulators. To obtain a realistic result, we take notice of the fact that commonly used (passive) modulators do not have a gain. The modulated lightwave is thus expressed as

$$P_o = P_i\{1 - m - m\cos(\omega t + \phi)\}$$

The instantaneous peak power of the lightwave does not exceed that of the input P_i. And, since the power of the lightwave never becomes negative, the modulation index m must be smaller than $\frac{1}{2}$: $|m| \leq \frac{1}{2}$ ($m = \frac{1}{2}$ means 100% modulation).

If this modulated lightwave is again modulated in reverse polarity (ϕ being replaced by $\phi + \pi$) with another modulator cascade connected, we have

$$P_o = P_i \left\{ 1 - 2m + \frac{m^2}{2} - \left(\frac{m^2}{2}\right) \cos 2(\omega t + \phi) \right\} \quad (1)$$

The lightwave was modulated at twice the fundamental frequency 2ω with the ω component being cancelled.

This modulation technique is extended to an arbitrary number of modulators. The n modulators are now connected in cascade so that the output light power is expressed by

$$P_o = P_i \prod_{k=1}^{n} \left\{ 1 - m - m \cos\left(\omega t - \frac{2k\pi}{n}\right) \right\} \quad (2)$$

where the phases of each modulator were shifted by the same value to achieve a total phase shift of 360°.

Noting that the nth root of 1 is $\exp(j2k\pi/n)$, we write as

$$x^n - 1 = \prod_{k=1}^{n} \left\{ x - \exp\left(\frac{j2k\pi}{n}\right) \right\}$$

Substituting $x = \exp(\alpha + j\omega t)$ into the above, and taking the square of its absolute value and dividing it by $2\exp(n\alpha)$, we obtain

$$\cosh(n\alpha) - \cos(n\omega t) = 2^{n-1} \prod_{k=1}^{n} \left\{ \cosh \alpha - \cos\left(\omega t - \frac{2k\pi}{n}\right) \right\}$$

If we set $\cosh \alpha = (1 - m)/m$, we have $\exp(\pm\alpha) = (1 - m \pm \sqrt{1 - 2m})/m$. With the aid of the above relations, equation (2) is calculated to be

$$P_o = P_i \cdot A\{1 - M \cos(n\omega t)\}$$

where the resultant amplitude is equal to

$$A = \frac{\{(1 - m + \sqrt{1 - 2m})^n + (1 - m - \sqrt{1 - 2m})^n\}}{2^n}$$

and the modulation index is

$$M = \frac{2m^n}{\{1 - m + \sqrt{1 - 2m})^n + (1 - m - \sqrt{1 - 2m})^n\}}$$

This tells us that the nth order harmonic modulation is attained with the components at the fundamental and any other harmonic frequencies are cancelled. It is seen that if the modulation index m of each modulator is 100% ($m = \frac{1}{2}$), then the nth order harmonic modulation M also becomes 100%. Because of the passive nature of optical modulators, the magnitude of the optical power decreases to $P_i/2^{2n-1}$. On the other hand, if the modulation index m of each modulator is kept small, the modulation index of the nth-order harmonic decreases to about $2(m/2)^n$, without decaying the optical power itself. In the intermediate case, both the modulation index and the power decrease. The attenuation of power may be compensated for by employing an optical source with high power, or by inserting optical amplifiers. There is a way to enhance the modulation index, which will be reported elsewhere.

3 High-frequency Demodulation

Normally the frequency response of optical detectors is lower than that of optical modulators, so that the high-frequency signal modulated in the previous section may not be detected directly, unless some measures are taken.

3.1 Direct demodulation

Consider an n–i–p photo-diode, as shown in Figure 8.4.1. A lightwave is injected into this diode from the left, where a DC bias electric field is applied across it. If we assume that all of the light power $\varphi h \nu$ (φ denotes the photon number per second) has been absorbed at the interface $x = 0$ between n and i, electron-hole pairs will be produced whose number is equal to φ per second. The electrons accelerated by the electric field immediately enter the heavily doped n region, losing their velocity. On the other hand, the holes travel rightwards with velocity v_0, inducing an external current $i = \varphi e$.

When the light is modulated at ω_s such that

$$\varphi = \varphi_0 + \varphi_s e^{j\omega_s t} + \varphi_s^* e^{-j\omega_s t}$$

with $\varphi_s = |\varphi_s| e^{j\phi_s}$, the induced external current is calculated to be

$$i = \varphi_0 e + \varphi_s e e^{j\omega_s(t - \pi/\omega_c)} \operatorname{sinc}\left(\frac{\omega_s}{\omega_c}\pi\right) \tag{3}$$

where $\omega_c = 2\pi v_0/L$, and L stands for the length of the diode. The factor $\operatorname{sinc}(\omega_s \pi/\omega_c)$ represents the frequency characteristic of demodulation due to the carrier transit across the diode. The frequency response is reduced to zero at the frequency $f_s = f_c \equiv v_0/L$. Beyond this frequency, demodulation is achievable by using the method as outlined in the following subsection.

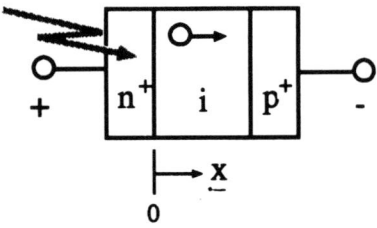

Figure 8.4.1
n–i–p photo-diode

3.2 Frequency-conversion demodulation

Next the local voltage is applied to the diode at a frequency ω_l superimposed on the DC bias. The electric field induced in the diode is written

$$E = E_0 + E_l e^{j\omega_l t} + E_l^* e^{-j\omega_l t} \qquad (4)$$

The lightwave modulated at frequency ω_s is illuminated onto the left surface of the diode. The number of holes produced, $p(0)$, is found from the relation $\varphi e = i = p(0)ev$, $v = \mu E$, or

$$p(0) = \frac{\varphi}{v} = \frac{(\varphi_0 + \varphi_s e^{j\omega_s t} + \varphi_s^* l e^{-j\omega_- t} + \cdots)}{v_0}$$

where $l = E_l/E_0$ and $\omega_- = \omega_l - \omega_s$. The velocity of these holes will be modulated by the local field $E_l e^{j\omega_l t}$. The density of holes in the diode may thus be written

$$p = \bar{p} + p_s e^{j\omega_s t} + p_- e^{j\omega_- t} + \cdots + c.c. \qquad (5)$$

We substitute equations (4) and (5) into the continuity equation

$$\frac{\partial p}{\partial t} = -\mu_p E \frac{\partial p}{\partial x} + D_p \frac{\partial^2 p}{\partial x^2} - \frac{p - p_0}{\tau_p}$$

The re-combination term usually may be ignored, as the life time τ_p of the holes is longer compared with the period of the frequencies concerned. For the sake of simplicity we will also neglect the diffusion D_p.

In order for the equation above to hold regardless of time, the coefficients of each time dependence (exponential functions) must be balanced between the right- and left-hand sides. For the ω_s and ω_- components we have

$$j\omega_s p_s = -\mu E_0 \frac{\partial p_s}{\partial x} - \mu E_l \frac{\partial p_-^*}{\partial x}$$

$$j\omega_- p_- = -\mu E_0 \frac{\partial p_-}{\partial x} - \mu E_l \frac{\partial p_s^*}{\partial x}$$

The solution of the above is written

$$p_s(x) = e^{-j\beta_a x} \left\{ p_s(0) \left(\cos \beta_b x - j\frac{\beta_d}{\beta_b} \sin \beta_b x \right) - p_-^*(0) \frac{jc_{12}}{\beta_b} \sin \beta_b x \right\}$$

$$p_-^*(x) = e^{-j\beta_a x} \left\{ p_-^*(0) \left(\cos \beta_b x + j\frac{\beta_d}{\beta_b} \sin \beta_b x \right) - p_s(0) \frac{jc_{21}}{\beta_b} \sin \beta_b x \right\}$$

where

$$\beta_a = \frac{\omega_s - \omega_-}{2v_0}, \quad \beta_d = \frac{\omega_s + \omega_-}{2v_0} \quad \beta_b = \sqrt{\frac{\beta_d^2 - |l|^2 \omega_s \omega_-}{v_0^2}}$$

$$c_{12} = -\frac{l}{1 - |l|^2} \frac{\omega_-}{v_0}, \quad c_{21} = \frac{l}{1 - |l|^2} \frac{\omega_s}{v_0}$$

With this solution, the current at ω_s is given by $i_s(\omega_s) = pev|_{\omega_s} = p_s(x)e\mu E_0 + p_-^*(x)e\mu E_l$. The current output from the diode at ω_s is obtained by integrating it throughout the diode

$$\bar{i}(\omega_s) \approx \varphi_s ee^{-j\omega_s\pi/\omega_c} \operatorname{sinc} \frac{\omega_s}{\omega_c}\pi \tag{6}$$

which is an approximate expression in the case of a small excitation $|l|^2 < 1$. In a similar way, the frequency-converted signal current flowing out of the diode is obtained as

$$\bar{i}(\omega_s) \approx -\varphi_s^* el \frac{\omega_-}{\omega_l} \left(e^{-j\omega_-\pi/\omega_c} \operatorname{sinc} \frac{\omega_-}{\omega_c}\pi - e^{j\omega_s\pi/\omega_c} \operatorname{sinc} \frac{\omega_s}{\omega_c}\pi \right)$$

From this result it is found that the frequency-converted demodulated signal increases in proportion to the local signal amplitude $l = E_l/E_0$ applied to the diode. This is in good agreement with experiments [5], as is also shown in Figure 8.4.2.

As was derived in equation (6), the output signal of the direct demodulation vanishes entirely at the frequency $f_s = f_c = v_0/L$, where the last term in the equation above is also zero. Since $\omega_- \neq \omega_s$ in general, the first term is not zero. This signifies that optical demodulation is possible at the converted frequency at ω_-, even when the photo-diode is unable to detect the signal at ω_s directly because of the frequency characteristic due to carrier transit. If we define the gain to be the ratio of the frequency-converted signal ω_- to the directly demodulated signal ω_s, this gain becomes infinitely large at this frequency f_c.

Figure 8.4.2 shows the experimental result. It is seen that the frequency-converted signal exceeds the directly detected signal when the local power is higher than a certain value.

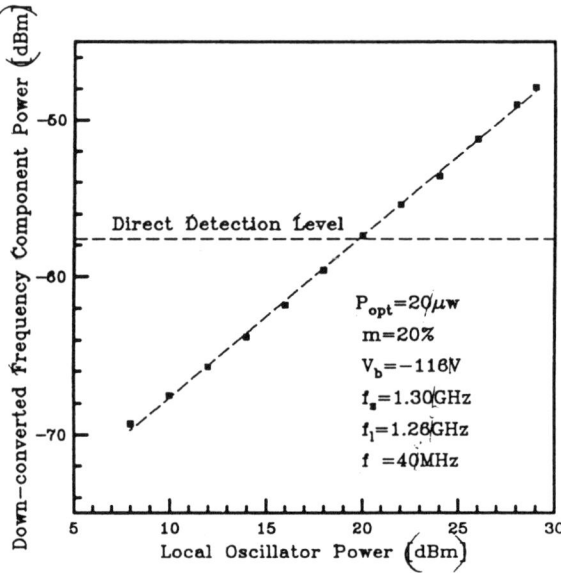

Figure 8.4.2
Experimental result of the frequency converted output signal versus local power

3.3 Demodulation with the aid of optical modulators

When a lightwave modulated at the signal frequency ω_s is applied to an optical modulator excited at ω_l, various frequency components appear according to

$$(1 + m \cos \omega_s t)(1 + l \cos \omega_l t)$$
$$= 1 + m \cos \omega_s t + l \cos \omega_l t + \left(\tfrac{1}{2}\right) ml \cos(\omega_l + \omega_s)t + \left(\tfrac{1}{2}\right) ml \cos(\omega_l - \omega_s)t$$

In the case when ω_s is higher than the cutoff frequency of the detector, the frequency $\omega_l - \omega_s$ can be chosen so that it falls below the cutoff, by selecting the local frequency ω_l to be properly higher than ω_s. In this manner, the signal ω_s is demodulated with a conventional photo-detector. When ω_s is higher than the workable frequency of the optical modulator, the modulation scheme described in section 2 is applicable.

4 Subcarrier Multiplexing

The high-frequency modulation and demodulation techniques described above are now applied to build a subcarrier multiplexed optical communication system to enhance the bandwidth capability. In the process of frequency multiplication no signal distortion occurs, as is seen in the mathematical formula (1). The signal amplitude included in the frequency ω or the phase ϕ is simply multiplied by a constant factor 2 in the process of frequency doubling. This means that the modulation index is increased, as is the case with FM generation through Armstrong's method. It is also noted that the increased frequency deviation is retained unchanged in the process of optical demodulation, the same as heterodyne receiving. This may be an advantage, for the modulation index in the first stage can be made low. A digital scheme of FSK or PSK is applicable, needless to say.

4.1 Frequency multiplexed modulation

An example of a frequency multiplexed optical modulation scheme is shown in Figure 8.4.3, in the case of three channels for brevity. As in a conventional system, a laser diode LD at the bottom is modulated by a baseband signal Ch. 1. The modulated lightwave from this diode may be expressed as $P_1(t) = \tfrac{1}{3} + m_1 \cos \omega_1 t$, where ω_1 is the frequency of the baseband signal itself or a subcarrier which is FM or PM modulated.

The lightwave from another laser diode LD is modulated by another subcarrier at ω_2 containing another signal Ch. 2. Since the frequency ω_2 is chosen higher than the feasible modulation frequency of the laser diodes, an external modulator MOD workable at this subcarrier frequency is used. The output of the modulated lightwave is expressed by $P_2(t) = \tfrac{1}{3} + m_2 \cos \omega_2 t$. The light wave from the third laser diode is modulated with a set of cascaded optical modulators at a subcarrier frequency ω_3 to yield $P_3(t) = \tfrac{1}{3} + m_3 \cos \omega_3 t$.

The combined total light output from this system of modulators becomes

$$P_1 + P_2 + P_3 = 1 + m_1 \cos \omega_1 t + m_2 \cos \omega_2 t + m_3 \cos \omega_3 t + \cdots$$

where the lightwave was normalized to its own average value.

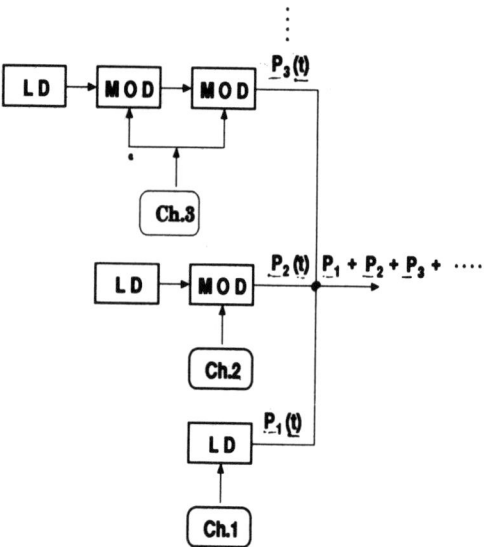

Figure 8.4.3
Frequency multiplexed modulator system

Following the same procedure as explained above, it may now be understood that the number of subcarriers or channels can be increased to a due amount.

4.2 De-duplexed demodulation

Figure 8.4.4 shows an example of de-duplexed demodulation scheme. A lightwave modulated at two subcarrier frequencies ω_1 which is lower than the cutoff frequency of (avalanche) photo-diodes PD and ω_2 which is higher than the cutoff frequency of them is split into two and injected into two optical modulators separately. The two modulators are excited in reverse polarity at a local frequency ω_l which is higher than twice the cutoff frequency of the photo-diodes. The output lightwave from the modulators is written as

$$\left(1 + \tfrac{1}{2}\cos\omega_1 t + \tfrac{1}{2}\cos\omega_2 t\right)(1 \pm \cos\omega_l t) = 1 + \tfrac{1}{2}\cos\omega_1 t + \tfrac{1}{2}\cos\omega_2 t$$
$$\pm \tfrac{1}{4}\cos(\omega_l - \omega_1)t \pm \tfrac{1}{4}\cos(\omega_l - \omega_2)t \pm \tfrac{1}{4}\cos(\omega_l + \omega_1)t \pm \tfrac{1}{4}\cos(\omega_l + \omega_2)t \pm \cos\omega_l t$$

where the upper sign corresponds to the upper optical modulator and the lower to the lower modulator. Because the frequencies ω_2, $\omega_l - \omega_1$, $\omega_l + \omega_1$, $\omega_l + \omega_2$ and ω_l are higher than the cutoff frequency of the photo-diodes from the assumption above, the components at these frequencies will not appear in the output terminals of the photo-diodes. Therefore the output signal at terminal ⓤ from the upper photo-diode is equal to $\tfrac{1}{2}\cos\omega_1 t + \tfrac{1}{4}\cos(\omega_l - \omega_2)t$ and that at terminal ⓛ from the lower photo-diode is $\tfrac{1}{2}\cos\omega_1 t - \tfrac{1}{4}\cos(\omega_l - \omega_2)t$. If the signal from terminal ⓤ is summed with that from terminal ⓛ, the baseband signal $\cos\omega_1 t$ in Ch. 1 is obtained, while the signal from ⓤ with that from ⓛ subtracted, produces the signal in Ch. 2, which is frequency converted

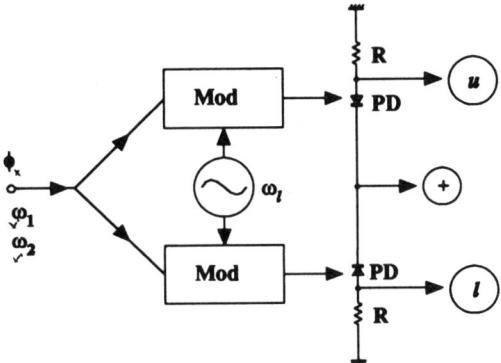

Figure 8.4.4
Demodulator system

at $\omega_l - \omega_2$. In this manner the signals on two subcarriers ω_1 and ω_2 are demodulated separately.

4.3 De-multiplexed demodulation

Frequency de-multiplexed demodulation involving more than two subcarriers is realized using the procedure explained in what follows. With the modulation of the baseband signal or subcarrier Ch. 1 omitted, the lightwave modulated with many subcarriers is divided by a star divider into the same number of branches as that of subcarriers at frequencies ω_n, and each lightwave is then injected into a photo-diode. Each photo-diode is excited with a local signal at frequency ω_{ln}. Each local frequency should be chosen such that the frequency differences $|\omega_{ln} - \omega_n|$ will fall within the baseband frequency. The subcarrier signals are demodulated in this fashion being converted into a baseband frequency in each photo-diode.

This technique may, in some aspects, resemble that of presently developed subcarrier multiplexed optical demodulation, but differs in that a frequency region higher than the cutoff of photo-diodes is utilized.

5 Conclusion

We have shown that a lightwave can be modulated and demodulated at frequencies higher than the feasible frequencies of the conventional optical modulators and demodulators. On the basis of this fact, an example of a frequency multiplexing and de-multiplexing scheme has been proposed to attain a wider bandwidth while using conventional optical components. This scheme may find application in a local area network because the bandwidth is extended without introducing highly advanced technologies like a coherent optical system requiring high stability of optical frequency.

References

[1] J. Nees, S. Williamson and G. Mourou, *Appl. Phys. Lett.*, **54**(20) (15), 1962-1964 (1989).
[2] Onodera, Awai, M. Nakajima and Ikenoue, *Appl. Opt.*, **23** (1), 118-123 (1984).
[3] Y-K. Choi, T. Fujita, M. W. Yee and M. Nakajima, *4th Optoelectronics Conference (OEC'92)*, July 1992, 16B4-3.
[4] Y-K. Choi and M. Nakajima, *IEICE Technical Report*, **OQE93**, 37-42 (1994).
[5] Y.K. Choi, Y.M. Wai and M. Nakajima, *Electron. Commun. Jpn*, Part 2, **75** (3), 20-26 (1992).
[6] M. Nakajima and F.C.V. Mendis, *International Symposium on Ultrafast and Ultra-parallel Optoelectronics*, 2-3, July 1994.

8.5

Ultralong-distance Fiber-optic Transmission and Multiplexing

Shinji Yamashita and **Takanori Okoshi**

Abstract

The present state of the art in the conventional optical fiber communication systems is not advanced enough to realize ultralong distance fiber-optic transmission and multiplexing. The need for the 3R repeater spoils the intrinsic simplicity of the system, and the present DD receivers do not have good frequency selectivity and cannot fully utilize the potential ultra-high capacity of an optical fiber. To overcome these difficulties, the authors have been engaged in research on optical fiber amplifiers which can compensate for the optical loss of the fiber, as well as coherent optical fiber communications which feature excellent frequency selectivity.

1 Introduction

Optical fiber communications are recognized as playing a leading role in future communications network, and have already been put into commercial operation in domestic and international trunk lines. The present state of the art in these systems, however, is not enough to realize ultralong-distance fiber-optic transmission and multiplexing. In ultralong systems, the 3R (reshaping, retiming, regeneration) repeaters accompanying O/E and E/O conversions are used at every 50–100 km to compensate for the loss of the fiber, which, however, spoils the intrinsic simplicity of the optical fiber communications. Furthermore, the present direct-detection (DD) receivers can provide only poor frequency selectivity, making it difficult to realize a densely multiplexed system.

To overcome these difficulties, the authors have been engaged in research on optical fiber amplifiers which can compensate for the optical loss of the fiber, as well as coherent optical fiber communications which feature excellent frequency selectivity.

2 Ultralong-distance Transmission Using Fiber Amplifiers

2.1 Erbium-doped fiber amplifier (EDFA)

Erbium-doped fiber amplifiers (EDFAs) have attracted researchers' attention and have been studied intensively and explosively because of their excellent features, such as high gain, high efficiency, high saturation output, low noise, and low coupling loss to optical fibers [1]. A number of experiments of long-span optical fiber communication systems using EDFAs have been reported; at present, ultralong transmissions over 10 000 km have already been demonstrated.

Figure 8.5.1 shows a typical construction of an EDFA with forward pumping. It consists of a wavelength-division multiplexing (WDM) coupler to mix the pump with the signal, a pumping LD (0.98 or 1.48 μm), polarization independent isolators at the input and output to prevent possible oscillation, and an erbium-doped fiber (EDF) with a proper length, a proper erbium density, and a proper doped-region size. An optical bandpass filter is inserted to remove the wideband amplified spontaneous emission (ASE) outside of the signal bandwidth and the rest of the pump.

The characteristics of fiber amplifiers are described by the expanded STT (Shimoda, Takahashi and Townes) theory [2]. The output power from an amplifier is expressed as

$$P_{\text{out}} = G(L)P_{\text{in}} + 2hfG(L)N_{\text{sp}}(L)\Delta f \tag{1}$$

where Δf is the spontaneous emission bandwidth, L is the total length of the amplifier, and $G(L)$ and $N_{\text{sp}}(L)$ are the gain and spontaneous emission factor of the amplifier in total. The first term represents the amplified signal, and the second the ASE. The beat between the amplified signal and the ASE is the dominant noise after detection if the gain is high and the bandwidth is small. This signal–ASE beat-noise-limited SN ratio is worse than the shot-noise-limited SN ration by $2N_{\text{sp}}$. For an ideal amplifier, N_{sp} is unity; thus the ideal noise figure (NF) of an EDFA is 2 (3 dB).

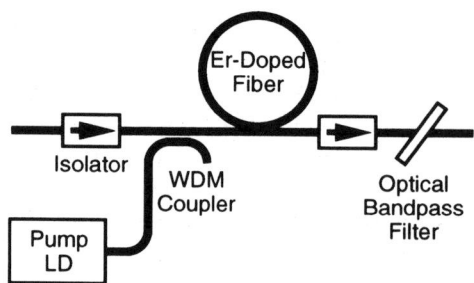

Figure 8.5.1
Typical construction of an erbium-doped fiber amplifier (EDFA)

The parameters G and N_{sp} are expressed using local parameters at a position (x, y, z) (gain and loss coefficient g and γ, spontaneous emission factor n_{sp}, and normalized intensity distribution function $\psi(x, y)$), as

$$G(z) = \exp\left[\int_0^z \int_{-\infty}^{\infty} \int_{-\infty}^{\infty} \psi(x, y)(g(x, y, z') - \gamma(x, y, z'))\,dxdydz'\right] \quad (2)$$

$$N_{sp}(z) = \int_0^z \int_{-\infty}^{\infty} \int_{-\infty}^{\infty} \psi(x, y) \frac{g(x, y, z')n_{sp}(x, y, z')}{G(z')}\,dxdydz' \quad (3)$$

Local parameters are obtained from simple rate equations describing erbium ion densities in the lower and upper states, N_1 and N_2, as

$$g = (\sigma_s^e + \sigma_s^a)N_2 - \sigma_s^a \rho \quad (4)$$

$$n_{sp} = \frac{\sigma_s^e N_2}{(\sigma_s^e + \sigma_s^a)N_2 - \sigma_s^a \rho} \quad (5)$$

where σ_s^e and σ_s^a are the emission and absorption cross-sections, respectively, ρ is the total erbium density,

$$N_2 = \rho \frac{(\alpha_s I_s/(1+\alpha_s)I_{sat}) + (\alpha_p I_p/(1+\alpha_p)I_{th})}{(I_s/I_{sat}) + (I_p/I_{th}) + 1}, \quad (6)$$

I_s and I_p are the signal and pump intensities, respectively, I_{sat} and I_{th} are the signal saturation and pump threshold intensities, respectively, and α_s and α_p are the ratios of absorption to emission cross-sections at signal and pump wavelengths, respectively. Details of the theory will be found in Reference [2].

The broken curves in Figure 8.5.2 show examples of the gain and the NF of an EDFA calculated in the above way as a function of the input signal power for two signal wavelengths, 1.553 μm and 1.535 μm [2]. (Solid curves will be discussed in

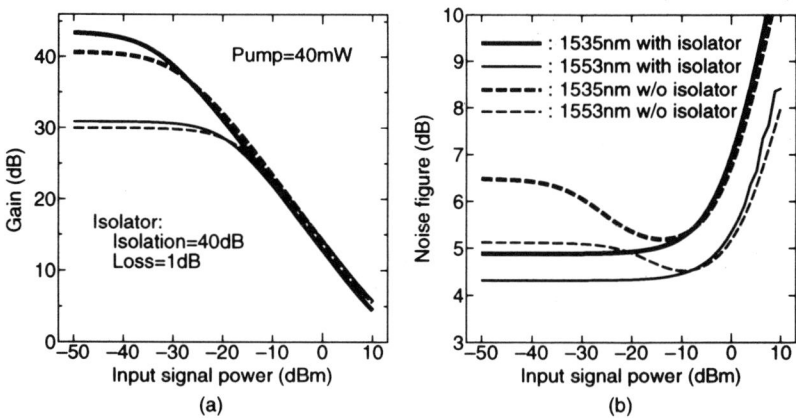

Figure 8.5.2
Calculated (a) gain and (b) NF of the EDFA without (broken curves) and with (solid curves) a midpoint isolator

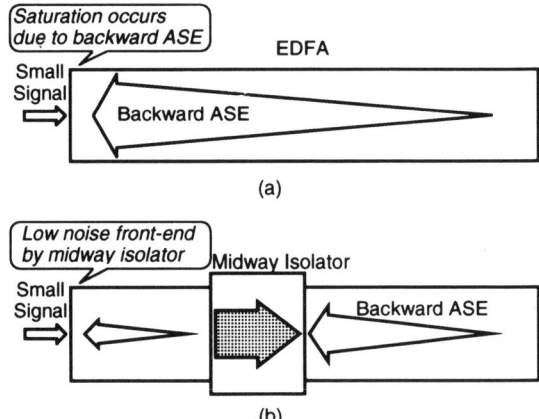

Figure 8.5.3
Degradation of the noise figure due to backward ASE and elimination of backward ASE by a midpoint isolator

subsection 2.2.) It is found that a gain greater than 30 dB and a NF below 5 dB can be obtained with relatively low (40 mW) pump power. It should be noted that NF degradation is observed at both large (0-10 dBm) and small (-50--30 dBm) input signal regions, and that the NF is minimized at a medium input signal level, in this case around -10 dBm. It is accounted for as follows. The NF, as equation (3) implies, is almost determined by the value of n_{sp} near the input end of the EDFA where the gain $G(z)$ is small. When the large signal is input, saturation (enhancement of I_s/I_{sat}) occurs near the input end due to the signal itself, and n_{sp} is enhanced through equations (6) and (5). When the small signal is input, contrarily, the ASEs become larger than the signal, in particular the backward ASE becomes dominant near the input end and causes saturation, as shown in Figure 8.5.3(a).

2.2 Factors degrading the characteristics of an EDFA and their improvement

An EDFA has almost ideal characteristics as an optical amplifier, as described above. However, some factors degrade its characteristics: the loss due to optical couplers and isolators; terminal reflection from input or output end; and the backward-propagated ASE noise discussed above.

As to the loss, only that at the input end degrades the noise characteristics because, once amplified, the signal and ASE decay at the same rate due to the loss, thus keeping the SN ratio unchanged. Therefore, low-loss couplers and isolators should be used, particularly at the input end.

Since an EDFA is a traveling-wave (TW) type device, reflections at both ends are harmful and in the worst case even make it oscillate. The authors have made an analysis on the degradation induced by the terminal reflections using the same method as in subsection 2.1 [3]. It has been shown that the reflection at the input end seriously affects the noise characteristics because the backward ASE is reflected at the input end to become excess noise. Reflections at the input end smaller than −50 dB at a signal wavelength of 1535 nm, and smaller than −36 dB at 1553 nm, are required to suppress the NF deterioration below 0.5 dB. An isolator at the input end has been shown to be indispensable for suppressing such NF deterioration.

The backward-propagated ASE causes degradation of the NF in the small input signal region, as shown in Figure 8.5.3(a). If the backward-propagated ASE is eliminated near the input end, better noise characteristics are expected. The authors have proposed a novel method to suppress the backward-propagated ASE, and made a detailed analysis of the novel EDFA [4]. The basic construction of the proposed EDFA is simple. An isolator is inserted at the 'midpoint' of an active EDF to protect the amplification of the backward ASE near the input end, as shown in Figure 8.5.3(b). The solid curves in Figure 8.5.2 show the calculated gain and NF for the cases with and without the midpoint isolator, as a function of the input signal power.

With the midpoint isolator, the NF degradation at small input signal regions is removed as expected, and a low and flat NF is obtained, as shown by solid curves in Figure 8.5.2(b). Moreover, the small signal gain is also improved with the midpoint isolator, as shown in Figure 8.5.2(a); this is because unwanted amplification of the backward ASE is effectively suppressed. More detailed analyses show that a 1.1–3.2 dB improvement for the gain and 0.9–1.8 dB for the NF are expected, and that optimum position of the isolator is one-third to one-half of the EDFA length from the input end. It is noted that a high-quality isolator is not necessary in this method; an ordinary isolator with 20 dB isolation and 2 dB loss has been found satisfactory.

2.3 *Factors limiting the transmission distance and its elongation*

By cascading EDFAs as optical 1R repeaters, ultralong-distance transmission can be realized. In such a system, factors limiting the transmission distance are accumulation of ASE, chromatic dispersion, and nonlinear effects [5].

In the system with EDFAs, the ASE from each EDFA is accumulated to degrade the SN ratio and limit the transmission distance. This limit depends on the repeater gain G and number of repeaters K, and can be improved by reducing G and increasing K. The optimum system is a distributed amplification system where $K = \infty$ and $G = 1$, that is, the loss is compensated by gain everywhere. Such a system could be realized with a low-erbium-doped fiber (LEDF) system; a detailed analysis of such a novel EDFA has been made using the same method as in subsection 2.1 [2]. The result showed that nearly ideal characteristics are realized in this system with bi-directional pumping.

A single-mode optical fiber has a certain amount of chromatic dispersion (∼17 ps/km·nm) at 1.55 μm. This can be reduced in a dispersion-shifted optical fiber, but cannot be reduced to zero in ultralong-distance transmissions. The effect of the dispersion itself is a linear one, and thus can be compensated for by an optical equalizer in a DD system, or an electrical equalizer in a coherent system. However, the synergistic interaction between the dispersion and nonlinear Kerr effect makes compensation impossible and the waveform distortion more serious.

Various nonlinear effects exist in an optical fiber into which high optical power is launched. Major problems in communication systems are stimulated Brillouin scattering (SBS) and the optical Kerr effect. SBS limits the maximum optical power in the fiber below a few milliwatts, but can be avoided by some methods [5]. The authors have proposed methods using spread-spectrum (SS) modulations [6] and in-line isolators in

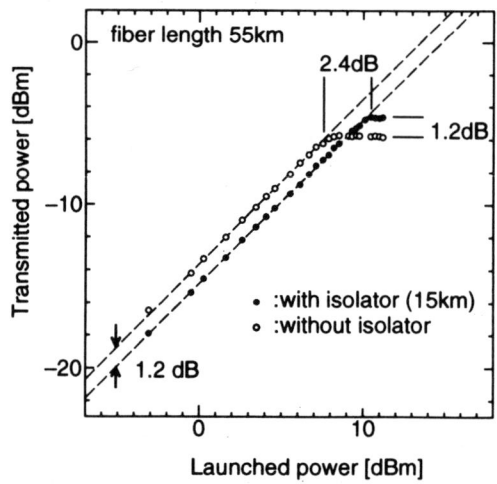

Figure 8.5.4
Output power as a function of input power measured with (solid circles) and without (open circles) as in-line isolator

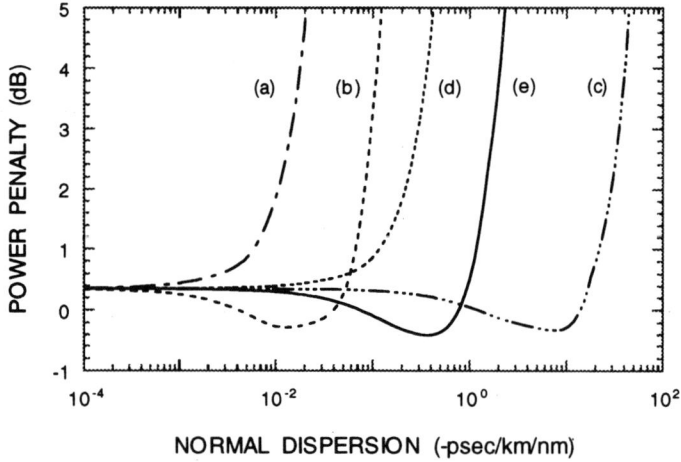

Figure 8.5.5
Penalty due to the distortion after 10 000 km transmission as a function of the value of the dispersion; (a) without equalizer, (b) with a single equalizer, (c) with distributed equalizers at every 50 km, (d) with distributed imperfect equalizers, and (e) with distributed imperfect equalizers and an adjustable equalizer

the system [7]. Figure 8.5.4 shows the output optical power as a function of input power, measured with (solid circles) and without (open circles) an in-line isolator. The SBS threshold is found to be successfully improved by 2.4 dB, and the saturation output power by 1.2 dB.

The most serious problem in ultralong-distance transmission is the Kerr effect. It has a very small nonlinear coefficient (3.2×10^{-20} m^2/W), but accumulates and cooperates with the dispersion to seriously distort the waveform as the transmission distance becomes longer. This effect depends severely on the magnitude of the dispersion. The authors have proposed and demonstrated through computer simulation that the insertion of optical dispersion equalizer(s) can reduce the waveform distortion [8].

Figure 8.5.5 shows the penalty due to the distortion after 10 000 km transmission (50 km repeater spacing) of a 10 Gbit/s signal, as a function of the magnitude of the dispersion. A single equalizer at the receiving end (curve (b)) can improve the distortion a little, whereas the distributed equalizers at every 50 km (curve (c)) improve the distortion considerably. A compensation mismatch of 5% in every equalizer makes the improvement less (curve (d)), but an adjustable equalizer at the receiving end can compensate for the mismatch (curve (e)). Thus, the system with distributed equalizers is much more tolerant of fiber dispersion, and can realize an ultralong-distance transmission more easily than by a conventional system.

3 Multiplexing Using Coherent Techniques and Fiber Amplifiers

3.1 Coherent optical fiber communications

The present optical fiber communication systems employ intensity modulation/direct detection (IM/DD). The transmitter laser diode is directly on-off modulated, and the signal is directly detected by an avalanche photo-diode (APD), as shown in Figure 8.5.6(a). This system is simple and robust, but suffers from low receiver sensitivity and poor frequency selectivity. The sensitivity can be enhanced by using a fiber pre-amplifier, whereas the selectivity is difficult to enhance because a sharp optical filter comparable with an electrical one can never be obtained. Coherent optical fiber communications have the advantages of high receiver sensitivity and high frequency selectivity [9]. The amplitude, frequency, or phase of the carrier are modulated directly or externally, and the signal is mixed with high-power local (LO) light, and the beat between the signal and LO is detected by a photo-diode (PD), as shown in Figure 8.5.6(b). The intermediate-frequency (IF) signal obtained, typically a few GHz, is fed to the heterodyne demodulator to obtain a baseband signal.[1] By scanning the LO frequency and using a sharp electrical IF filter, the coherent scheme enables us to realize an extremely dense frequency-division multiplexing (FDM) system.

---

[1] In a homodyne scheme, the baseband signal is obtained directly by the first detection.

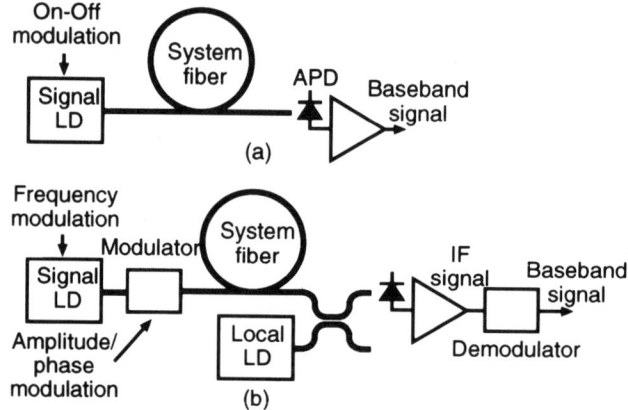

Figure 8.5.6
(a) An IM/DD system and (b) a coherent system

3.2 Reduction of the beat noise from EDFA by using coherent techniques

Using a coherent receiver together with a fiber pre-amplifier is a favorable combination for a phase- and/or polarization-diversity receiver, or an integrated coherent receiver which often suffers from insufficient LO power and a lossy diversity and/or integrated circuitry in the receiver. In such receivers, the LO and ASE make their beat the dominant noise (LO-ASE beat-noise-limited case). In this case, the ideal NF is still 3 dB. However, when the LO power is insufficient, the signal–ASE and ASE–ASE beat noises predominate over the LO–ASE beat noise, thus degrading the SN ratio. The authors have shown and demonstrated that these are common-mode noises for a balanced mixer, and hence can be suppressed by using a balanced receiver [10]. Figure 8.5.7 shows the spectra measured in a single (Figure 8.5.7(a)) and a balanced (Figure 8.5.7(b)) receiver

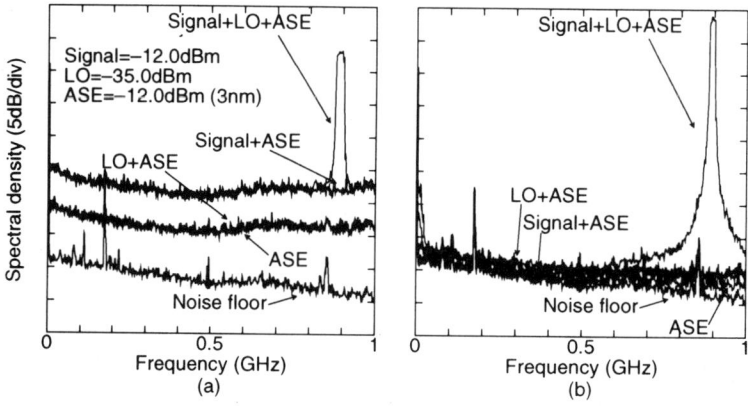

Figure 8.5.7
Spectra measured in (a) a single and (b) a balanced receiver when the LO power is small

when the LO power is small. The SN ratio is found to be limited by the signal–ASE beat noise in a single receiver, whereas it is improved to be nearly the LO–ASE beat-noise-limited SN ratio in a balanced receiver.

When a heterodyne receiver is used in a FDM system, an 'image' signal interferes with the 'real' signal and degrades the receiver performance, if the LO frequency is chosen at the middle of two channels. The situation is the same with the case when the signal with a wideband ASE noise is detected by a heterodyne receiver. The image band ASE causes the same amount of beat noise with the inevitable real band noise. If the image band beat noise could be eliminated, the NF would be improved by 3dB, and hence a NF of 0 dB would be realized. An image-rejection receiver is suitable for this purpose.

Another solution to this problem is the use of a double-stage phase-diversity (DSPD) receiver, proposed by the authors, which features the spectrum-unfolding of a homodyne-detected signal by second-stage phase-diversity frequency up-conversion [11,12]. Figure 8.5.8 shows how the real and image beat-noise spectra folded into the baseband can be unfolded in the second IF stage. The in-phase (I) and quadrature (Q) homodyne-detected signals using a 90° optical hybrid are up-converted by the second electrical LOs which have a 90° phase difference. The real spectra in the two up-converted signals are in-phase, whereas the image spectra are out-of-phase. By adding the two up-converted signals, the image spectra are eliminated, and hence the SN is improved by 3 dB.

Figure 8.5.9 shows the spectra measured after (a) up-conversion and (b) addition. In Figure 8.5.9(a), the signal (at 0.2 GHz) to noise (at 0.6 GHz) ratio is estimated to be 22.5 dB. By adding the two up-converted signals in Figure 8.5.9(b), a 2.8 dB S/N improvement is observed as well as a 17.9 dB image signal suppression. The authors have also observed a sensitivity improvement of about 2.5 dB by measuring the bit-error rate in DSPD and heterodyne receivers.

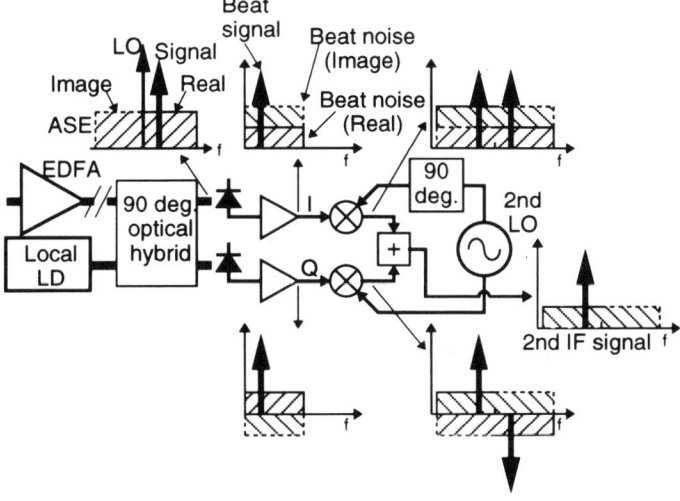

Figure 8.5.8
Folded real and image beat-noise spectra and their unfolding in a DSPD receiver

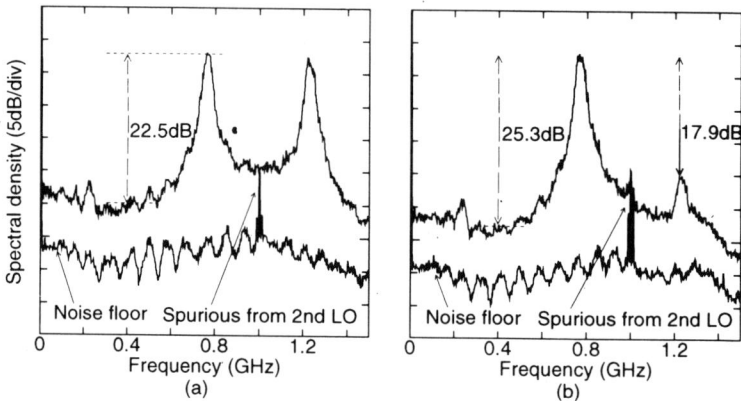

Figure 8.5.9
Spectra measured after (a) up-conversion and (b) addition

References

[1] S. Nakagawa, M. Nakazawa, K. Aida and K. Hagimoto, *Optical Amplifiers and Their Applications*, Ohm Publishing Co. Ltd, Tokyo, 1992 [in Japanese].
[2] S. Yamashita and T. Okoshi, *Trans. IEICE*, **J75-B-I** (5), 263–272 (May 1992) [in Japanese]. (Translation: *Electron. Commun. Jpn*, **76** (8), 1–13 (Aug. 1993).)
[3] S. Yamashita and T. Okoshi, *Electron. Lett.*, **28** (14), 1323–1324 (July 1992).
[4] S. Yamashita and T. Okoshi, *Photon. Technol. Lett.*, **4** (11), 1276–1278 (Nov. 1992).
[5] T. Okoshi, *Optoelectronics Conference (OEC'92)*, July 1992, no. PL-1.
[6] A. Hirose, Y. Takushima and T. Okoshi, *J. Opt. Commun.*, **12** (3), 82–85, (Sept. 1991).
[7] Y. Takushima and T. Okoshi, *Electron. Lett.*, **28** (12), 1155–1157 (June 1992).
[8] J. Nakagawa, K. Hotate and T. Okoshi, *J. Opt. Commun.*, **15** (6), 202–207 (Dec. 1994).
[9] T. Okoshi and K. Kikuchi, *Coherent Optical Fiber Communications*, KTK/Kluwer, Tokyo/Dordrecht, 1988.
[10] S. Yamashita and T. Okoshi, *Electron. Lett.*, **28** (21), 1970–1972 (Oct. 1992).
[11] T. Okoshi and S. Yamashita, *J. Lightwave Technol.*, **8** (3), 376–384 (Mar. 1990).
[12] S. Yamashita and T. Okoshi, *J. Lightwave Technol.*, **12** (6), 1029–1035 (June 1994).

8.6

Control of Fiber Characteristics

Yutaka Sasaki

Abstract

The optimum design of stress-applying parts (SAPs) in a polarization-maintaining optical fiber (PMF) of a PANDA profile, being the abbreviation of polarization-maintaining and absorption-reducing one, is theoretically presented by an analysis dealing with the off-diagonal terms of the permittivity tensor arising from the shear stress inherently caused by the SAPs. The opitimally designed PANDA fiber can make an improvement of about 40 dB on polarization crosstalk compared with the conventional one.

1 Introduction

Polarization-maintaining optical fibers (PMFs) [1-3], which can preserve a polarization state despite bends, twists, and environmental changes, are essential for fiber-optic sensing systems [4] and long-length optical transmission systems, such as coherent optical transmissions and active transmissions with optical nonlinear effects.

Several different types of PMFs are now available by a variety of fabrication techniques [5]. These fibers are most applicable for carrying linear-polarized light, because high birefringence is provided in the fibers to preserve polarization in a stable state against environmental influences. Polarization crosstalk of −42 dB/km was achieved in a short-length PMF, in which low-loss characteristics are not so important [6]. Furthermore, polarization crosstalk of −34 dB/km was obtained in a long-length PMF with low loss of 0.22 dB/km [7].

PMFs are now required to have lower polarization crosstalk compared with −40 dB/km now commercially available in some applications such as fiber-optic sensing systems. Its value, for an example, is required to be less than about −70 dB/km in the high performance interferometer fiber gyroscope. To improve the polarization-maintaining ability of PMFs, their polarization mode-coupling must be adequately suppressed. The magnitude of the mode-coupling increases with an increase in the quantity of the minor component power of the electrical field which arises from the structural fluctuation along the fiber length and the structural composition in the fiber cross-section. The mode-coupling on the structural fluctuation by bending stress has already been analyzed by Okamoto *et al.* [8]. Subsequently, an analysis of the structural composition, using the off-diagonal terms of the permittivity tensor [9] arising from the inner shear stress, was carried out by Sadar and Wong [10]. The analysis theoretically demonstrated the power of the minor component in the PMF of bowtie profile to be comparable or less compared with that of the conventional single-mode fiber. But the minor component power of the PMF of PANDA profile, namely the PANDA fiber, which has merits of excellent polarization-maintaining performance and easy design and fabrication, has not yet been analyzed.

Here, the off-diagonal terms of the permittivity tensor in the fiber cross-section of the PANDA fiber is introduced. Then, the polarization crosstalk degradation by the minor component caused by mode-coupling is investigated. As a result of this, the optimum design of SAPs of the PANDA fiber is obtained.

2 Fiber Parameters

A cross-sectional view of the PANDA fiber is shown in Figure 8.6.1. The diameters of the core and cladding are represented by 2a and 2b, respectively, and the relative refractive index difference between the core and cladding, $(n_1 - n_2)/n_2$, is represented by Δ where n_1 and n_2 are the refractive indices of the core and cladding, respectively. The half distance between the stress applying parts (SAPs) is represented by t and

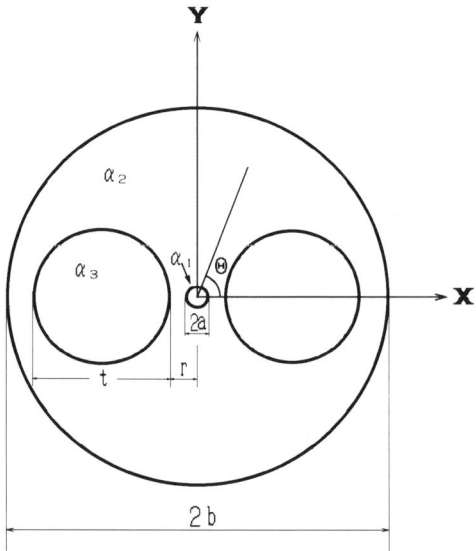

Figure 8.6.1
Cross-section of a PANDA fiber

Table 8.6.1
PANDA fiber parameters used

Young's modulus: E	7830 (kg/mm^2)
Poisson's ratio: ν	0.186
Thermal expansion coefficient:	
α_1 (core)	9.24×10^{-7} (°C^{-1})
α_2 (cladding)	5.40×10^{-7} (°C^{-1})
α_3 (SAP)	2.12×10^{-6} (°C^{-1})
Optoelastic coefficient:	
C_1	7.420×10^{-6} (mm^2/kg)
C_2	4.104×10^{-5} (mm^2/kg)
Temperature difference: ΔT	-767 (°C)
Refractive index:	
n_1	1.4632
n_2	1.4585
Core radius: a	5.0 (μm)
Cladding radius: b	62.5 (μm)

the diameter of the SAPs is presented by r. The thermal expansion coefficients of the core, the cladding and the SAPs are presented by α_1, α_2 and α_3, respectively. The core region is germanium-doped silica with GeO$_2$ percentage molar concentration ρ_{Ge}, the SAP region is boron-doped silica with B$_2$O$_3$ percentage molar concentration ρ_B, and the cladding is pure silica. As for the fiber's principal axes, the axis in the direction of SAPs is denoted as the x-axis, and that normal to it is denoted as the y-axis. The projection angle 2Θ, where Θ is the azimuthal angle from the x-axis, is the angle subtended by each SAP at the core center.

The PANDA fiber parameters used here are listed in Table 8.6.1, where ρ_{Ge} is 3.5 mol% and ρ_B is 15 mol%.

3 Analysis

3.1 Off-diagonal term in the permittivity tensor

The x-polarized fundamental mode HE$_{11}^x$ of the PANDA fiber shown in Figure 8.6.1 is analyzed here, which propagates along the fiber length and generates a major x component and a minor y component. Thus, the only off-diagonal terms present in the permittivity tensor are n_{xy}^2 and n_{yx}^2 [8]. These off-diagonal terms are defined as

$$n_{xy}^2 = n_{yx}^2 = 2 \cdot n_0 \cdot \{C_2 - C_1\} \cdot \tau_{xy}$$

where n_0 is the stress-free refractive index profile of the fiber in question, C_1 and C_2 are the optoelastic coefficients [11,12], and τ_{xy} denotes the shear stress inherently caused by the SAPs. The generalized plane strain (GPS) model [13], assuming elastic homogeneity in the thermal stress analysis, is implemented. Moreover, in the homogeneous model implanted, Young's modulus E is taken to be 7830 kg/mm^2 and Poisson's ratio ν to be 0.186 throughout the cross-section of the fiber.

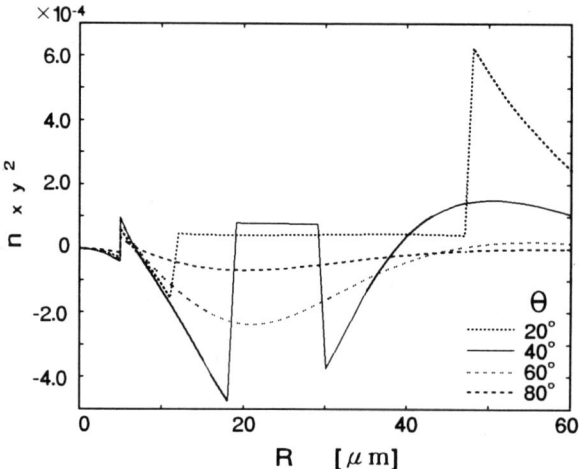

Figure 8.6.2
Off-diagonal term n_{xy}^2 versus radial coordinate R for various angular coordinates Θ in a PANDA fiber, where the normalized SAP diameter is $t/b = 0.6$

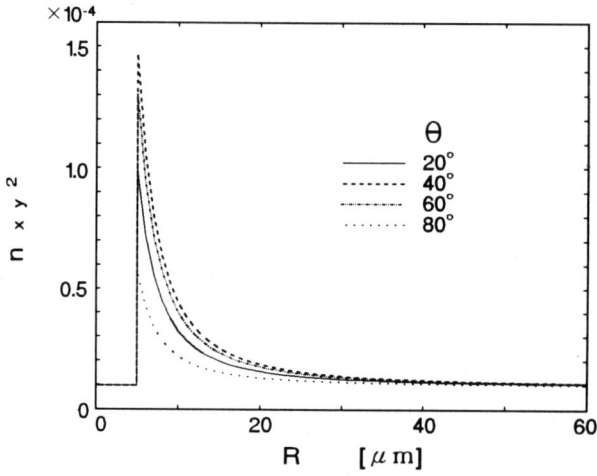

Figure 8.6.3
Off-diagonal term n_{xy}^2 versus radial coordinate R for various angular coordinates Θ in a single mode fiber, where the normalized SAP diameter $t/b = 10^{-5}$ is used

The off-diagonal term n_{xy}^2 versus the radial coordinate R for various angular coordinates Θ is shown in Figure 8.6.2. It is found that in the directions $\Theta = 20°$ and $\Theta = 40°$, where the SAPs are located, the off-diagonal terms change variously and largely, but in the directions $\Theta = 60°$ and $\Theta = 80°$, where the SAPs are not located, the off-diagonal terms change only slightly. It is sure that $n_{xy}^2 = 0$ in $\Theta = 0°$ and $\Theta = 90°$ not being illustrated in the figure. In comparison with the off-diagonal term of a PANDA fiber, that of a single-mode fiber, is shown in Figure 8.6.3, where the single-mode fiber can be considered as the PANDA fiber with $t/b = 10^{-5}$. It is not present in the core and its maximum value is gained at the core–cladding boundary. Furthermore, n_{xy}^2 in the cladding decreases exponentially with an increase in the radial coordinate R. It is found that in the cladding, $|n_{xy}^2|$ of the single-mode fiber is smaller than that of the PMF, but at the core–cladding boundary, that of the single-mode fiber is rather large compared with that of the PMF.

3.2 Mode-coupling coefficient

The mode-coupling problem at random is solved by Marcuse [14] and the power coupling coefficient h between the powers of the x-component and the y-component is given by

$$h = \int_{-\infty}^{\infty} \langle \Gamma(Z)\Gamma(Z-\ell) \rangle \cdot \exp[-j(\beta x - \beta y)\ell]\, dl$$

where the symbol $\langle \cdot \rangle$ indicates an ensemble average and the mode-coupling coefficient is shown as follows

$$\Gamma(Z) = \omega\varepsilon_0 \int_0^{2\pi} \int_0^{\infty} \boldsymbol{E}_x^* \boldsymbol{X}(Z) \boldsymbol{E}_y\, r\, dr\, d\theta$$

where β_x and β_y denote the propagation constants and $\boldsymbol{E}_x$ and $\boldsymbol{E}_y$ denote the electric fields of the fiber. $\boldsymbol{X}(z) = (\tilde{\boldsymbol{K}} - \boldsymbol{K})$ represents the small deviations of the permittivity tensor $\boldsymbol{K}$. Furthermore, Γ is given by

$$\Gamma(Z) = \hat{\Gamma} \cdot f(Z)$$

where $\hat{\Gamma}$ is constant and $f(z)$ presents the fluctuation of the parameter along the fiber length. The autocorrelation function $R(\ell) \equiv \langle f(Z) \cdot f(Z-\ell) \rangle$ is shown as follows,

$$R(\ell) = \sigma \cdot \exp\left(\frac{-|\ell|}{L_c}\right)$$

where σ is the rms deviation and L_c is the correlation length.

The optimum condition for the polarization-maintaining ability of a PANDA fiber realizes complete suppression of the mode coupling and is shown by

$$\Gamma(Z) = 0$$

This condition is satisfied by $\Gamma_i(z) = 0$, where $\Gamma_i(z)$ is the mode-coupling coefficient in the ith quadrant of the orthogonal coordinate system because $\Gamma(z)$ is represented by

$$\Gamma(Z) = \sum_{i=1}^{4} \Gamma_i(z)$$

Moreover, $\Gamma(z) = 0$ is satisfied by $\Gamma_1(z)$ since $\Gamma_1(z) = \Gamma_3(z)$, $\Gamma_2(z) = \Gamma_4(z)$, and $\Gamma_1(z) = -\Gamma_2(z)$ according to the structural symmetry in the fiber cross-section. That is, the optimum condition $\Gamma(z) = 0$ is replaced by $\Gamma_1(z) = 0$. Therefore, the characteristics of $\Gamma_1(z)$ are analyzed in detail here.

The mode-coupling coefficient $\Gamma_1(z)$ versus the normalized SAP diameter t/b for various normalized half distances r/a between SAPs is shown in Figure 8.6.4. When

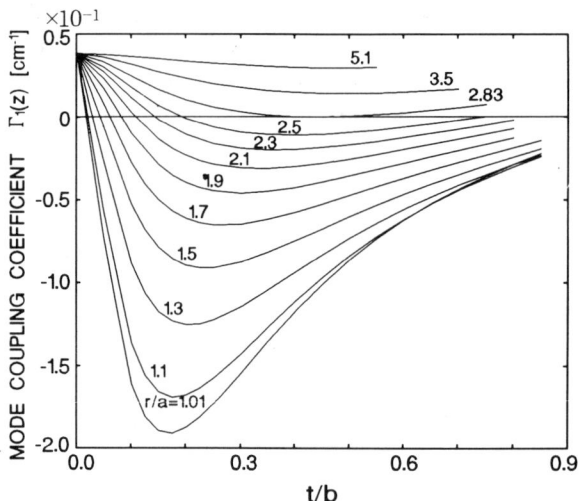

Figure 8.6.4
Mode-coupling coefficient $\Gamma_1(z)$ in the first quadrant versus normalized SAP diameter t/b

the SAP parameters take $t/b = 0.44$ and $r/a = 2.83$, $\Gamma_1(z)$ is equal to zero, that is, the magnitude of the mode-coupling is zero. Therefore this shows that the PANDA fiber is optimally designed under those SAP parameters, which is discussed in detail in section 4.

3.3 Polarization crosstalk

Thus, the power coupling coefficient h is given by

$$h = \left| \frac{(k/2)(C_2 - C_1)\gamma_S}{\xi} \right|^2 \cdot 2\sigma_0^2 \frac{L_C}{1 + (\Delta\beta \cdot L_C)^2}$$

where $\Delta\beta = \beta_x - \beta_y$, ξ is the constant gained by the polarization crosstalk measured experimentally, and

$$\gamma_S = 4\sqrt{\frac{\varepsilon_0}{\mu_0}} A^2 \left\{ n_1 \int_{\text{core}} \tau_{xy} J_0^2\left(\frac{u}{a}r\right) dS + n_2 \int_{\text{clad}} \tau_{xy} \frac{J_0^2(u)}{K_0^2(w)} K_0^2\left(\frac{w}{a}r\right) dS \right\}$$

$$A = \frac{w}{V J_1(u)} \sqrt{\frac{\sqrt{\mu_0/\varepsilon_0}}{2\pi a^2 n_1}}$$

where u and w are the normalized transversal propagation constants of the core and cladding, respectively, n_1 and n_2 are the core and cladding indices, respectively, and V is the normalized frequency.

The mode-coupling problem at random is shown as follows

$$\frac{dP_\nu}{dz} = -q_\nu P_\nu + \sum_{\mu=1}^{N} h_{\nu\mu}(P_\mu - P_\nu)$$

where P_ν is the power of the ν-mode, q_ν is the attenuation constant, $h_{\nu,\mu}$ is the power coupling coefficient between the ν- and μ-modes. N is the number of total modes. The fundamental modes, HE_{11}^x and HE_{11}^y, only propagate in a PMF and $P_x(z)$ and $P_y(z)$ are shown as follows

$$P_x(z) = [\cosh(hz) + \zeta \sinh(hz)] \cdot \exp(-hz)$$

$$P_y(z) = [\sinh(hz) + \zeta \cosh(hz)] \cdot \exp(-hz)$$

where $P_x(0) = 1$, $P_y(0) = \zeta$, $h_{\nu\mu} = h_{\mu\nu} = h$, and $q_\nu = 0$. Therefore the polarization crosstalk $\tilde{\eta}(z) \equiv P_y(z)/P_x(z)$ is shown by the power coupling coefficient h as follows

$$\tilde{\eta}(Z) = \tanh(h \cdot Z)$$

Moreover, η is defined as the polarization crosstalk in a 1 km length. The polarization crosstalk is conventionally expressed as follows

$$\eta(z) \equiv 10 \cdot \log \tilde{\eta}(z) = 10 \cdot \log \left[\frac{P_y(z)}{P_x(z)} \right]$$

In the case of a conventional PANDA fiber with $t/b = 0.6$ and $r/a = 1.3$, if the correlation length L_c is assumed to be 200 μm, evaluated from an investigation of the fiber loss mechanism, the polarization crosstalk can be experimentally given as -40 dB/km. Therefore, ξ is given by 4.0×10^{-6} since σ is about 10^{-8} m, corresponding to ten orders of the SiO_2 molecular distance, that is, the magnitude of the core diameter variation by fiber drawing.

The polarization crosstalk η versus the rms deviation σ with $\eta(0) = -\infty (\zeta = 0)$ is shown in Figure 8.6.5. The SAP parameter of t/b is taken to be 0.45 ($= 0.44 + 0.01$) because the deviation of SAP diameter, $(\Delta t)_{max}/b$, is known to be about 0.01 from experimental data. It is found that the PANDA fiber with $t/b = 0.45$ and $r/a = 2.83$ can make an improvement of 40–50 dB in the polarization crosstalk compared with that of the conventional PANDA fiber.

The polarization crosstalk η versus the rms deviation σ with $\eta(0) = -80$ dB, is shown in Figure 8.6.6. It is found that when σ is larger than 10^{-8} m, the deviation in

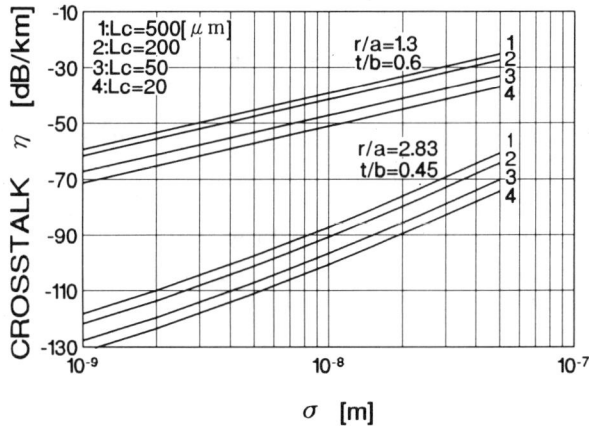

Figure 8.6.5
Polarization crosstalk η versus rms deviation σ with $\eta(0) = -\infty$

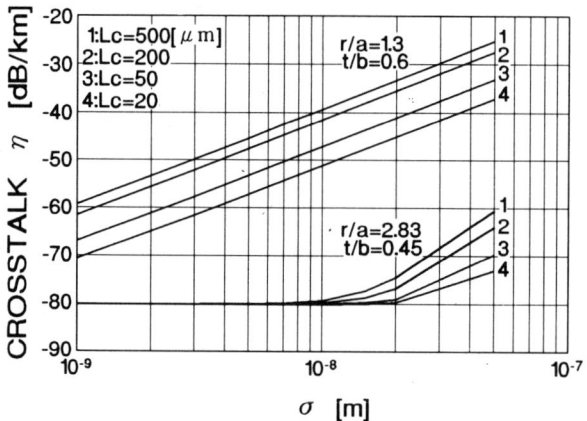

Figure 8.6.6
Polarization crosstalk η versus rms deviation σ with $\eta(0) = -80$(dB)

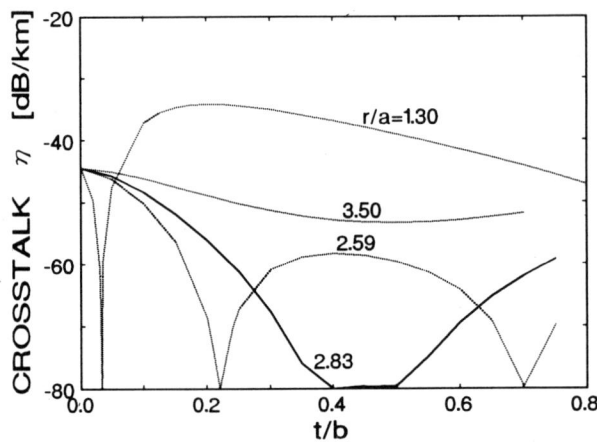

Figure 8.6.7
Polarization crosstalk η versus rms deviation σ with $\sigma = 1.0 \times 10^{-8}$ m and $L_c = 200$ μm

the fiber parameter causes mode-coupling and gives polarization crosstalk, but when σ is smaller than 10^{-8} m, incident conditions of linear-polarized give polarization crosstalk.

The polarization crosstalk η versus the normalized SAP diameter t/b, with $\sigma = 1.0 \times 10^{-8}$ m, $L_c = 200$ μm, and $\eta(0) = -80$ dB, is shown in Figure 8.6.7. It is found that the setting clearance of the SAPs of the PANDA fiber proposed here is not so strict and the fiber is easily fabricated.

4 Discussion

From Figure 8.6.4 the optimum design parameters for the PANDA fiber are given as the normalized SAP diameter $t/b = 0.44$ and the normalized half distance between SAPs,

Control of Fiber Characteristics

$r/a = 2.83$. This condition is found to be based on the fabrication conditions as follows. First, $t/b \leq 0.75$ under which PANDA fibers are easily drawn regardless of the B_2O_3 molar concentration. Secondly, t/b is required to give a large modal birefringence of more than 10^{-4}. Lastly, r/a is desired to be larger than 2 because of the escape from the perturbation of the electric field caused by the geometrical refractive index of the SAPs and from the strict setting clearance of SAPs in the vicinity of the core for fabrication.

It is clear in Figure 8.6.5 that there are three regions on r/a, where $t/b \leq 0.75$.

Region I : $r/a > 2.83$, where $\Gamma_1(z) \neq 0$.
Region II : $2.5 \leq r/a \leq 2.83$, where there are two (t/b) values satisfying $\Gamma_1(z) = 0$.
Region III : $0 \leq r/a < 2.5$, where there is only one (t/b) value satisfying $\Gamma_1(z) = 0$.

In Region I, the mode-coupling coefficient $\Gamma(z) \neq 0$ from $\Gamma_1(z) \neq 0$, that is, mode-coupling cannot be suppressed. In Region III, $\Gamma_1(z) = 0$ are satisfied by only one (t/b) value with varying r/a. But the (t/b) value is smaller than about 0.2 and the modal birefringence is smaller than 10^{-4}. Mode-coupling also cannot be suppressed. In Region II, $\Gamma_1(z) = 0$ is satisfied by two (t/b) values with varying r/a. It is found that the larger of these two values suppresses mode-coupling since it gives a modal birefringence of more than 10^{-4} (from Figure 8.6.8). Furthermore, when $t/b = 0.44$, the curve with $r/a = 2.83$ is tangent to the line $\Gamma(z) = 0$. This means that $\Gamma(z) \approx 0$ is accomplished easily even if t/b deviates slightly from 0.44. Therefore, $t/b = 0.44$ and $r/a = 2.83$ are gained as the optimum design parameters of the PANDA fiber.

The modal birefringence B versus the normalized SAP diameter t/b is shown in Figure 8.6.8. The modal birefringence of the PANDA fiber proposed here with $t/b = 0.44$ and $r/a = 2.83$ is 1.7×10^{-4}. This modal birefringence is found to be half of

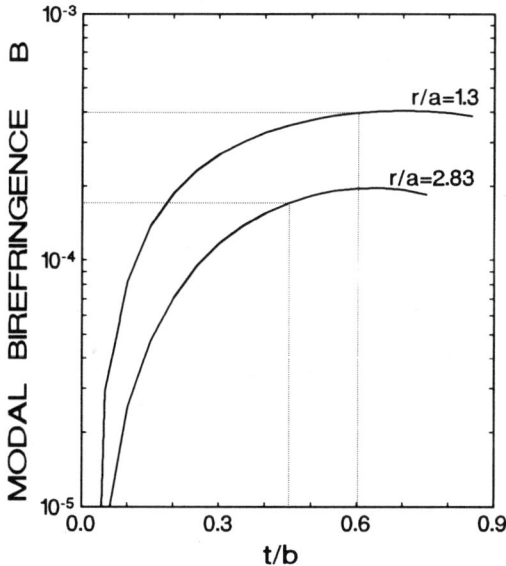

Figure 8.6.8
Modal birefringence B versus normalized SAP diameter t/b

the modal birefringence of the conventional PANDA fiber which has $t/b = 0.6$ and $r/a = 1.3$. It is also found that this modal birefringence is large enough to overcome the modal birefringence of 3.2×10^{-7} caused by a bending radius of 40 mm in the interferometer fiber gyroscope.

5 Conclusion

The off-diagonal terms of the permittivity tensor arising from the inner shear stress in the fiber cross-section of the PANDA fiber has been introduced. The degradation of polarization crosstalk by the minor component power in the PANDA fiber has been shown. As a result, the optimum design of the SAP parameters of the PANDA fiber, which is established when the magnitude of the mode-coupling relating to the structural composition becomes zero, has been obtained. The optimally designed PANDA fiber with $t/b = 0.44$ and $r/a = 2.83$ can bring about an improvement in polarization crosstalk of 40–50 dB although it has a birefringence of one-half or less compared with that of the conventional PANDA fiber.

References

[1] V. Ramaswamy, I. P. Kaminow, P. Kaiser and W.G. French, *Appl. Phys. Lett.*, **33** (9), 814–816 (Nov. 1978).
[2] H. Matsumura, T. Katsuyama and T. Suganuma, in *Proc. 6th ECOC*, York, U. K., 1980, pp. 49–52.
[3] Y. Sasaki, K. Okamoto, T. Hosaka and N. Shibata, in *Tech Dig. 5th OFC*, Pheonix, AZ, 1982, paper THCC6, pp. 54–56.
[4] R. A. Bergh, H. C. Lefevre and H. J. Shaw, *J. Lightwave Technol.*, **LT-2** (2), 91–107 (Apr. 1984).
[5] J. Noda, K. Okamoto and Y. Sasaki, *J. Lightwave Technol.*, **LT-4** (8), 1081–1099 (Aug. 1986).
[6] T. Hosaka, Y. Sasaki, K. Okamoto and J. Noda, *Trans. IECE Jpn*, **J67-C** (10), 741–748 (Oct. 1984).
[7] Y. Sasaki, T. Hosaka, M. Horiguchi and J. Noda, *J. Lightwave Technol.*, **LT-4** (8), 1097–1102 (Aug. 1986).
[8] K. Okamoto, Y. Sasaki and N. Shibata, *IEEE J. Quantum Electron.*, **QE-18** (11), 1890–1899 (Nov. 1982).
[9] C. Vassallo, *J. Lightwave Technol.*, **LT-5** (1), 24–28 (Jan. 1987).
[10] J.E. Sader and D. Wong, *IEEE J. Quantum Electron.*, **28** (6), 1533–1538 (June 1992).
[11] Y. Namihira, *J. Lightware Technol.*, **LT-3** (5), 1078–1083 (Oct. 1985).
[12] W. Primak and D. Post, *J. Appl. Phys.*, **30** (5), 779–788 (May 1959).
[13] D. Wong and P.L. Chu, *Int. J. Optoelectron.*, **6** (1), 65–82 (Jan. 1991).
[14] D. Marcuse, *Theory of Dielectric Optical Waveguides*, Academic Press, New York and London, 1994.

9
DIGITAL OPTICAL COMPUTING

9.1

Optical Parallel Computing Systems and Parallel Algorithms: Experimental Demonstration

Yoshiki Ichioka, Tsuyoshi Konishi and Jun Tanida

Abstract

In this section we describe an attractive goal of research on optical parallel computing: an optical parallel computing system and a native parallel processing technique. One aspect of the system, is a pure optical parallel array logic system (P-OPALS) which has been developed. To clarify the principle of P-OPALS, the basic concepts of optical digital computing and our approach to optical digital computing are briefly summarized. Then the results of an experimental demonstration of P-OPALS are described. Several fundamental operations are executed on the experimental P-OPALS, illustrating its potential capability.

1 Concept of Optical Digital Computing

Optical digital computing is a research area attempting the effective use of the physical characteristics of light for massively parallel information processing based on a digital computing scheme [1,2]. Among the various approaches to optical computing, a digital scheme has distinct advantages. Namely, fruitful resource and methodology accumulated in computer sciences such as logic design, system architecture, and programming theory, can be utilized in device development, system construction, and operation programming.

The major advantages of a digital computing system are flexibility in computational precision, resistance against noise, inherent correctness of computation, and easy fabrication for implementing devices [3]. In addition, a unified methodology can be used to construct a digital computing system: combining several kinds of logic gates to make logic circuits, connecting the logic circuits to build functional modules, and organizing the modules to form a system. This methodology is effective even for the optical computing experimental, though specific consideration is required to utilize the characteristics of the optics. This is one of the important advantages of optical digital computing. As an extension to the research, an optical digital computing system is considered to be attractive and reasonable goal of optical computing.

An optical digital computing system is expected to be useful in various fields of massively parallel information processing. Image processing, large-scale scientific computation, database management, and simulation of highly parallel processes are promising applications of an optical digital computing system. Although these applications are also the targets of highly parallel electronic computers, optical schemes have advantages in the degree of parallelism and system complexity.

2 Optical Array Logic and OPALS

The authors have developed a generalized computing paradigm for optical digital computing called optical array logic (OAL) [4]. Figure 9.1.1 shows a schematic diagram of OAL. Two binary images are processed by logical neighborhood operation and the result is provided as one binary image. In logical neighborhood operation, the value of an individual pixel, $c[i, j]$, is determined by the a combination of the values of the pixels at the corresponding location, $a[i, j]$ and $b[i, j]$, and those of the neighboring pixels, $a[i + m, j + n]$ and $b[i + m, j + n]$ $(m, n = -L, \ldots, L)$ on the two input images. By logical expression, the neighborhood operation in OAL is described as follows:

$$c[i, j] = \sum_{k=1}^{K} \prod_{m=-L}^{L} \prod_{n=-L}^{L} f_{m,n;k}(a[i + m, j + n], b[i + m, j + n]) \tag{1}$$

where L and K indicate the size of the neighborhood area and the number of product terms, respectively; $f_{m,n;k}(a, b)$ means any one of 16 two-variable binary logical functions assigned to the individual set (m, n, k). For all combinations of (i, j), the same operation is executed in parallel.

In the processing procedure of OAL, the input image is converted into a coded image composed of spatial patterns shown in the coding rule. The coded image is correlated with the operation kernel consisting of several delta functions located at the grid points, and the result is sampled for every other pixel along the vertical and horizontal directions. The above procedure is repeated for different operation kernels and the results are collected with inverted-OR operation to produce the output image. Since the operation in OAL can be specified by the operation kernels in the correlation, a combination of the operation kernels is used to describe the processing executed in OAL. For programming utility in OAL, a specialized programming language has been developed [4].

As a conceptual architecture of an optical digital computing system based on OAL, we have proposed the optical parallel array logic system (OPALS) [5]. Figure 9.1.2 shows a schematic diagram of OPALS. It consists of the OAL processor with a feedback system for iterative processing. Processing and data transfer are achieved in parallel. Several versions of OPALS have been proposed with various techniques [6,7]. Although the optoelectronic hybrid version, or the H-OPALS [7], is considered to be the practical

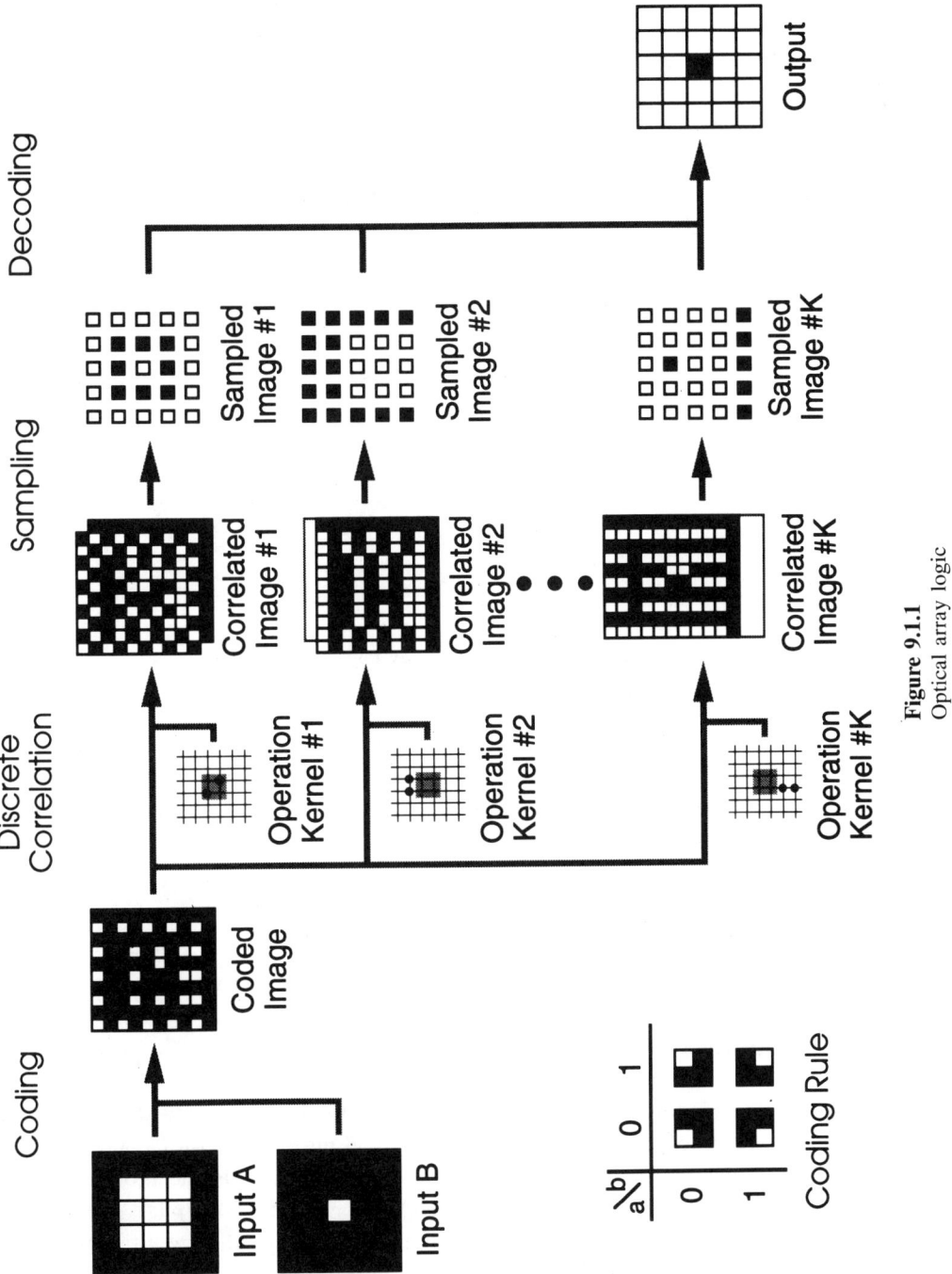

Figure 9.1.1 Optical array logic

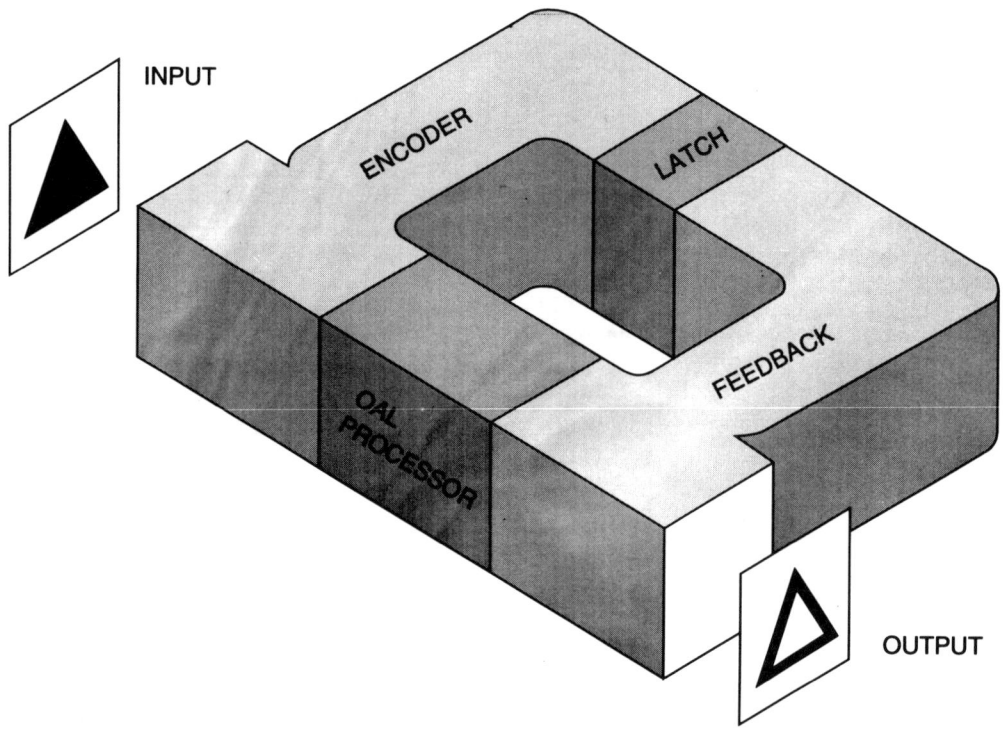

Figure 9.1.2
OPALS (optical parallel array logic system)

one, the potential capability of optical computing cannot be fully utilized with the current immature technologies in optoelectronics. Therefore, the authors have attempted to develop a pure optical version of OPALS, called P-OPALS, to investigate the capability of optical digital computing.

3 Pure Optical Parallel Array Logic System

3.1 System overview

P-OPALS is an instance of OPALS in which the optical procedural techniques are effectively used to achieve extensive parallelism and flexible connectivity. In P-OPALS, a large amount of data is processed with simple implementation. Of course, a combination of optics and electronics provides more flexibility and more powerful capabilities than the pure optical system. However, we consider that simplicity is the most important factor in system construction to increase the performance, reliability, and productivity of the system. If desired, the simplified system can be modified with the help of electronics.

P-OPALS executes equivalent operations of the original OPALS with complete parallel data flow. All information located on image data is processed and transferred in parallel with optical techniques. In P-OPALS, various techniques are adopted for effective implementation, *e.g.* single-input optical array logic (S-OAL), birefringent encoding/decoding [8], latch-and-amplification, and optical discrete correlation. As a result, the information capacity of its optical system can be utilized effectively, which enables us to achieve massively parallel processing.

3.2   Single-input optical array logic

S-OAL is a simplified version of OAL. Figure 9.1.3 shows a schematic diagram of S-OAL. In S-OAL, a single binary image is processed with logical neighborhood operations according to the following operation:

$$c[i, j] = \sum_{k=1}^{K} \prod_{m=-L}^{L} \prod_{n=-L}^{L} g_{m,n;k}(a[i+m, j+n]) \tag{2}$$

where L and K indicate the size of the neighborhood area and the number of product terms, respectively; $g_{m,n;k}(a)$ means any one of four one-variable binary logical functions assigned to the individual set (m, n, k). The processing procedure is the same as that of OAL except the number of inputs and a difference in the coding rule.

In comparison with OAL, the number of optical components can be reduced to implement S-OAL. In terms of processing efficiency, S-OAL requires more product term operations than OAL, where the number of product terms, K, corresponds to the required number of iterations. However, the number is estimated to be twice the OAL for the worst case. Usually, there is no noticeable difference in processing efficiency between OAL and S-OAL. In addition, S-OAL provides the possibility to increase the information capacity of the system. Since the coding rule of S-OAL does not care about data continuity in the vertical direction, much more data can be mapped along the vertical direction than the horizontal direction.

3.3   Birefringent encoding/decoding

Birefringent encoding/decoding [8] is an optical technique for image encoding and decoding in S-OAL. Figure 9.1.4 explains the principle of birefringent encoding on one pixel. Using this technique, the birefringent phenomenon appearing in a uniaxial crystal is utilized. The horizontally polarized probe beam is converted into either a horizontally or vertically polarized beam according to the control signal applied to the spatial light modulator. The converted beam enters the birefringent crystal in which both states of the polarization beams propagate in different directions. Behind the crystal, the probe beam reaches different positions according to the control signal. Consequently, either of the code patterns for 0 and 1 in S-OAL is generated optically. Note that the technique can be applied to multiple data arranged on an image plane and executed in parallel with an optically addressable spatial light modulator (OASLM). The same procedure can be used for the decoding process.

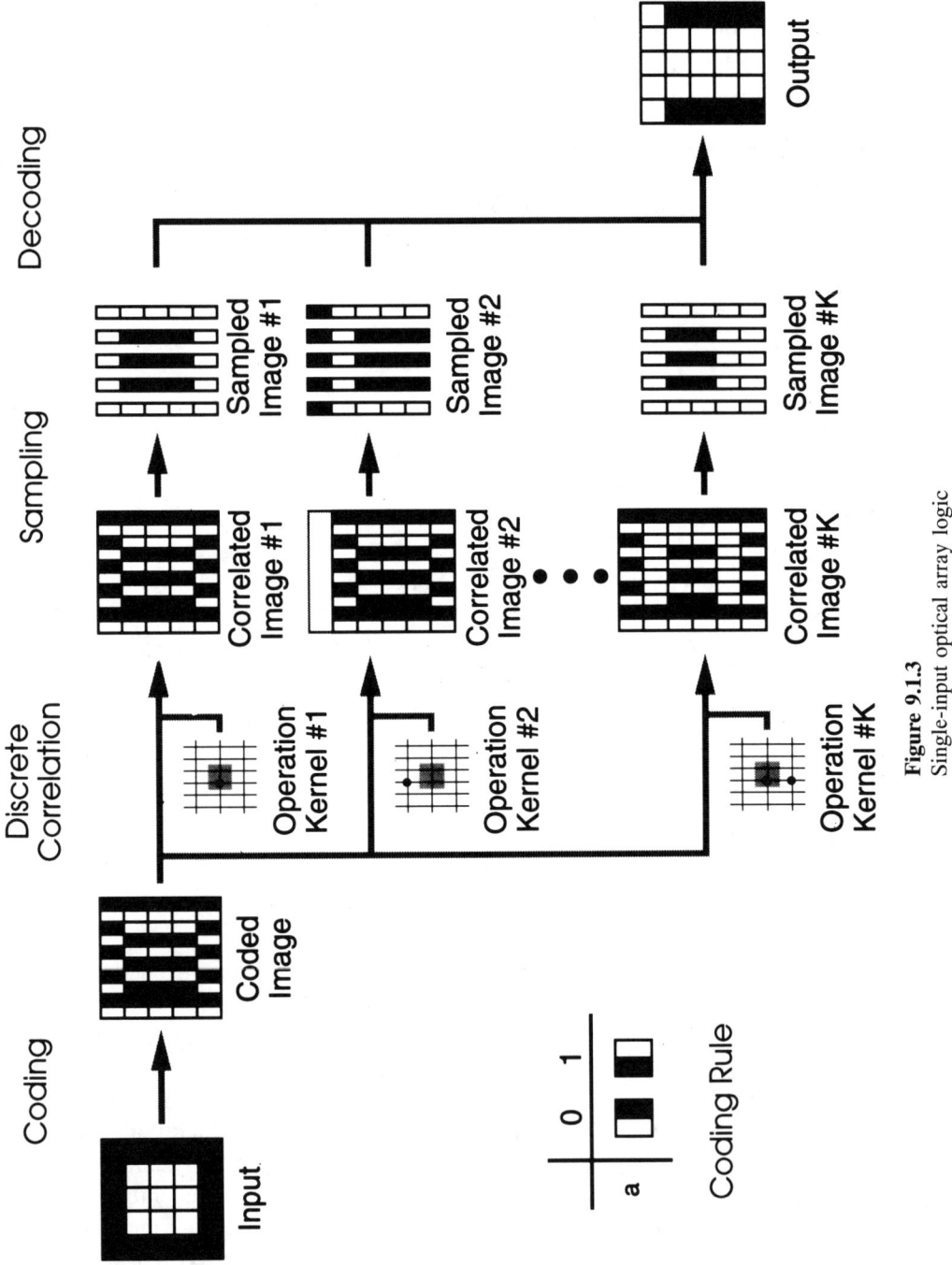

Figure 9.1.3
Single-input optical array logic

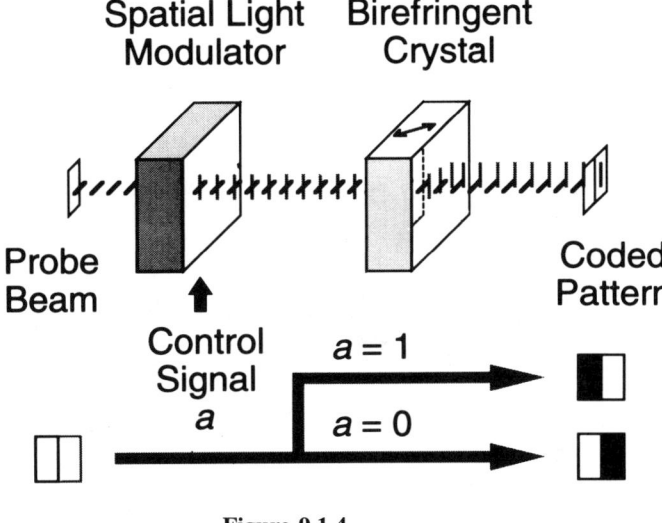

Figure 9.1.4
Birefringent encoding

3.4 Latch and amplification

The latch-and-amplification technique is useful for synchronous and cascade operation of the system. To obtain system stability, synchronous operation with a clock signal is effective, and to compensate for power loss in the cascade operation, signal amplification is required. Thus after coding and correlation, the processed images are latched onto the OASLMs and read out by a bright light. As an additional advantage of the technique, signal saturation caused by the nonlinear characteristics of the OASLM regulates the signal level so that it avoids noise accumulation.

3.5 Optical discrete correlation

Optical discrete correlation is one of the fundamental processes used in digital optical computing. For P-OPALS, the capability of fast reconfiguration of the kernel, large power throughput, and large information capacity are important requirements for the correlator. Optical shadow casting, multiple imaging, split-and-shift imaging, and holographic optical elements can be used for the purpose.

4 Experimental Results

An experimental version of P-OPALS was constructed to clarify the practical issues in system construction and to explore the potential capability of the system. Figure 9.1.5 shows the optical setup of the experimental P-OPALS. In the system, two ferroelectronic liquid crystal light valves (LAPS-SLM by Seiko Instrum. Inc.) [9] are used as the OASLMs. As the birefringent crystal, a 1 mm thick calcite plate is used, which provides about 0.1 mm lateral shift between different polarization states of the incident beams.

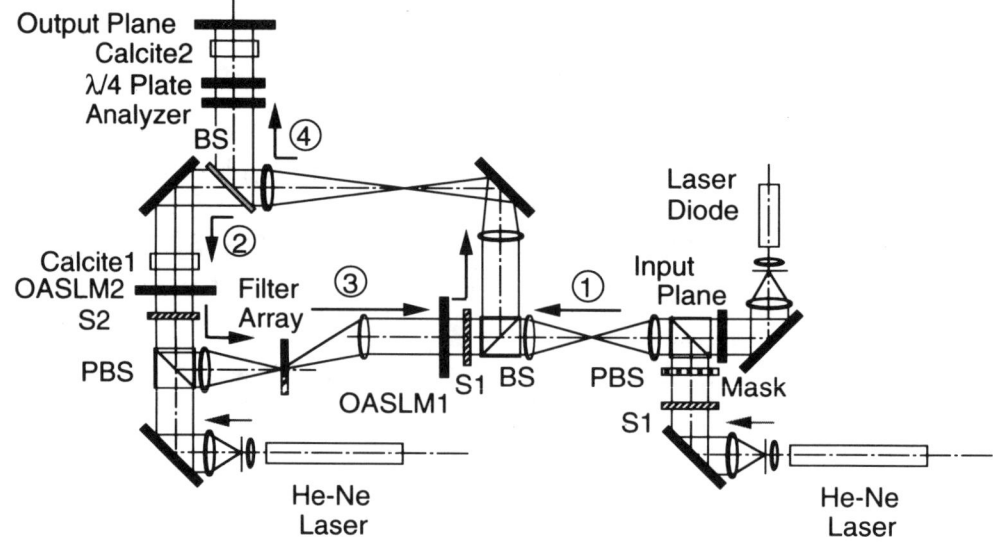

Figure 9.1.5
Optical setup of experimental P-OPALS

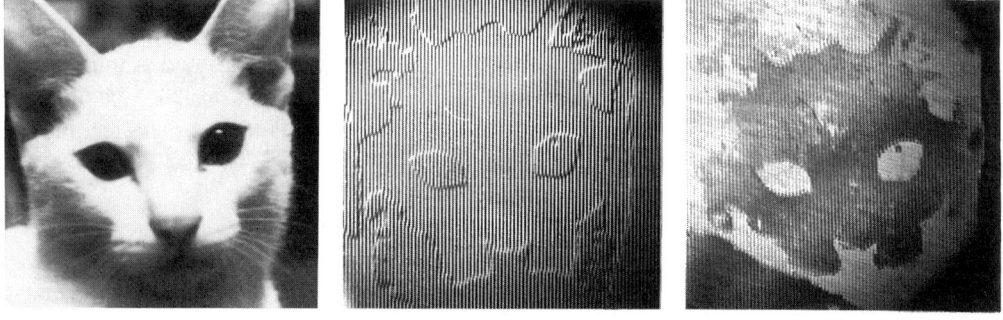

Figure 9.1.6
Experimental results on P-OPALS: (a) input image, (b) encoded image, and (c) output image of inversion

The optical system consists of cascaded 4-f systems. Discrete correlation is implemented by coherent spatial filtering with a holographic filter recorded on photopolymer material. The effective processing area is about 20×20 mm^2. Magnification of all imaging systems in the system is unity.

The experimental system is operated in four phases: (1) data latch, (2) encoding, (3) command execution, and (4) decoding. The data flow in the system is indicated by the circled numbers in Figure 9.1.5. The optical path is controlled by the laser diode configuration and the shutters S1 and S2. In the encoding phase, the input image is read out with a vertical strip pattern generated by a Ronchi grating with 5 lp/mm frequency.

Figure 9.1.6 shows the experimental results of the system. A gray image, Figure 9.1.6(a), is set at the input plane and converted into the coded image,

Figure 9.1.6(b). The output image of the logical inversion is shown in Figure 9.1.6(c). As seen from the figure, the desired result can be obtained. In the processing, 220 ms is required to complete the operation, which is mainly determined by the response of the OASLMs.

The total number of data points in the experimental P-OPALS is estimated by the resolution of the internal optical system. With the USAF test chart, we observed the value as 17.95 lp/mm. Since the horizontal data density is determined by the spatial frequency of the Ronchi grating for encoding, the total number of data points is estimated as 100×360. Although this value is not large compared with conventional optical information processing systems, it is a remarkable one in the demonstration of optical digital computing systems.

5 Conclusion

In this section we have reported an experimental demonstration of a digital optical computing system called P-OPALS. P-OPALS has the capability to execute an arbitrary logical neighborhood operation based on single-input optical array logic. The constructed system can handle 100×360 data points at 4.5 Hz per frame in parallel. To improve the performance, OASLMs with fast response, high contrast ratio, and good uniformity are strongly desired. Since P-OPALS executes processing described by OAL, various applications can be implemented on the system. For example, image processing [10], inference processing [11], and database management [12] have been coded in OAL programs. Therefore, it can be expected that various types of parallel processing is achievable optically on OPALS.

References

[1] D.G. Feitelson, *Optical Computing*, MIT Press, Cambridge, MA, 1988.
[2] A.D. McArley, *Optical Computer Architectures*, John Wiley & Sons, 1991.
[3] A.A. Sawchuk and T.C. Strand, *Proc. IEEE*, **72** (7), 758 (1984).
[4] J. Tanida and Y. Ichioka, *Int. J. Opt. Comput.*, **1** (2), 113 (1990).
[5] J. Tanida and Y. Ichioka, *Appl. Opt.*, **25** (10), 1565 (1986).
[6] J. Tanida and Y. Ichioka, *Appl. Opt.*, **26** (18), 3954 (1987).
[7] J. Tanida, D. Miyazaki and Y. Ichioka, *Proc. SPIE.*, **1806**, 568 (1992).
[8] J. Tanida, J. Nakagawa and Y. Ichioka, *Appl. Opt.*, **27** (18), 3819 (1988).
[9] S. Yamamoto, R. Sekura, J. Yamanaka, T. Ebihara, N. Kato and H. Hoshi, *Proc. SPIE*, **1211**, 273 (1990).
[10] S. Kakizaki, J. Tanida and Y. Ichioka, *Appl. Opt.*, **31** (8), 1093 (1992).
[11] M. Iwata, J. Tanida and Y. Ichioka, *Appl. Opt.*, **31** (26), 5604 (1992).
[12] M. Iwata, J. Tanida and Y. Ichioka, *Appl. Opt.*, **32** (11), 1987 (1993).

9.2

Parallel Optoelectronic Computing System

Masatoshi Ishikawa

Abstract

A massively parallel processing system which employs a reconfigurable diffractive element to perform a shift-invariant optical interconnection among electronic processing elements (PEs) is described. Each PE in the system is so compact that more than 4000 PEs can be integrated into one chip for direct coupling with array-type optical devices in parallel. The optical interconnection is constructed by using a phase modulation type spatial light modulator to be reconfigurable to realize general purpose parallel processing. In this section the design concept of the system and the PE, an experimental scaled-up model of a PE array, and the optical interconnection are shown.

1 Introduction

There is growing interest in integrated two-dimensional optical parallel processing systems using two-dimensional optical devices such as laser diode arrays and photo detector arrays. Processing architectures for utilizing such devices are intrinsically based on massively parallel processing with parallel optical input and output. It is expected that high performance parallel computing systems can be realized by using such two-dimensional optical devices and a fine grain electronic processing element (PE) array and can be applied to high-speed visual image processing for robot vision, automated visual inspection, pattern recognition, and so on.

Most conventional image processing systems consist of a CCD camera and an image processor connected through the video signal which is generated by scanning circuits

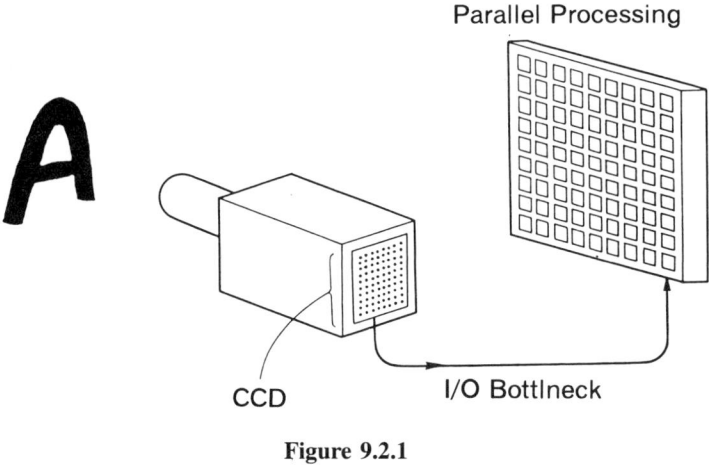

Figure 9.2.1
I/O bottleneck

in the CCD camera. Although this type of two-dimensional data transmission is useful for simple interconnection between the camera and an image processor, there is an essential limitation of processing speed due to the bottleneck of the video signal, as shown in Figure 9.2.1. In other words, the processing speed of the conventional image processing is limited to the 33.3 ms of the frame rate as long as the video signal is used for the interconnection, even if a high-speed image processor is used. The bottleneck of the interconnection is called the 'I/O bottleneck'.

To eliminate the I/O bottleneck, a perfectly parallel processing architecture in which two-dimensional pattern information is directly transferred to a two-dimensional parallel PE array is proposed [1,2]. Conceptual diagrams of the architecture are shown in Figure 9.2.2. An architecture of a hierarchical processing system with optical interconnections is shown in Figure 9.2.2(a). Each layer in this architecture consists of the same module with a two-dimensional PE array and optical I/O devices and is optically interconnected with neighboring layers. Since the processing of the each layer is independent of the processing of the other layers, different programs can be carried out at each layer. Figure 9.2.2(b) shows a feedback type of architecture. This architecture can implement the hierarchical architecture shown in Figure 9.2.2(a) by using a time-sharing algorithm which carries out the processing of each layer in a time sequential order to be equivalent to Figure 9.2.2(a).

However, these types of processing architectures require a one-to-one full connection between optical I/O devices and the PE array. Macro scale electronic wiring technology cannot implement such high-density interconnections. Only micro scale wiring technology such as LSI technology and flip chip bonding technology can implement such interconnections. However, the PEs used in conventional parallel processing systems are too large to integrate into one chip. In addition, fine grain PEs and nonflexible interconnections between PEs are used to limit the generality of the processing. Therefore, in such architectures, an essential problem is how to integrate PEs and optical input/output with general processing.

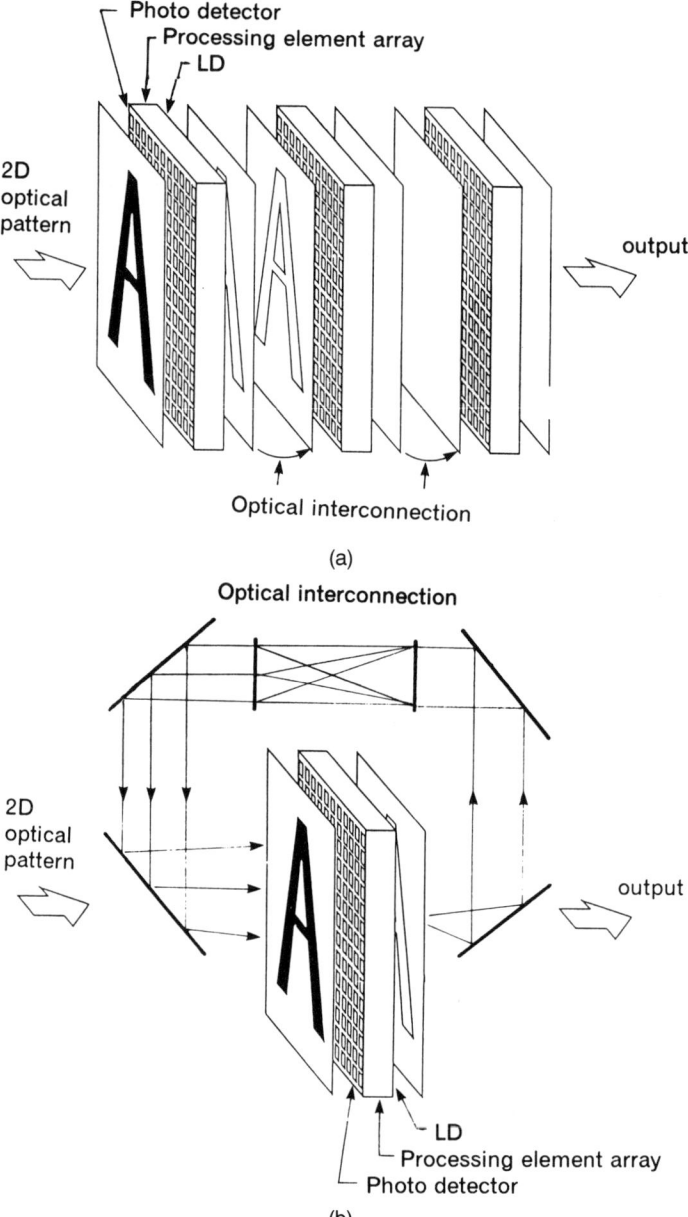

Figure 9.2.2
Hierarchical optoelectronic processing system: (a) hierarchical type; (b) feedback type

In this section the design concept and processing architecture of the system based on the architecture, an experimental system as a scale-up model, a reconfigurable optical interconnection using an optimized computer-generated hologram and a phase modulation type spatial light modulator, and basic examples of visual image processing are described.

2 System Architecture

From the viewpoint of integrated circuits, or smart sensors, much research into the integration of both detectors and processing circuits has been done [3,4]. Mead [3] developed a silicon retina which realizes the function of early vision integrating photo transistors, a resister network, and active circuits. The circuit is designed as special purpose analog circuits and does not have general purpose processing capabilities.

MIT has a vision chip project focused on analog integrated circuits with photo detectors [4]. In the project, circuits to detect the position, moments, and other scalar values for two-dimensional patterns are developed and integrated into one chip or a combination of several chips.

However, the main circuits of these vision chips are fixed and have special purposes without the generality of processing. Utsugi and Ishikawa [5] proposed a learning method of resistor networks for the adaptation of a coordination system between the vision system and the real world.

On the other hand, a combination of optical computing and the learning capabilities of neurocomputing gives an effective tool for visual perception. Ishikawa et al. [6,7] developed an optical associative memory system, called the optical association, with learning capabilities using microchannel spatial light modulators (MSLMs) and combined with conventional optical processing such as optical Fourier transform [8]. The systems can carry out direct processing for two-dimensional pattern information using optical parallel computing and can realize high performance in special purpose pattern processing by using learning capabilities. However, they are adaptable only for the input pattern and are not programmable for general processing.

Ishikawa et al. have developed a general purpose PE to solve the problem [1,2]. The PE has been designed to be so compact that more than $64 \times 64 = 4096$ PEs could be integrated into one chip by using present VLSI technology. An experimental scale-up model with $64 \times 64 = 4096$ PEs has also been developed. The system is called SPE-4k (sensory processing elements — 4k).

In the following sections we describe the design concept and configuration of the PE and the SPE-4k.

2.1 Processing element (PE)

To implement the optoelectronic parallel processing system shown in Figure 9.2.2, Ishikawa et al. designed a compact and general purpose PE. The block diagram of the PE is shown in Figure 9.2.3.

Each PE has three 8 bit registers (A: accumulator, T: template, W: weight), one arithmetic logical unit (ALU, 1 bit), and one 4 bit multiplier, as shown in Figure 9.2.3. The processing architecture of the ALU is based on bit serial processing which is slow in comparison with bit parallel processing, but does not require so many gates. It also has major advantages for integration and variable bit length processing. The functions of the ALU include AND, OR, exclusive OR, addition, subtraction, multiply (4 bits × 4 bits), and combinations of these basic functions such as weighted sum for the calculation of correlations.

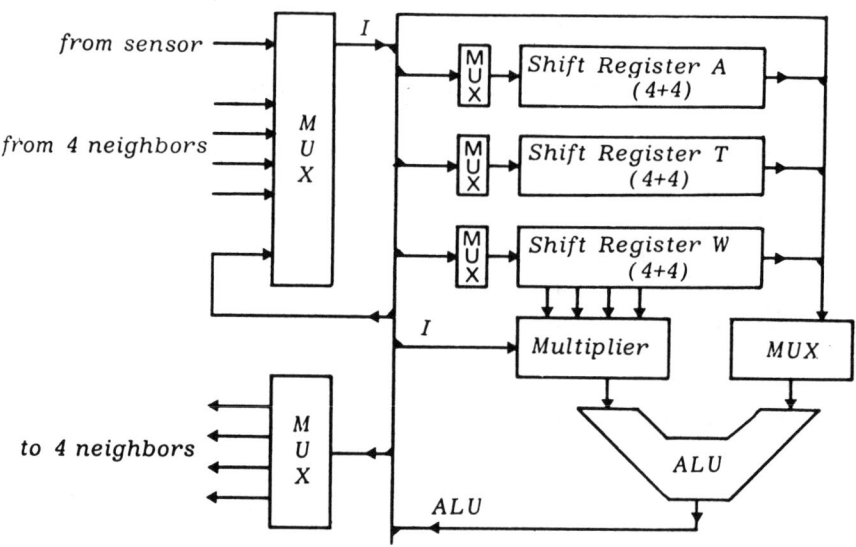

Figure 9.2.3
Block diagram of a processing element

The instruction set of the PE is rearranged from 32 bit direct horizontal-type microcode to four kinds of 10 bit microinstruction by classifying the instructions. The three static instructions are held in instruction registers until the next change of instructions, because these types of instructions are almost always fixed for the period of a set of processing. By this method, not increasing the number of instructions in practice, equivalent control with horizontal microprogramming is realized by small gates of the PE and small bit length of the instruction.

Most important for the specification of the PE is the number of gates. In the result, the PE is implemented by using 337 gates. The processing cycle time is typically 44 ns, with a maximum of 87 ns.

2.2 SPE-4k

An experimental optoelectronic processing system SPE-4k which has matrix positioned $64 \times 64 = 4096$ (4k) PEs was constructed using LSI gate array in which eight PEs ($8 \times 337 = 2696$ gates) and a common part (274 gates) are implemented. The system is a scale-up model of one-chip optoelectronic processing layer shown in Figure 9.2.2. The structure of the SPE-4k is shown in Figure 9.2.4 and an overview of SPE-4k is shown in Figure 9.2.5. The LSIs are arranged between the front LED array and the back PTR array.

The SPE-4k system uses I/O processor type of control from a personal computer and the development of software may be carried out on the computer (Figure 9.2.4). However, the processing speed is limited by the speed of the parallel I/O of the computer. The cycle time of the system is 10 μs at present, but the SPE itself can work in a 100 ns cycle time. Considering the 8 bit integer addition as a basic operation

Parallel Optoelectronic Computing System

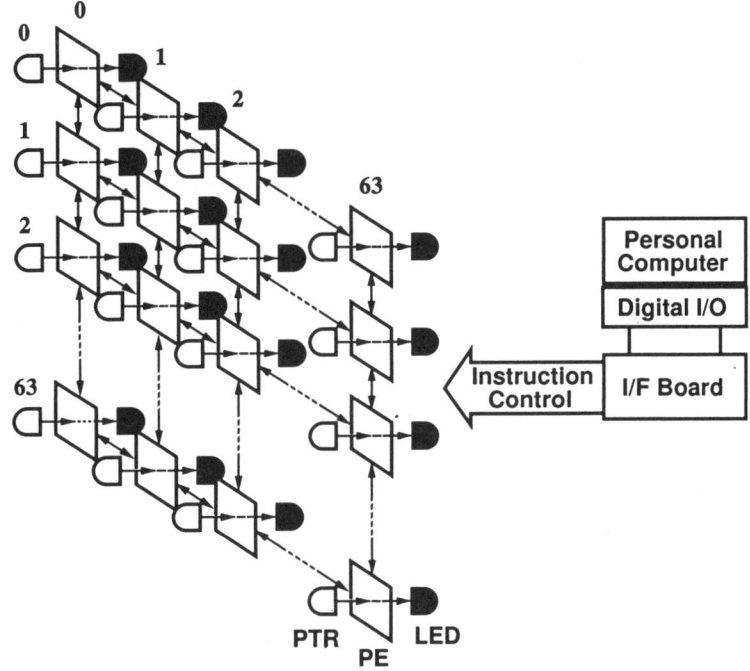

Figure 9.2.4
Block diagram of SPE-4k

Figure 9.2.5
SPE-4k

Table 9.2.1
Processing time of applications

Processing	Number of steps*	Processing time	
		Present	Maximum speed
Edge detection	33	330 μs	3.3 μs
Skeletonization	149	15 ms	150 μs
Detection of moving object	7	70 μs	0.7 μs
Trace	4	40 μs	0.4 μs
Poisson's equation	125	250 ms	2.5 ms

*Number of microinstructions.

of processing for the evaluation of the speed of SPE-4k, total 32MOPS (mega operations per second) by the present system and 3.2GOPS (giga operations per second) at the maximum speed of the system are obtained. The processing speed is much higher than general image processors or parallel processing systems.

2.3 Applications

To evaluate the processing performance of the SPE-4k, the following applications have been carried out on the system. The processing times of the following applications are shown in Table 9.2.1. The maximum speed column shows the performance of the SPE-4k at maximum speed (cycle time: 100 ns). As shown in Table 9.2.1, high-speed processing, of the order of microseconds, is obtained.

Two and four neighborhood edge detections to the 1 bit input patterns are implemented as logical operations. Simple skeletonization for 1 bit input data using four or eight neighbors is implemented by pattern matching with input data and judgment of edge using prepared two and four neighborhood patterns of edge for four neighbors and eight neighbors, respectively. Rotated patterns are also matched with the templates. Ten times iteration of this operation is made on the SPE-4k.

The edge detection of moving objects is implemented by using the time derivative of 1 bit input data. The trace of moving objects using logical OR operation between memory data and temporal input data is also implemented. Sampling time of detection, which should be determined by the velocity of the objects, is available as 0.7 μs and 0.4 μs at the maximum speed of the SPE-4k.

As an example a solution of the simple partial differential equation, the two-dimensional Poisson's equation is solved in 8 bit fixed point operation for 1 bit input patterns. Since the solution is obtained as a converged pattern of a diffusion equation associated with Poisson's equation, the solution is obtained after 200 iterations of a converging operation for a difference equation derived from the diffusion equation.

3 Reconfigurable Optical Interconnection

The PE has electronic interconnects with four neighbors, but other interconnects require iterative operations which take long processing times when only electronic interconnects

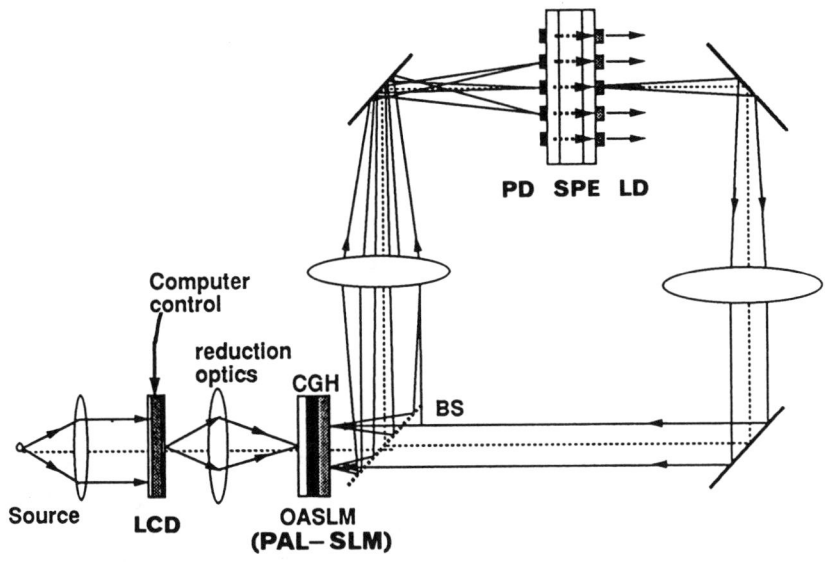

Figure 9.2.6
Parallel processing system with reconfigurable optical interconnects

are used. To directly connect between arbitrary PEs, Kirk *et al.* have designed a reconfigurable optical interconnection [9] and showed the basic experimental results [10] using a phase modulation type spatial light modulator, PAL-SLM (parallel aligned spatial light modulator) [11].

The block diagram of the system is shown in Figure 9.2.6. The system is a typical example of the feedback type architecture shown in Figure 9.2.2(b). The interconnection subsystem is based on a computer-generated hologram (CGH) optimized by using a simulated annealing algorithm. Therefore it realizes a shift invariant interconnection limited by the specification of the optical devices. The reconfigurability of the interconnection is most important for realizing general processing. It is implemented by rewriting the CGH on the PAL-SLM by using a liquid crystal display controlled by a personal computer. Consequently, the practical functions of the interconnection are obtained and demonstrated.

Recently, Ishikawa *et al.* have developed a whole system and some algorithms for parallel processing, including use of the reconfigurable interconnection, and demonstrated the system behavior [2]. In the system a surface-emitting laser diode array [12] is used for the interconnection and the operation of matrix–vector product is demonstrated.

4 Conclusion

A massively parallel processing system using an architecture for optoelectronic computing and reconfigurable diffractive interconnects is proposed and implemented. The system has general purpose computing capabilities using compact programmable processing elements and the reconfigurable interconnection. The design concept of the system can lead to an integrated optoelectronic parallel computing system.

References

[1] M. Ishikawa, A. Morita and N. Takayanagi, *Optical Computing Technical Digest 1993*, **7**, Optical Society of America, Washington, D.C., 1993, pp. 272-275.
[2] M. Ishikawa, *Proc. Optical Computing '94, Inst. Phys. Conf. Ser.* No. 139: Part I, 41-46 (1995).
[3] C. Mead, *Analog VLSI and Neural Systems*, Addison-Wesley, 1989.
[4] J.L. Wyatt, D.L. Standley and W. Yang, *1991 IEEE Int. Conf. on Robotics and Automation*, 1991, pp. 1130-1135.
[5] A. Utsugi and M. Ishikawa, *Neural Networks*, **4**, 81-87 (1991).
[6] M. Ishikawa, N. Mukohzaka, H. Toyoda and Y. Suzuki, *Appl. Opt.*, **28**, 291-301 (1989).
[7] M. Ishikawa, N. Mukohzaka, H. Toyoda and Y. Suzuki, *Appl. Opt.*, **29**, 289-295 (1990).
[8] H. Toyoda, N. Mukohzaka, Y. Suzuki and M. Ishikawa, *Appl. Opt.*, **32**, 1354-1358 (1993).
[9] A. Kirk, T. Tabata, M. Ishikawa and H. Toyoda, *Opt. Commun.*, **105**, 302-308 (1994).
[10] A. Kirk, T. Tabata and M. Ishikawa, *Appl. Opt.*, **33**, 1629-1639 (1994).
[11] N. Yoshida, N. Mukohzaka, T. Hori, H. Toyoda and T. Hara, *Spatial Light Modulator and Applications Technical Digest 1993*, **6**, Optical Society of America, Washington, D.C., 1993, pp. 97-99.
[12] M. Kajita, T. Numai, K. Kurihara, T. Yoshikawa, H. Saito, Y. Sugimoto, M. Sugimoto, H. Kasaka, I. Ogura and K. Kasahara, *Jpn. J. Appl. Phys.*, **33**, 859-863 (1994).

9.3

Temporal Coding Logic Array and its Application to Hybrid Computing

Toyohiko Yatagai

Abstract

A new type of a digital parallel logic gate array technique is proposed, based on a temporal coding method. A space variant logic gate array is able to be implemented by this technique. This technique enables us to make edge detection of binary input patterns and feature extraction. Next an optical computing system combined with the temporal coding system and a neural computing system is proposed. The preprocessing is done by the digital system and then pattern classification is done by a three layer neural network system based on the back-propagation learning algorithm. Some experimental results are described.

1 Introduction

Some effort to develop optical computing systems has been made in this decade. Optical digital, optical analog, and neural computing architectures have been discussed. These architectures have their own capabilities and possibilities in information processing. For example, the optical digital architecture is suitable for parallel logical and image processing [1,2], whereas optical neural computing is suitable for pattern classification and recognition [3,4]. Here an optical computing system combined with a digital system and a neural computing system is proposed. In the digital parallel processing system based on a temporal coding method, the edge detection of binary input patterns and feature extraction is made, and then pattern classification is done by a three-layer neural network system based on the back-propagation learning algorithm.

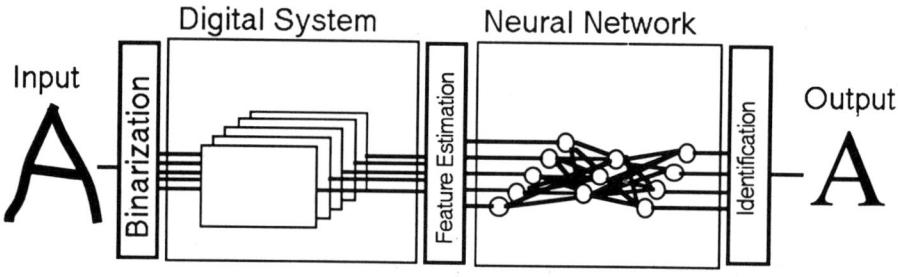

Figure 9.3.1
Concept of a hybrid optical computing system

2   General Aspects of Hybrid Computing

In general, optical neural systems are known to have adaptive processing abilities for large amounts of input data. But many practical optical neural systems based on the matrix-vector multiplication architecture are limited by the scale of parallelism, because of the spatial resolution of optical systems and spatial light modulators. In addition, when the bipolar synaptic weights are represented by optical intensity, they are divided into positive and negative unipolar weights. This means that twice the number of synaptic weights is needed. To solve this problem, as an alternative method we introduce optical feature extraction techniques. The feature extraction can map the input data into a new feature space. The data in the feature space are subjected to input data for an optical neural system. If we can design a suitable mapping system, the ability and the function can be increased and the size of the optical neural system can be reduced.

Figure 9.3.1 shows an example of a hybrid computing system for recognizing characters. In this system inputed characters are processed by an optical digital system as a preprocessor, and then the processed results are pipe-lined to an optical neural system for pattern classification. In the preprocessing part, the feature extraction of the input data and the data mapping into feature space are done by optical digital computing. The temporal coding logic array technique was developed for this purpose.

3   Temporal Coding Space-variant Parallel Logic Operation

A new method for optical space-variant logic operation is developed for the preprocessor of a hybrid optical computing system. The proposed method, which is called the temporal method, is based on the temporal coding and temporal gating of encoded input patterns, corresponding to spatial coding [5–7] and an operation mask, respectively. A fundamental system for verifying our method of the space-variant logic operation has been constructed. We also discuss the expected capabilities in a larger system of the space-variant parallel logic operation (SVPLO).

The principle of the temporal method for the SVPLO is based on the temporal coding of two input patterns and a temporal gate for performing the different logic operational outputs, as shown in Figure 9.3.2. Consider a simple example of performing the SVPLO

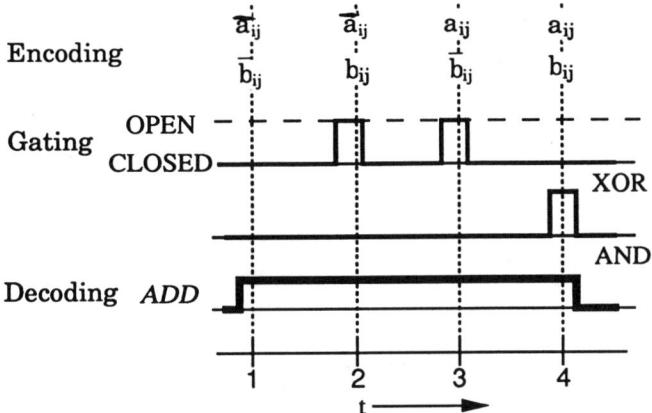

Figure 9.3.2
Example of space-variant logic operations based on the temporal method. All the logic operations are performed in four sequential steps. The temporal coding patterns $C(t)$ are obtained from two binary input patterns. The operational output O is obtained by adding four transient outputs $T(t)$ applied at the temporal gates $F(t)$ to the temporal coding patterns $C(t)$

for the input patterns $A = \{a_{ij},\ i, j = 1, 2, \ldots, N\}$ and $B = \{b_{ij},\ i, j = 1, 2, \ldots, N\}$ consisting of $N \times N$ square elements. The temporal coding represents temporally a pair of elements a_{ij} and b_{ij}. In the case of the logic operation of two binary variables, four pairs of elements are represented by four sequential steps, since the element has logical states 0 and 1. The logical values are optically represented by black and white, dark and bright, or opaque and transparent states. Four temporal coding patterns $C = \{C(t),\ t = 1, 2, 3, 4\}$ are obtained by the input pattern A or its inverted pattern $\bar{A}$, and the input pattern B or its inverted pattern $\bar{B}$. The temporal coding sequence is shown in Figure 9.3.3. The (i, j) the element of the tth temporal coding patterns $C(t)$, $c(t)_{ij}$, is bright at only one time in the four sequential steps. For example, in the case of the elements $a_{ij} = 0$ and $b_{ij} = 1$, the coded element $c(t)_{ij}$ is bright at the second step ($t = 2$) and dark at the other steps. The temporal coding is performed by superimposition of the input a_{ij} or its inverted input $\bar{a}_{ij}$ with the input b_{ij} or its inverted input $\bar{b}_{ij}$. The (i, j)th elements of the temporal coding, $c(t)_{ij}$, are obtained as

$$c(1)_{ij} = \bar{a}_{ij} \cdot \bar{b}_{ij} \tag{1}$$

$$c(2)_{ij} = a_{ij} \cdot \bar{b}_{ij} \tag{2}$$

$$c(3)_{ij} = \bar{a}_{ij} \cdot b_{ij} \tag{3}$$

$$c(4)_{ij} = a_{ij} \cdot b_{ij} \tag{4}$$

where $\cdot$ denotes superimposition corresponding to the logical AND operation.

To perform the space-variant logic operation, we use the temporal gate $f(t)_{ij}$, which is arranged to perform the desired logic operation on the desired element position and constitutes four temporal gates $F = \{F(t),\ t = 1, 2, 3, 4\}$. The temporal gate $f(t)_{ij}$ with a specific temporal structure, controls the decoded output by gating the temporal coding $c(t)_{ij}$. In Figure 9.3.2 the cases for the logical exclusive OR (XOR) and the logical AND

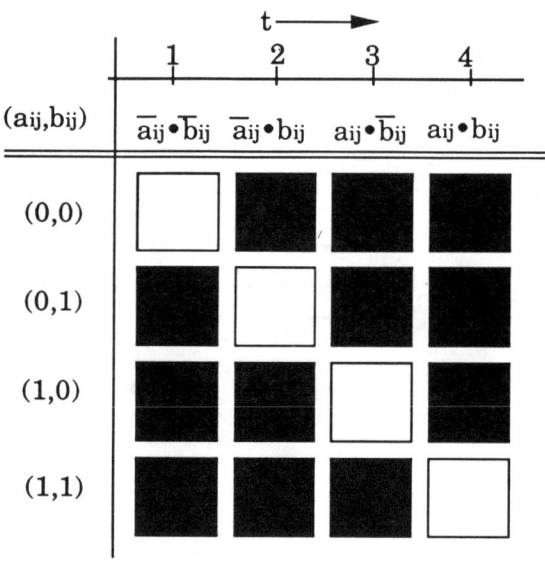

Figure 9.3.3
Temporal coding sequence. Each element is ON at only one time in the four sequential steps. In an optical implementation, the temporal coding is performed by superimposition of the input pattern A or its inverted pattern A with the input pattern B or its inverted pattern B. Temporal gates of all 16 logic functions. The temporal gate controls the operational output by selecting opaque or transparent for the pixel of the temporal coding patterns

are shown. By applying the temporal gate $f(t)_{ij}$ to the temporal coding $c(t)_{ij}$ at four sequential steps, four transient output $t(t)_{ij}$, making the transient pattern $T = \{T(t), t = 1, 2, 3, 4\}$, are obtained. The transient output $t(t)_{ij}$ is obtained by superimposition of the temporal coding $c(t)_{ij}$ with the temporal gates $f(t)_{ij}$ as follows

$$t(t)_{ij} = f(t)_{ij} \cdot c(t)_{ij}, \quad t = 1, 2, 3, 4 \tag{5}$$

The transient output $t(t)_{ij}$ is bright at only one time or dark all the time in the four sequential steps. Figure 9.3.4 shows the temporal gates for all 16 logic functions. For example, the temporal gate $f(t)_{ij}$ of the logical XOR is CLOSED in the first and fourth steps, and OPEN in the second and third steps.

A decoded output o_{ij} making the decoded output pattern O is performed by adding the four transient outputs $f(t)_{ij}$ as follows

$$\begin{aligned} o_{ij} &= t(1)_{ij} + t(2)_{ij} + t(3)_{ij} + t(4)_{ij} \\ &= f(t)_{ij} \cdot \bar{a}_{ij} \cdot \bar{b}_{ij} + f(t)_{ij} \cdot a_{ij} \cdot \bar{b}_{ij} + f(t)_{ij} \cdot \bar{a}_{ij} \cdot b_{ij} + f(t)_{ij} \cdot a_{ij} \cdot b_{ij} \end{aligned} \tag{6}$$

where $+$ denotes temporal addition. In the pattern formulas, equation (6) is rewritten as

$$O = F(1) \cdot \bar{A} \cdot \bar{B} + F(2) \cdot A \cdot \bar{B} + F(3) \cdot \bar{A} \cdot B + F(4) \cdot A \cdot B \tag{7}$$

Table 9.3.1 shows a comparison between the spatial method and the temporal method in the space-variant technique. The spatial method requires four times the number of elements of the input pattern, and the temporal method requires the same number as

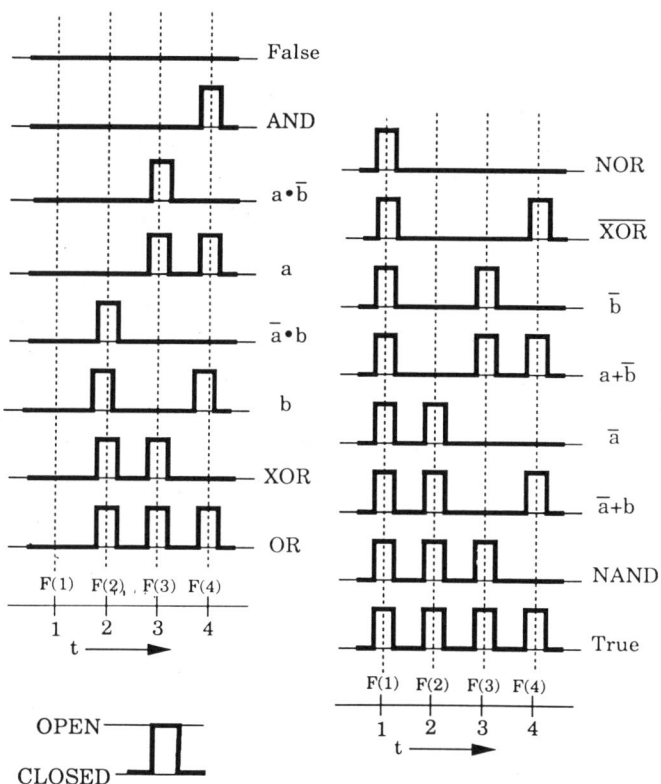

Figure 9.3.4
(a) Fundamental system constituent of the space-variant logic operation based on the temporal method.
(b) Schematic diagram of the experimental system. The computer simulates the time integral device's properties, such as linearity, with saturation and nonlinearity thresholding

Table 9.3.1
Comparison between the spatial method and the temporal method in the space-variant technique

	Spatial method	Temporal method
Number of elements	$P \times 4$	P
Operation time	$T_c + T_g + T_d$	$(T_c + T_g + T_a) \times 4$
Main operation		
(coding)	Pattern conversion	Pattern inversion
(decoding)	Pattern conversion	Pattern addition

the elements. The operation times, T_s and T_t, of the spatial method and the temporal method, respectively, are

$$T_s = T_{sc} + T_{sg} + T_{sd} \tag{8}$$

and

$$T_t = (T_{tc} + T_{tg} + T_{td}) \times 4 \tag{9}$$

where T_c, T_g, and T_d are the times of coding, gating (masking), and decoding.

The spatial method needs a pattern conversion for each element such as controlling bright and dark of the four subelements, while the temporal method needs only a pattern inversion. In the decoding, the spatial method should restore an original element from a bright subelement, because the element size of the decoded output is different from that of the original input. The decoding of the temporal method, on the other hand, is the temporal addition without a complex pattern transformation. The temporal method leads to easy and inexpensive implementation from simple processes and a high cascadability because of no change in size and arrangement of the output pattern. In general, it is thought that the parallel computing system is composed of many cascaded logic operating systems. Therefore the proposed technique is very important in logic operation systems since it can make a direct cascaded connection between the logic operation systems.

4   Three-layer Neural Network [8,9]

A three-layer neural network of 4 input neurons, 4 hidden layer neurons, and 5 output neurons is implemented in the optical neural system. The back-propagation learning rule is used. Consider a three-layer neural network model as shown in Figure 9.3.5. This system consists of an input layer, a hidden layer, and an output layer. The values of the ith neuron in the input layer, the jth neuron in the hidden layer, and the kth neuron in the output layer are denoted by I_i, H_j, and O_k, respectively. The weights between the ith neuron of the input layer and the jth neuron of the hidden layer, and

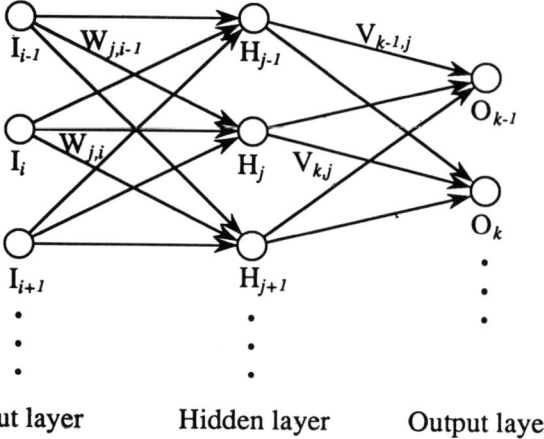

Figure 9.3.5
Three-layer neural network model

between the jth neuron of the hidden layer and the kth neuron of the output layer are denoted by W_{ji} and V_{kj}, respectively. According to Rumelhart [8] the values of the hidden layer neurons are given by

$$\dot{H}_j = f\left(\sum_i W_{ji} I_i + \theta_j\right) \tag{10}$$

and the values of the output layer neurons are given by

$$O_k = f\left(\sum_j V_{kj} H_j + \gamma_k\right) \tag{11}$$

where θ_j and γ_k are offsets of each neuron in the hidden and output layer, respectively, and f is a sigmoid function. Using the actual value O_k and the target input T_k, the sum of the squared errors E is given by

$$E = \tfrac{1}{2} \sum_k (T_k - O_k)^2 \tag{12}$$

To obtain the optimum weights, we use the back-propagation learning rule. As the error of equation (12) decreases, the error signal in the output layer is given by

$$\delta_k = \frac{(T_k - O_k)^*2}{u^* O_k (1 - O_k)} \tag{13}$$

and an error signal in the hidden layer is given by

$$\delta'_j = \frac{\delta_k V_{kj}^* 2}{u^* H_j (1 - H_j)} \tag{14}$$

where u is the gradient of the sigmoid function. By using the error signal of equation (13) or (14), the modification value ΔV_{kj} of the weights V_{kj} is given by

$$\Delta V_{kj} = \alpha \delta_k H_j \tag{15}$$

and the modification value ΔW_{ji} of the weights W_{ji} is given by

$$\Delta W_{ji} = \alpha \delta'_j I_i \tag{16}$$

where α is the learning rate.

5 Hybrid Computing System

The hybrid computing system developed is shown in Figure 9.3.6. This system consists of an optical digital computing system and an optical neural computing system, which are connected by an electronic computer. In the digital computing part, CRT and LCTV (liquid crystal TV) are used to display temporally coded patterns and operation selection patterns, respectively. Superimposition of the coded patterns and the operation selection patterns are imaged onto a MSLM (micro-channel plate spatial light modulator). The MSLM carries out temporal addition corresponding to decoding.

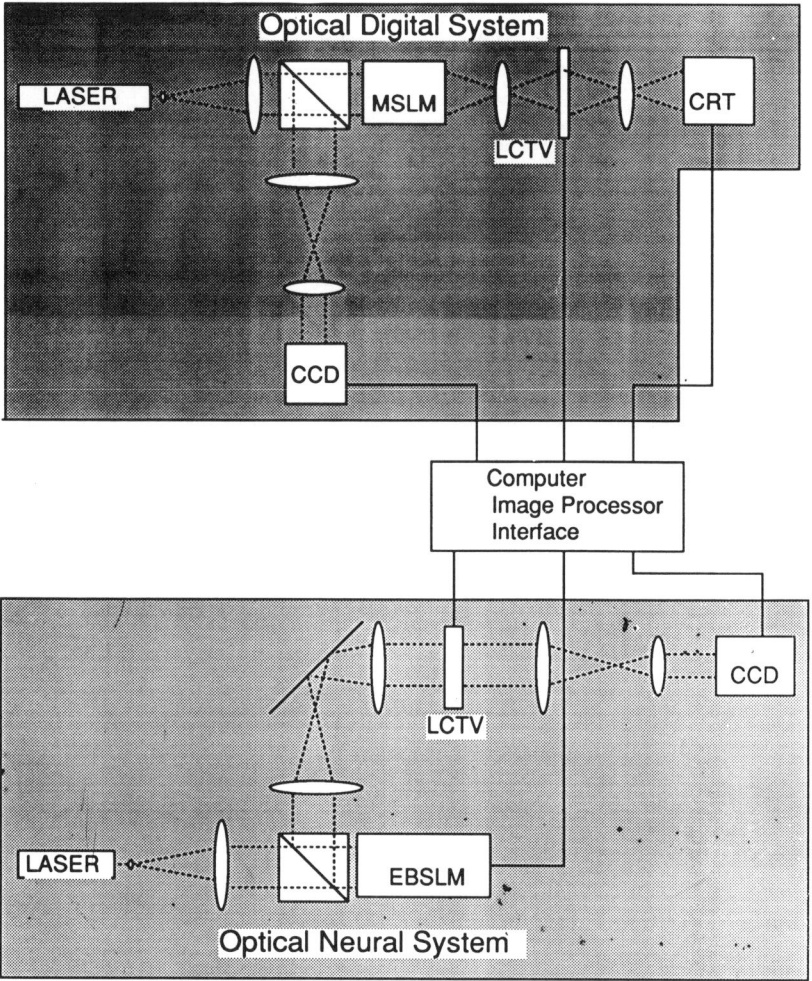

Figure 9.3.6
Developed hybrid optical computing system

The optical neural network part consists of an EBSLM (electron beam spatial light modulator) in which the weights are stored, a LCTV which displays the output values of the input and hidden layer neurons, and a CCD camera. The recorded weights in the EBSLM are read out by a He-Ne laser light. The superimposed images of the weights displayed on the EBSLM and the output of the neurons displayed on the LCTV are detected by the CCD camera. Computer controls of the EBSLM and the LCTV perform nonlinear processing of neurons.

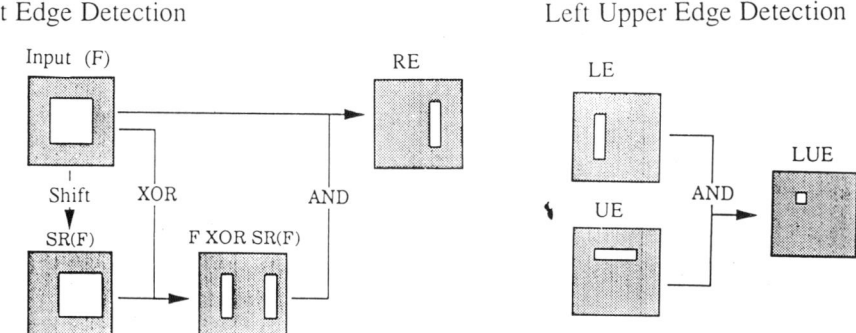

Figure 9.3.7
Edge detection using SVPLO

Table 9.3.2
Detected features corresponding to the pixel numbers of edges

Input	Total	RE	UE	RUE	LUE
M	320	25	71	7	8
N	243	21	63	1	14
T	129	20	26	0	1
Y	137	21	38	6	8
Z	203	46	26	11	2

Table 9.3.3
Neuron output for input characters

Input	Neuron output				
	1	2	3	4	5
M	0.963	0.046	0.004	0.000	0.000
N	0.028	0.985	0.004	0.021	0.000
T	0.043	0.000	0.934	0.029	0.044
Y	0.002	0.015	0.023	0.957	0.000
Z	0.000	0.000	0.044	0.000	0.965

In the preprocessing system of the digital computer, the edge information of the input characters is extracted by the temporal coding space-variant parallel logic operation technique mentioned above. The edge of the binary pattern is detected by Exclusive OR operation (XOR) between a boundary pattern and its shifted pattern and AND operation, as shown in Figure 9.3.7. In our experiment there are four features, i.e. right edge (RE), upper edge (UE), right upper edge (RUE), and left upper edge (LUE).

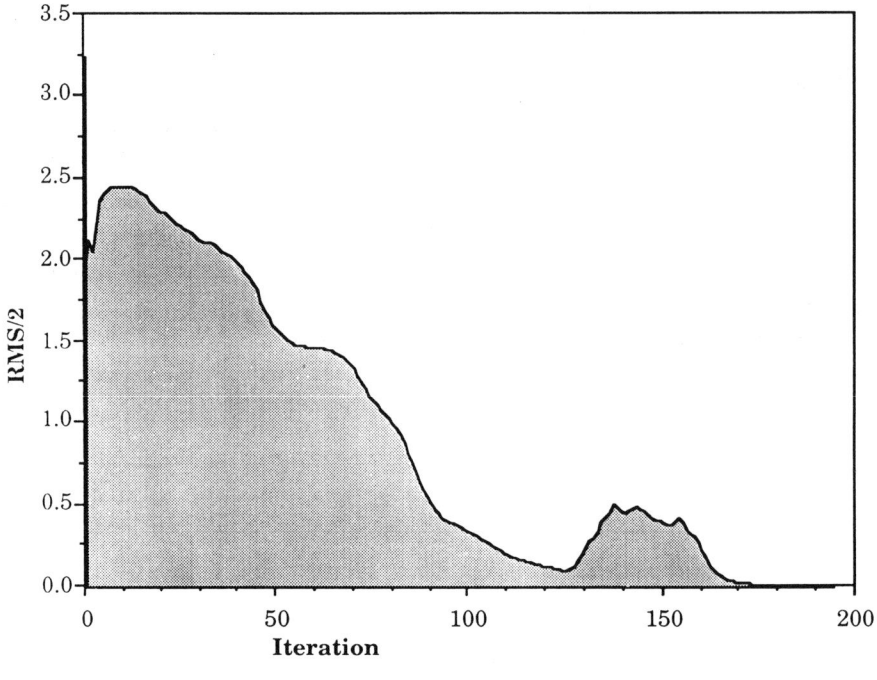

Figure 9.3.8
Change of errors vs. learning iteration

The input characters for recognition are *M*, *N*, *T*, *Y*, and *Z*. The features defined by the number of detected edge pixels are given in Table 9.3.2. Figure 9.3.8 shows the change in the error for learning iteration times. Table 9.3.3 shows the output of neurons for five input characters.

6 Conclusion

We have described the temporal method for an optical space-variant logic operation, which was based on temporal coding and temporal gating. With regard to large-scale parallelism, the space-variant logic operation executes different operations in parallel. We also have proposed a hybrid computing system, combined with a digital system and a neural computing system. In the digital computing part of this system the feature extraction of the input data is performed, and the features estimated are subjected to the optical neural computing part. Such a hybrid computing architecture is expected to be useful for developing practical and advanced optical computing systems.

References

[1] A. Huang, *IEEE Proc.*, **72**, 780 (1984).
[2] T. Yatagai, *Appl. Opt.*, **25**, 1571 (1986).
[3] N.Y. Farhat and D. Psaltis, A. Prata and E. Paek, *Appl. Opt.*, **244**, 1469 (1985).
[4] M. Ishikawa, N. Mukohzaka H. Toyoda and Y. Suzuki, *Appl. Opt.*, **29**, 289 (1990).
[5] S. Fukushima and T. Kurokawa, *IEEE Photon. Technol. Lett.*, **3**, 682 (1991).
[6] J. Tanida and Y. Ichioka, *J. Opt. Soc. Amer.*, **73**, 800 (1983).
[7] T. Yatagai, *Opt. Lett.*, **11**, 270 (1987).
[8] D.E. Rumelhart, J.L. McClelland and the PDP Research Group, *Parallel Distributed Processing*, **1**, MIT Press, Cambridge, MA, 1986, Chapter 8, p. 318.
[9] N. Kasama, Y. Hayasaki, T. Yatagai, M. Mori and S. Ishihara, *Jpn J. Appl. Phys.*, **29**, L1565 (1990).

10
NEURAL COMPUTING

10.1

Neural Network Learning: Generalization and Over-learning

Hidemitsu Ogawa

Abstract

A framework for discussing the learning problem for multi-layer feedforward neural networks is introduced from the point of view of an inverse problem. It naturally leads us to the concept of optimal generalization and methods for choosing training data and the optimal number of hidden units and for constructing optimally generalizing neural networks. It also leads us to the concept of J-over-learning and methods for choosing training data for preventing over-learning.

1 Introduction

This section discusses the learning problem for multi-layer feedforward neural networks. Since Rumelhart *et al.* proposed the error back-propagation (BP) algorithm, multi-layer neural networks have been widely used. However, it has been observed in many fields that a decrease in error over the training set does not mean a decrease over novel data and in fact may lead to lower generalization. This phenomenon is called *over-learning*.

Although there has been much discussion about preventing over-learning [1-7], most of the work uses the same least mean squares (LMS) criterion as used in standard BP. However, this criterion only measures the error over the data set; it does not evaluate the error outside of the training set.

That means that when we use the LMS criterion, any responses to novel inputs are allowed for a neural network. Therefore, the LMS criterion is not the proper place to begin discussing the generalization ability of a neural network.

To discuss the generalization problem, we should start from a criterion that evaluates directly the function of a neural network. From this point of view, we propose a framework for discussing the generalization problem [8,9]. It naturally leads us to the concept of optimal generalization and J-over-learning. It also leads us to methods for constructing optimally generalizing neural networks and for preventing over-learning.

2 Optimally Generalizing Neural Network

In this section we are concerned with multi-layer feedforward neural networks whose number of input and output units are L and 1, respectively. Assume that its output unit has a linear activation function. Let D be a subset of the L-dimensional Euclidean space R^L. A neural network can be expressed as a function from D to R, denoted by $f_0(x)$ with x an L-dimensional real vector in D.

The learning problem is to construct a neural network by using a training set so that $f_0(x)$ becomes the best approximation to a desired function $f(x)$. The training data is given as a set of M input vectors, say $\{x_m : 1 \leq m \leq M\}$, and corresponding desired output values, say $\{y_m : 1 \leq m \leq M\}$, where $y_m = f(x_m)$. Let us denote by y the vector consisting of $\{y_m : 1 \leq m \leq M\}$. The point x_m and the value y_m can be called the mth sample point and the corresponding sample value of the desired function f, respectively.

Let $\mathcal{H}$ be the set of all functions f to be approximated by neural networks. Assume that $\mathcal{H}$ is a Hilbert space with the reproducing kernel $K(x, x')$. The reproducing kernel $K(x, x')$ is the bivariate function defined on $D \times D$ which satisfies the following two conditions [10]:

(i) For any fixed x', $K(x, x')$ is a function in $\mathcal{H}$.

(ii) For any function f in $\mathcal{H}$ and for any x' in D, it follows that

$$(f(x), K(x, x')) = f(x'), \tag{1}$$

where the left-hand side of equation (1) denotes the inner product in $\mathcal{H}$.

Once the training set $\{x_m : 1 \leq m \leq M\}$ is fixed, the vector y is uniquely determined from f. Then we can introduce an operator A which transforms f to y:

$$y = Af. \tag{2}$$

The operator A becomes a linear operator even when we are concerned with nonlinear neural networks.

Now we can say that the learning problem is the problem of obtaining an estimate f_0 to f from y in the model (2). It is equivalent to the problem of obtaining an operator B, which provides f_0 from y:

$$f_0 = By. \tag{3}$$

This problem belongs to the general class of inverse problems. To solve it we can use deep results obtained from the field of inverse problems [11,12].

The space $\mathcal{H}$ is of infinite dimension in general, whereas neural networks with a fixed number of hidden units can express only functions which belong to a subset of $\mathcal{H}$. Hence, we need a criterion for choosing an operator B which provides the best approximation f_0 to the desired function f of $\mathcal{H}$. Such a criterion is called a *learning criterion*.

When an operator B is discussed in the context of a learning criterion J, B is called a *J-learning*. An operator B satisfying J is called a *proper J-learning* and denoted by $A^{(J)}$. Since $A^{(J)}$ is not always unique, let $A\{J\}$ be the set of all $A^{(J)}$. The function f_0 given by an $A^{(J)}$ is called the *J-optimally generalizing neural network* or *J-OGNN* for short.

Three examples of learning criteria are listed below.

(1) *Memorization criterion, J_M*. Find B that minimizes the objective function

$$J_M[B] = \|ABy - y\|^2 \tag{4}$$

for each $y = Af$, where $\|\cdot\|$ is the norm in R^M.

(2) *Wiener criterion, J_W*. Find B that minimizes the objective function

$$J_W[B] = E_f \|By - f\|^2 \tag{5}$$

where $\|\cdot\|$ is the norm in $\mathcal{H}$ and E_f is the expectation taken over $\{f\}$.

(3) *Projection criterion, J_P*. Find B such that By is equal to the orthogonal projection of f onto the range $\mathcal{R}(A^*)$, where A^* is the adjoint operator of A. The J_P-learning will be discussed in the next section in detail.

3 Projection Generalizing Neural Network

3.1 Projection generalization

The memorization criterion J_M does not guarantee the generalization ability of a neural network. The Wiener criterion J_W provides the best approximation in the averaged sense with respect to $\{f\}$. In this section we shall introduce the concept of projection learning which provides the best approximation f_0 to an individual f.

Whenever we use a linear operator B for reconstructing f_0, the range of B becomes a closed subspace of $\mathcal{H}$. Hence 'the best approximation' implies that f_0 is the nearest point to f in a subspace, say $\mathcal{X}$, of $\mathcal{H}$, i.e. the orthogonal projection of f onto $\mathcal{X}$. The following lemma is essential.

Lemma 1. Let $\mathcal{X}$ be a closed subspace of $\mathcal{H}$. For any function f of $\mathcal{H}$, its orthogonal projection onto $\mathcal{X}$ is computable from $y = Af$ if and only if $\mathcal{X}$ is included in $\mathcal{R}(A^*)$.

Lemma 1 implies that $\mathcal{R}(A^*)$ is the largest subspace in which we can obtain the best approximation from y. Hence, we shall concentrate our attention on the subspace $\mathcal{R}(A^*)$ hereafter. Thus, we have the projection criterion J_P mentioned in section 2. The function f_0 given by J_P is called the *J_P-optimally generalizing neural network* (J_P-OGNN) or the *projection generalizing neural network* (PGNN).

3.2 Design of a training set

By using the reproducing kernel $K(x, x')$ of $\mathcal{H}$, we shall introduce the so-called sampling function ψ_m as

$$\psi_m(x) = K(x, x_m). \tag{6}$$

The range $\mathcal{R}(A^*)$ is the subspace spanned by $\{\psi_m : 1 \leq m \leq M\}$. Since $\mathcal{R}(A^*)$ is the subspace in which the best approximation is obtained, the larger $\mathcal{R}(A^*)$ is, the better it is. The largest $\mathcal{R}(A^*)$ is given when $\{\psi_m : 1 \leq m \leq M\}$ becomes linearly independent. Let K be the $M \times M$ matrix with elements $K_{ij} = K(x_i, x_j)$. The matrix K is the Gram matrix of $\{\psi_m : 1 \leq m \leq M\}$. Then we have

Theorem 1 (Design of a training set). If and only if the training set $\{x_m : 1 \leq m \leq M\}$ is chosen in such a way that the matrix K is nonsingular, the range $\mathcal{R}(A^*)$ becomes the largest subspace.

3.3 Construction of a PGNN

If we are free from the sigmoidal activation function, then any multi-layer feedforward neural network can be constructed by a two-layer feedforward neural network. It is written as the L-N-1 network, which means that the two-layer network has L input units, N hidden units, and one output unit.

An input–output relation of the nth hidden unit is denoted by $u_n(x)$ and is called a basis function, where x is an L-dimensional real vector whose lth element is denoted by ξ_l. Conventional feedforward neural networks have the following basis functions:

$$u_n(x) = \sigma\left(\sum_{l=1}^{N} w_{nl}\xi_l\right), \quad 1 \leq n \leq N, \tag{7}$$

where $\sigma(\cdot)$ is the so-called sigmoidal activation function and w_{nl} is a weight which connects the lth input unit to the nth hidden unit. In this section, however, we are free from such type of basis functions. Our standpoint is that the basis functions $\{u_n : 1 \leq n \leq N\}$ should be chosen for some proper reason.

The linear output unit has weights w_n connected to each hidden unit. The vector consisting of $\{w_n : 1 \leq n \leq N\}$ is denoted by w. Finally, the input–output relation $f_0(x)$ of the network is given by

$$f_0(x) = \sum_{n=0}^{N} w_n u_n(x). \tag{8}$$

There are two standpoints for using neural networks. Let $\mathcal{S}$ be the subspace spanned by the basis functions $\{u_n : 1 \leq n \leq N\}$. The first standpoint is that we try to find the best approximation of $f \in \mathcal{H}$ in $\mathcal{S}$. In this case, $\mathcal{S}$ should be included in $\mathcal{R}(A^*)$ because of Lemma 1. Otherwise, we cannot obtain it from y. The second standpoint is that we try to find the best approximation of $f \in \mathcal{H}$ in $R(A^*)$. In this case we use a neural network to represent the best approximation. Hence, $\mathcal{S}$ should include $R(A^*)$. Otherwise, we cannot represent all elements in $R(A^*)$. In this section we take the second standpoint.

To construct a PGNN we have to decide the number N of hidden units, the basis functions $\{u_n : 1 \leq n \leq N\}$, and the weight parameter w so that equation (8) holds. Let

$$U = \sum_{n=1}^{N} (e_n \otimes \overline{u_n}), \tag{9}$$

where e_n is the N-dimensional vector consisting of zero elements except the element m equal to 1 and $(e_n \otimes \overline{u_n})$ is the Schatten product defined by

$$(e_n \otimes \overline{u_n})f = (f, u_n)e_n. \tag{10}$$

Let $A\{i, j, \ldots, l\}$ be the set of operators T which satisfy conditions $(i), (j), \ldots, (l)$ among the following four conditions:

(1) $ATA = A$,
(2) $TAT = T$,
(3) $(AT)^* = AT$,
(4) $(TA)^* = TA$.

An operator T of $A\{i, j, \ldots, l\}$ is called an $i, j, \ldots, l$-inverse of A and denoted by $A^{(i,j,\ldots,l)}$. There exists a unique operator T which satisfies the above four conditions. It is called the Moore–Penrose generalized inverse of A and denoted by $A^\dagger$.

Theorem 2 (Construction of PGNN). A neural network becomes a PGNN if and only if

(i) The number N of hidden units is chosen in such a way that $N \geq M$.
(ii) The basis functions $\{u_n : 1 \leq n \leq N\}$ are chosen in such a way that S includes the subspace spanned by $\{\psi_m : 1 \leq m \leq M\}$.
(iii) The weight parameter w is chosen in such a way that $w = Wy$, where $W = (U^*)^\dagger A^{(1,4)}$.

4 Over-learning

4.1 J-over-learning

When the memorization criterion J_M is discussed in the context of the generalization problem, we can say that J_M is used as a substitute for some 'true' criterion. The essential point of over-learning lies in the relation between those two criteria. In this section we discuss the case where a criterion J', not limited to the memorization criterion, is used as a substitute for a true criterion J.

Definition 1 (*J-over-learning*). If a proper J'-learning $A^{(J')}$ does not satisfy a criterion J, then $A^{(J')}$ is said to cause J-over-learning.

The definition of J-over-learning naturally leads us to the following concepts:

Definition 2 (*Admissibility*).

(i) *Nonadmissibility*. If all proper J'-learnings cause J-over-learning, i.e. if it follows that

$$A\{J\} \cap A\{J'\} = \phi, \tag{11}$$

then it is said that *J does not admit J'*.

(ii) *Partial admissibility*. If there is at least one proper J'-learning which does not cause J-over-learning, i.e. if it follows that

$$A\{J\} \cap A\{J'\} \neq \phi, \tag{12}$$

then it is said that *J partially admits J'*.

(iii) *Admissibility*. If all proper J'-learnings do not cause J-over-learning, i.e. if it follows that

$$A\{J\} \supset A\{J'\}, \tag{13}$$

then it is said that *J always admits J'* or, in short, that *J admits J'*.

(iv) *Complete admissibility*. If J always admits J' and vice versa, i.e. if it follows that

$$A\{J\} = A\{J'\}, \tag{14}$$

then it is said that *J completely admits J'*.

From the over-learning point of view, the relation

$$A\{J\} \subset A\{J'\} \tag{15}$$

has no more meaning than partial admissibility. However, it may still have significance within the context of the learning problem. For example, if $J' = J_M$, then it implies that all proper J-learnings $A^{(J)}$ can give correct responses to the training data. This does not always hold in general (see Theorem 4).

4.2 *Projection learning versus memorization learning*

Let us consider the case that the memorization criterion J_M is used as a substitute for the projection criterion J_P.

Theorem 3. The following two cases occur.

(i) *Complete memorization*. Any proper projection learning can always give correct responses to the training data, i.e. it follows that

$$A\{J_P\} \subset A\{J_M\}. \tag{16}$$

(ii) *Complete admissibility*. The projection criterion J_P completely admits J_M, i.e. it follows that

$$A\{J_P\} = A\{J_M\} \tag{17}$$

if and only if
$$\mathcal{N}(A) = \mathcal{H} \quad \text{or} \quad \mathcal{N}(A) = \{0\} \tag{18}$$
with $\mathcal{N}(A)$ the null space of A.

4.3 Wiener learning versus memorization learning

Let us consider the case that the memorization criterion J_M is used as a substitute for the Wiener criterion J_W.

Theorem 4. The following four cases occur.

(i) *Partial admissibility.* J_W partially admits J_M.

(ii) *Admissibility.* J_W always admits J_M, if and only if
$$\mathcal{N}(A) \supset \mathcal{R}(R) \quad \text{or} \quad \mathcal{N}(A) = \{0\}. \tag{19}$$

(iii) *Complete memorization.* A proper Wiener learning gives correct responses to the training data, if and only if
$$\mathcal{N}(A) + \mathcal{R}(R) = \mathcal{H}. \tag{20}$$

(iv) *Complete admissibility.* J_W completely admits J_M, if and only if
$$\mathcal{N}(A) = \mathcal{H}$$
or
$$\mathcal{N}(A) = \{0\} \quad \text{and} \quad \mathcal{R}(R) = \mathcal{H}. \tag{21}$$

5 Conclusions

A framework for discussing the learning problem for multi-layer feedforward neural networks has been introduced from the point of view of an inverse problem. Within this framework the concepts of optimal generalization, J-over-learning, and relevant concepts like admissibility have been introduced. The problems of choosing the training data to achieve better generalization and to prevent over-learning have been solved. Methods for constructing optimally generalizing neural networks have also been obtained. The method can be easily extended to any J-OGNN.

References

[1] E.B. Baum and D. Haussler, *Neural Computation*, **1**, 151–160 (1989).
[2] Y. Kimura, *Trans. IEICE, Jpn*, **J73-D-II**, 840–847 (1990) [in Japanese].
[3] T. Poggio and F. Girosi, *Proc. IEEE*, **78**, 1481–1497 (1990).
[4] D.B. Schwartz, V.K. Samalam, S.A. Solla and J.S. Denker, *Neural Computation*, **2**, 374–385 (1990).

[5] J. Sietsma and R.J.F. Dow, *Proc. ICNN'88*, IEEE Int. Conf. on Neural Networks, 1988, pp. 325-333.
[6] S. Suzuki and H. Kawahara, *Trans. IEICE, Jpn*, **J75-D-II**, 637-645 (1992) [in Japanese].
[7] Y. Wada and M. Kawato, *Trans. IEICE, Jpn*, **J74-D-II**, 955-965 (1991) [in Japanese].
[8] H. Ogawa and K. Yamasaki, *Proc. ICANN'92*, Int. Conf. on Artificial Neural Networks, 1992, pp. 215-218.
[9] H. Ogawa, *Proc. ICIIPS'92*, Int. Conf. on Intelligent Information Processing & System, **2**, 1992, pp. 1-6.
[10] S. Bergman, *The Kernel Function and Conformal Mapping*, American Mathematical Society, 1972.
[11] H. Ogawa, *J. IEICE, Jpn*, **71**, 491-497, 593-601, 739-748, 828-835 (1988) [in Japanese].
[12] H. Ogawa, E. Oja and J. Lampinen, *Proc. IEEE*, Int. Conf. on Systems Engineering, 1989, pp. 93-97.

10.2

Integrated Image Processing System Using an Optical Neural Network

Nagaaki Ohyama, Masahiro Yamaguchi and **Joong-Sun Lee**

Abstract

The integration of the artificial neural network and the electronic computing system is studied. A method to control the neural network using external constraints is investigated to combine the intuitive and logical processing systems, and the integrated system is applied to the image processing. The optical parallel processing is used for high-speed and high-performance computing.

1 Introduction

Information processing by an artificial neural network is promising for various computing applications, because highly adaptable computation is possible at high speed by a parallel system such as an optical processor. On the other hand, conventional electronic computers have very high ability in logical matters. Therefore, the purpose of this research is to integrate the intuitive processing by a neural system and the logical processing by a Von Neumann computer to realize high-performance and high-speed computing, and to apply the integrated system to image processing using the optical parallel processing technology.

In this section, the concept of the integrated computing is described and a method to combine the neural and logical systems is introduced. In the character reader system, the word dictionary is employed as logical constraints, resulting in the improvement of the recognition rate [1]. The technique of the integrated system is also applied to the image retrieval system, and it becomes possible to suppress or to induce specified patterns in the image memory. A large-scale optical parallel processing system is constructed in the experiment to investigate a method for integration.

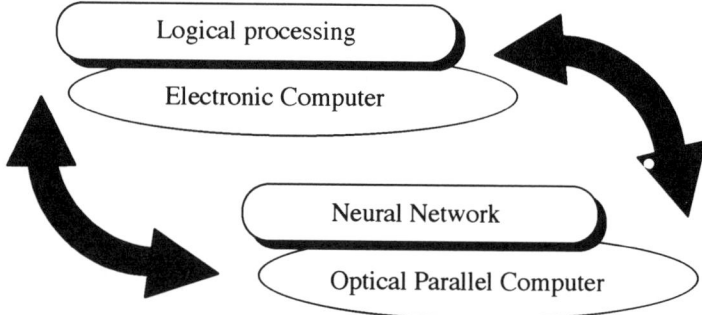

Figure 10.2.1
The concept of integrated computing

2 Concept of Integrated Computing

Although information processing by a neural network is suitable for human interfaces, conventional computers are superior in programming and logical processing. Thus, for efficient and flexible computing, logical processing by conventional computer should be involved in the neural network. To realize the integrated computing system shown in Figure 10.2.1, it is required to develop a method to control the neural network by an external system.

An external condition can easily be introduced into the Hopfield-type [2] network because the energy function (Liapunov function) is defined and the effect can be reasonably understood. Let us consider the states of the neurons as a vector v and the connection as a matrix T, then the dynamics of the network can be expressed as

$$v' = \Phi\{Tv + w\} \quad (1)$$

where Φ denotes the nonlinear output function and v' is the next states of the neurons. Thus, the energy function $E(v)$ becomes

$$E(v) = -\tfrac{1}{2}v^t Tv - w^t v \quad (2)$$

where w is the external input as a bias vector. From equation (2) it is easy to understand that the stable state of the network can be controlled by changing the external input vector w. In the case when a step function is used as the output function, varying the threshold levels can substitute the external input vector.

When a fixed input vector is applied, the energy function still holds as quadratic, while the stable state changes. So, to avoid convergence of undesirable patterns, the external input should be given so that such patterns become unstable. On the other hand, to induce a specific state, the energy function can be modified by the external input so as to generate a strong stable state.

3 Character Recognition

When reading characters, we can guess the meaning of words even if some of the characters are hidden or mistaken, because we can work out the incorrect characters

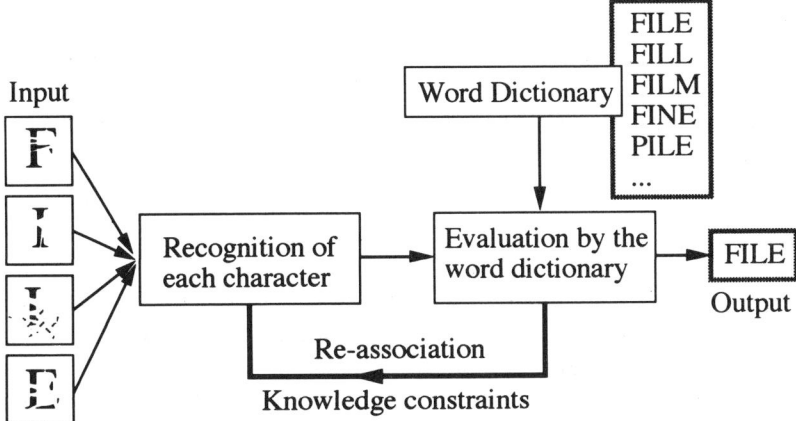

Figure 10.2.2
Schematic of an intelligent associative memory. A degraded word is input and the word 'FILE' is obtained using the associative memory and the word dictionary

using the knowledge of a word dictionary. To realize the human attribute of character reading using artificial neural networks, it is effective and practical to apply integrated computing. Namely, each character in a word is processed by the neural associative memory, and the results from the neural processes are corrected and verified by the logical processing system.

Figure 10.2.2 shows a schematic of a character reading system by integrated processing, called intelligent associative memory [1]. When severely distorted input patterns are independently processed, they cannot be correctly recognized. The results of all the characters in a word are evaluated with using the 'word dictionary'. If the resultant word is not found in the dictionary, the suspicious character is recognized again, with a constraint to suppress the erroneous character and to generate one of the correct words. If the output word of the associative memory is found in the dictionary, the correct word is obtained as the output. Therefore, although the associative memory employed in this system has a small-scale network, intelligent processing becomes possible with knowledge from a logical system.

The intelligent associative memory for character recognition is also applied to the ID numbers printed on a medical X-ray film. For the efficient and advanced information system of medical image data, it is required to manage the film image as digital information, and the films should be read by a film scanner. Since the patient's name and ID number are usually printed on the X-ray film, the digitized image can be automatically filed by using the technology of character recognition. The most important matter in this application is to avoid misreading, because incorrect identification is very dangerous in medical applications. For the practical system of automatic ID reading, the utilization of knowledge is powerful in suppressing misrecognition. In this application, the database of lists of patients in a hospital's information system can be used as the dictionary. Moreover, in most hospitals patient ID numbers include an error check code, which can be also employed as a logical constraint. When one of the results of recognition conflicts with the logical constraints, the second candidate is picked up and verified with the constraints. If the constraints cannot be satisfied, the film is provisionally held as an unreadable one. According to the above procedure, the system has high reliability to overcome the problem of misreading. Even though a synthetic discriminant filter is used instead of a neural network for character recognition [3], the integrated computing concept is utilized and is very effective in this application.

4 Image Associative Memory

Integrated computing is also applied to the image associative memory, which can be used for image content retrieval. Figure 10.2.3 shows a schematic of the large-scale image associative memory developed in this work. When a distorted image or part of an image is used as input, the most similar image among the stored patterns is obtained as the output. In Figure 10.2.3, dot-products between the input and stored patterns are calculated to give correlation results, and after nonlinear processing of the correlation values, each stored image is recalled with weights given by the corresponding correlation value. The reconstructed images are summed to generate the output, which is converted again by another nonlinear function and fed back into the input stage. Repeating the above process, the system can associate an image that is close to the input pattern.

The system shown in Figure 10.2.3 is almost equivalent to the Hopfield-type associative memory, in so far as the nonlinear function Ψ can be regarded as linear.

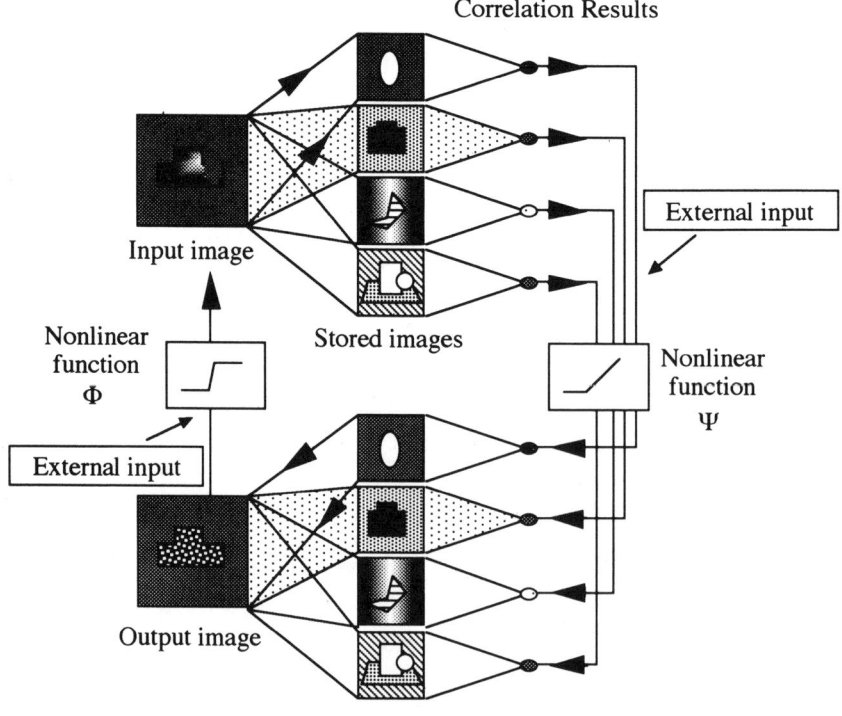

Figure 10.2.3
The image associative memory system

Thus, the method for controlling the network by external inputs can be directly applied. In the system shown in Figure 10.2.3, the external inputs can be applied to both the image pattern and the index of the stored image, respectively.

To reveal the effect of the external input, the stable states of the network are analyzed. Although the stable states of a Hopfield-type associative memory have already been described [4], the effect of the external input has not been considered up to now. In the following discussion the memory vectors are assumed to be orthogonal. For simplicity, consider the outer product rule for learning, then we can write

$$T = \sum_{k=1}^{M} \frac{1}{\|\mathbf{v}^k\|^2} \mathbf{v}^k \mathbf{v}^{kt} \tag{3}$$

where M is the number of memory vectors $\mathbf{v}^k$ ($k = 1, \ldots, M$).

If we disregard the nonlinear output function, the stable state of the neural network $\mathbf{v}_s$ must satisfy

$$\mathbf{v}_s = T\mathbf{v}_s + \mathbf{w} \tag{4}.$$

Obviously, when $\mathbf{w} = \{0\}$, all the linear combinations of $\mathbf{v}^k$ are stable states. To see the effect of the introduction of an external input $\mathbf{w}$, we separate the component within the subspace spanned by the stored pattern vectors from $\mathbf{w}$ and $\mathbf{v}_s$ as

$$\mathbf{w} = \sum_{k=1}^{M} \omega^k \mathbf{v}^k + \Delta_w \tag{5}$$

$$\mathbf{v}_s = \sum_{k=1}^{M} \alpha^k \mathbf{v}^k + \Delta_v \tag{6}$$

where $\omega^k = \mathbf{w}^t \mathbf{v}^k / \|\mathbf{v}^k\|^2$, $\alpha^k = \mathbf{v}_s^t \mathbf{v}^k / \|\mathbf{v}^k\|^2$, and Δ_w and Δ_v denote the components orthogonal to the subspace of the stored patterns in $\mathbf{w}$ and $\mathbf{v}_s$, respectively. Substituting equations (5) and (6) into equation (4), the stability condition becomes

$$\omega^k = 0, \text{ for all } k,$$

$$\Delta_w = \Delta_v \text{ and}$$

$$\alpha^k = \frac{\mathbf{v}_i^t \mathbf{v}^k}{\|\mathbf{v}^k\|^2} \tag{7}$$

where $\mathbf{v}_i$ denotes the initial state of the neurons. Therefore, ignoring the nonlinear effect, it can be noted that

(1) the association diverges when the input pattern has a component within the subspace spanned by the stored pattern vectors, and
(2) there arises a spurious stable state by the component orthogonal to the subspace of stored patterns in the input bias vector.

Actually, taking into account the nonlinear function, such as a step function, the stable states must be the lattice points, where each neuron has 0 or 1, or -1 or 1. Accordingly, supposing $\omega^{k_0} \neq 0$ and $\omega^k = 0$ ($k \neq k_0$), the state $\mathbf{v}_s = \mathbf{v}^{k_0}$ becomes a global minimum of the energy function of equation (2). In this case the network would always converge to v^{k_0} regardless of the initial probe vector if the external input is large. Moreover, a spurious stable state would appear by Δ_w, in the case when the nearest lattice point from v'_s is not placed in the subspace of stored patterns, where

$$v'_s = \sum_{k=1}^{M} \alpha^k v^k + \Delta_w \qquad (8)$$

Hence, considering the nonlinear function, we can claim that,

(1) the stability is increased or reduced if the external input vector has a positive or negative component that lies in the subspace spanned by a partial set of the memory patterns, and
(2) the component orthogonal to the subspace of stored vectors causes a spurious stable state if the component is too large.

It is possible, by applying an external input, to release the false state and to suppress or to induce a specific pattern; however, from the above discussions, a weak input vector within the subspace of memory vectors is preferable as an external input to avoid convergence to the spurious state.

To realize high-speed processing in the image associative memory, an optical parallel system is developed based on the holographic memory technique. All memory patterns are recorded in an array of Fourier transformed holograms, and two of the holograms are used in both correlating and recalling phases, in the optical system shown in Figure 10.2.4. In the correlation phase, the reconstructed light from the hologram

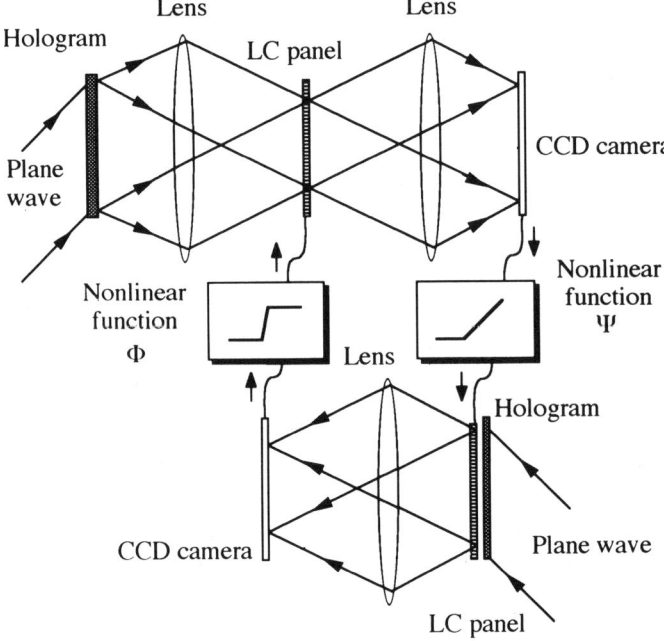

Figure 10.2.4
Optical setup for the image associative memory

illuminates the liquid crystal (LC) panel, and inner products between the input and stored images are simultaneously obtained on the CCD camera. In the recalling step, each stored pattern reconstructed from the hologram is weighted by the pattern on the LC panel. In this optical system, in spite of the fact that dynamic modification of the connection weights is difficult, a number of images can be stored by sequentially exposing images in the hologram.

In the experiment, the controllability of the associative memory by the external input is examined. Thirty binary images in 256×220 pixels are stored, where two of the stored images are shown in Figure 10.2.5(a,b). The image of Figure 10.2.5(c), which comprises the upper and lower halves of the images in Figure 10.2.5(a,b), is introduced in the associative memory. The output image is none of the stored patterns but as in Figure 10.2.5(d), if no external input is applied. As soon as an external input is applied on the upper or lower half of the input, one of the stored images corresponding to the weighted part is obtained as in Figure 10.2.5(e,f). The external input works as the reliability of the input image in this operation.

Hence, the image associative memory presented here is suitable for an information retrieval system from image content, because good man–machine interface is provided by the technology of integrated computing. The reliability of the input can be easily employed, and in addition, logical constraints can be applied to restrict specific patterns. Furthermore, the content retrieval system can be connected to the retrieval system based on a keyword list [5], using the external inputs as an interface between them. Keywords concerned with an image are presented as well as a key image pattern, and the reliability

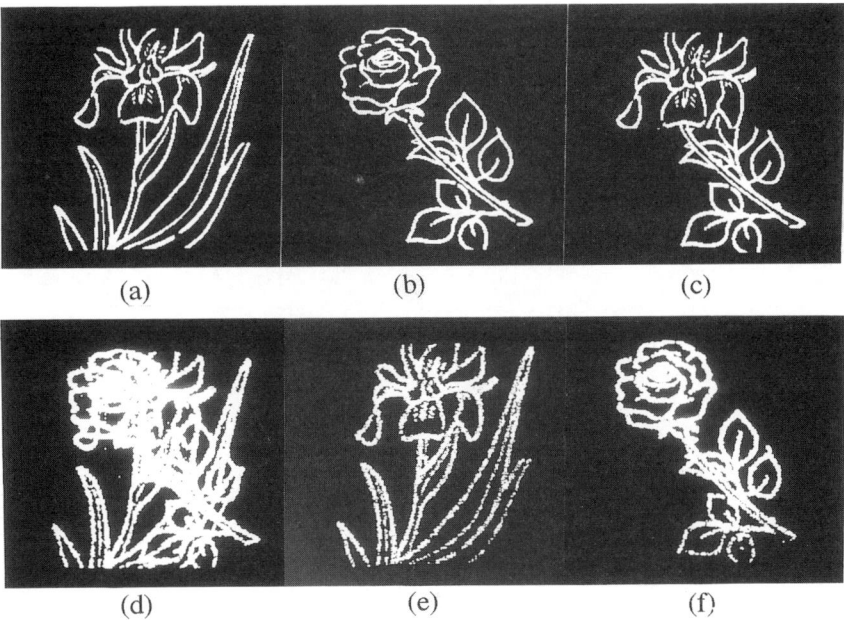

Figure 10.2.5
Experimental results. (a, b) Two of the stored patterns. (c) Input image. Output images (d) without external input, with external inputs on (e) upper and (f) lower half of the image

of the keywords and the pattern can be also utilized for image retrieval. If an inadequate image is obtained, the image should be re-associated by suppressing the association of the inadequate output, resulting in the desired output image which satisfies both the keywords and the key image conditions given by the operator.

5 Conclusion

The concept of integrated computing is promising in the application of image processing, such as a character recognition and an image retrieval system. The optical parallel processor is connected to the electronic computer to realize an efficient and intelligent system for large-scale image processing. A technique to control the neural network by external constraints is presented and the effect of the external input is described. By applying the technique of integrated computing, a functional image memory is realized. The experimental system presented in this section could be applied to an image retrieval system with good human interface.

References

[1] X. Lu, M. Yamaguchi, N. Ohyama, T. Honda, M. Oita, S. Tai and K. Kyuma, *Opt. Commun.*, **90**, 165–172 (1992).
[2] J.J. Hopfield, *Proc. Nat. Acad. Sci.*, **79**, 2554–2558 (1982).
[3] K. Yoshimoto, M. Yamaguchi, N. Ohyama, T. Honda, M. Konishi, M. Kuranishi, Y. Yamanaka and M. Matsui, *Med. Imag. Technol.*, **11** (3), 367–368 (1993) [in Japanese].
[4] S.V.B. Aiyer, M. Niranjan and F. Fallside, *IEEE Trans. Neural Networks*, **1**, 204–215 (1990).
[5] T. Kondoh, M. Yamaguchi, N. Ohyama, M. Oita, S. Tai and K. Kyuma, *Proc. 23rd Joint Conference on Imaging Technology*, 1993, pp. 199–202 [in Japanese].

10.3

Real Time Optical Correlator with a $Bi_{12}SiO_{20}$ Crystal

Katsuyuki Okada, Koji Ito, Toshio Honda and **Jumpei Tsujiuchi**

Abstract

This section discusses some techniques for improving the performance of an optical joint transform correlator with a photo refractive crystal. One of the improvement is a reduction in the aberration of the optical system. To accomplish this, a new design of a joint transform correlator is presented in which two separate transform lenses are used for the Fourier transform of the object and the reference images, and they are then joined on a Fourier plane. The second problem is due to the thickness of the crystal. An equation for the difference between the thickness of the crystal and the size of the Fourier plane is derived and, to reduce the size of the optical system, an optical system with a magnification optics is presented. Another problem is the degradation of the signal-to-noise ratio of the correlation image caused by the nonlinear response of the photo refractive crystal. This nonlinearity is due to photo carrier accumulation in the low light intensity area on the crystal, and the saturation of diffraction efficiency at the high light intensity area. To reduce the nonlinear effect, a technique is proposed for increasing the dark current of the crystal by flat white light illumination. The effectiveness of the new optical system and flat light illumination are verified experimentally by measuring the signal-to-noise ratio of the correlation images.

1 Introduction

Searching and recognizing objects are basic operations in associative memories and computer vision. For these purposes, correlation is one of the most important procedures. This operation, however, needs enormous computation. For example, taking a correlation of two 1000×1000 pixel images with a TV frame ratio needs at least a few giga-operations per second (10^9 operations/s), and this can hardly be treated with a conventional digital computer. On the other hand, optical methods have inherent advantages in the ability of parallel processing and high-speed operation.

One well-known optical technique is a holographic correlator [1]. This technique enables us make a correlation between the image of a object and the recorded images in a hologram at *the speed of light*. The drawback of this system, however, is difficulty of changing references that need to be recorded in the hologram.

Another technique is a joint transform correlator. Here the object and reference images are placed side by side on the input plane, and are Fourier transformed simultaneously on a recording device, such as a photo refractive crystal [2–4] or a TV camera [5]. This pattern is read out by another laser light and transformed again to make a correlation pattern. Since the two images are set in an image plane, rather than a Fourier plane, they can be changed easily with an electrically addressable spatial light modulator, such as a liquid crystal spatial light modulator.

We have been researching into an optical joint transform correlator with a photo refractive crystal. Although this type of correlator has the advantage of high parallelism of the optical method, and can perform the correlation operation in an almost TV frame interval, the performance of the system is restricted by the limitation of practical components for constructing the correlator. One of the limitations is the aberration of the lens used to transform the input patterns, and another is the thickness of the crystal. The other source of error is the nonlinear response of the photo refractive crystal. This causes degradation of the signal to noise ratio of the correlation pattern.

In this section we discuss the degradation of the correlation obtained by an optical joint transform correlator with a photo refractive crystal. In section 2 the basic configuration and the theoretical introduction of a joint transform correlator is presented. In section 3 the influence of the aberration of the optical system and a new design of the optical system to reduce the aberration is shown. In section 4 the influence of the thickness of the crystal is discussed, and a relation is derived between the thickness and the size of the crystal to be able to ignore the influence of the thickness of the crystal. In section 5 the degradation of the signal-to-noise ratio is discussed, and a technique is proposed to linearize the response of the crystal. In section 6 some experimental results are shown to verify the effectiveness of the new design of the optical system and the linearization of the response of the crystal by measuring the signal-to-noise ratio of the obtained correlation. In section 7 a conclusion is given.

2 *A Basic Optical Joint Transform Correlator*

Figure 10.3.1 shows the basic configuration of an optical joint transform correlator with a photo refractive crystal. On the input plane of the system two transparencies, one an image of an unknown object $O(x+h,y)$ and the other a set of reference images $R(x-h, y)$, are placed side by side and are illuminated by a parallel coherent light from an Ar^+ laser. The lens L_1 placed behind the images makes a Fourier transform of these two images on its focal plane. The intensity distribution I_F on the plane is

$$I_F \propto |\Im[O(x, y) + R(x, y)]|^2 = |\Im[O(x, y)]|^2 + |\Im[R(x, y)]|^2$$
$$+ \Im[O(x, y)]\Im[R(x, y)]^* + \Im[O(x, y)]^*\Im[R(x, y)] \quad (1)$$

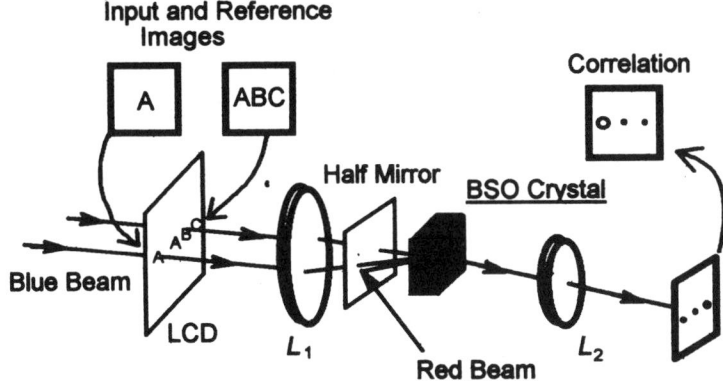

Figure 10.3.1
Schematic diagram of a basic joint transform correlator with a photo-refractive crystal

where $\Im[\cdot]$ denotes the Fourier transform and an asterisk denotes the complex conjugate. This intensity distribution produces refractive index modulation in the photo refractive crystal.

A light beam from a He–Ne laser illuminates the crystal through a half mirror, and the diffracted beam, which corresponds to the third term in equation (1), is transformed by another lens to create the correlation pattern between the object and the reference images.

The photo refractive material we used is a $Bi_{12}SiO_{20}$ (BSO) crystal. This crystal is sensitive only to light in the short wavelength region and not to light in the long wavelength region. Thus the processes of writing by the Ar^+ laser and reading by the He–Ne laser can proceed independently, and the correlation of the images can be obtained in real time [2–4].

3 Separate Transform Lenses Configuration

One of the disadvantages of the design of Figure 10.3.1 is the need for a large Fourier transform lens to cover the two sets of input images. This causes an increase in the system's weight and pushes up its cost. In addition, the aberration of the large lens influences the intensity of the correlation peak. Assuming that the spherical aberration is dominant in the aberration, the wavefront aberration W can be expressed by a fourth-order polynomial

$$W = ar^4 + br^2 \qquad (2)$$

where $r (0 \leq r \leq 1)$ is the normalized radial coordinate on the pupil plane of the transform lens. To minimize the wavefront aberration, usually the coefficient of defocus b is set opposite to the coefficient of the fourth-order a, that is

$$b = -a \qquad (3)$$

If we set the offset of the input image as $\frac{1}{2}$, and set the size of the image as 1, then the maximum wavefront aberration in the image is $\pm a/8$.

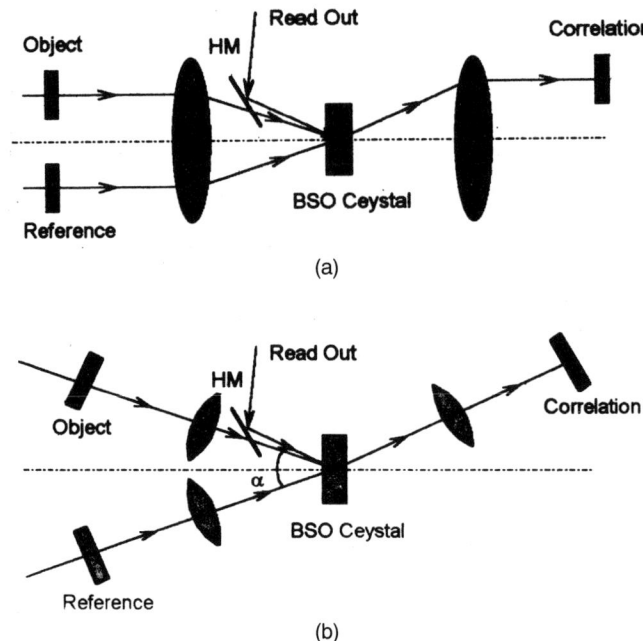

Figure 10.3.2
Designs of a correlator of (a) a conventional one transform lens system, and (b) a separate transform lens system

On the other hand, if the image can be placed on the optical axis, the range of r can be reduced to $\frac{1}{2}$ to $-\frac{1}{2}$. The wavefront aberration then becomes $\pm a/128$, or the aberration can be reduced by a factor of 16. Equation (1) indicates that the operation needs only to take a Fourier transform of two input images and then add the transformed functions in the Fourier plane. In this sense, it is not necessary to take a Fourier transform by one lens.

On the basis of these considerations we propose a modified version of the optical system for a joint transform correlator. Figure 10.3.2(b) shows a schematic diagram of the new design. In this system, two separate lenses transform the object and the reference images, and then both results fall on the BSO crystal. As mentioned before, the aberration is reduced by a factor of 16 compared with the basic configuration of Figure 10.3.2(a).

4 Influence of the Crystal Thickness

Another problem of the optical system is caused by the angular selectivity of the thick hologram recording material. Although a precise analysis of the influence of the crystal thickness has been reported [6,7], here we make a simple approximation in which the reference image is assumed to be small. In this case the influence of the thick recording material becomes similar to the reconstruction of a volume hologram. The intensity of the diffracted correlation pattern $I(x_c)$ is weighted by the square of a sine function, i.e.

$$I(x_c) \propto \text{sinc}^2\left(\frac{hD}{n\lambda f^2}x_c\right) \qquad (4)$$

where x_c is the coordinate in the correlation plane, D is the thickness of the crystal, f is the focal length of the transform lens, h is the offset of the images in the input plane, λ is the wavelength, and n is the refractive index of the crystal.

For a temporal criterion to reduce the influence of the thickness of the crystal, we assumed that the size of the correlation plane L_c should be smaller than the width of the sine function, that is

$$L_c < \frac{n\lambda f^2}{hD} \qquad (5)$$

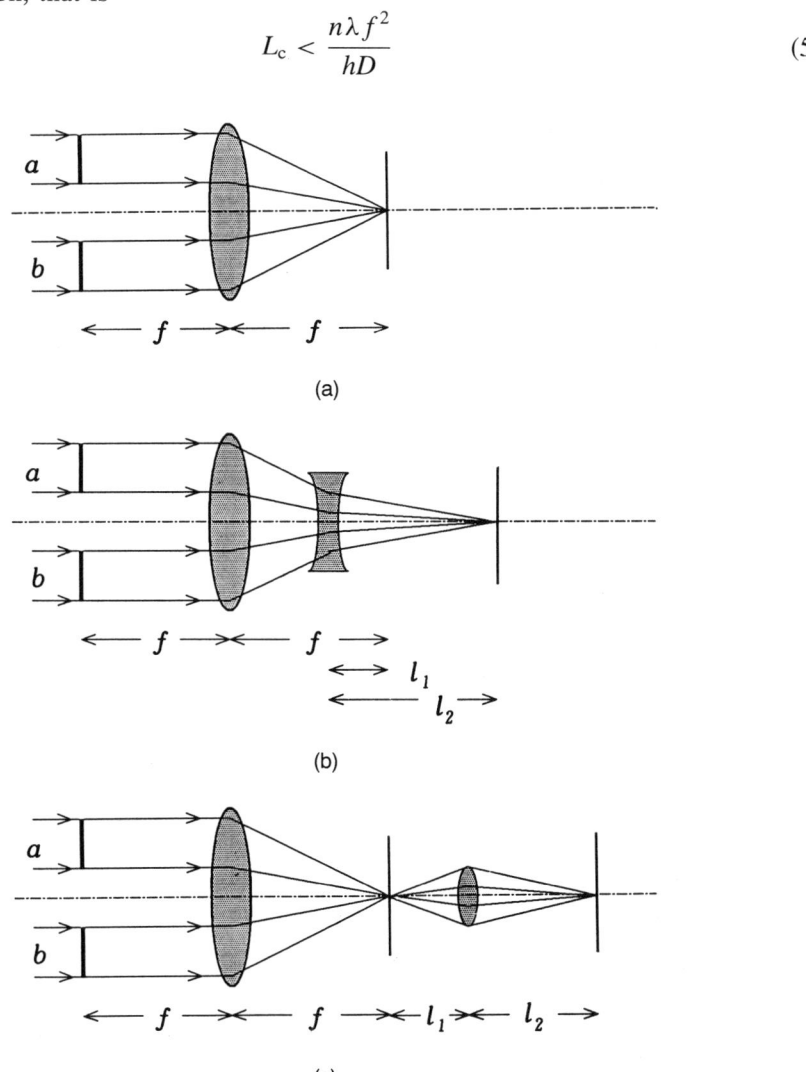

Figure 10.3.3
Magnification of the Fourier transform plane with a concave lens (b), and with a convex lens (c)

Then we have the following relation for the focal length of the transform lens:

$$f > \sqrt{\frac{DhL_c}{n\lambda}} \tag{6}$$

From equation (6) it is conceived that there are two ways to reduce the influence of the thickness of the BSO crystal. One is to reduce the offset of the images h in the input plane or, the same thing, to reduce the size of the images, but these are difficult due to the size and the resolution of the spatial light modulator. Another method is to lengthen the focal length of the Fourier transform lens f. The long focal length raises another problem regarding the size of the optical system. The length of the correlator needs to be four times the focal length, but a large optical system tends to be influenced by air turbulence.

To solve this problem, magnification optics can be used to lengthen the focal length effectively. There are two ways to magnify the Fourier plane. One is to employ a concave lens [6,7], and the other is to employ a convex lens [4]. In Figure 10.3.3, optical systems with a magnification lens are shown, and the magnification ratio M is decided as

$$M = \frac{l_2}{l_1} \tag{7}$$

The focal length of the transform lens is lengthened by this factor.

5 Nonlinear Response of the Crystal

In principle, a joint transform correlator gives the exact correlation pattern between an object and the reference images. This is true even in an optical correlator, as long as the operation of equation (1) is performed exactly. The main assumption of the equation is that the diffraction efficiency of the crystal should be in proportion to the light intensity of the joint transform pattern. If this assumption fails, the result is no longer a correct correlation.

On the Fourier plane of a joint transform correlator, most of the incident light converges at the center of the Fourier plane, and a weak high-frequency spectrum part reaches the area around the peak. If the response of the crystal is linear, the diffraction efficiency at the peak will be high, and low in the surrounding area. This situation would be true in a short period just after the object light is turned on and the index modulation at the center peak does not reach the saturation value.

After this period, however, the index modulation at the center peak reaches and stays at the saturated value, whereas in the area around the peak, photo carrier accumulation continues to reach modulation at the saturation value. Therefore, although the period is different, the index modulation, and thus the diffraction efficiency, will finally become almost uniform on the crystal. This situation is far from the linear operation assumed in equation (1), and greatly reduces the signal to noise ratio.

The light intensity variation on the Fourier plane influences the local strength of the static electric field in the crystal. In the experiment, a DC electric field is applied across the crystal to increase the diffraction efficiency, and if its strength is completely uniform in the crystal, the diffraction efficiency would be in proportion to the light intensity. The conductivity of the crystal, however, is influenced by the carrier density, and thus by the light intensity, and the local electric field in inversely proportional to the conductivity of the area. Therefore, the strength of the electric field, and the diffraction efficiency, will be low at the center peak, where the conductivity is low, and become high in the area around the peak, where the conductivity is high. This effect causes a nonlinear response of the crystal, too.

To reduce these effects, it is effective to flatten the light intensity on the crystal. In a holographic correlator variation of the light intensity can be controlled by the reference light intensity, but, in a joint transform correlator, this technique cannot be used. We propose a new technique which adds a flat incoherent light illumination to the crystal to homogenize the light intensity. This flat illumination restricts the accumulation of a photo carrier around the center peak on the crystal, and makes the crystal response linear. If the intensity of the illumination is too strong, total diffraction efficiency may be decreased, but it can be controlled by monitoring the intensity of the joint transform pattern.

There may be another method to linearize the response, i.e. to increase the dark current of the crystal by doping some impurities in the crystal. It is, however, not easy to obtain a crystal with proper conductivity, and the response of the crystal cannot change during the operation, so we used the former technique in the experiments.

6 Experimental Results

Figure 10.3.4 shows the entire system of an optical joint transform correlator used for the experiments. The unknown object and the reference images are displayed separately on liquid crystal spatial light modulators (SLMs) and transformed by two Fourier transform lenses, L_3 and L_4, with focal lengths of 200 mm. In general, it is not easy to

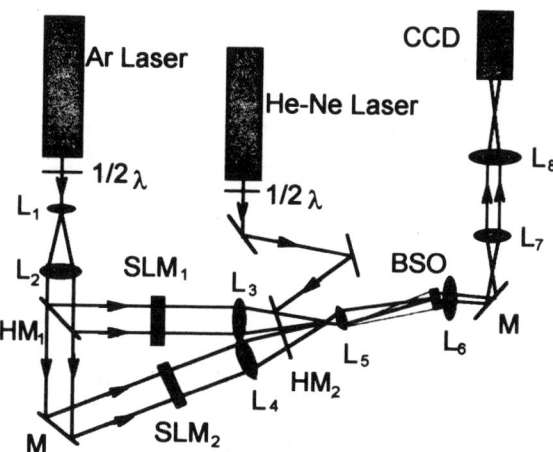

Figure 10.3.4
Full system of the real time optical correlator used for the experiments

obtain a well-corrected concave lens, but it is easy to obtain a convex lens, because such lenses are used for microscope objectives. Thus, the optical system with a convex lens seems more practical, and we adopted this configuration for the experiments.

The transformed patterns are superimposed on the Fourier plane and magnified by a microscope objective lens. In the optical system of Figure 10.3.4, the thickness of the crystal D is 3 mm, the refractive index n of the BSO crystal is 2.65, the wavelength λ is 0.515 μm, and the size of the correlation plane L_c and the offset h are 25 mm. Then the focal length of the transform lens f should be larger than 1.2 m from the relation of equation (6). The focal lengths of L_3 and L_4 are 200 mm and the magnification ratio is set to 15 in the experiment, then the effective focal length becomes 3 m, which is larger than the above value. A half mirror combines the light beam from a He–Ne laser to the transformed pattern, and it illuminates the crystal. The lens L_6 just behind the crystal compensates for the quadratic phase factor introduced by the magnification lens, and the diffracted light by the BSO crystal is demagnified and transformed by lenses L_7 and L_8 to make correlation pattern on a CCD camera.

A small computer controls all the systems, i.e. taking an image of the unknown object by a TV camera, cutting and pasting the interesting part of the object image onto the SLM, calling back the reference images from the computer memory and showing them on the SLM, taking the correlation pattern with a TV camera, and searching for the maximum peak in the correlation pattern. These operations are completed in almost real time.

In the first experiment the effect of the flat illumination is examined. For comparison of the obtained correlation, the signal-to-noise ratio of the correlation is defined as the ratio of the peak intensity to the standard deviation of the sidelobes, as shown in Figure 10.3.5. Figure 10.3.6(a) shows the object and the reference images, in which five characters are arranged in a vertical column for the object image and five are aligned in a horizontal column for the reference. The correlation of these images has 5×5 peaks, as shown in Figure 10.3.6(c); each is an autocorrelation of the character. Figure 10.3.7 shows the signal-to-noise ratio versus intensity of the flat light illumination, where the signal-to-noise ratios for all 25 peaks are averaged. This graph shows that the signal-to-noise ratio is improved as increasing the intensity of flat light illumination.

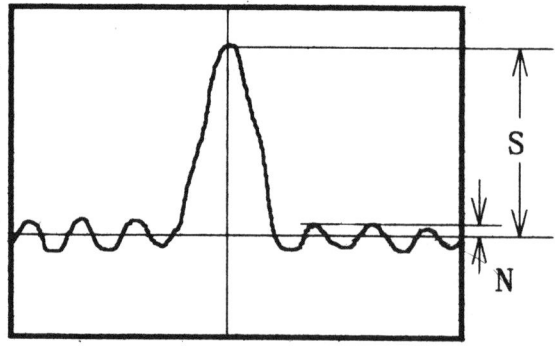

Figure 10.3.5
Signal-to-noise ratio of the correlation image

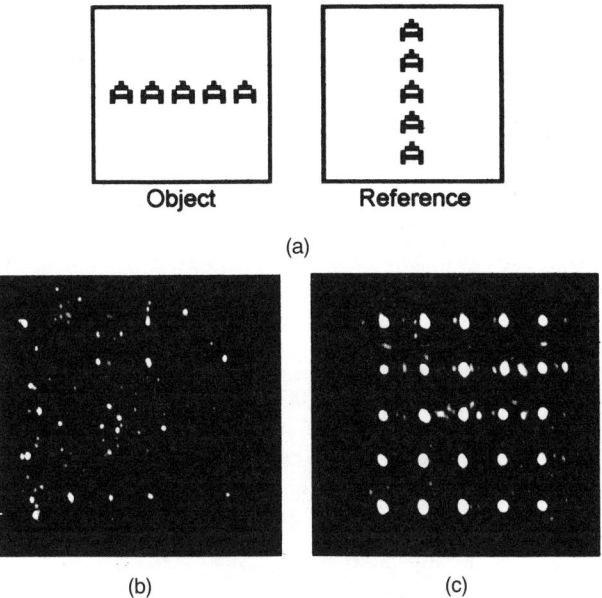

Figure 10.3.6
Matrix like correlation for character recognition used for signal to noise ratio measurement of the correlation image. (a) shows the object and the reference images, (b) shows the result without back light, and (c) shows 25 peaks of autocorrelation obtained with back light illumination

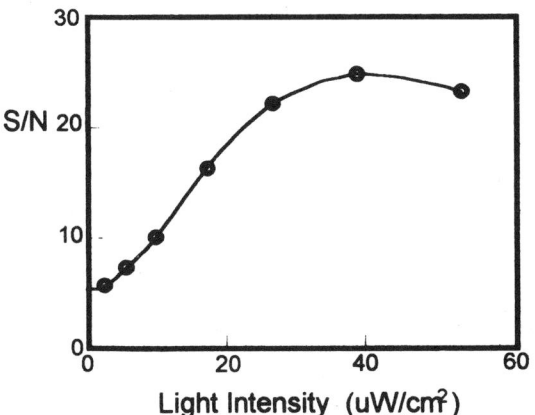

Figure 10.3.7
Result of signal-to-noise ratio versus back light intensity

Figure 10.3.8(a) shows another example of the object and the reference images, in which five different characters are arranged in vertical and horizontal lines. Five by five correlation peeks corresponding to the correlation will be obtained, as depicted in the Figure 10.3.8(b). Figure 10.3.8(c) shows the result of the correlation. Five bright autocorrelation peaks are shown in the photograph with some dimmer cross-correlation peaks.

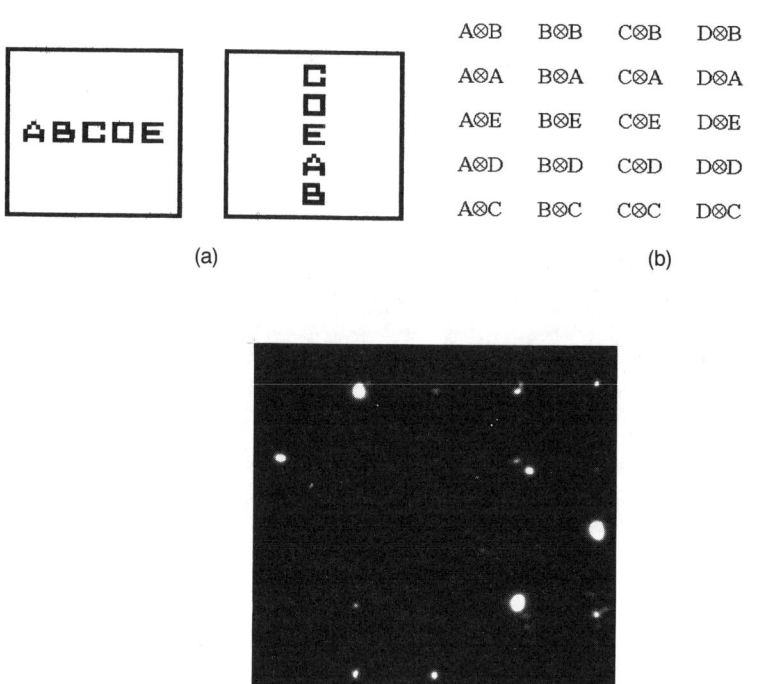

Figure 10.3.8
Results of experiments for character recognition. (a) are the object and the reference images, (b) is the expected correlations, and (c) is the experimental result

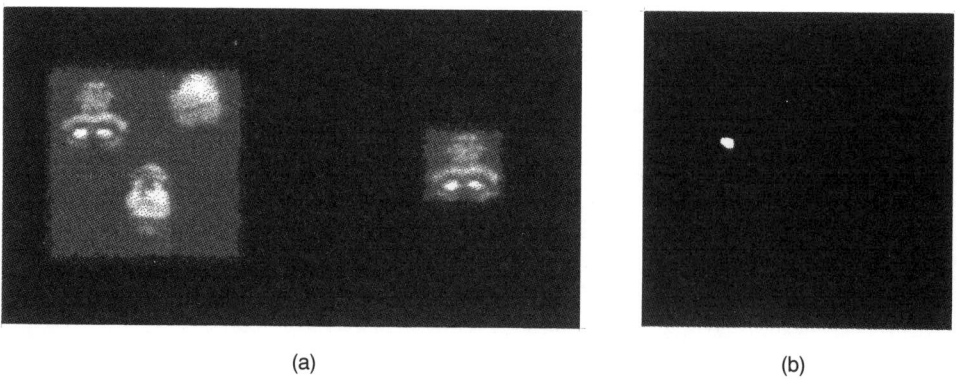

Figure 10.3.9
Results of experiments for half-tone images. (a) are the object and the reference images and (b) is the experimental result

Figure 10.3.9 shows the case of half-tone images. (a) shows the object and the reference images, and (b) shows the resultant correlation, in which a strong correlation peak can be recognized at one of the reference images corresponding to the object image.

7 Conclusion

In this section some improvement in the performance of an optical joint transform correlator with a photo refractive crystal has been discussed. To reduce the practical limitations of an optical system, a new configuration of a correlator is presented. In the system with two transform lenses, aberration of the lens is minimized, and by employing a magnification lens, a compact optical system is realized. Another problem of the correlator is degradation of the signal-to-noise ratio of the correlation image caused by the nonlinear response of the photo refractive crystal. This nonlinearity is due to photo carrier accumulation in the low intensity area on the joint transform plane, and saturation of the diffraction efficiency at the high intensity area. To reduce the nonlinear effect, a technique is proposed in which the dark current of a crystal is increased by flat white light illumination. Experimental results show the effectiveness of the flat light illumination by measuring the signal-to-noise ratio of the correlation images, and shows that the correlator recognizes objects in real time from a set of images.

The main advantage of this system is flexibility in the references, and this enables us to compare an object with various images sequentially. For future work, we are considering making this system flexible for scale change and orientation of the object. One of the methods is to take advantage of the system's real time operation feature, changing the scale and the orientation of objects and/or references until the brightest correlation peak is obtained. By these improvements of the system, an optical correlator would be a powerful tool for image recognition and associative memory.

References

[1] E.G. Paek et al., *Opt. Eng.*, **26**, 428 (1987).
[2] K. Okada and J. Tsujiuchi, *Interim Report of UUO Project*, **1**, 91 (1992).
[3] H. Rajbenbach, S. Bann, P. Refregier, P. Joffre, J.P. Huignard, H.S. Buchkremer, A.S. Jenson, E. Rasmussen, K.H. Brenner and G. Lohman; *Appl. Opt.*, **31**, 5666–5674 (1992).
[4] K. Okada, K. Ito, T. Honda and J. Tsujiuchi, *Opt. Rev.*, in press.
[5] B. Javidi and J.L. Horner, *Appl. Opt.*, **28**, 1027–1032 (1989).
[6] J.A. Davis, M.A. Waring, G.W. Bach, E.A. Lilly and D.M. Cottrell, *Appl. Opt.*, **28**, 10 (1989).
[7] T. Francis S. Yu, S. Wu, S. Rajan and D.A. Gregory, *Appl. Opt.*, **31**, 2416–2418 (1992).

10.4

Dynamics of Complex Neural Fields in a Phase-conjugate Resonator

Mitsuo Takeda and **Takaaki Kishigami**

Abstract

A new model of a neural network, whose state variables and synaptic weights can take complex values and whose dynamics are governed by a Hopfield-like energy function, is introduced. The model is referred to as *a phase-conjugate neural network* because the neurons have the function of phase conjugation and their dynamics have a close analogy with those of self-oscillation generated in a phase-conjugate resonator. A physical interpretation of the model is given, which shows that the optical gain medium in the resonator should have the function of phase conjugation in order for the generated complex optical fields to have an energy function that decreases monotonically with the time evolution of the fields. The results of experiments and computer simulations are presented which demonstrate the behaviors of the complex neural fields predicted by the theory.

1 Introduction

In all-optical neural networks implemented by using coherent light [1–3], the states of neurons and the synaptic weights are represented, respectively, by complex optical fields and complex amplitude transmission functions which have both amplitude and phase information [4,5]. In addition, the complex neural fields in such networks have dynamics that are continuous both in time and space. Recently, Noest [6] has proposed a complex neuron model called the *phasor neuron model*, and has shown that a Hopfield-like energy function [7] exists when the synapses have Hermitian symmetry ($T_{mn} = T_{nm}^*$). This model, however, is not compatible with a physical system of optics because it does not allow amplitude variations, and more importantly, because the law of light propagation,

Helmholtz's reciprocity theorem, demands symmetric synaptic weights ($T_{mn} = T_{nm}$) rather than Hermitian symmetry as required by the Noest model. This section reviews our recent research [8] on the dynamics of a complex neural network model. We propose an alternative model that is continuous in time and space, allows amplitude variations, and has symmetric synaptic weights to satisfy Helmholtz's reciprocity theorem. We show that the complex neural fields of our model have a close analogy to optical fields generated in a phase-conjugate resonator. On the basis of this analogy, we predict the existence of a Hopfield-like energy function that governs the dynamics of self-oscillating optical fields in a certain class of phase-conjugate resonators. We present the results of experiments and computer simulations that demonstrate the behaviors of the complex neural fields predicted by the theory.

2 Dynamics of the Phase-conjugate Neuron Model

Let the state of the complex neural field at time t and point r be $V(r, t)$, and its *internal* state be $u(r, t)$. Denoting the synaptic weight for the signal from the neural field at $\hat{r}$ to that at r by $T(r, \hat{r})$, the dynamics of our model is expressed by

$$\tau \frac{\partial u(r,t)}{\partial t} = -\alpha u(r,t) + \int_{-\infty}^{\infty} T(r,\hat{r})V(\hat{r},t)\mathrm{d}\hat{r}, \tag{1}$$

$$V(r,t) = g(|u(r,t)|)\frac{u^*(r,t)}{|u(r,t)|} \tag{2}$$

where $g(\cdot)$ is a nondecreasing real function, an asterisk denotes a complex conjugate, and τ and α are, respectively, a time constant and a damping parameter (which are assumed to be real parameters). We can show [8] that, for symmetric synapses $T(r,\hat{r}) = T(\hat{r},r)$, our complex neural network has a Hopfield-like energy function

$$E = -\frac{1}{2}\int_{-\infty}^{\infty}\int_{-\infty}^{\infty}\Re[T(r,\hat{r})V(r,t)V(\hat{r},t)]\mathrm{d}r\mathrm{d}\hat{r} + \alpha\int_{-\infty}^{\infty}\int_{0}^{|V(r,t)|}g^{-1}(s)\mathrm{d}s\mathrm{d}r \tag{3}$$

which decreases monotonically with the time evolution of the complex neural field: $\Re$ denotes the real part and s is a real parameter for integration.

3 Physical Interpretation

The dynamics of our complex neural network has a close analogy to that of optical field oscillation generated inside a phase-conjugate cavity with nonlinear gain given by degenerate four-wave mixing, as shown in Figure 10.4.1 [9,10]. We consider the complex amplitude of the optical field generated at point r inside the crystal as the state of neuron $V(r, t)$, and associate with the internal state of the neural field the complex amplitude of the diffraction grating formed inside the crystal by interference between the pump beam and the beams $T(r,\hat{r})V(\hat{r}, t)$ emitted from other points at $\hat{r}$ inside the crystal and reflected back into the crystal by a reflector at the other side of

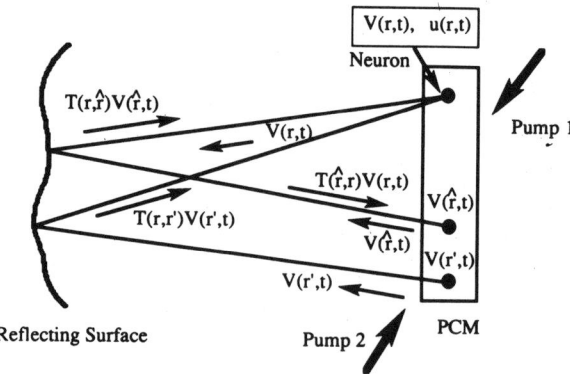

Figure 10.4.1
Analogy of the phase-conjugate neuron model to the dynamics of optical fields generated in a phase-conjugate resonator

the cavity. Regarding the transmission function $T(r, \hat{r})$ as the synaptic weight between the two points at r and $\hat{r}$, we have symmetric synapses $T(r, \hat{r}) = T(\hat{r}, r)$ from the Helmholz reciprocity theorem. By these associations, equation (1) can be regarded as a writing process of the grating, where the complex grating amplitude $u(r, t)$ increases in proportion to the sum of the writing fields $\int_{-\infty}^{\infty} T(r, \hat{r}) V(\hat{r}) d\hat{r}$. The damping term $-\alpha u(r, t)$ can be interpreted as an erasing process which the grating undergoes at the same time. Likewise, equation (2) can be considered as a read-out process of the grating by the pump beam. This read-out process produces a phase-conjugated beam whose amplitude is transformed by a nondecreasing function $g(\cdot)$ which may incorporate the effects of possible nonlinear gain saturation and/or thresholding. We should note the important role played by the phase conjugation. If the phase conjugation were not included, the system would form, in general, an unstable cavity, and the optical field inside the cavity would not converge into a stable mode even when sufficient gain is provided. Since a monotonic decrease in the energy function means that the complex neural field must converge into some stable state, the phase conjugation function plays an important role in the existence of the energy function. In other words, the existence of the energy function in the proposed model is guaranteed by the fact that a phase-conjugate mirror can form a stable cavity irrespective of the shape of the mirror in its counter part (of course, part of the reflected light needs to return to the phase-conjugate mirror) [11,12].

4 Energy Function and Total Intensity of the Oscillating Fields

It is of interest to examine if we can observe the predicted energy function of the complex optical neural fields, equation (3), by experiments. It generally is not possible to specify the synaptic weights $T(r, \hat{r})$ for all possible optical ray paths between the distributed neurons. We therefore eliminate them by substituting equations (1) and (2) into equation (3), and obtain [8]

$$E = -\frac{1}{2} \int_{-\infty}^{\infty} \left\{ |V(r,t)| \left[\tau \frac{\partial}{\partial t} g^{-1}(|V(r,t)|) + \alpha g^{-1}(|V(r,t)|) \right] - 2\alpha \int_{0}^{|V(r,t)|} g^{-1}(s) ds \right\} dr \quad (4)$$

This relates the energy function to the modulus of the field amplitude $|V(r,t)|$. In most cases where the oscillation grows rather slowly, both the field amplitude and the grating amplitude remain small for some time period after the start of oscillation, so that $|V(r,t)|, |u(r,t)| \ll 1$ for $0 \leq t \leq t_w$. In such a weak-field limit, we may expect that the gain function can be approximated by a linear function $g(|u(r,t)|) \approx a|u(r,t)|$, and we have

$$g^{-1}(|V(r,t)|) \approx a^{-1}|V(r,t)| \quad (5)$$

where a is a positive constant. Substituting equation (5) into equation (4), we have

$$E = -\frac{\tau}{4a} \frac{dI}{dt} \quad (6)$$

where I is the total intensity or the power of the optical field defined by

$$I(t) = \int_{-\infty}^{\infty} |V(r,t)|^2 dr \quad (7)$$

Thus we have shown that, in the weak-field limit, the energy function is proportional to the time derivative of the sign-reversed total intensity of the fields. In other words, the energy function (with its sign reversed) is proportional to the rate of the intensity growth of the oscillating beam. Since we have already shown that $dE/dt \leq 0$, we have

$$\frac{d^2 I}{dt^2} = -\frac{4a}{\tau} \frac{dE}{dt} \geq 0 \quad (8)$$

which states that the total intensity $I(t)$ grows as a downward convex function of time. Finally, we point out again that, since the total intensity of the oscillating beam $I(t)$ is a measurable quantity, the behavior of the energy function in the weak-field region can be observed by experiments.

5 Dynamics of Simulated Complex Neural Fields

Computer simulations were performed for a one-dimensional cavity to demonstrate that the complex neural field of the proposed model converges into one of its eigen modes by decreasing the value of its energy function. As a reflecting surface, we assumed a parabolic mirror with a radius of curvature $r = 500$ mm (focal length $f = 250$ mm). The phase-conjugate mirror is located at the focal point of the mirror so that the synaptic weight represented by the transmission function for the light propagation via the reflecting mirror becomes (apart from a constant factor) $T(x, \hat{x}) = \exp(-ikx\hat{x}/f)$. The field reflected back to the crystal is given by the Fourier transform $\int_{-\infty}^{\infty} T(x, \hat{x}) V(\hat{x}) d\hat{x}$. The amplitude is transformed by a sigmoid function $g(|u|) = \tanh(|u|/T)$ (with $T =$

240) to incorporate the effect of gain saturation. The effective size of the phase-conjugate mirror is determined by the diameter of the pump beams, which was chosen as 0.72 mm. The wavelength is 0.5 μm, $\alpha = 1$, and $\tau = 1$. For the initial state, we used a weak random field whose intensity and phase distributions are shown in Figure 10.4.2(a). The neurons first change their states so as to have a smoother phase distribution, as shown in

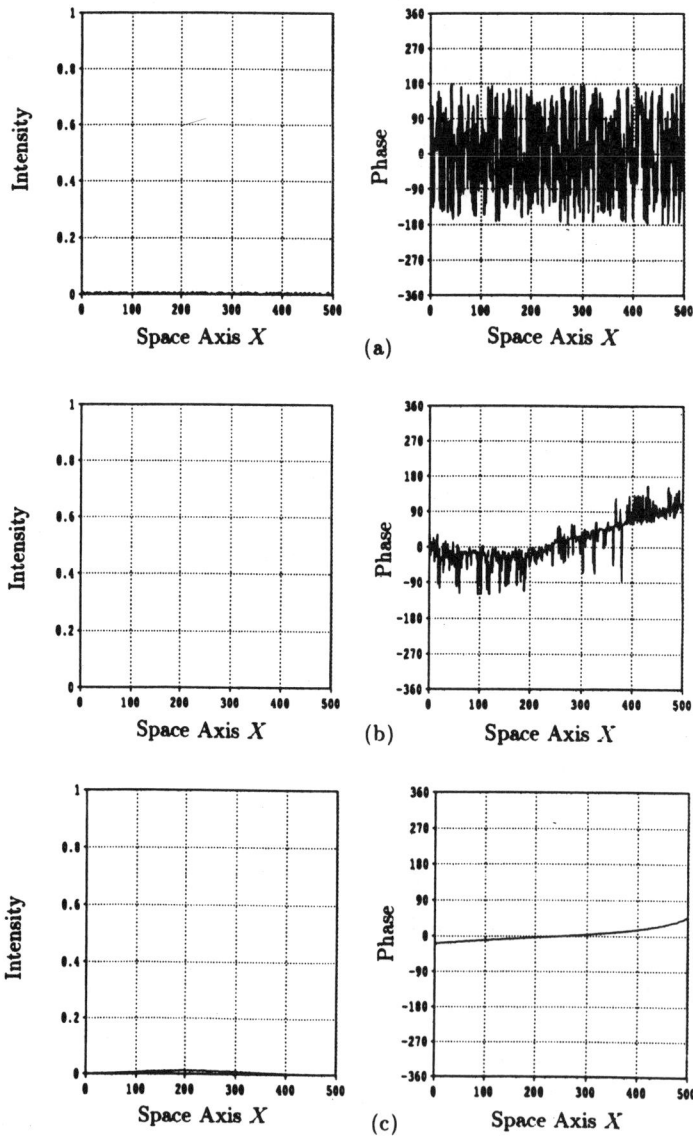

Figure 10.4.2
An example of simulated complex neural fields that converge to a mode like a fundamental Hermite–Gaussian mode. Figures on the left and right show, respectively, the intensity distributions and the phase distributions at the initial state [(a)], after 1 time unit [(b)], and after 15 time units [(c)]

Figure 10.4.2(b) and Figure 10.4.2(c). They show the intensity distributions (left) and the phase distributions (right) after 1 time unit [(b)], and after 15 time units [(c)], where the time unit is defined as a period during which all the neurons renew their states. Physically, this process of phase smoothing may be interpreted as being performed by spatial lowpass filtering. The Fourier transform kernel of the synaptic weights

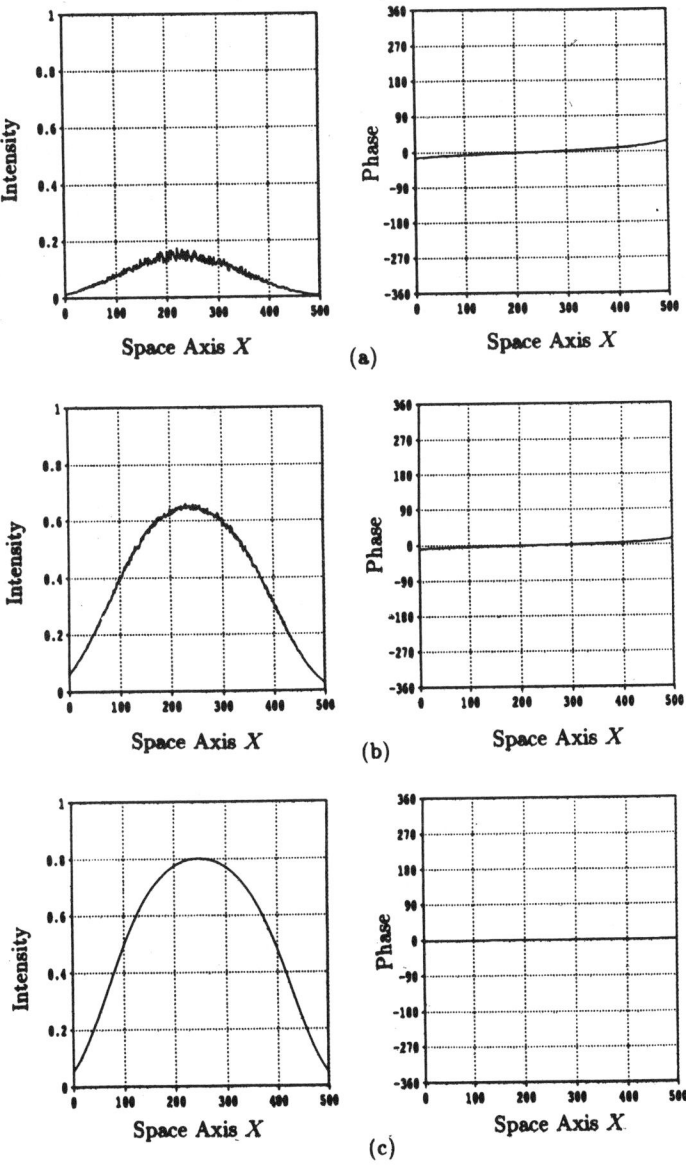

Figure 10.4.3
An example of simulated complex neural fields that converge to a mode like a fundamental Hermite–Gaussian mode. Figures on the left and right show, respectively, the intensity distributions and the phase distributions after 20 time units [(a)], after 26 time units [(b)], and after 50 time units [(c)]

produces a spatial frequency spectrum distribution of the neural fields over the PCM (phase-conjugate mirror) from which they originated. Only the very low spatial frequency components of these spectrum distributions will be returned as a phase-conjugate beam because we have limited the effective size of the PCM. Since the spatial frequency spectrum distribution is most sensitive to the phase distribution of the original fields, we may consider it natural that the phase of the complex neural fields first tries to take a smoother spatial distribution. Once the smooth phase distribution is achieved, the fields become more and more concentrated onto the effective area of the PCM, and the intensity starts to increase. The behavior of the simulated complex neural fields was found to be in agreement with that predicted from the physical picture described above; Figure 10.4.3 shows the intensity distributions (left) and the phase distributions (right) after 20 time units [(a)], after 26 time units [(b)], and after 50 time units [(c)]. In this example, the fields converged into a mode that looks like a fundamental Hermite–Gaussian mode as shown in Figure 10.4.3(c). Illustrated in Figure 10.4.4(a) is the time variation of the (sign-reversed) total intensity $-I(t) = -\int_{-\infty}^{\infty} |V(x,t)|^2 dx$, which starts to saturate at around 25 time units. In Figure 10.4.4(b), the solid line shows the energy function which reduces its value monotonically as predicted by theory. The broken line shows the time derivative of the sign-reversed total intensity $-dI(t)/dt$ scaled by a factor $\tau/4a$. Note that, as predicted by equation (6), it can well represent the energy function (apart from the constant scale factor) in the weak-field region that appears to last until 20 time units. This means that we can observe the behavior of the energy function in the weak-field region experimentally by detecting the total intensity of the fields and computing its sign-reversed time derivative. In another example where we increased the weak-field gain to $T = 200$ and the effective size of the PCM to 0.90 mm, the field distribution converged to a higher order mode with sign-reversed twin peaks and a phase jump by π. Here again it was found that the energy function was well approximated by the time derivative of the sign-reversed total intensity $-dI(t)/dt$ in the weak-field region.

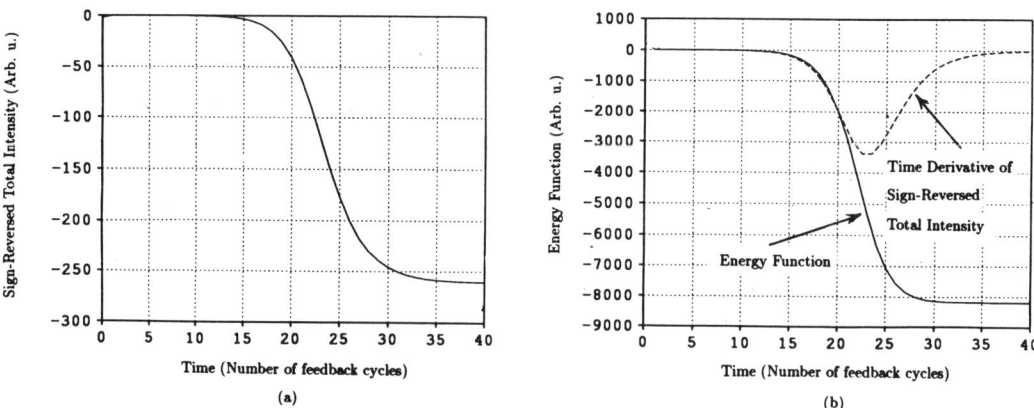

Figure 10.4.4
(a) Sign-reversed total intensity. (b) The energy function (solid line) and the time derivative of the sign-reversed total intensity (broken line), both exhibiting similar behaviors in the weak-signal region which lasts for 20 time units

6 Experiments

On the basis of the analogy that we have found between the complex neural fields of our phase-conjugate neuron model and the optical fields generated in a phase-conjugate resonator, we carried out experiments to observe the behavior of the energy function in a physical system, as shown in Figure 10.4.5. The phase-conjugate resonator is formed by combining a conventional mirror M_4 with a PCM which consists of a $BaTiO_3$ crystal and a pair of pump beams, pump1 and pump2, mutually counter-propagating from mirrors M_2 and M_3. The optical power for the pump beams is supplied by an argon-ion laser operating at a wavelength of 514.5 nm. Experiments of self-oscillation using such degenerate four-wave mixing have already been reported by many people. It should therefore be emphasized that our aim is not to demonstrate the self-oscillation itself but to observe the predicted behaviors of the energy function that we have associated with the optical fields in the phase-conjugate resonator. Likewise, studying the behavior of the total intensity of the oscillating beam may not be of interest by itself, since it may have also been done by many people. However, we consider that the significance of our experiments lies in that they are conducted with the understanding that the Hopfield-like energy function of the optical fields in the phase-conjugate resonator can be observed through measurement of the total intensity of the oscillating fields and the computation of its sign-reversed time derivative. Returning to Figure 10.4.5, we observe the field intensity distribution in the PCM by a CCD camera focused on the $BaTiO_3$ crystal, and take the intensity distribution into a frame memory for display and analysis. At the same time, we detect the total intensity of the oscillating field with a photodetector and take it into a personal computer to compute its sign-reversed time derivative. Figure 10.4.6 shows an example of the intensity growth of the optical fields observed by the CCD camera, which resembles the example of the computer simulations depicted in Figures 10.4.2 and 10.4.3. Figure 10.4.7(a) shows the sign-reversed total intensity of the optical

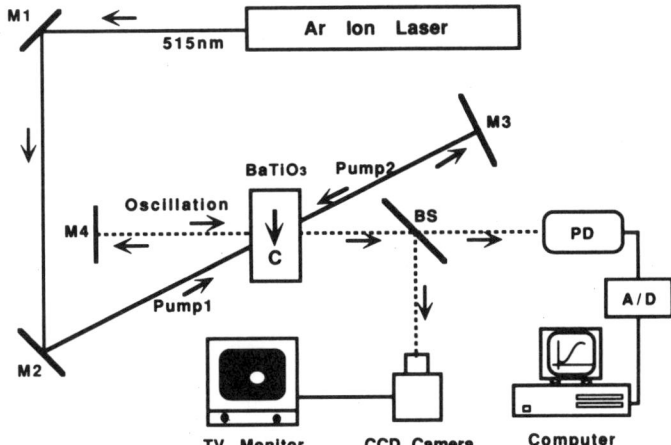

Figure 10.4.5
Experimental set-up for the observation of the Hopfield-like energy function of the optical fields generated in a phase-conjugate resonator

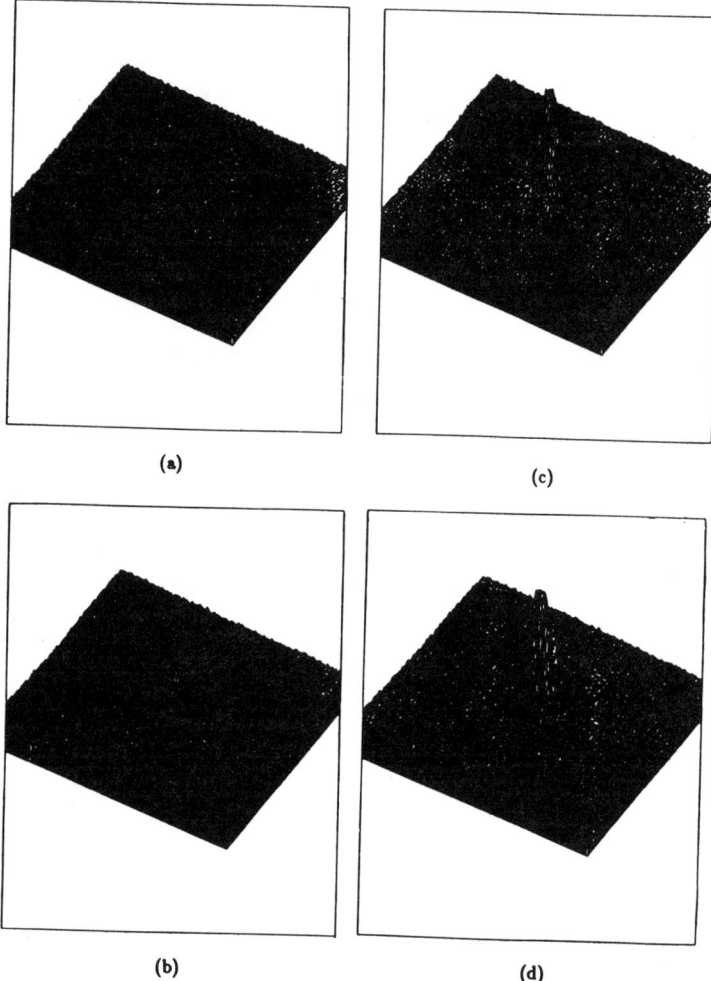

Figure 10.4.6
Intensity growth of optical fields in a phase-conjugate resonator: (a) after 80 s, (b) after 100 s, (c) after 150 s, and (d) after 240 s

fields which resembles the result of simulations shown in Figure 10.4.4(a). Its time derivative is shown in Figure 10.4.7(b). Ignoring the noisy fluctuations enhanced by differentiation, we can see that it decreases monotonically in the weak-field region which appears to last for approximately 150 s after the start of pumping. According to equation (6) and the result of the computer simulation shown in Figure 10.4.4(b), we consider that the derivative of the sign-reversed total intensity shown in Figure 10.4.7(b) represents the Hopfield-like energy function of the optical fields in the region where the signal is small.

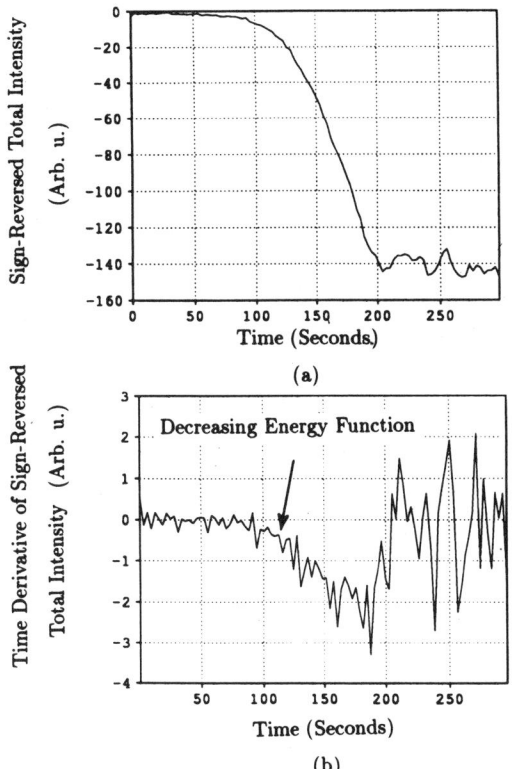

Figure 10.4.7
(a) Sign-reversed total intensity of the optical fields, and (b) its time derivative representing the energy function that decreases monotonically in the weak-field region (ignore the noisy fluctuations enhanced by differentiation)

7 Conclusion

We have reviewed our recent research on the dynamics of a complex phase-conjugate neural network model with a Hopfield-like energy function. We have pointed out that the dynamics of our complex neural network model have a close analogy with the dynamics of self-oscillation generated in a phase-conjugate resonator. From the physical interpretation of the model we have found that the optical gain medium should have a phase-conjugate property in order for the generated optical fields to have the Hopfield-like energy function that decreases monotonically with the time evolution of the fields. We have shown that, in the weak-field limit, the energy function can be approximated by the time derivative of the sign-reversed total intensity of the fields, and is observable by experiments. We have conducted experiments and computer simulations, and demonstrated the behaviors of the complex neural fields predicted by the theory.

References

[1] D.Z. Anderson, *Opt. Lett.*, **11**, 56–58 (1986); D.Z. Anderson and M.C. Erie, *Opt. Eng.*, **26**, 434–444 (1987).
[2] B.H. Soffer, G.J. Dunning, Y. Owechko and E. Marom, *Opt. Lett.*, **11**, 118–120 (1986).
[3] A. Yariv and S. Kwon, *Opt. Lett.*, **11**, 186–188 (1986); A. Yariv, S. Kwon and K. Kyuma, *Appl. Phys. Lett.*, **48**, 1114–1116 (1986).
[4] D. Psaltis, D. Brady and K. Wagner, *Appl. Opt.*, **27**, 1752–1759 (1988).
[5] G.R. Little, S.C. Gustafson and R.A. Senn, *Appl. Opt.*, **29**, 1591–1592 (1990).
[6] A.J. Noest, in *Neural Information Processing Systems*, D.Z. Anderson, Ed., American Institute of Physics, New York, 1988, pp. 584–591; A.J. Noest, *Phys. Rev. A*, **38**, 2196–2199 (1988); A.J. Noest, *Europhys. Lett.*, **6**, 469–474 (1988).
[7] J.J. Hopfield, *Proc. Nat. Acad. Sci. USA*, **81**, 3088–3092 (1984).
[8] M. Takeda and T. Kishigami, *J. Opt. Soc. Am.*, A **9**, 2182–2191 (1992); M. Takeda and T. Kishigami, *Proc. SPIE*, **2039** *Chaos in Optics*, 1993, 314–322.
[9] A. Yariv and D.M. Pepper, *Opt. Lett.*, **1**, 16–18 (1977).
[10] J. Feinberg and R.W. Hellwarth, *Opt. Lett.*, **5**, 519–521 (1989).
[11] A. Yariv, *Introduction to Optical Electronics*, CBS College Pub., New York, 1985, p.511.
[12] J. Feinberg and K.R. MacDonald, in *Photorefractive Materials and Their Applications II*, P. Günter and J.-P. Huignard, Eds., Springer, Berlin, 1989, p.154.

11
MULTI-DIMENSIONAL OPTICAL SENSING AND PROCESSING

11.1

Display Systems of Autostereoscopic 3D Images

Joji Hamasaki

Abstract

The section discusses the image flippings common for autostereoscopic images composed of sampled views, advancements in 3D TV displays, data compression schemes for a variety of 3D displays, and reflection characteristics of deformable mirror devices for a new 3D camera scheme.

1 Introduction

An autostereoscopic image is a 3D image the corresponding aspect of which can be seen whenever an observer changes his direction of viewing. The autostereoscopic character is necessary for interactive observation because an observer moves around interactively to understand the shape of objects in his natural daily life. The quality of a 3D image depends on spatial resolution, temporal resolution, brightness, latitude and chromaticity, noisiness, the field of view, *the autostereoscopic character, the depth of object space, image flipping, phantom image*, and *the viewing zone width and depth*. The last four factors are specific to an autostereoscopic image. The quality depends also on the needs of the vertical parallax.

An autostereoscopic image has been displayed with techniques belonging to three major methods: *the lens-plate method, holography*, and *the volume scanning method*. Lenticular and slit array methods and fly's-eye-lens and pin-hole array methods, including Integram, follow the same principle of the lens-plate method in reconstructing 3D images. All of these major methods need to handle a huge amount of data, because so many pairs of eyes, either of many observers or due to the motion of an observer, are looking at an object from different directions.

Let us review just how huge the amount of data is for the major methods. Assuming that the pairs of eyes are looking at the whole field of view uniformly, one can confine the quality of a 3D image into three factors, namely *the spatial resolution, the field of view,* and *the viewing zone width*. Then, the bandwidth necessary for analogue circuits of a display belonging to a major method can be compared. It can be shown that the necessary bandwidth for a lenticular display is proportional to that of a sampled holographic display, but the latter is 16 times greater than the former, even for images without vertical parallax. A numerical example of 40 cm × 40 cm × 20 cm (depth) object space with 500 × 500 lateral resolutions and 30 frames/s rate, and viewed at 100 cm distance from 40 cm wide viewing zone, the minimum bandwidth amounts to 94 MHz, which is 25 times greater than the bandwidth necessary for a 2D display. Among images with both horizontal and vertical parallax, the smallest bandwidth is obtained with a volume scanning display. For the same numerical example, this amounts to 375 MHz. The greater the bandwidth is, the greater the resolution necessary to display an image and the computational burden necessary to generate picture data. For these reasons, the lenticular display is the major subject of this investigation.

This section is divided into three parts: (1) the sampling errors of a lenticular image, which are specific to autostereoscopic images composed of sampled views; (2) advancements in 3D TV displays; and (3) data compression schemes including a 3D camera with which the compressed data for a variety of 3D displays will be obtained.

2 *Sampling Errors of Lenticular 3D Images*

2.1 *Flippings of 3D images composed of many sampled views*

A *view-point* is a projection center through which a view of an object scene is obtained or displayed. *Resolution-points* are the centers of many pixels of a view on the screen. To reduce the amount of data to be displayed with electronic devices, either with lenticular displays or sampled holographic displays, it is necessary to sample view-points. A ray that traverses a view-point and a resolution-point is called *a characteristic ray*.

The sampling errors of a 3D image, which is composed of a limited number of views, are the difference between two image positions: one that is reconstructed by characteristic rays generated with the display system, and the other that an observer at an arbitrary position should see geometrically. Since a characteristic ray is the central ray of its associated ray bundle with a finite spread, the rays that actually enter an observer's pupils are rays in the ray bundle belonging to the same view-point of a characteristic ray. Because the rays entering his pupils are different from characteristic rays, the position of the image that he should see geometrically is different from the image that is reconstructed by the intersection of the characteristic rays. When the observer moves while looking at a displayed 3D image, the rays entering his pupils suddenly belong to the next view-point and then cause a dicontinuous change in the position of the image. The discontinuity occurs in both the lateral and depth directions.

Display Systems of Autostereoscopic 3D Images

Table 11.1.1
Symbols explaining flippings of 3D images composed of many views

C_d : View-plane of a display	c : Horizontal separation of view-points
C : Observer's view plane	e : Horizontal pitch of resolution-points
E : Resolution-plane (a screen) of the display	P : Observer's interocular distance
F_d : Distance from the plane C_d to the plane E	$W = (c/e)/|F_d/F - 1|$: size of a 'View'
F : Distance from the plane C to the plane E	$\rho = (F_d/c)/(F/P)$: stability parameter

When the discontinuity exceeds the limit of the human visual system, it causes flippings for the observer.

The 3D image space is divided horizontally into many diamond-shaped regions (called *(binocular) Views*) by discontinuity lines which depend on the position of the eye. When the observer moves, the network of discontinuity lines also moves. Using the symbols explained in Table 11.1.1, the size of Views is proportional to the W in the retina coordinates, and if $|\rho|$ is smaller than unity, the depth discontinuity becomes infinite. Empirically, the relation $F_d/c = 30$, that is, an interocular image at 1 m distance of viewing, is a good choice to obtain stable observation within a viewing zone having a resonable depth. It is also important to design the screen of the display to alleviate a sudden decrease in brightness on the discontinuity lines. The phenomenon of Views has been precisely examined experimentally [1].

2.2 Scattering of a 3D image point due to discrete resolution-points

Another phenomenon of scattering of an image point, namely the sampling errors due to resolution-points, also occurs for a reconstructed 3D image. This phenomenon resembles the quantizaion error due to discrete pixels for a 2D image point, but it causes blurring

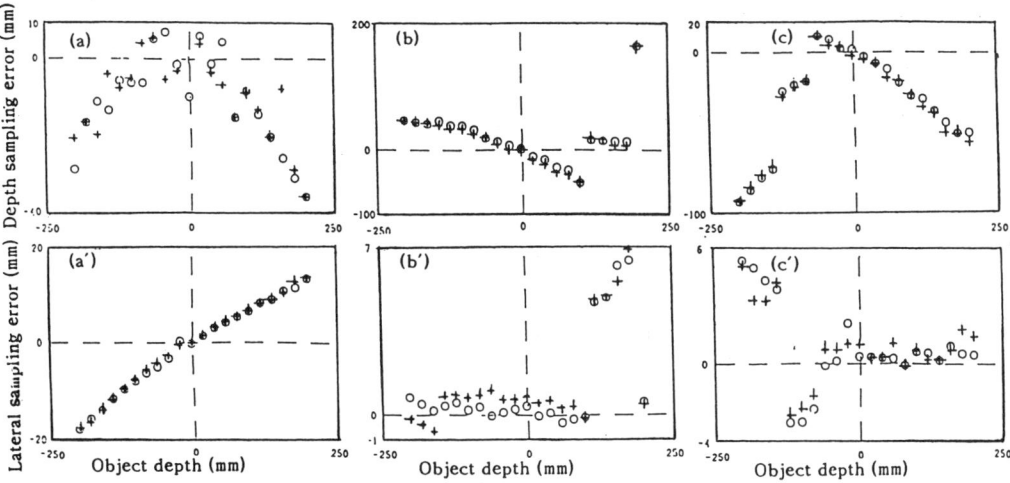

Figure 11.1.1
Comparison between experimental sampling errors (crosses) and theoretically predicted sampling errors (circles). Abscissa : Depth coordinate in the object space. Ordinate : Sampling error. (a, a') : $P = 100$ mm, $F = 1200$ mm; (b, b') : $P = 160$ mm, $F = 1200$ mm; (c, c') : $P = 160$ mm, $F = 800$ mm

of a 3D image point apart from the screen of a display. The ray belonging to a view-point and focusing on an object point is a ray in a bundle centered at a characteristic ray traversing to the same view-point and a certain resolution-point. The reconstructed ray bundle from this resolution-point has a characteristic ray as its central ray, which is different from the ray focusing on the object point. For this reason, a cluster of image positions are generated depending on which resolution-points are used at the reconstruction. When viewing an image point, the observer feels that the image point is blurred.

Figure 11.1.1 exemplifies the comparison between theoretical predictions (crosses) and experimental values (circles) of the view-point sampling errors. Large jumps indicate the discontinuity of View borders. Small wiggles in the theoretical predictions indicate scattering of the image points. Experimental values follow well the wiggles of the theoretical predictions.

3 Advancements in Autostereoscopic Lenticular TV Displays

3.1 Improvements of the autostereoscopic lenticular TV display using a CRT with beam indices

To improve the 3D TV display, a new high-resolution and high-brightness CRT with beam indices behind its convex face plate was made. With this CRT, displayed images became much brighter than the previous experiment. Figure 11.1.2 shows a block

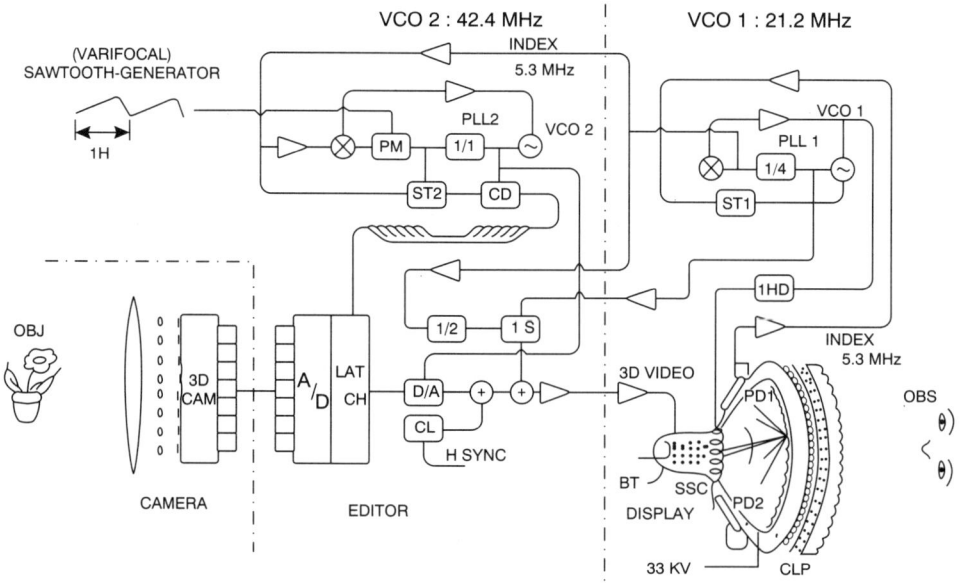

Figure 11.1.2
Block diagram of a direct-view-type autostereoscopic 3D TV display using a beam index CRT. PLL1 : to register the reproduced pixels with the index stripes. PLL2 : to vary the positional phase of pixels to register with lenslets

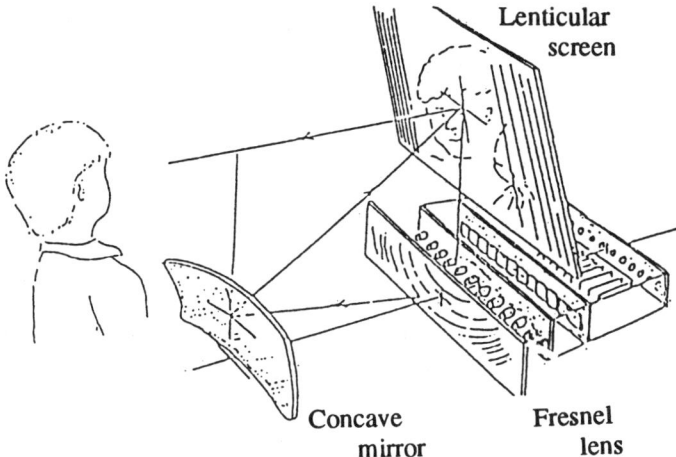

Figure 11.1.3
Schematic of a projection-type 3D TV display. Twelve elementary projectors have parallel axes

diagram of a 3D TV using the CRT [2], where two PLLs (phase-lock-loops) are used. The improvements include the scheme of *the index stabilization pulses* with which the phase of the VCO becomes almost independent of the fluctuating width of the index signal [3], the scheme of *an electronic equivalent of a varifocal lens* in the sense of varying electronically the distance F_d between the CRT screen and the view plane of the display [2], and a *simplified structure of a composite lens-plate* [4].

3.2 A projection type 3D TV display using an array of LCD panels

A projection-type display using inverted real images from many view points is one of the most promising candidates for an autostereoscopic 3D TV display. Figure 11.1.3 shows a scheme of an experimental 3D TV display of this type [5]. A one-dimensional array of 12 small LCD projectors is the core of this display. An elementary projector consists of a halide lamp for illumination, a condensor lens behind the LCD panel, and a projection lens. The optical axes of projectors are aligned in parallel. An off-axis Fresnel lens and a convex mirror are used for the transformation optic, with which separation of the projection lenses approaches an adequate value for an observer, and also all projected images merge on a lenticular screen. An observer in front of the screen sees a 3D image.

4 Data Compression and Associated 3D Camera Scheme for Autostereoscopic Images

4.1 Data compression of autostereoscopic images

The intermediate data common to all 3D TV displays are the 3D coordinate data and brightness data of object surfaces [6]. A 3D TV camera needs to have a camera processor which extracts the intermediate data. A 3D TV display needs to have a display processor, with which the picture data necessary for the display are computed from the intermediate data. The two schemes have been demonstrated for extracting 3D coordinate data and brightness data of object surfaces.

Analyzing texture pattern composed of horizontal line images of many views

When horizontal line images of the same line number of horizontally and densely equi-spaced views are arranged in the vertical direction in order of viewing, a texture pattern, which is composed of many overlapped bands with various slopes, is generated in the (m, n) space, where m and n are the indices of resolution-points and view-points, respectively, as depicted in Figure 11.1.4(a). The edge of the band corresponds to the intersection of the contour of 3D image and the horizontal plane determined by the line number. The slope corresponds to the depth of the point on the contour. From this figure another diagram is generated, as depicted in Figure 11.1.4(b), in which the abscissa and the ordinate are the horizontal and the depth coordinates in a normalized 3D space, respectively. The equi-value lines of the normalized variance of brightness seen from all views are drawn on this diagram. It is shown that the brightness constancy of an object point is reliable for finding the depth only if the point is on the contour. In the region between the two contour points on the same surface, it has been found that a small and systematic difference in the distributions of brightness is a much more reliable feature for finding the surface shape. Experiments to find the 3D coordinates of object surfaces have been demonstrated [7,8].

'Evolution strategy' in finding depth data from views

'The evolution strategy (ES)' [9] is a mathematical technique to optimize a large set of parameters of a system according to values of *an error function* which describes the

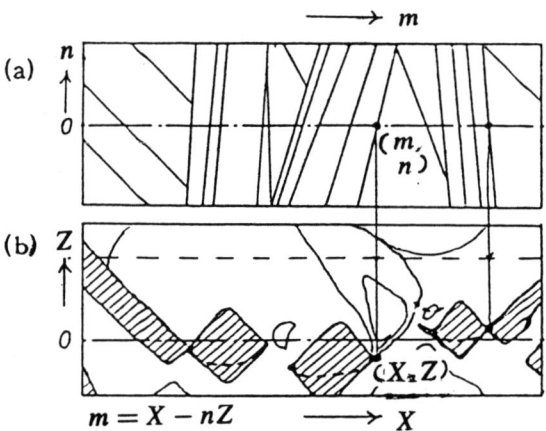

Figure 11.1.4
Analyzing texture pattern. (a) Texture pattern with various slopes. (b) Equi-variance diagram. Hatched areas have very small variance. Dashed curves show object surface shapes

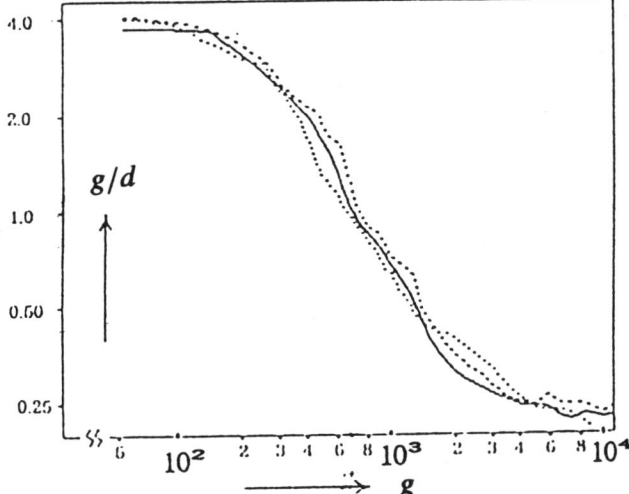

Figure 11.1.5
The rms error of matching (ES method). Abscissa : *g*, number of generations, ordinate : *gld*; rms error (bits/pixel), both in logarithmic scales

optimization. To find depth data from views experimentally, we adopted *the simplest interpretation* of *the assumption of the brightness constancy*. Then, the problem reduces to optimization of the displacements (disparities) of the records (in views) of object points to obtain the best match between the grey levels in views. These displacements are the parameters to be found by means of the evolution strategy.

Special considerations have been made on choosing *the error function, the control structure* adopting 'the one-fifth-rule' to adjust the step length of random shifts, and *the random number generator* itself. Also, *the booster* is found to be very important: it eliminates the peaks of the error distribution function where the matchings are poor. Furthermore, *the median filtering* is useful in eliminating extraordinary small values of the error function of some parent generations [10].

Figure 11.1.5 shows a typical evolution curve: the abscissa is the logarithm of the generation count *g*, and the ordinate is the rms error of matching (designated by *gld*) between the grey level functions, *f* and *mf*. After 200 generations, most of the peaks in the error distribution function *ef* were eliminated, and the progress of evolution was accelerated. At the 800th generation, the *gld* became below 1 bit (actually 0.75 bits) per pixel and an approximate shape of the object appeared on the displacement function *df* curve. At the 4000th generation, the form of *df* was mostly completed, and the *gld* became less than 0.25 bits per pixel. The time of computation is comparable with the other method [8].

4.2 Need for new camera technologies for communication

Empirically, 3D coordinates and brightness can be extracted from parallax views, but a huge amount of computational effort is necessary to find the correct correspondences between views because approximate depths are unknown from views having a very deep and clear focus. A compromise is needed between the complexity of the camera system and the allowable computational burdens of its processor. A practical TV camera to be further investigated should follow the principles of the 3D camera called 'Datagraphy' [11]:

(1) A volume scanning method should be incorporated with parallax processing. With this scheme an approximate depth is obtained.
(2) A whole 3D scene should be reconstructed from partial 3D surface images of an object.

In other words, a new camera scheme must optically more closely simulate our eye system. Both the lens adjustment and the accommodation actions of a 3D camera can be simultaneously performed with a pair of *deformable mirror devices* (DMDs) [12].

A new scheme of a 3D camera facilitating the extraction of depth data has been proposed, using DMDs [13,14]. Since a DMD has both deflection and focusing characteristics, it can direct the lines of sight of the constituent 2D cameras of the scheme to part of an object surface to obtain its depth data, and then scan the whole field of view successively. Furthermore, primary color data should be simultaneously obtained using three images of the adjacent orders of the diffracted waves from a DMD.

4.3 Reflection characteristics of DMDs

A DMD is a device consisting of an array of millions of tiny metallic plane mirrors (having 10 μm size, for example), integrated on a silicon LSI driver array. The mirror structure is made by micro-machining. The mirrors are suspended in air by fine metallic beams, and tilted and displaced by electrostatic force. Figure 11.1.6(a) and (b) explain two structures of DMD mirrors. The mirrors are on 2D lattice points, and on a plane where no driving voltages are applied.

When a plane wave is incident on a tilted tiny mirror, its reflected wave has a finite directivity according to the theory of diffraction from a finite aperture. Looking at the mirror from a fixed pupil, the intensity is modulated by tilting the mirror, and the phase is modulated by displacing the mirror. When a plane wave is incident on the large-scale array of mirrors, many orders of diffracted waves are generated. But, the intensities and the phases of the diffracted waves are composed of wavefronts from tiny mirrors, which are individually controlled in general. The main features of the array characteristics are explained with a 1D Fourier analysis under the paraxial approximation.

The symbols for the analysis are explained in Table 11.1.2. The x-axis is chosen in the plane of the array. The array of mirrors and the details of a mirror centered at $x = np$ are depicted in Figure 11.1.7(a) and (b), respectively. The reflection angle ψ_n is the angle of a reflected wave if the mirror has an infinite size in the same plane as the mirror. The suffix n is dropped if the values are the same for all n's.

Two conditions for the monochromatic incident wave are considered. (1) A wave having locally plane wavefronts, and (2) a wave having spherical wavefronts. The

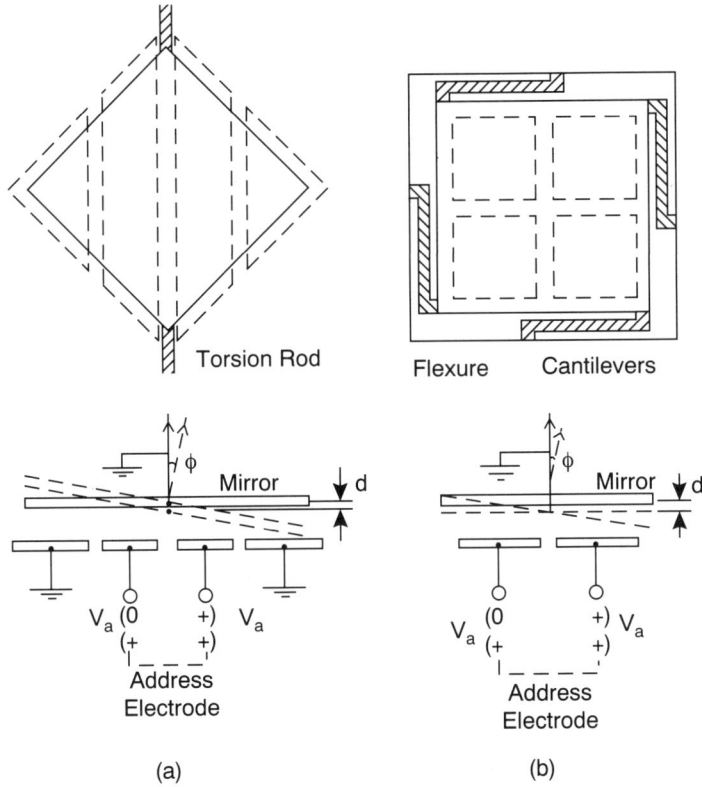

Figure 11.1.6
Structures of a mirror of a DMD (a) 45 degree torsion beam type; (b) flexture cantilever type

reflection characteristics with the necessary control conditions of DMDs are also summarized in Table 11.1.2 [14,15]. It is shown in this table that the ideal characteristic of a plane mirror and a spherical mirror are represented by the product of the sinc function $F_1(\xi; \sigma)$, which represents the diffraction from the aperture of a mirror, and the array factor $F_2(\xi; \rho)$, which represents infinite orders of diffraction from a periodic array of sources. To obtain the ideal characteristics of a spherical mirror, the mirrors must be displaced according to the parabolic displacement condition and also tilted according to the linear tilt condition.

If the tilting angle ϕ_n and the displacements d_n are the same for all mirrors, the array becomes *an echellet-like grating*. The array can be used as a plane mirror with a step-wise controllable tilting angle. If the mirror displacements are also controllable, the array can be used as a plane mirror having a continuously variable tilt.

For varifocal mirror characteristics it is evident that *a phase-type Fresnel zone plate lens* is more suitable for DMDs in a varifocal mirror application than *an amplitude-type Fresnel zone plate lens*. But both are of limited use if *the zone parameter τ* is not extremely small: for the phase type, $N < \tau^{-1}$, and for the amplitude type, $N < (7/8)(\tau)^{-1/2}$. To utilize the full aperture of DMDs for a varifocal mirror, it is necessary to have both displacement control and tilting angle control.

Table 11.1.2
Reflection characteristics of DMDs: $p = 0.01$ mm, $\lambda = 0.0005$ mm, $2N + 1 = 1000$, $z_o = 1000$ mm

Symbols:
$2N + 1$: Total number of tiny mirrors in the x-direction
n : Index of a mirror in the x-direction ($n = -N, -N + 1, \ldots, -2, -1, 0, 1, 2, \ldots, N - 1, N$)
m : Order of diffraction, ϕ_n : Mirror tilt angle, θ_n : Incident angle
p : Pitch of the array, d_n : Mirror displacement, ψ_n : Reflection angle
z_o, x_o : Coordinates of the center of spherical wavefronts
λ : Wavelength $\mu_n = \phi_n/\lambda$, $\rho_n = \theta_n/\lambda$, $\sigma_n = \psi_n/\lambda$, $\gamma_n = d_n/\lambda \cdot \theta_n + \psi_n = 2\phi_n$
ξ : Spatial frequency with respect to the x-coordinate, $2\alpha p$: Size of tiny mirrors $0 < \alpha \leq 1/2$

Conditions:
Synchronization with diffraction order : $\phi_{(m)} = (1/2)m\lambda/p$.
Linear tilt condition : $\phi_n = \phi_o - np/(2z_o)$, $\phi_o = (1/2)(\theta_o + \psi)$, $\theta_o = x_o/z_o$, $\tau = p^2/(\lambda z_o)$: 'Zone parameter'
$\mu_n = \mu_o - n\beta$; $\mu_o = (1/2)(\rho_o + \sigma)$, $\beta = \tau/(2p)$. k_n : integers
Linear displacement condition : $\gamma_n = 2\mu np + k_n$, k_n : integers
Parabolic displacement condition : $\gamma_n = -(\tau/2)n^2 + 2\mu np + k_n$

Functions:
$F_1(\xi; \sigma) = 2\alpha p \sin[\pi(\xi - \sigma)2\alpha p]/[\pi(\xi - \sigma)2\alpha p]$. : Sinc function
$F_2(\xi; \rho) = \sin[(2N + 1)\pi(\xi + \rho)p]/\sin[\pi(\xi + \rho)p]$. : Array factor
$F_3(\xi; \rho_o, \tau) = 1 + 2\Sigma(n = 1, N) \exp(j\pi n^2 \tau) \cos(2\pi nv)$; $v = (\xi + \rho_o)p$.
$F_4(\xi; \rho_o, \tau) = 2\alpha p\Sigma(n = -N, N) \exp(j2\pi nv) \sin[\pi(v - n\tau)2\alpha]/[\pi(v - n\tau)2\alpha]$; : 'Frequency parameter'
$v = (\xi - \sigma_o)p = (\xi + \rho_o - 2\mu)p$

Reflection characteristics:
Plane wave incidence :   Step-wise title control : Synchronization with diffraction order.
$F(\xi) = F_1(\xi; \sigma)F_2(\xi; \rho)$.   Continuous tilt control : Linear displacement condition.
Spherical wave incidence :
Ideal varifocal mirror :   Parabolic displacement condition and the linear tilt condition.

$F(\xi) = F_1(\xi; \sigma)F_2(\xi; \rho_o)$ (if $|\tau| \ll 1$)
Without displacement control : Linear tilt condition. $F(\xi) = F_1(\xi)F_3(\xi; \rho_o, \tau)$
Without tilt control : Parabolic displacement condition. $F(\xi) = F_4(\xi; \rho_o, \tau)$

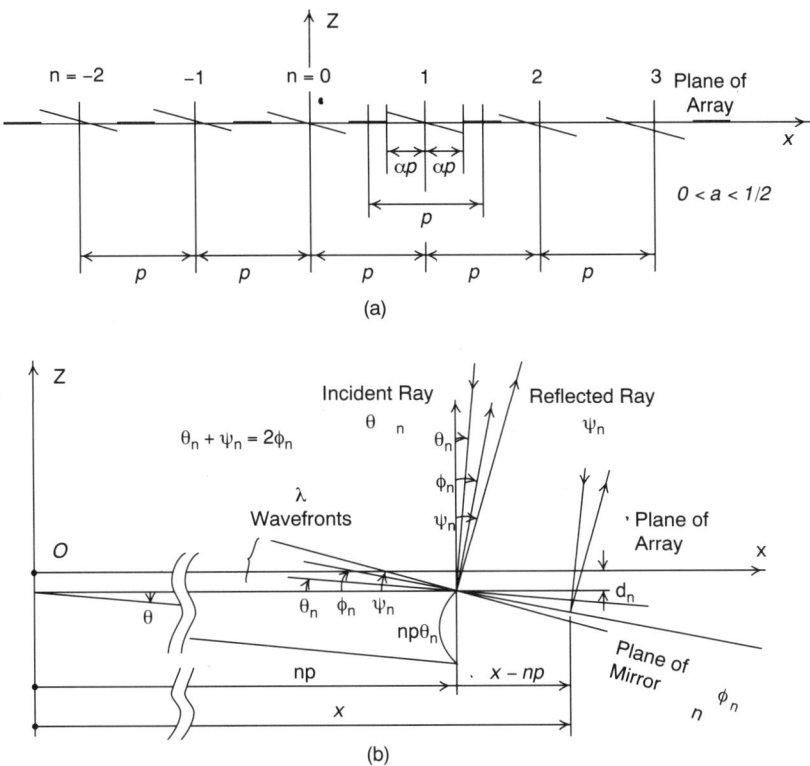

Figure 11.1.7
(a) Array of mirrors of a DMD. Pitch : p. (b) Details of a tiny mirror of a DMD. Center at $x = np$, $z = -d_n$. Tilting angle : ϕ_n

5 Concluding Remarks

The work summarized here mostly concerns autostereoscopic lens-plate displays using a screen of moderate size. Further work is needed to improve the quality of the displayed images and to establish a 3D camera system with which the data common for a variety of 3D displays are obtained.

References

[1] J. Hamasaki, '92 3Dmt Conference, Montreal, Canada, 1992, to be published.
[2] M. Okada, J. Hamasaki, S. Utsunomiya and O. Takeuchi, *Proc. First International Symposium on 3D Images*, Paris, France, 1991, pp. 145–153.
[3] J. Hamasaki, M. Okada, S. Utsunomiya, O. Takeuchi, *SID 91 Digest*, Anaheim, CA, 1991, pp. 844–847.
[4] J. Hamasaki, M. Okada and H. Sakaki, *Digests of Technical Meeting of UUO*, Hakodate, 1992, pp. 61–64.
[5] M. Okada, S. Utsunomiya, J. Hamasaki and T. Sonehara, *Proc. International Symposium on 3D Technology and Arts*, Tokyo, 1992, pp. 117–124.

[6] J. Hamasaki, *Proc. Japan Display '92*, 1992, pp. 295-298.
[7] T. Fujii, J. Hamasaki and R. Ishima, *Proc. First International Symposium on 3D Images*, Paris, France, 1991, pp. 173-181.
[8] T. Fujii, J. Hamasaki and M. Pusch, *Proc. International Symposium on 3D Technology and Arts*, Tokyo, 1992, pp. 171-178.
[9] I. Rechenberg, "Evolutions strategie" Problemata 15, Frommann-Holzboog, F. Frommann Verlag, Stuttgart-Bad Cannstatt (1973).
[10] M. Pusch, J. Hamasaki and T. Fujii, *Proc. Japan Display '92*, 1992, pp. 299-302.
[11] M. Rioux, G. Godin and F. Blais, *'92 3Dmt Conference*, Montreal, Canada, 1992, to be published.
[12] L.J. Hornbeck, *Proc. SPIE*, **1150**, *Spatial Light Modulator and Applications III* (Invited Critical Reviews), San Diego, CA, 1989.
[13] J. Hamasaki, *Proc. Fourth European Workshop on Three-dimensional Television*, Rome, Italy, 1993, pp. 25-28.
[14] J. Hamasaki, *Proc. First TAO Symposium on 3D Image Communication Technologies*, Tokyo, 1993, pp. S4.3.3-S4.3.12.
[15] J. Hamasaki, *Digests of Technical Meeting of UUO*, Miyajima, 1993, pp. 65-68.

11.2

Selective Image Retrieval by Synthesis of the Coherence Function

Kazuo Hotate

1 Introduction

Optical information processing, which is expected to have the potential to process information two-dimensionally or three-dimensionally by utilizing parallelism, has been studied widely in recent years (for example, see references in Reference [7]). One effective scheme for optical information processing is to utilize optical coherency. We have proved that the visibility of the interference pattern as a function of the path length difference between two arms in the interferometer, or the shape of the coherence function, can be synthesized by utilizing the direct frequency modulation of a laser diode [1]. First, the optical coherence function with a delta-function-like shape was synthesized, which was then successfully applied to the reflectometry for diagnosing optical devices and/or circuits [1-4]. Extending this method, we have applied the synthesis of the coherence function to 2D or 3D optical information processing systems [5-9].

In what follows we first discuss the basic nature of the coherence function, and then show the synthesis of the delta-function-like shape and the notch shape coherence function. By adopting those functions, several information processing functions are shown to be realized. The performance deterioration factors in the synthesis of the coherence function are discussed as well as the countermeasures. Selective extraction and masking of the 2D image from 3D optical information is demonstrated in experiments. Moreover, the improvements in the reflectometry using the same manner is also shown, in which the performance deterioration factors are successfully compensated [2,10,11].

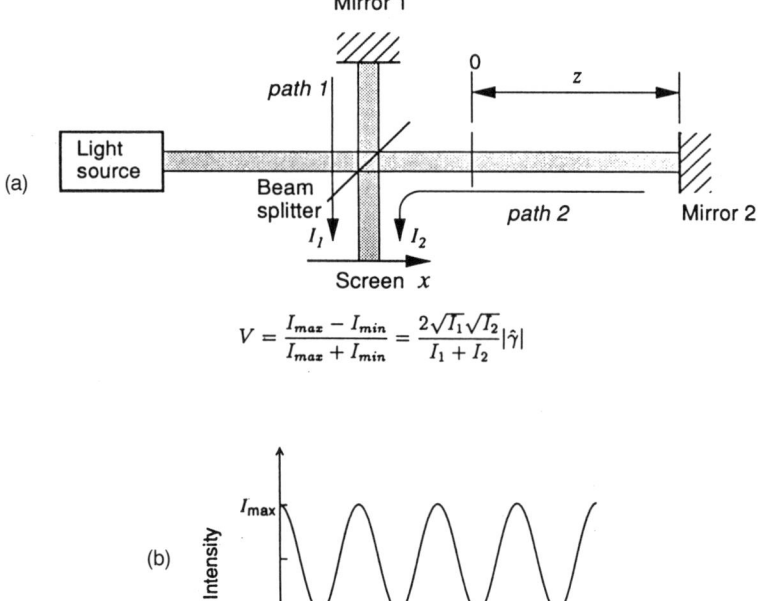

Figure 11.2.1
Optical coherence function. (a) Interferometer with path length difference z; (b) interference pattern [7]

2 Synthesis of the Coherence Function

To explain the optical coherence function, let us consider the Michelson interferometer shown in Figure 11.2.1. In Figure 11.2.1, $2z$ is defined as the optical path length difference between the two arms. The visibility V of the interference pattern is defined by the equation shown in the figure, where I_{max} and I_{min} indicate the local maximum and minimum values, and I_1 and I_2 indicate the intensity when path 2 or path 1 is blocked, respectively. In this equation, γ represents the complex coherence function. The visibility V is proportional to $|\gamma|$. In general, $0 < |\gamma| < 1$, and $|\gamma|$ is called the 'degree of conference' [12]. The coherence function is fixed when the light source is given, that is, the complex coherence function γ is calculated as the Fourier transformation of the power spectrum of the source.

On the contrary, we have proposed and proved that the coherence function can be synthesized by modulating a laser frequency with an appropriate waveform. To do so, we use a high coherence tunable semiconductor laser.

When the laser frequency is modulated using the waveform shown in Figure 11.2.2(a), the power spectrum of the laser has N frequency pairs with the frequency spacing $f_s, 2f_s, 3f_s, \ldots, Nf_s$, with the center frequency f_0 in the sense of time averaging. In this case, the degree of coherence $|\gamma|$ is calculated the Fourier

Selective Image Retrieval by Synthesis of the Coherence Function

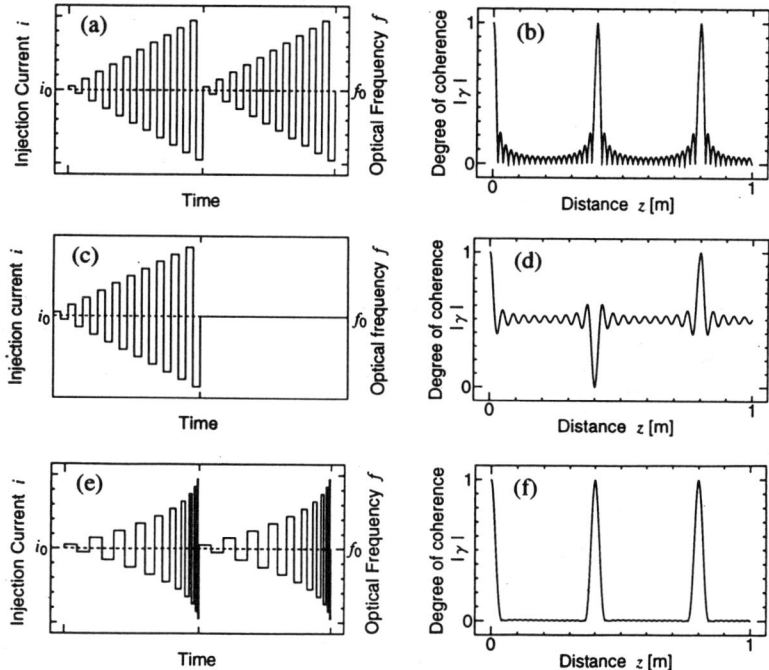

Figure 11.2.2
Synthesis of the coherence function. (a) Modulation waveform to synthesize the delta-function-like coherence function, and (b) the synthesized coherence function [5-7]. (c) Modulation waveform to synthesize the notch-shaped coherence function, and (d) the synthesized coherence function [7,8]. (e) Modulation waveform to realize the Hamming window, and (f) the synthesized coherence function [7]

transformation, as shown in Figure 11.2.2(b) [5-7], where the frequency pairs $N = 10$. Under frequency modulation with this waveform, there is no multi-frequencies just one frequency at a time. We have proved that the optical coherence function shown in Figure 11.2.2(b) can be synthesized, in the time average sense, even under this condition [3].

The coherence function $|\gamma|$ becomes a periodic delta-like function when the number of the frequency pairs N is large enough. This means that light with the frequency modulation shown in Figure 11.2.2(a) can interfere only when the specific optical path length difference exists. The position of the coherence peak can be swept by changing the amplitude of the modulation waveform, f_s. This delta-function-like optical coherence function is suitable for a variety of information processing, as shown below. The spatial resolution defined as the FWHM of $|\gamma|$ is inversely proportional to the tuning range in the waveform, or Nf_s. As N increases, the better the spatial resolution becomes. The tuning range of 150 GHz corresponds to a resolution of about 1 mm.

By combining this frequency modulation with a constant frequency as shown in Figure 11.2.2(c), the peak of the delta-function-like coherence function can be canceled. Consequently, the notch-shaped coherence function is also synthesized, as shown in Figure 11.2.2(d) [7,8].

Figures 11.2.2(b) and 11.2.2(d) tell us that the degree of coherence has a ripple when the number of frequency pairs N is finite. To suppress the ripple, it is effective to use a frequency window, such as the Hamming Window. By modifying the modulation waveform, as shown in Figure 11.2.2(e), such windows can also be realized. The ripple is suppressed, as shown in Figure 11.2.2(f).

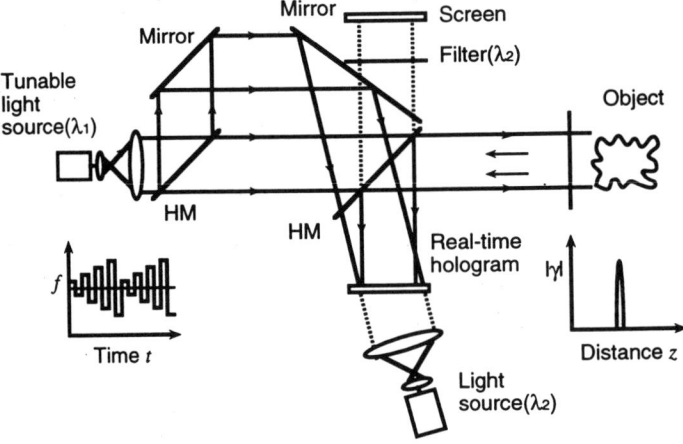

Figure 11.2.3
Conceptual drawing of an optical information processing system by synthesis of the coherence function [5-8]

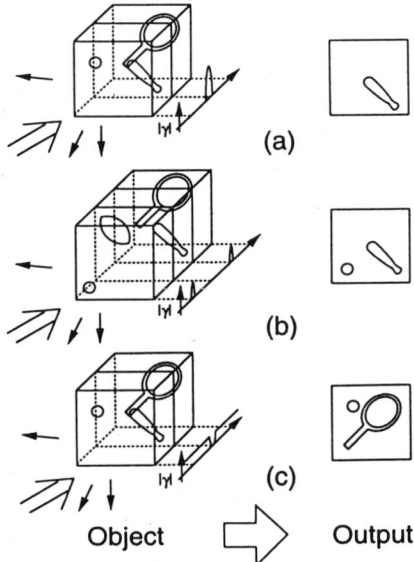

Figure 11.2.4
Information processing functions obtained by the system. (a) Selective extraction; (b) addition; and (c) selective masking of 2D information [5-8]

3 Optical Information Processing System

A conceptual drawing of an optical information processing system by synthesis of the coherence function is shown in Figure 11.2.3 [5–7]. A laser beam is divided into the reference wave and an object wave after the beam expander. Both waves are incident on the same plane indicated as the hologram in Figure 11.2.3. When the coherence function is synthesized to have a delta-function-like shape, only the reflected wave at the plane corresponding to the peak of the coherence function can interfere with the reference wave.

The modulation parameters can be set so that there exists only one peak in the object. In this case we have selective interference having information corresponding to only one plane. To pick up the interference component, holography is used. The information is reconstructed on the screen by another laser light. In this way the selective extraction of 2D information from a 3D object is carried out, as shown in Figure 11.2.4(a) [5–7]. The addition of two 2D information is also performed, as shown in Figure 11.2.4(b).

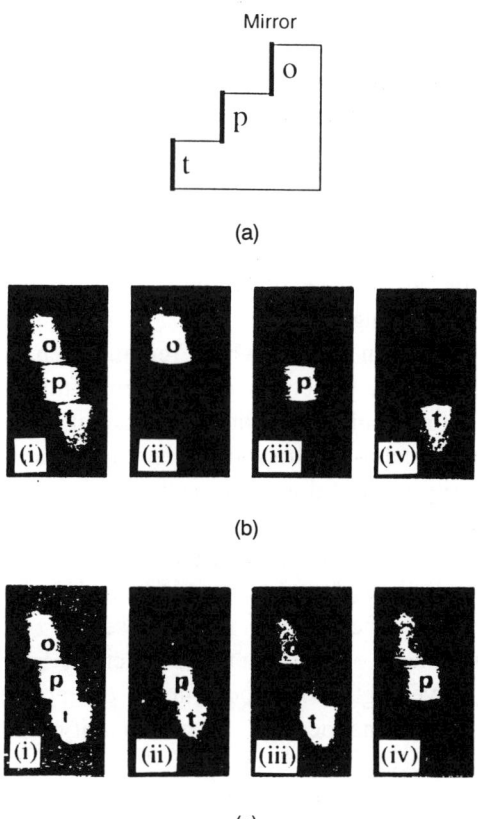

Figure 11.2.5
Experimental results. (a) Simplified 3D object; (b) selective extraction; and (c) selective masking [5–8]

When the coherence function is synthesized to have a notch shape, the object wave, except that from the plane corresponding to the position of the notch, can interfere with the reference wave. Therefore the selective masking of 2D information can also be carried out, as shown in Figure 11.2.4(c) [7,8].

Figure 11.2.5 shows the experimental results of selective extraction [5-7] and selective masking [7,8] of 2D information from a 3D object. In the experimental setup a silver halide holographic plate is used. Three mirrors are set as the 3D object, as shown in Figure 11.2.5(a). Each mirror is placed at a distance of 44, 40, and 36 cm from the reference place, and has the letter 'o', 'p', and 't', respectively, for identification. Figures 11.2.5(b) and (c) show the experimental results, that is, the reconstructed images of the holography. Figure 11.2.5(b-i) and 11.2.5(c-i) are the images from the holograms recorded without modulation. All the letters 'o', 'p' and 't' can be seen. Figures 11.2.5(b-ii)-11.2.5(b-iv) and Figures 11.2.5(c-ii)-11.2.5(c-iv) are the images from the holograms recorded with modulation to synthesize the delta-function-like shape and the notch shape coherence functions, respectively. The selective extraction and selective masking of the 2D image from 3D optical information have clearly been demonstrated.

4 *Improvement of the Performance*

In the above experiments the lasers with a wavelength of about 700 nm were used as the light source. The wavelength was selected taking into consideration the sensitivity of the holographic plate. The lasers with such a wavelength have relatively poor tuning characteristics, which restricts the spatial resolution. On the contrary, multi-electrode DFB/DBR semiconductor lasers have good tuning characteristics, which have been developed mainly in the wavelength regions of 1.3 μm and 1.5 μm.

The performance deterioration factors which remain even in the DFB/DBR lasers have been discussed, such as the effect of parasitic amplitude modulation in changing the driving current, and the nonlinearity and the nonflat response in the frequency modulation [2-4,8]. For the delta-function-like coherence function, it was clarified that the effect of the nonlinearity is most dominant, and a countermeasure was found [2-4]. The modulation waveform can be modified to compensate for the nonlinearity [2]. For the notch-shaped one, the shift in the center frequency in the modulation region with respect to the constant region was found to be dominant [8].

We have also studied the reflectometry to diagnose the optical waveguide and/or fiber devices by using synthesis of the coherence function, in which a three-electrode DFB laser of 1.55 μm wavelength with good tuning characteristics has already been used [2,4]. The compensation manner for the nonlinearity, in which the modulation waveform is modified, has successfully been applied to this system [2,4]. Figure 11.2.6 shows the experimental setup. The oscillation frequency is controlled by an injection current having the waveform shown in the figure. The coherence function of the periodic delta-function-like shape along the optical path is synthesized, peak point of which is swept by changing the current amplitude. The distribution of back-reflection can directly be obtained as the square-law-detector (SQD) output. A balanced detection scheme is

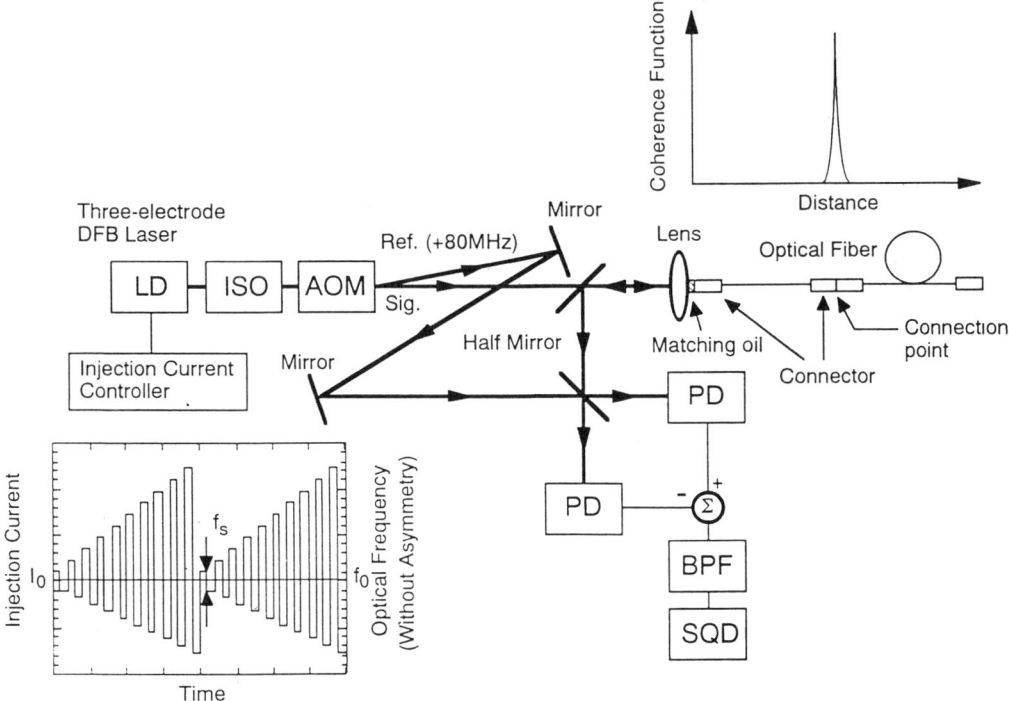

Figure 11.2.6
Experimental setup of the optical coherence domain reflectometry by synthesis of the coherence function [2]

adopted to improve the signal-to-noise ratio. In direct frequency modulation of a three-electrode DFB laser, an asymmetrical change in oscillation frequency with respect to the center frequency f_0 is induced by spatial hole burning. Therefore the injection-current waveform must be asymmetrical in order to obtain a symmetrical optical frequency change. We have found a method to measure the asymmetry by using the SQD output itself [2]. Then, the modulation waveform is modified to compensate for the asymmetry, as is illustrated in Figure 11.2.6.

Figure 11.2.7 shows the experimental results of the reflection position and the reflectivity at the fiber connector obtained in the setup shown in Figure 11.2.6 [2]. A resolution of 2 mm was demonstrated, which agrees well with the theoretical value.

Recently, we have proposed a novel way to scan the coherence peak or notch, in which the phase of the reference wave is modulated to be proportional to the modulated frequency [10,11]. Figure 11.2.8 shows the scan of the coherence peak by changing the phase modulation parameter [11]. In this method the modulation waveform for the LD does not need to be changed. Therefore the compensation described above becomes quite easy. Moreover, the phase modulation itself can also be used to compensate for the nonlinearity in the LD frequency modulation [11]. Experimental results of the reflectometry have already been obtained [11].

A photonic/video hybrid system for information processing has also been proposed and demonstrated. When holography is used to distinguish the interference component

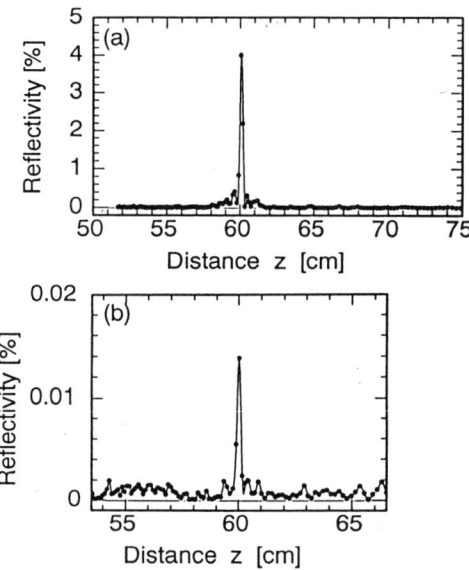

Figure 11.2.7
Improvement of the spatial resolution by the compensation of the asymmetrical nonlinearity in the frequency modulation of LD: reflection position and the reflectively at the fiber connector measured by the optical coherence domain reflectometry by synthesis of the coherence function. (a) Without and (b) with the connection [2]

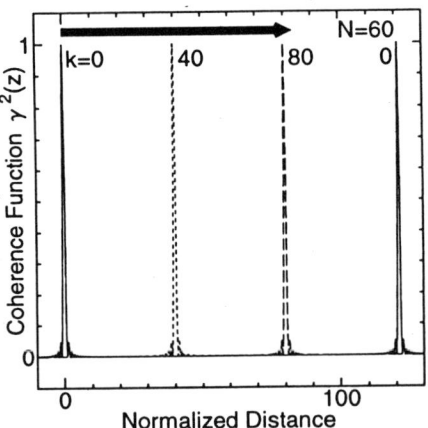

Figure 11.2.8
Scanning the coherence function by phase modulation [11]

from the others, the signal and the reference wave must have some angle with each other to produce the interference component as the diffracted wave. This results in degradation of the spatial resolution. On the contrary, the photonic/video hybrid system can avoid this problem [9]. Experimental results of the selective extraction of 2D information have already been demonstrated [9].

5 Conclusion

Optical information processing by synthesis of the coherence function has been proposed and studied. Selective extraction and selective masking of 2D information from a 3D object have been successfully demonstrated. The performance deterioration factors have been analyzed and countermeasures have been found. The reflectometry applying the synthesis of the coherence function has also been studied. It has been shown that a millimeter or higher of spatial resolution can be realized when using the tunable DFB LD.

References

[1] K. Hotate and O. Kamatani, *Electron. Lett.*, **25**, 1503–1505 (1989).
[2] O. Kamatani and K. Hotate, in *Proc. Conference on Optical Fiber Communication/International Conference on Integrated Optics and Optical Fiber Communication, OFC/IOOC'93*, San Jose, 1993, ThA4.
[3] K. Hotate and O. Kamatani, *IEEE J. Lightwave Techn.*, **11**, 1701–1710 (1993).
[4] O. Kamatani and K. Hotate, *IEEE J. Lightwave Technol.*, **11**, 1854–1862 (1993).
[5] K. Hotate and T. Okugawa, *Optoelectronics Conference, OEC'92*, Chiba, 1992, 16B2-4.
[6] K. Hotate and T. Okugawa, *Opt. Lett.*, **17**, 1529–1531 (1992).
[7] K. Hotate and T. Okugawa, *IEEE J. Lightwave Technol.*, **12**, 1247–1255 (1994).
[8] T. Okugawa and K. Hotate, *Opt. Rev.*, **1**, 8–11 (1994).
[9] T. Okugawa and K. Hotate, in *Proc. Optoelectronics Conference, OEC'94*, Chiba, 1994, 13E1-3.
[10] K. Hotate and T. Saida, in *Proc. SPIE Distributed and Multiplexed Fiber Optic Sensors IV*, **2294**, San Diego, 1994, 2294-03.
[11] K. Hotate and T. Saida, in *Proc. International Conference on Optical Fiber Sensors, OFS-10*, Glasgow, 1994, 534–537.
[12] M. Born and E. Wolf, *Principles of Optics*, Fifth Edition, Pergamon Press, 1975, Chapter X.

11.3

Ultraweak Biophoton Imaging and Information Characterization

Humio Inaba

Abstract

We have pursued the research and development of highly sensitive technology for detecting, analyzing, and processing extremely weak optical signals such as those not detectable by the human eye, especially those called biophoton emission. Biophoton is a new concept representing ultraweak photon emission phenomena closely related to a variety of life processes and biological activities, observable quite generally in nature. Hence these phenomena are considered to be a novel source of biological and vital information originating from microscopic and macroscopic systems, such as biomolecular species, cellular organelles, cells, tissues, organs, and living systems regardless of differences in biological hierarchy. This section reports and discusses recent progress and results of our biophoton emission research with emphasis on two-dimensional imaging and an approach to the quantum biophoton statistics to analyze the quantum optical properties for information characterization.

1 Introduction

Remarkable advances have been achieved over the last two decades in the field of optical electronics and photonics to develop various techniques for quantitative measurement and analysis not only of the intensity but also of the spectral and spatial distributions of extremely weak light not detectable by the human eye. Light emission phenomena closely related to a variety of biological activities and vital functions can be classified into two main categories [1,2], according to the emission intensity, as shown in Figure 11.3.1. In the lower part of this figure the number of photons/s·cm^2 and the converted optical power corresponding to the three wavelengths of 0.25 μm, 0.5 μm, and 1 μm are indicated for reference.

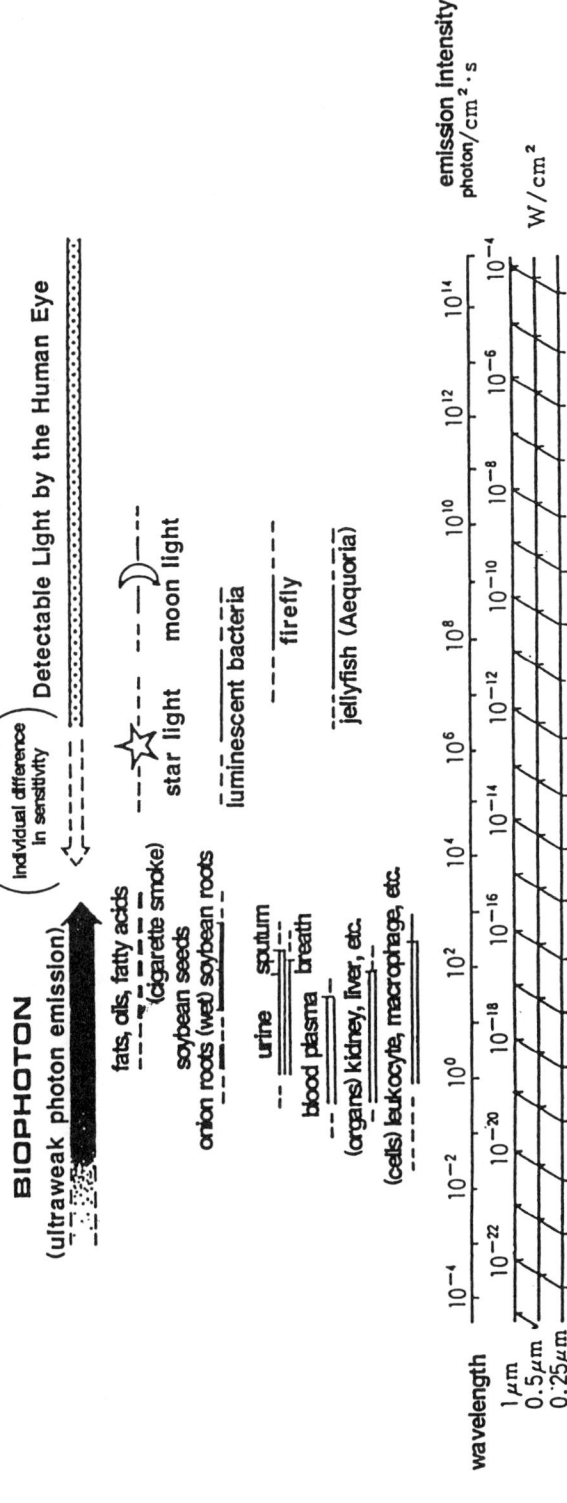

Figure 11.3.1
Comparison of typical intensity distributions of various light emission phenomena closely associated with life functions and biological activities: ultraweak biophoton emission and bioluminescence

The first category is bioluminescence, which is relatively intense and normally can be detected even by the naked eye. Typical examples of bioluminescence are the well-known cases observed in the firefly, luminescent bacteria, jellyfish (Aequoria), and the like. We know the mechanisms and specific substances responsible for this type of rather strong light emission, based on research begun almost a hundred years ago. Their characteristics and purpose, such as mating signals, appear to be rather well documented.

On contrast, the second category is extremely weak light emission originating spontaneously from living systems, organs, tissues, cells, cellular organelles, and related biomolecular materials regardless of differences in biological hierarchy. This ultraweak photon emission, termed simply biophoton emission [2–4], is generally so weak, ranging in intensity from perhaps a few to about 10^3–10^4 photons/s·cm², that it cannot be detected by the naked eye or even by means of ordinary optical detectors. Specific light emitters or sources and concrete mechanisms of biophoton emission are not yet well understood, although it is considered that this phenomenon occurs quite generally in nature in conjunction with a wide variety of processes of life and living functions, as illustrated on the left-hand side of Figure 11.3.1, for example.

In an effort to study the essential role of biophoton emission and to clarify its basic mechanisms and information in actual living systems and materials noninvasively, nondestructively, and without the use of any photosensitizers, we have implemented several types of highly sensitive measuring systems [1–8]. They include extremely low-noise photon counters for detection, computer-based spectral analyzers using a set of sharp cut-off filters or a polychromater incorporating a two-dimensional photon counter, two-dimensional imaging systems, and systems to measure the quantum statistical properties.

In this section we report recent progress and results of our biophoton emission studies with emphasis on two-dimensional imaging and an approach to quantum biophoton statistics.

2 Two-dimensional Imaging of Ultraweak Biophoton Emission Phenomena

Biochemically generated electronically excited states are ubiquitous at all levels of the biological hierarchy and the biophoton that such states emit can be utilized to probe the underlying processes and activities involved. The photon counting method offers, as is well known at present, a reasonable approach to the measurement and study of ultraweak photon emission processes. The advent of the microchannel plate and image processor, moreover, has made possible two-dimensional photomultiplier tubes. These devices greatly extend the usefulness of the photon counting technique, adding the versatility of an imaging capability.

Figure 11.3.2 shows a typical block diagram of highly sensitive ultraweak photon imagery systems developed in our laboratory. This system is based on a modified video

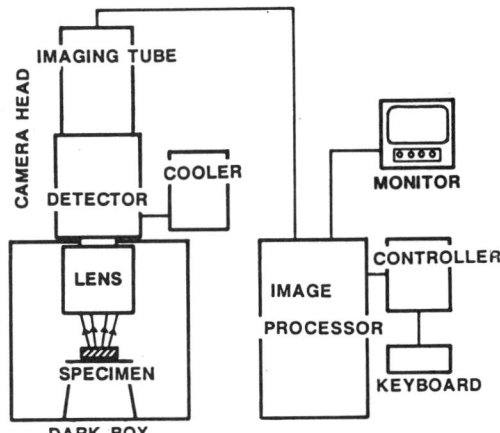

Figure 11.3.2
A block diagram of the two-dimensional photon counting system for ultraweak biophoton emission imaging

intensified microscope (Hamamatsu Photonics K.K.) together with a specially arranged sample chamber for controlling the temperature, humidity, and environmental gases. The detector assembly consists of a sensitive photocathode and a microchannel plate. The spectral response of the photocathode is about 370–700 nm and is cooled to $-20\,°C$ to reduce the dark noise count rate to about 4 counts/s. The two-dimensional positions of the incident photons are computed by an image processor and displayed on a monitor after suitable integration.

As one of samples, etiolated soybeans were prepared by germinating soybeans in darkness. Then a pair of cotyledons of one of the dark-adapted soybeans after 4–5 days of germination had been carefully separated, and one half of the sample soybean including its intact radicle and plumule was set in a sample cell with a small amount of 0.01 mM KCl solution [4]. Figure 11.3.3(a) is the biophoton emission pattern and (b) illustrates schematically the soybean sample used for clarity. The total exposure time required was 93 min, and the total number of photoelectrons, represented by dots on the display, was 67 788 with an average rate of 12.1 counts/s.

It should be noted here that this emission pattern was obtained by keeping the specimen in complete darkness in stable conditions for a long period of time (5 days) in order to eliminate the possible effects of delayed fluorescence due to chlorophyll and others, and undesirable physiological excitement caused by external stimuli such as light exposure and physical injury. Although the average count rate of the biophoton emission is relatively low, remarkable emission is observed in the segment of hypocotyl, the junctional region between radicle and plumule, where active cell division is taking place for growth. In contrast, emission from the plumule itself or from the cotyledon is much less. It is believed that this biophoton emission is closely related to the biochemical and biophysical processes of cell multiplication and successive growth.

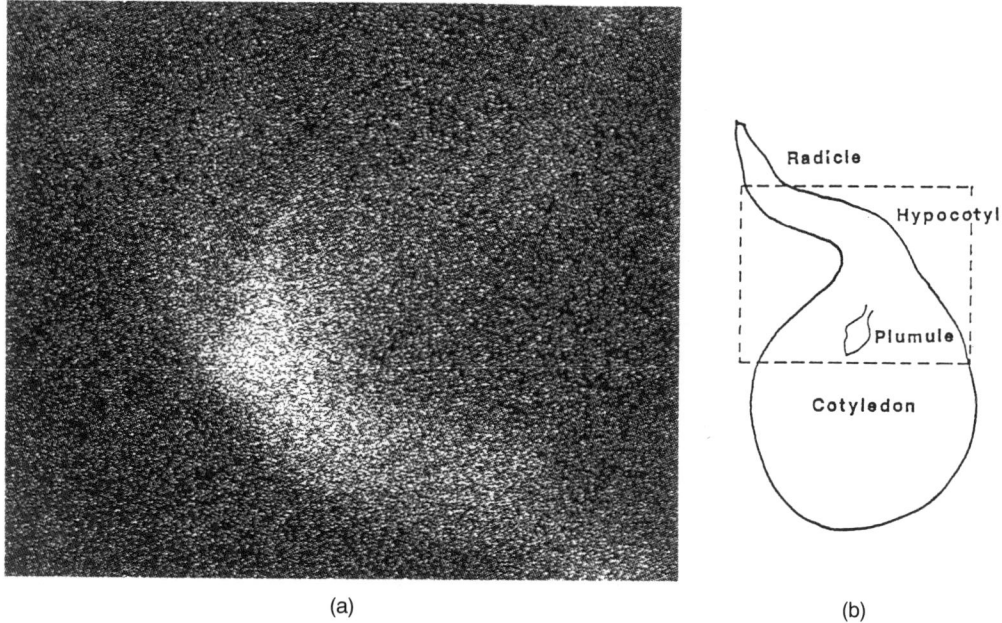

Figure 11.3.3
(a) Ultraweak biophoton emission image of mitosing soybean after 5 days of germination, and (b) schematic drawing of the sample soybean. The area included in the image is indicated by the dashed line

To examine the effect of physical injuries on the ultraweak biophoton emission from the living systems, we used soybeans and adzuki beans which were adapted to darkness for 3-4 days (for etiolation) at room temperature after several hours of imbibition [6]. The injuries were cuts to the cotyledons of the seedlings in the form of crosses 2-3 mm deep and 5-10 mm long. Figure 11.3.4 shows the emission patterns from an adzuki bean; (a) before injury, and (b) and (c) taken 15 min and 2 h after injury, respectively. The detected time was 1 h in all cases. A photograph of the specimen taken immediately after measurement is shown in Figure 11.3.4(d). In Figure 11.3.4(b), the emission intensity centered around the injured region is about 7-20 times higher than the other regions [6]. Consequently, the emission intensity was also higher and more widespread in the root system away from the injury site. However, in Figure 11.3.4(c) it can be seen that the emission pattern is attenuated in both regions. The simultaneous increase in photon emission far away from the injury region could offer clues to the mechanism of information transfer in the living plant system.

Furthermore, we found that the biophoton emission intensity, along with its emission pattern, varies on injury, even in the case of animals. Figure 11.3.5(a) shows a photograph of a mouse with a wound of about 1 cm in diameter made by operation on shaved skin on its back, and (b) displays the two-dimensional image obtained immediately after the operation which did not exhibit any distinct emission pattern or localization. We took 90 min for each biophoton image in this experiment [8]. The image was clearer on the third day (48 h) after injury, as is seen in Figure 11.3.5(c),

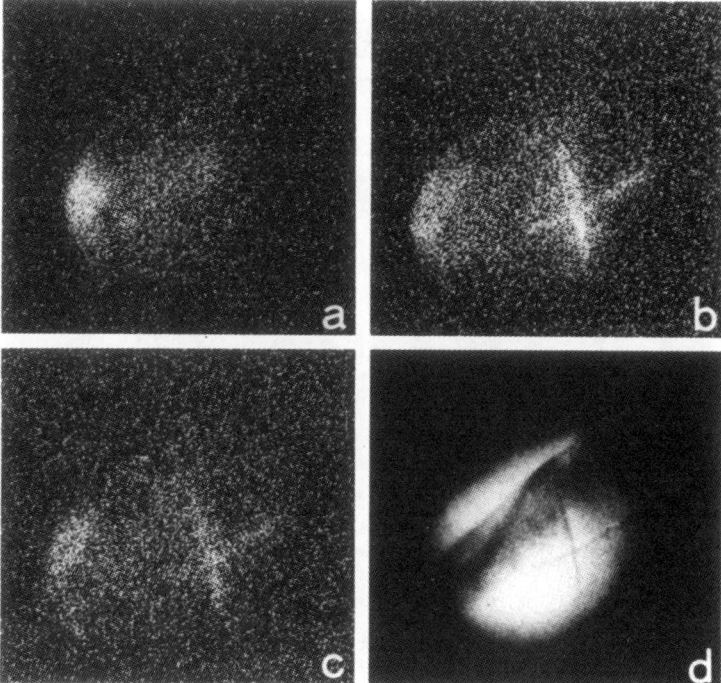

Figure 11.3.4
Ultraweak biophoton emission images of germinating adzuki bean. (a) Its image before injury, (b) and (c) 15 min and 2 h after injury, respectively, and (d) photograph of the same specimen used in this experiment. The cut can be clearly seen as the cross on the cotyledon

and resembled the shape of the wound, while the emission intensity was continuously rising, corresponding to activation of the immune system as examined by other medical techniques. The emission intensity was at a maximum between the third and the fifth days after the operation, as observed in Figures 11.3.5(c) and 11.3.5(d). From the sixth day of the injury the emission intensity began to decrease, and on the ninth day after injury (when the wound had almost healed), the image in Figure 11.3.5(e) did not show any clear pattern, as in Figure 11.3.5(b). The emission intensity returned to normal levels once the scab had fallen off.

3 Approach to the Quantum Statistical Properties of the Biophoton

To investigate the essential roles and fundamental significance of the biophoton field in science and technology, as recognized to be biophotonic information carrier, it would be of great importance and a challenge to study the statistical properties of the biophoton fields from the view-point of quantum optics. In general, the statistical properties of the photon field are related to those of the underlying, primary excitation processes that lead to optical emission. In the area of interest, namely that of weak fields and instantaneous single photon emissions, the coherence and statistical properties of the

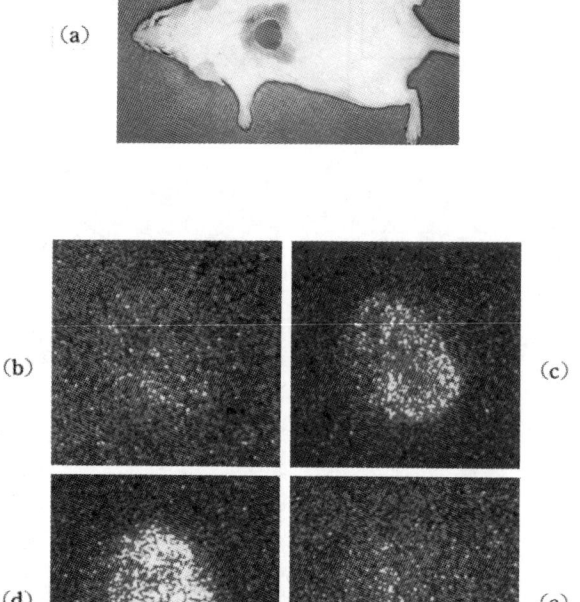

Figure 11.3.5
Ultraweak biophoton emission images showing the time-span of a wound healing on an injured mouse back. (a) Photograph of the mouse with a wound of roughly 1 cm in diameter, (b) its image taken 15 min after the operation, (c) 2 days, (d) 4 days, and (e) 8 days after injury, respectively

primary excitations are transferred directly to those of the field [9]. This general property has been demonstrated experimentally by the observation of nonclassical light in the Franck–Hertz effect, where the anti-bunched, sub-Poisson character of a space-charge-limited electron beam is transferred to the photon field emitted from Hg vapor [10].

The primary excitations at all levels of the biological hierarchy arise from a wide variety of biophysical events, and all such events are considered ultimately to be coupled chemically to significant biophysical processes, although most events remain uncharacterized as yet. However, the experimental results shown briefly above as well as observed evidence published previously by us [1–8] could support this view. Also these quantitative measurements evidently validate that in many cases ultraweak photon emission is localized in the heterogeneous and highly structured biomolecular environment of the organism, and that the emission processes are sensitive to structure and dynamics at the molecular level. Subsequently, this insight and idea motivated our attempt to measure precisely the quantum optical properties of ultraweak, endogenous optical fields, that is, biophoton emission. There has been no report, to our best knowledge, of direct measurements of such properties of biophoton fields.

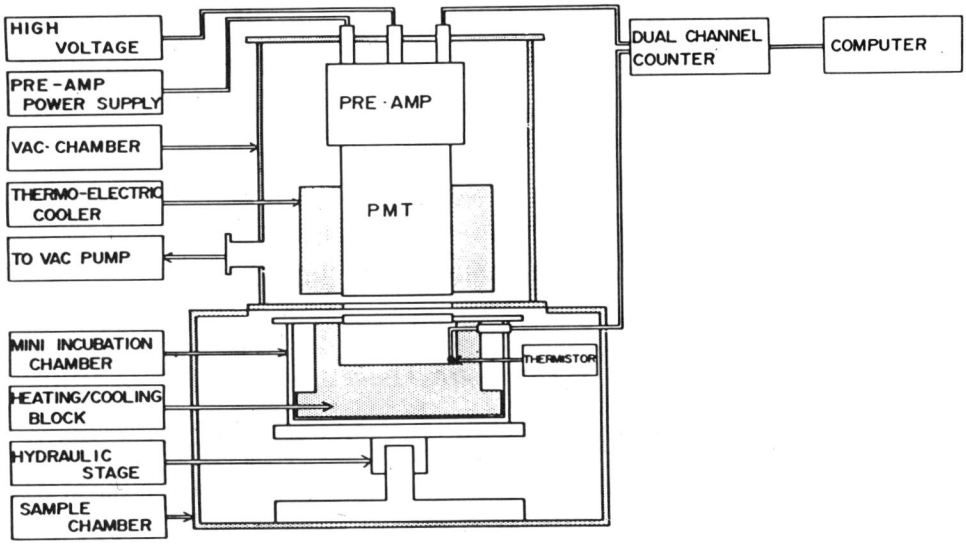

Figure 11.3.6
Schematic diagram of the extremely highly-sensitive, vacuum-isolated photon counting system

Figure 11.3.6 shows schematically the block diagram of an extremely highly-sensitive, vacuum-isolated photon counting system designed and constructed in our laboratory for these purposes [5]. To achieve ultimately the design goals such as quantum-limited low noise operation, absolute stability of the residual noise process, and maximum practicable sensitivity, a vacuum-isolated configuration was conceived in which the detector and its first preamplifier are thermally isolated in a high vacuum. Isolation in a high vacuum enables cooling of the photomultiplier tube to extremely low temperatures (approximately $-50°$ to $-70\,°C$, depending on tube type) beyond which no further decrease in noise is obtained. This technique allows maximum exploitation of the inherent capability of the photomultiplier tube. Furthermore, this isolating configuration carries with it other benefits; the vacuum cell conventionally employed just before the photomultiplier tube is eliminated and reflection losses are thus reduced. Also, extremely short sample-to-photocathode distances become possible, significantly enhancing sensitivity simply by the geometrically favorable detection geometry for efficient light collection. The absolute stability of the noise statistics is further insured in this system by critical stabilization of the associated electronics. In addition to the vacuum housing, our system shown in Figure 11.3.6 is equipped with a very large, light-tight experimental chamber with provision for the introduction in the dark of liquids and gasses and for external electrical connections or the introduction of light beams.

For biophoton statistics experiments, photon counting distributions, $P_n(T)$, are constructed by counting the number of possible locations in the pulse-to-pulse sequence of a time window of width T covering exactly n photoelectron counts. The second-order intensity correlation function, $g^{(2)}(\tau)$, is derived from the same data set by exploiting the formal equivalence between the correlation function and the two photon conditional probabilities. A selected photomultiplier tube (Hamamatsu R375) was used in a high vacuum and cooled to $-45\,°C$, which gave an average dark count rate of 7.8 counts/s with optimal high voltage and discriminator settings. As the photon sources for the measurement, we employed suspensions of isolated chloroplasts, whole detached leaves of spinach, and a preparation of human polymorphonuclear neutrophil leukocytes together with an AlGaAs semiconductor diode laser for comparison.

In Figure 11.3.7 a preliminary result of the spinach leaf analysis is shown by closed circles in comparison with that of the semiconductor laser exhibiting a standard Poisson distribution, presented by the solid line. These data of biophoton emission from a spinach leaf were obtained after 2 h of dark adaptation at $30\,°C$ in the sample chamber of our photon counting system [11]. This period of dark adaptation ensures that the delayed fluorescence associated with the photosynthetic units of the plant has decayed away leaving only the residual, nearly constant ultraweak biophoton emission. We adopted the procedure used by Teich and Saleh [10] in which the photon counting distribution of the optical field to be analyzed is compared with a known Poisson standard laser source. Then corrections for instrumental effects such as dead time and noise were made for the reference light after adjusting both intensities to the same count rate of 20 counts/s.

The result shown in Figure 11.3.7, although quite preliminary, appears to suggest that the magnitude of the measured effect is of the same order as that observed for the

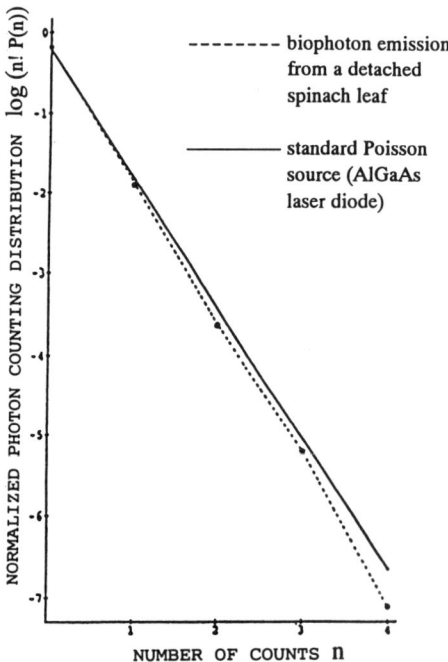

Figure 11.3.7
Normalized photon counting distribution of biophoton emission from a spinach leaf, after 2 h of dark adaptation, corrected for instrumental effects such as dead time and noise by comparison with a standard Poisson source (an AlGaAs diode laser) which provided the Poisson distribution, as shown by the solid line

first time in the generation of sub-Poisson light in the Franck–Hertz experiment [10]. However, a quantitative assessment should be provided by the Fano factor, which is the most sensitive measure of departure of a photon counting distribution from Poisson behavior, and detailed analyses of biophoton counting distributions could be expected to investigate its quantum optical properties including the possibility of a sub-Poisson field.

Finally, it would be of interest to add some comments on the quantum statistical features of biophoton fields in connection with our new findings in the two-dimensional images of ultraweak biophoton emission. It is known theoretically that in the information transmission conveyed by the coherent light corresponding well to the classical electromagnetic wave followed by direct detection with an ideal optical detector without internal noise, an average number of photons of nearly 21/s is necessary to keep the signal error rate to the order of 10^{-9} which is demanded traditionally in the communication channel [12]. This situation means that the minimum required number of photons is approximately 10/s in this short noise limit. To overcome this standard quantum limit, the quantum states of electromagnetic waves called squeezed states, photon number states, and amplitude-squeezed states were proposed and are being studied. In these nonclassical states information can be extracted from the one observable with reduced quantum noise by preserving the Heisenberg uncertainty principle with an increased quantum noise for the conjugate observable. Especially, the photon number state providing the sub-Poisson statistics is completely determined by its photon number and the photon number fluctuation of the signal wave can be reduced to zero in the ideal case so that the signal error rate in the direct detection is thought to be made as small as possible without any limitation of the signal-to-noise ratio.

On the other hand, we have found in Figure 11.3.3 that the average photoelectron number of 12.1 counts/s is detected, verifying the tissue-specific localization of biophoton emission to the hypocotyl and radicle regions having higher metabolic activity. Subtracting a noise count rate of about 4/s in this measurement, a few tens/s of photons are approximately estimated to be emitted from these localized regions in the germinating soybean. Also, similar estimation in Figure 11.3.4(a) leads to about the same or a little less average number of biophotons which are generated from the root tip region in the case of a normal adzuki bean without injury.

It was pointed out theoretically that the information which can be transmitted by a single photon is finite under the condition of a large amount of thermal photons, and there exists an upper limit on the information capacity by any selection of quantum states, modulation types, and detection schemes [12]. This limit corresponds to about 46 bit/photons for a wavelength of 1.5 μm at 300 K. Taking into account this achievable information capacity, therefore, emitted biophotons in the range of approximately 10–40/s could be inferred, enough to transfer a part of the necessary information such as prompting active cell division, multiplication, and successive growth leading to ontogenesis in the normal state. Furthermore, once abnormal or dangerous states like injury and damage happen, the number of emitted biophotons increases, as described in the previous section, as though they carry the information remotely and quickly. In addition, from the aspect of a quantum-mechanical information channel as discussed above, it could be conjectured that the biophoton field conveying various kinds of inherent information is advantageous to take the photon number states associated with antibunching and sub-Poisson statistics. Consequently, the precise measurements to confirm such characteristics seem to be extremely interesting and appealing because direct transfer of the coherence and statistical features of the underlying, primary excitations takes place basically to those of the photon field, and vice versa, namely, statistics governing endogenous biophysical processes are in principle experimentally accessible via those biophoton fields.

4 Conclusion

This section reviewed and discussed our recent approaches to two-dimensional imaging of ultraweak biophoton phenomena and to the quantum optical properties of biophoton fields for their information characterization. We also included an attempt to illustrate briefly the breadth of scientific disciplines on which the biophoton research may bear. We have demonstrated, for the first known time, the feasibility of measuring the quantum statistical properties of ultraweak photon emission from biological sources and materials. A vast amount of original experimental work and analyses awaits us in this newly developing area, simply called biophoton, to verify its fundamental significance and to establish biophotonic information science and technologies along with their applications in the future.

References

[1] H. Inaba, Y. Shimizu, Y. Tsuji and A. Yamagishi, *Photochem. Photobiol.*, **30** (2), 169 (1979).
[2] H. Inaba, *Modern Radio Science 1990*, J.B. Andersen, Ed., Oxford Univ. Press, 1990, p. 163, and references cited therein.
[3] H. Inaba, *Experientia*, **44** (7), 550 (1988).
[4] R.Q. Scott, M. Usa and H. Inaba, *Appl. Phys. B*, **48** (2), 183 (1989).
[5] B. Devaraj, R.Q. Scott, P. Roschger and H. Inaba, *Photochem. Photobiol.*, **54** (2), 289 (1991).
[6] S. Suzuki, M. Usa, T. Nagoshi, M. Kobayashi, N. Watanabe, H. Watanabe and H. Inaba, *J. Photochem. Photobiol. B*, **9** (2), 211 (1991).
[7] T. Nagoshi, N. Watanabe, S. Suzuki, M. Usa, H. Watanabe, T. Ichimura and H. Inaba, *Photochem. Photobiol.*, **56** (1), 89 (1992).
[8] M. Usa, B. Devaraj, M. Kobayashi, M. Takeda, H. Ito, M. Jin and H. Inaba, *Proc. Third Int. Conf. of Int. Soc. on Optics within Life Sciences: Opt. Methods in Biomed. and Environ. Sciences (OWLS III)*, Tokyo, 10-14 April, Elsevier Sci. Publishers, 1994, p. 3.
[9] M.C. Teich, B.E.A. Saleh and J. Perina, *J. Opt. Soc. Am. B*, **1** (3), 366 (1984).
[10] M.C. Teich and B.E.A. Saleh, *J. Opt. Soc. Am. B*, **2** (2), 275 (1985).
[11] R.Q. Scott, *Final Research Reports of Inaba Biophoton Project, ERATO*, Research Development Corporation of Japan, 1991, p. 1.
[12] E.g. Y. Yamamoto and H.A. Haus, *Rev. Mod. Phys.*, **58** (4), 1001 (1986).

11.4

Super-parallel Fourier-transform Spectral Imaging

Kazuyoshi Itoh

Abstract

Spectral imaging techniques dedicated to the ultrafast measurement of spatial and spectral information of a distant object are described. The ultrafast technique combines a multiple-imaging optical system with a conventional Fourier spectrometer. The basic sequential system with a normal imaging system is introduced first and the principle of the ultrafast system is presented along with a compact spectral imaging system based on a polarization interferometer with a wedge-shaped liquid crystal. An object attached to the rotating shaft of an electric motor was successfully reconstructed by using the single flash of a strobe light.

1 Introduction

The term 'spectral imaging' is now common in the literature. This term may be defined as simultaneous acquisition of spatial and spectral information from remote radiative objects. Various techniques have already been used for spectral imaging in many fields such as astronomy and remote sensing. Much effort has been made so far to develop new efficient instruments for measuring spectral images [1]. However, few techniques have been suggested so far to the author's knowledge for the spectral imaging of very fast phenomena. In what follows, very fast techniques that are based on super-parallel signal detection are described. The reader will understand that spectral imaging of optical phenomena much faster than a microsecond is readily achievable by the suggested techniques.

2 Fast Techniques for Spectral Imaging

Fourier transform spectroscopy is one of the most efficient methods to measure spectral information, and has been widely used, especially in infrared regions. If we use a two-dimensional detector array in place of the point detector in the Fourier spectrometer, we may obtain a Fourier spectrometer array. Such an approach to spectral imaging has already been tried by Simmons and Cowie [2] and a similar idea has been proposed by Gay and Mekarnia for astronomical speckle interferometry [3]. A holographic approach has also been suggested by Lindegren and Dravins [4]. In these approaches, each element of the detector array detects an interferogram associated with a small area on the surface of the object. Such methods are expected to have high resolving powers in both the spectral and spatial regions [5,6] and high signal-to-noise ratios [7]. We call this class of methods Fourier-transform spectral imaging in the image plane (FTSI/I).

Except for the holographic technique, FTSI/I substantially realizes an array of Fourier spectrometers that measures the spectrum at every point on the surface of an object in parallel. Each elementary Fourier spectrometer obtains an interferogram as a time series. Thus, the time of data collection is a limiting factor of the observation of time-varying phenomena. The interferometric methods to be discussed in this section obtain interferograms at all positions on the surface of an object simultaneously [8,9]. The data necessary for spectral imaging are detected in parallel. The interferometric data are expanded in a two-dimensional detection plane.

In the next section, the technique of Fourier-transform spectral imaging in the image plane is briefly reviewed. The principle of multiple-image Fourier-transform spectral imaging is illustrated in section 4. In sections 4 and 5 two different techniques for the implementation of the multiple-image technique are described. Experimental results of these fast techniques are presented in section 6, and conclusions offered in section 7.

3 Fourier-transform Spectral Imaging in the Image Plane

The technique of Fourier-transform spectral imaging is classified into two types; the pupil plane [10,11] and the image plane [6,7] techniques. It has been shown that if the size of the entrance pupil is fixed, the image plane technique has a much better signal-to-noise ratio than the pupil plane technique [7].

We explain the principle of the image plane technique by referring to Figure 11.4.1. After passing through a lens (lens 1) and a beam splitter, the incident beam is split into two. The split beams make two images of the same object on the surface of the two mirrors in the interferometer. The two images are then superimposed by the beam splitter on the active area of the image detector via the second lens (lens 2). If these two images are exactly superimposed upon each other, they interfere. Let us assume that the two mirrors in the interferometer are strictly normal to the optical axis and a light beam in the interferometer from a certain point on the object makes an angle φ with respect to the optical axis. When one of the mirrors is translated along the optical axis by an amount $z/2$ from the origin of the zero path difference, the optical path

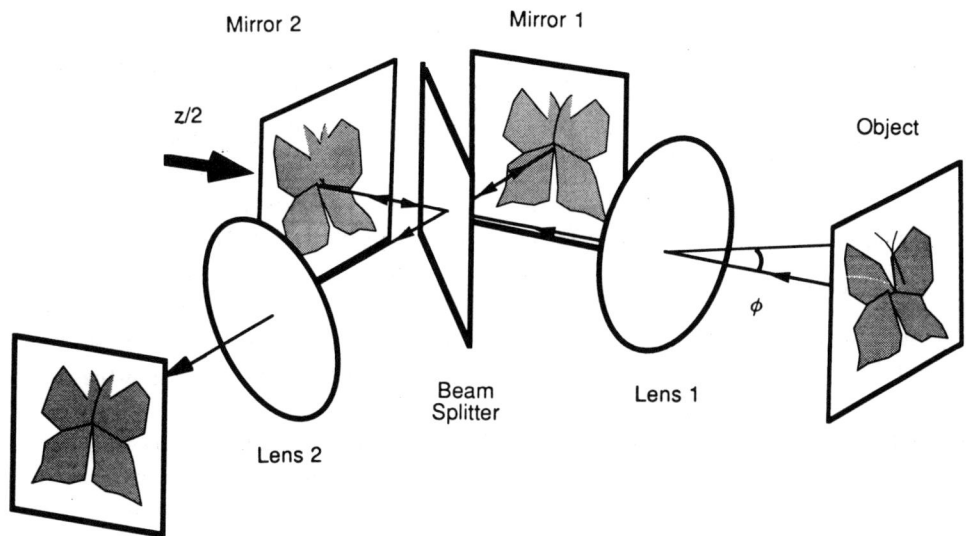

Figure 11.4.1
Principle of Fourier-transform spectral imaging in the image plane

difference between the light beams that make angle φ with the optical axis is given by

$$\Delta z = z \cos \varphi \tag{1}$$

Then the intensity at the detector element located by the position vector $\mathbf{r}$ can be written as

$$I(\mathbf{r}, z) = \int S(\mathbf{r}', k)[1 + \cos(2kz \cos \varphi)] dk \tag{2}$$

where k is the wave number and $S(\mathbf{r}', k)$ is the spectral density at the object point which is located at $\mathbf{r}'$. The two position vectors in the object and image planes are connected by the magnification M of the imaging system as $\mathbf{r}' = M\mathbf{r}$. Equation (2) is identical to that for the interferogram of Fourier-transform spectrometry. The important difference is that each element of the image detector collects an interferogram at a different location and detection is done in parallel. Thus, the interferometer acts as a two-dimensional Fourier spectrometer array. The spectral image, $S(\mathbf{r}', k)$, is reconstructed from the interferogram $I(\mathbf{r}, z)$ by inverse Fourier cosine transformation with respect to z. Note that a million elementary Fourier spectrometers operate simultaneously even if a commercially available standard CCD image sensor is used.

The spectral resolution, δk, and the maximum measurable wave number, $k_{\max}$, are given by the theory of Fourier spectrometry. At image points that are characterized by the beam inclination angle φ, they are given by

$$\delta k = \frac{1}{Nd \cos \varphi} \tag{3}$$

and

$$k_{\max} = \frac{1}{d \cos \varphi} \tag{4}$$

respectively, where N is the number of sampling points of the interferogram and d is the sampling interval. Note that the spectral resolution and measurable range are dependent on the location of the image point.

4 Principle of Multiple-image Fourier Transform Spectral Imaging [8]

A schematic illustration of the ultrafast method for spectral imaging is shown in Figure 11.4.2. Light emanating from an object passes through a lenslet array and is incident on the interferometer. The lenslet array forms multiple images of the object on the mirrors in the interferometer via a beam splitter. The light beams reflected on the mirrors are recombined by the beam splitter and make multiple images of interference signals on a two-dimensional image sensor through a lens. By tilting one of the mirrors on the arms of the interferometer, one can create path differences that depend on the locations on the mirror surface, say (x, y). We represent the path difference by $z(x, y; \nu)$. In general the path difference is dependent on the wave number, ν. By separating the multiple images of the interference signals in the rectangular array and piling them up according to their path differences, we get three-dimensional data that can be considered as a set of interferograms associated with every point on the surface

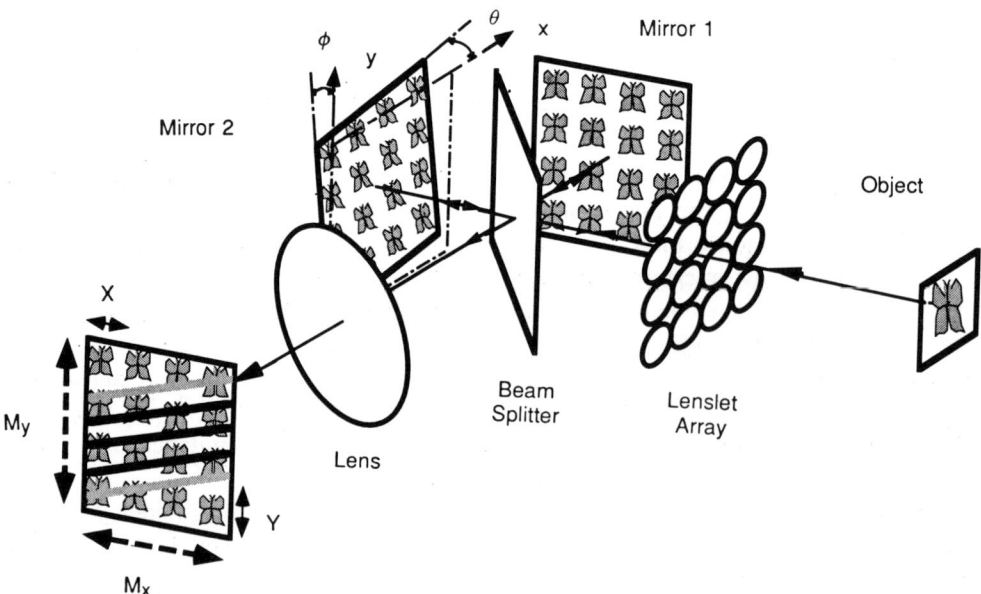

Figure 11.4.2
Principle of multiple-image Fourier-transform spectral imaging

of the object. This volume of three-dimensional data may be called an interferogram image. Let us take a Cartesian coordinates in this three-dimensional volume of data. Two of the three axes represent the spatial axes (ξ and η) and the other axis represents the axis of the path difference (z). For simplicity we assume that the magnification of the imaging system is unity. The interferogram observed at particular point P located at (ξ_P, η_P) in the interferogram image is given by

$$i(\xi_P, \eta_P) = \int S_P(\nu) \cos\{2\pi \nu z_i(\xi_P, \eta_P; \nu)\} d\nu \quad (i = 0, \ldots, N-1) \quad (5)$$

where N is the number of multiple images, $S_P(\nu)$ denotes the power spectral density at P, respectively, and $z_i(\xi_P, \eta_P; \nu)$ denotes the path difference for the light of wave number ν at point (ξ_P, η_P) in the ith cross-section of the interferogram image.

Let us consider the case of an ideal optical system. The path difference $z(x, y; \nu)$ at a point (x, y) on the image sensor can be written as

$$z(x, y; \nu) = 2(x \tan \theta + y \tan \phi) + z_0 \quad (6)$$

where θ and ϕ are inclination angles of the mirror M_2 with regard to the plane perpendicular to the optical axis in the respective directions of the x- and y-axes, and z_0 is the path difference at (x, y) = (0, 0). As is shown in Figure 11.4.2, suppose that the multiple images of the interference signals consist of $M_x \times M_y$ elementary images and the size of the elementary image is $X \times Y$. Then it can be shown that if the angles θ and ϕ satisfy the condition that $X \tan \theta = M_y Y \tan \phi$, sampling of the interferogram with equal intervals of path difference is realized.

Let us denote the interferogram at P by the vector I_P the elements of which are represented by $I_i(\xi_P, \eta_P)$ and let the original spectrum at this point be denoted by the vector $S_P = (S_P(\nu_0), \ldots, S_P(\nu_{N-1}))^T$, where N is the number of sampling points for the wave number. The interferogram and spectrum are related by the Fourier cosine transform. The vectors I_P and S_P are, therefore, related by $I_P = F_P S_P$, where the matrix F_P is a transformation matrix whose element $(F_P)_{ij}$ is expressed by $(F_P)_{ij} = \cos\{2\pi \nu_j z_i(\xi_P, \eta_P; \nu)\}$. If F_P has an inverse matrix, the spectrum can be reconstructed by multiplying equation (5) by the inverse matrix. In the case of an ideal optical system, the matrix F_P is identical to the matrix of the discrete Fourier-cosine transform.

The matrix F_P must be estimated before calculating the generalized inverse matrix. The matrix F_P can be estimated if the path differences $z_i(\xi_P, \eta_P; \nu_j)$ are known. If we assume that only one kind of glass is used in the interferometer, then $z_i(\xi_P, \eta_P; \nu_j)$ can be expressed by

$$z_i(\xi_P, \eta_P; \nu_j) = n(\nu_j) z_{gi}(\xi_P, \eta_P) + z_{ai}(\xi_P, \eta_P) \quad (7)$$

where $z_{gi}(\xi_P, \eta_P)$ is the path difference in the part of glass in the interferometer, $z_{ai}(\xi_P, \eta_P)$ is that in the air, and $n(\nu_j)$ is the refractive index of the glass. The path difference, $z_i(\xi_P, \eta_P; \nu_j)$, may be calibrated by using several interferograms measured at different spectral bands of known wave numbers $n_k (k = 1\text{-}K, K \geq 2)$ and an appropriate parametric model for $n(\nu)$. The distribution of the path difference can be estimated from the interference patterns of each band of the light observed through the interferometer. We can make simultaneous equations by substituting the value of $z_i(\xi_P, \eta_P; \nu_k)$ and $n(\nu_k)$ into equation (7) for each band. The unknown variables $z_{gi}(\xi_P, \eta_P)$ and $z_{ai}(\xi_P, \eta_P)$ can be estimated by solving these simultaneous equations.

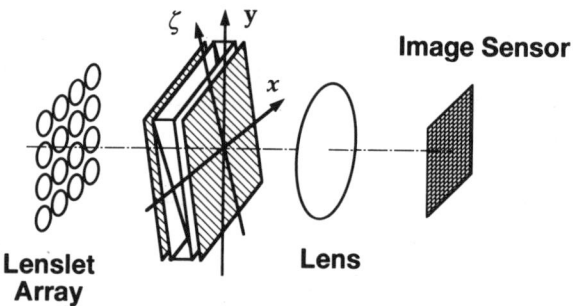

Figure 11.4.3
Principle of liquid-crystal polarization interferometry

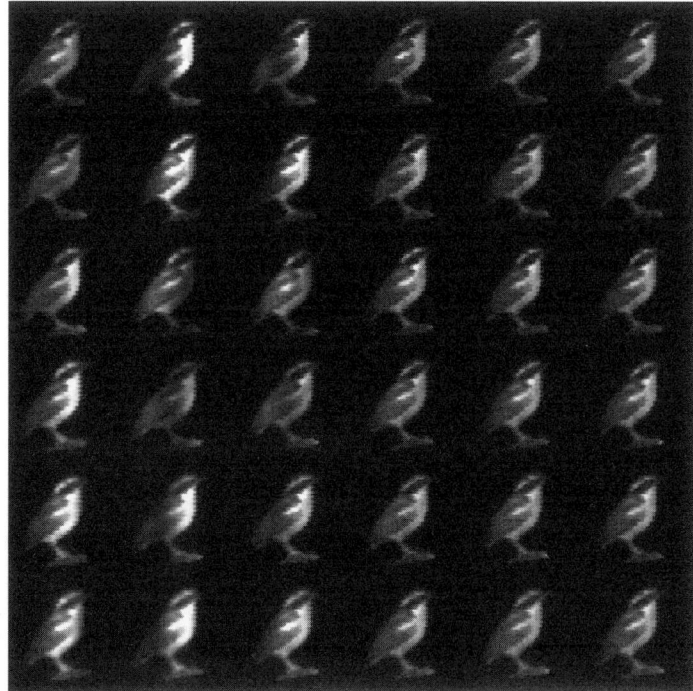

Figure 11.4.4
Multiple images of interference signals obtained by the Michelson interferometer

5 Liquid-crystal Polarization-interferometer System [9]

The Michelson interferometer in the Fourier-transform spectral imaging system described in the previous section can be replaced by another interferometer if the new interferometer can produce the position-dependent optical path difference. The wedged liquid crystal (LC) layer can produce position-dependent optical path differences in compact form. This LC polarization interferometer can be composed of two LC cells and two polarizers, as shown in the middle of Figure 11.4.3. The LC layer in each cell is sandwiched between two plates of glass that make a wedge-shaped gap. The wedged gaps are opposite each other. The glass plates are coated with a polymer and appropriately rubbed to align the LC molecules homogeneously. The rubbing directions of the two cells are perpendicular to each other. The two polarizers that are placed ahead of and behind the LC cells select light whose electric vectors make an angle 45° to the rubbing directions, The light polarized linearly by the first polarizer passes through the first LC layer. In the first layer, the light beam is split into ordinary and extra-ordinary rays. In the second layer the ordinary ray in the first layer changes to an extra-ordinary ray and vice versa. Thus the optical path difference is linearly dependent on the position, and the locations of the zero optical path difference form a straight line parallel to the apex of the wedge. Figure 11.4.3 shows the optical configuration of the proposed system using the interferometer. An array of lenslets creates multiple images of an object in the LC layer. Let these images form a 2D rectangular lattice in the $x-y$ plane. The LC cell is tilted so that the ζ-axis makes an angle with the x-axis. Since the optical path difference is linearly dependent on the distance from the apex of the wedge, the multiple images of the interference signals obtained behind the analyzer are equivalent to those discussed in the previous section. These multiple images of the

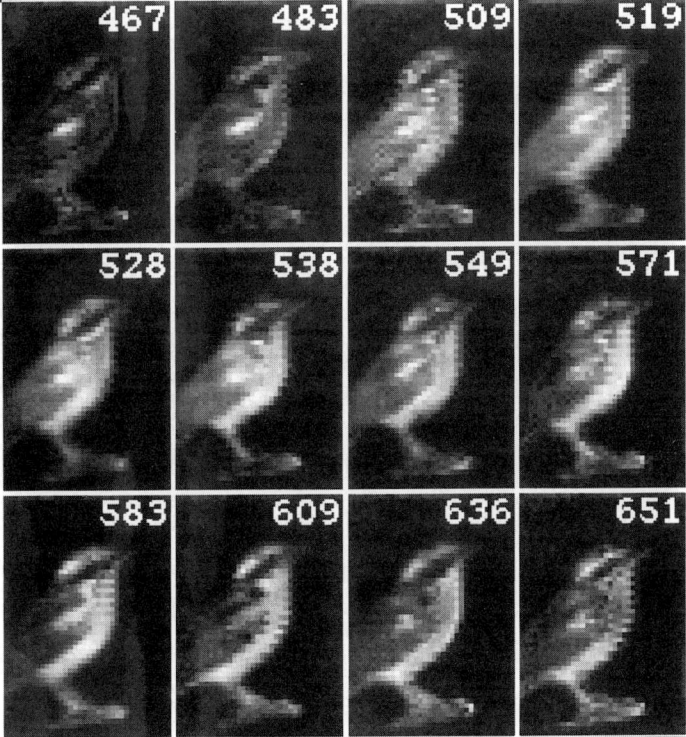

Figure 11.4.5
Cross-sections of the spectral image reconstructed by the multiple-image Michelson interferometer

interference signals are simultaneously detected by an image sensor in the same way as the basic multiple-image system.

6 Experimental Results

We fabricated a lenslet array with a rectangular aperture by using 8×8 lenslets. The central portion of the achromatic objective of a binocular telescope was cut to produce a small piece of lenslet with an aperture of 4×4 mm^2. The object used in this experiment was a picture of a bird with green wings, yellow chest and head, red abdomen, and white throat. The central part of the wings is blue. The object was attached to the shaft of an electric motor and rotated at a rate of approximately 30 rev/s. The object was illuminated by a single flash of a xenon lamp. The multiple images of the interference signals of the object produced by an 8×8 lenslet array were detected at the moment when the lamp flashed. The detected multiple images of the interference signals are partly shown in Figure 11.4.4. The white-light fringes are barely seen because the fringe spacing is quite long and the interferometer has considerable aberrations. However, the reader will find variations of details from image to image. The spectral image was reconstructed by using the generalized inverse matrix. The reconstructed spectral image has 32 cross-sections that are perpendicular to the wave number axis. A series of 12 cross-sections

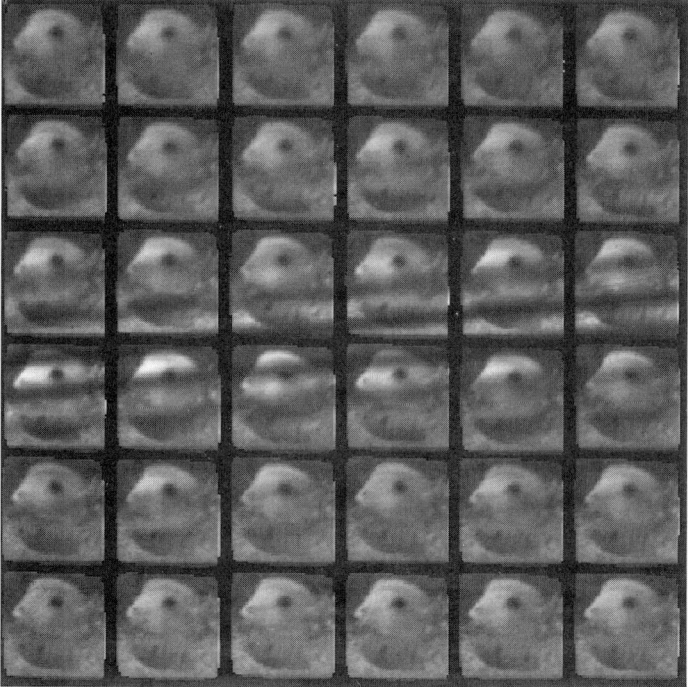

Figure 11.4.6
Multiple images of the interference signals obtained by the liquid-crystal polarization interferometer

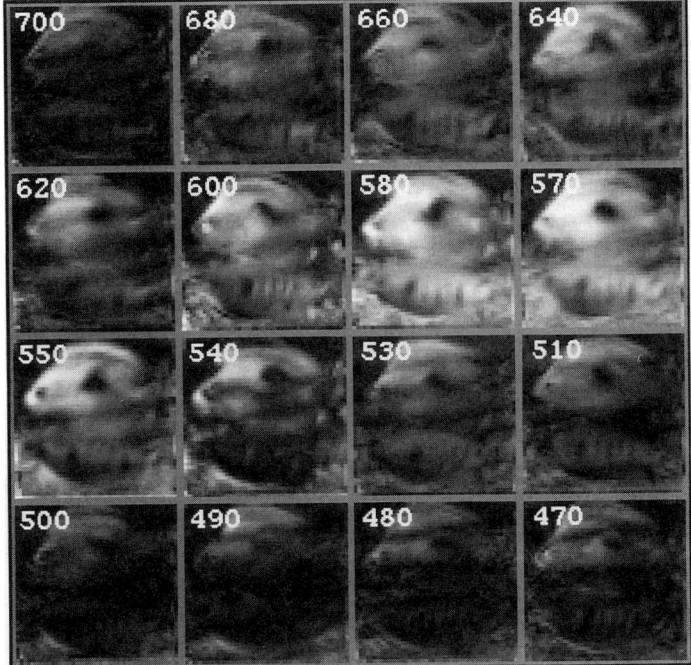

Figure 11.4.7
Cross-sections of the spectral image reconstructed by the liquid-crystal polarization interferometer

is displayed in Figure 11.4.5. Each cross-section contains 36×52 pixels. It is shown that the reconstructed spectral image is composed of areas of distinct colors.

We constructed an LC polarization interferometer consisting of two LC cells. The angle of the wedges of the two LC layers was 0.05°. These cells were filled with a nematic LC named E7 (BDH). The effective aperture size of the interferometer was approximately 50×50 mm^2. Figure 11.4.6 shows the array of interference images. The straight white-light fringes that appear in the central part of the picture indicate that the interferometer has little aberrations. The reconstructed spectral image is shown in Figure 11.4.7. Each small image consists of 45×45 pixels and represents the cross-section of the spectral image at the wavelength indicated in each image.

7 Conclusions

Two method suggested recently by the author's group to obtain spectral images of fast phenomena are presented. The combination of a lenslet array and an interferometer makes it possible to obtain instantaneously all data necessary for the reconstruction of spectral images. By the experiments it is shown that the spectral image of a rotating object can be measured and a liquid crystal can be used for the multiple-image interferometer. The liquid crystals are expected to ease fabrication of interferometers with wide apertures.

References

[1] G. Vane (Ed.), *Imaging Spectroscopy*, Proc. SPIE **834**, SPIE, Bellingham, Washington, DC, 1987.
[2] D. Simmons and L. Cowie, *CFHT Information Bulletin*, No.22, p.8 (1st semester 1990).
[3] J. Gay and D. Mekarnia, *J. Opt.*, **18**, 119-132 (1987).
[4] L. Lindegren and D. Dravins, *Astron. Astrophys.*, **67**, 241-255 (1978).
[5] K. Itoh, T. Inoue, T. Ohta and Y. Ichioka, *Opt. Lett.*, **15**, 652-654 (1990).
[6] T. Inoue, K. Itoh and Y. Ichioka, *Opt. Lett.*, **16**, 934-936 (1991).
[7] K. Itoh, T. Inoue and Y. Ichioka, *Proc. SPIE*, **1319**, SPIE, Bellingham, Washington, DC, 1990, pp. 370-371.
[8] A. Hirai, T. Inoue, K. Itoh and Y. Ichioka, *Meeting Digest of Topical Meeting of the ICO, Frontiers in Information Optics*, 1994, pp. 359-359.
[9] T. Inoue, A. Hirai, K. Itoh and Y. Ichioka, *Meeting Digest of Topical Meeting of the ICO, Frontiers in Information Optics*, 1994, pp. 360-360.
[10] K. Itoh and Y. Ohtsuka, *J. Opt. Soc. Am. A*, **3** 94-99 (1986).
[11] K. Itoh, T. Inoue and Y. Ichioka, *Appl. Opt.*, **29** 1625-1630 (1990).

11.5

Optical Heterodyne Spatio-temporal Polarimetry

Yoshihiro Ohtsuka

Abstract

This section describes the novel techniques capable of measuring the spatio-temporal principal stresses over a photoelastic sample and the change in the spatio-temporal orthogonal phases in a liquid crystal cell. Three techniques, using temporal, spatial, and spatio-temporal carrier frequency, are explained.

1 Introduction

The major objective of polarimetric studies is to discover some significant information about the material parameters of a bire-fringent sample of interest [1–4], which is obtained in general from knowledge of a polarized optical wave transmitted or reflected by such a sample. For example, photoelastic analysis permits us to make a mapping of the principal stress difference distributed over a photoelastic sample that is being loaded, whereas ellipsometry allows us to measure the complex refractive index and thickness of a thin film. In such polarimetric investigations, analysis of the state of polarization (SOP) plays a major role in the characterization of material parameters.

Roughly speaking, almost all the currently available polarimetric techniques can be divided into two classes: those for measuring only the time-dependent SOP and those for measuring only the space-dependent SOP. Still more generally, however, it is desirable that both the time- and space-dependent SOP can be determined simultaneously, although no polarimeter has been available so far for such simultaneous determination. This is mainly because of instrumentational difficulties with respect to the conventional polarimeters. For example, a rapid change in SOP cannot be followed by the mechanically movable optical components arranged in the polarimeter. Under these circumstances, the present authors have made a great efforts to exploit a novel polarimeter that allows simultaneous determination to be made.

This section aims to explain those techniques that enable us to measure the spatio-temporal principal stresses over a photoelastic sample and the change in spatio-temporal orthogonal phases along the principal axes in a liquid crystal cell. In what follows, three techniques [5-8] are successively interpreted.

2 Temporal Carrier Frequency Technique

The first technique introduces substantially the optical heterodyne detection processes; the major scheme for photomixing is shown in Figure 11.5.1. Emphasis should be given to the fact that this method needs a local oscillator beam of light consisting of orthogonal linearly polarized two-frequency components, ν_1 and ν_2, to be photomixed with a signal beam of light. Since there exists in general spatio-temporal irregularities in the birefringent parameters over a photoelastic sample, a circularly polarized signal beam of frequency ν_0 transmitted by the sample is transferred into a space- and time-dependent, elliptically polarized beam of light, the SOP of which is also a function of space and time. At every pixel of an arrayed photodetector over which photomixing takes place, the elliptically polarized signal beam is decomposed into orthogonal linearly polarized two-field components to be photomixed, respectively, with the counterpart ν_1 and ν_2 components of the local oscillator beam. It follows that the resultant photocurrent generated at every pixel involves the two beat-components at intermediate frequencies

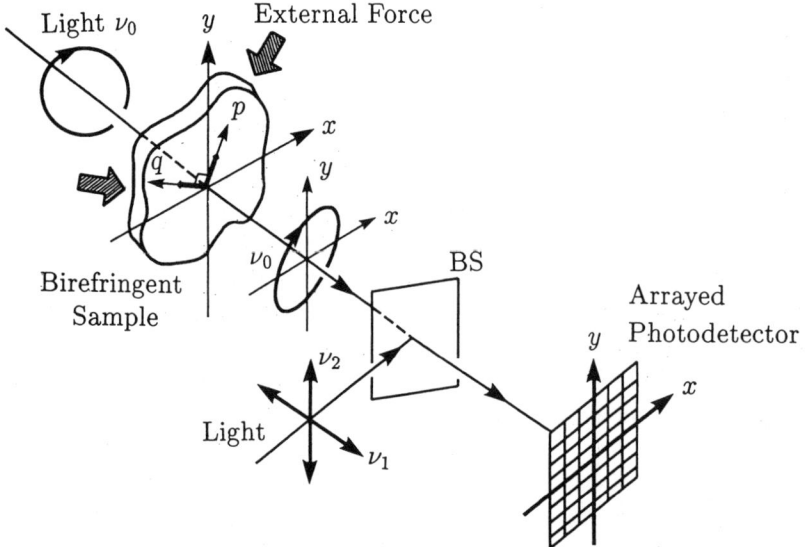

Figure 11.5.1
Working principle of an optical heterodyne polarimeter with temporal carrier frequencies

($\nu_0-\nu_1$) and ($\nu_0-\nu_2$), which possess the orthogonal decomposed two-field components of the elliptically polarized signal beam of light. These field components can be filtered in the frequency domain to offer some parameters for the determination of the spatial distribution of SOP varying with time. This polarimeter has the advantage of being able to simultaneously determine the changes in the two principal refractive indices without any other experiment as well as to measure the change in phase retardation over the birefringent sample.

The experimental demonstration is made by constructing a Mach–Zehnder type optical heterodyne interferometer. Its overall scheme is illustrated in Figure 11.5.2. A beam of light from a He–Ne laser source is first split at the beam splitter BS_1. The reflected beam of light works as a signal beam, which is transferred into a circularly polarized beam of light by a polarizer P and a quarter-wave plate (QWP) in such a way as to illuminate an epoxy photoelastic sample plate. This sample is imaged over an arrayed TV camera, as denoted by the dotted lines in the figure. The other beam of light that is transmitted by BS_1, proceeding toward the polarization beam splitter (PBS), works as a local oscillator beam. This beam is split into two orthogonal linearly polarized components at the PBS, which are modulated at different frequencies by means of the phase modulators PM_1 and PM_2, arranged in the optical paths between the PBS and the two reflection mirrors, M_1 and M_2. As a result, the returning beam of light combined at the PBS has orthogonal linearly polarized two-frequency components, ν_1 and ν_2, and travels toward the arrayed TV camera. It follows that the optical heterodyne detection processes for the signal and local oscillator beams are made over the TV camera.

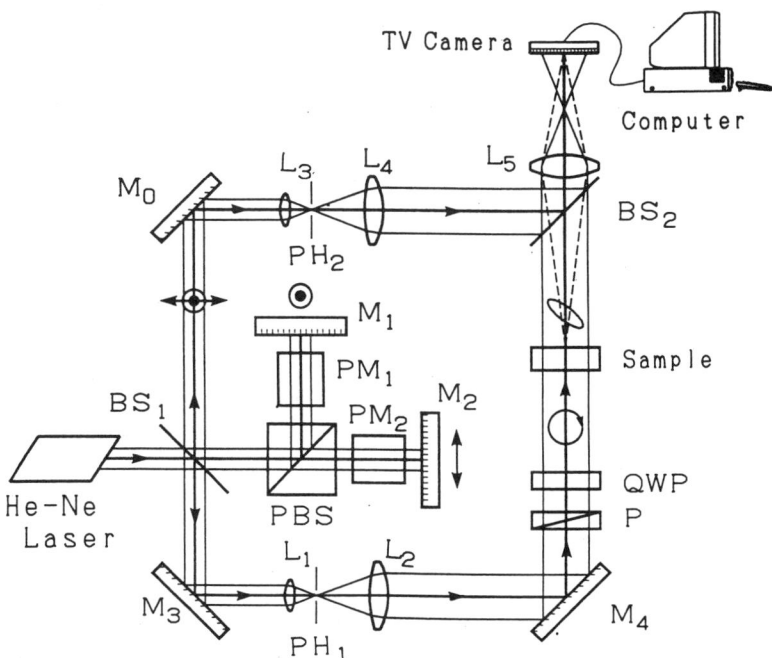

Figure 11.5.2
Optical system for optical heterodyne detection processes

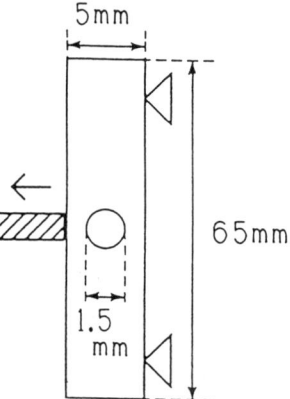

Figure 11.5.3
External view of a photoelastic sample. The stick for pushing is first in contact with the sample above the central hole

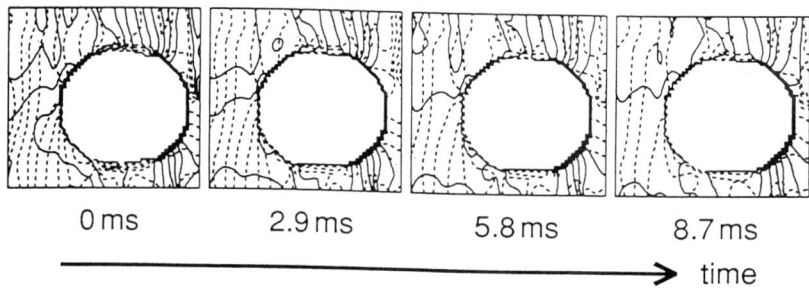

Figure 11.5.4
Time-sequential contour maps of the principal stresses around the central hole

The allowable frequency bandwidth of the TV camera used is 1033 Hz, and the filtering bandwidth is 172 Hz, which reduces the temporal resolution to 2.9 ms. Figure 11.5.3 shows the epoxy sample plate with a central hole where a stick first gives a pressing force. A decreasing force is thereby being loaded upon the sample for a while as the stick is taken out. The time-sequential contour lines for the orthogonal principal stresses are successively mapped, as shown in Figure 11.5.4. In all the maps the compressive and tensile forces remain in the regions, left and right to the circular hole, respectively. As the external force decreases, two successive countor lines spaced by 2.0×10^5 n/m^2 become more separated as time elapses.

3 Spatial Carrier Frequency Technique

In the second technique developed, a reference beam of light with orthogonal linearly polarized, two identical frequency components is introduced into an optical interferometer. Its major scheme for interference is shown in Figure 11.5.5, which is very similar to the scheme in Figure 11.5.1. For the present scheme the orthogonal linearly polarized

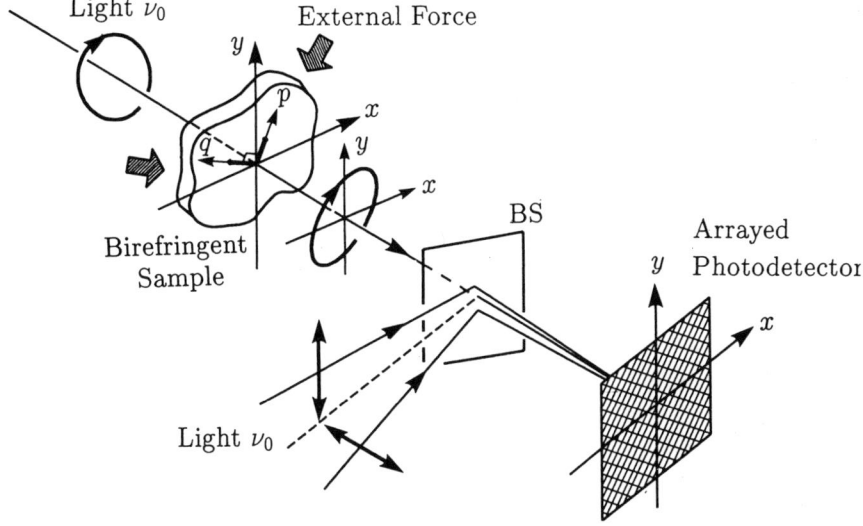

Figure 11.5.5
Working principle of an optical interferometric polarimeter with spatial carrier frequencies

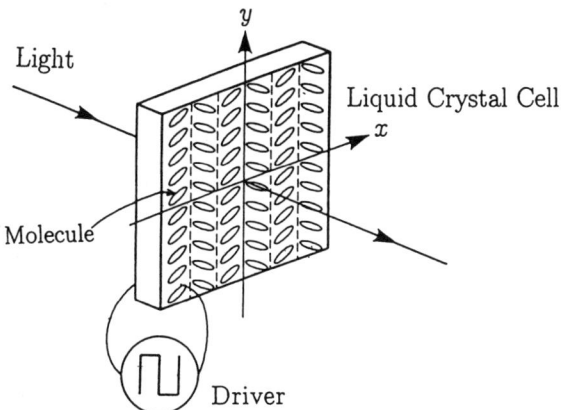

Figure 11.5.6
Schematic of a liquid crystal cell under operation

two-field components of the reference beam are tilted in the wavefront at different angles to each other in such a way that they interfere with the counterpart components of the elliptically polarized signal beam of light. As a result, a crossed interference fringe pattern is formed over the arrayed TV camera. Unlike the polarimeter described in the preceding section, the polarimeter constructed has no phase modulator, but an anti-ferroelectric liquid crystal cell is used to demonstrate its operation. This cell is shown schematically in Figure 11.5.6. The crossed interference fringe pattern, which fluctuates with time as an electric potential is applied to the liquid crystal cell, is taken successively at the frame rate of the TV camera into a computer. As mentioned in the preceding

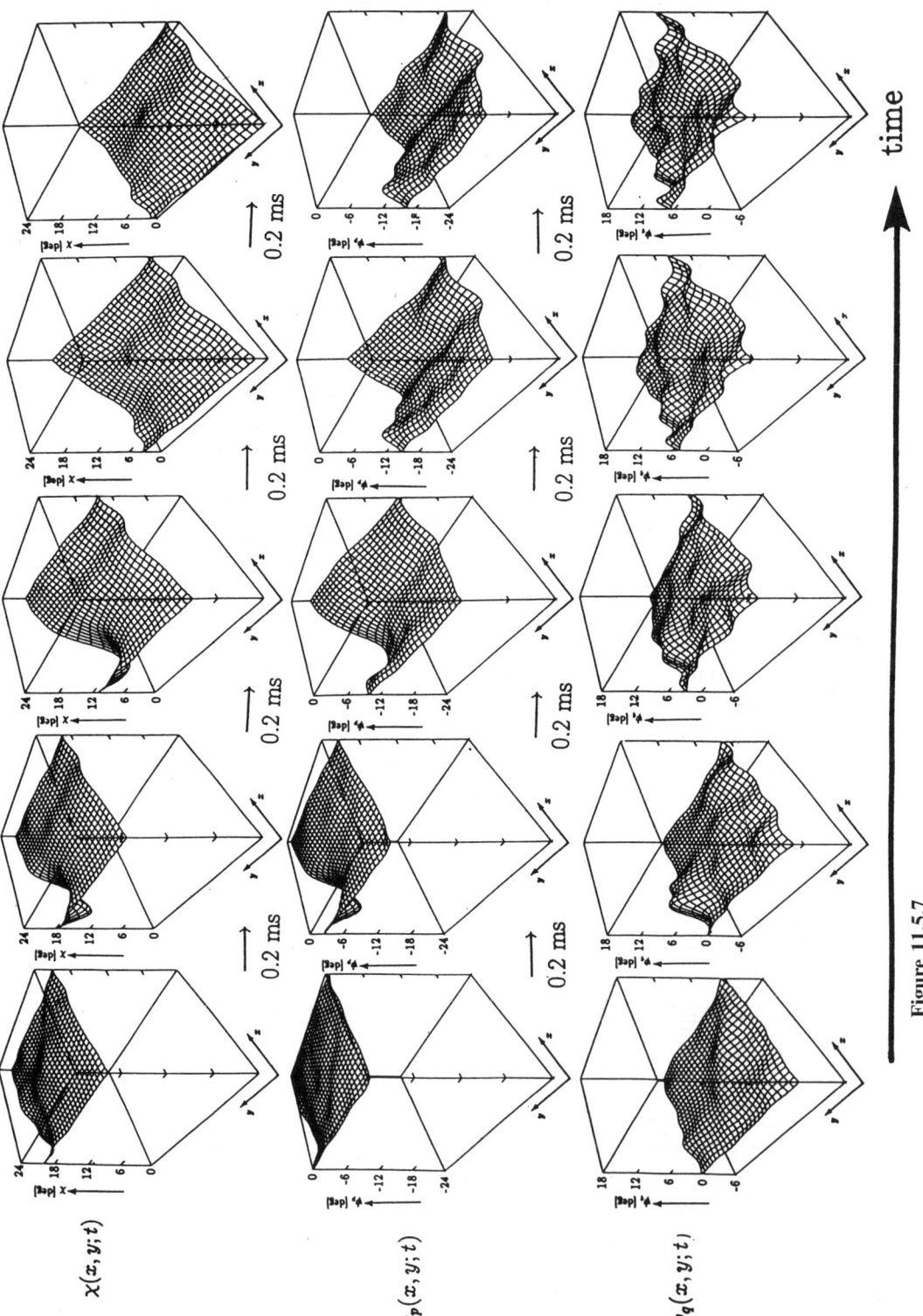

Figure 11.5.7
Time variations for the azimuth χ and the orthogonal phases ψ_p and ψ_q

section, the orthogonal two-field components of the elliptically polarized signal beam of light can be determined simultaneously from the interference fringe pattern recorded in the computer, which allows us to compute the spatial distribution of SOP varying with time. From knowledge of the SOP distribution, the azimuth $\chi(x, y; t)$ for an ellipse at a location (x, y) over the liquid crystal cell, and the changes in orthogonal phases $\psi_p(x, y; t)$ and $\psi_q(x, y; t)$ along the principal axes can be derived, respectively, for the signal beam of light. When a series of rectangular electric potentials [9,10] is applied to the liquid crystal cell, the SOP distributions are conveniently determined at every interval, in that the electric potential suddenly falls periodically. Such SOP distributions enable us to compute the above three parameters, as shown in Figure 11.5.7. Immediately after the rectangular electric potential falls to null, the azimuth $\chi(x, y; t)$ tends to zero in almost 1 ms at its falling edge, heavily depending upon the location on the cell. The phases $\psi_p(x, y; t)$ and $\psi_q(x, y; t)$ change conversely with each other from left to right in the figure, but the change in $\psi_p(x, y; t)$ is greater compared with $\psi_q(x, y; t)$.

4 Spatio-temporal Carrier Frequency Technique

The third technique introduces the spatio-temporal carrier frequencies into an interference fringe pattern, which is produced by a combination of the two polarimeters described in the preceding two sections. It is evident that this combination polarimeter is capable of broadening the bandwidth in the spatial and temporal frequency domains as compared with those in the above respective polarimeters. This fact is attractive for dynamic polarimetric studies in that a wider bandwidth is preferable.

In the experiment, the temporal and spatial bandwidths for the arrayed TV camera are assigned to be 1033 Hz and 100 cm^{-1}, respectively, and also the temporal and spatial bandwidth for filtering 500 Hz and 50 cm^{-1}, respectively. Under these conditions the temporal and spatial resolutions are, respectively, 1 ms and 0.1 μm for the polarization measurements. The epoxy sample plate, mentioned previously, is also used to demonstrate the normal operation of this polarimeter. The principal stress distributions as a function of elapsing time are shown in Figure 11.5.8. Note that the loaded force is being decreasing with elapsing time.

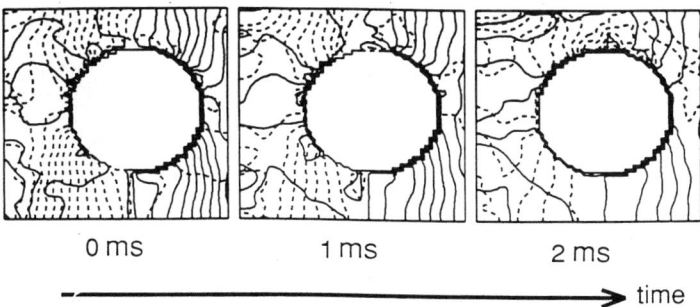

Figure 11.5.8
Time-sequential contour maps of the principal stresses around the hole

5 Concluding Remarks

The three polarimeters used allow us to use the spatio-temporal polarization analysis for a birefringent sample. A time-varying change in the SOP distribution can be determined by mapping the birefringent parameters of the sample within the filtering bandwidth of the electronic devices including an arrayed TV camera used. The spatio-temporal contour mapping is finely made with respect to a photoelastic epoxy plate and an antiferroelectric liquid crystal cell.

References

[1] M.M. Frocht, *Photoelasticity*, I & II John Wiley & Sons, New York, 1941, 1948.
[2] A. Kuske and G. Robertson, *Photoelastic Stress Analysis*, John Wiley & Sons, New York, 1974.
[3] J.F.S. Gomes, *Photoelasticity in Optical Metrology*, O.D.D. Soares, Ed., Martinus Nijhoff, Dordrecht, 1987 pp. 677.
[4] R.M.A. Azzam and N.M. Bashara, *Ellipsometry and Polarized Light*, North-Holland, Amsterdam, 1988.
[5] K. Oka, T. Takeda and Y. Ohtsuka, *J. Mod. Opt.*, **38** (8), 1567 (1991).
[6] Y. Ohtsuka and K. Oka, *Appl. Opt.*, **33** (13), 2633 (1994).
[7] K. Oka and Y. Ohtsuka, *Exp. Mech.*, **33** (1), 44 (1993).
[8] K. Oka, J. Ikeda and Y. Ohtsuka, *J. Mod. Opt.*, **40** (9), 1713 (1993).
[9] A.D.L. Chandani, T. Hagiwara, Y. Suzuki, Y. Ouchi, H. Takezoe and A. Fukuda, *Jpn J. Appl. Phys.*, **27** (5), L729 (1988).
[10] A.D.L. Chandani, E. Gorecka, Y. Ouchi, H. Takezoe and A. Fukuda, *Jpn J. Appl. Phys.*, **28** (7), L1265 (1988).

INDEX

aberration 527
absorption 217
absorption coefficient 402
Al/GaAs/AlAs 77
AlGaAs/InGaAs laser 289
alignment 21
all-optical neural networks 536
all-optical switch 6, 18
all-optical systems 4
amplifier spontaneous emission (ASE) 456
amplitude-squeezed states 579
anisotropic crystal 229
antiferroelectric LC 378
antiferroelectric liquid crystal cell 595
antiresonant condition 187
ARROW 185
ARROW-B 185
ASE 456
assembling 21
asymmetric coplanar strip line 350
attenuation constant 201
Auger recombination 264
autocorrelation function 469
autostereoscopic image 549

back propagation (BP) algorithm 418
backward degenrate four-wave mixing (BDFWM) arrangement 109
balanced receiver 462
band offset 266
band operation 352
band structure 263
basis function 512
beam divergence angle 201
beat-noise 456
biaxial strain 262
binocular view 551

bioluminescence 572
biophoton emission 572
birefringent encoding 481
bistable set-reset flip-flop 258
bottleneck 310
branching device 232
BSO crystal 527

CAD 198
cascaded modulation 446
Cerenkov-type SHG 148
CGSEL 259
CGT DFB laser 273
change in the error for learning iteration times 504
channel waveguide 148, 220
characteristic ray 550
characteristic temperature 344
chemical beam epitaxy 363
chemical-mechanical polishing (CMP) 391
chirp characteristic 56
chirp compensation 58
chirped gratings 273
chromatic dispersion 459
circular grating-coupled SEL (CGSEL) 259
CMP 391
coherent length 136
coherence multiplexing 48
coherent optical communication systems 442
coherent transient effects 157
coherent optical fiber communications 461
competitive learning algorithm 418
complex neural field 537
complex neuron model 536
confinement factor 201

construction of PGNN 513
controlled-spontaneous-emission diode 299
conventional QPM 136
correlation 525
correlator 525
Cotton-Mouton effect 227
coupled soliton 431
CPM dye laser 19

D-STOP 41
DAN 19
data compression of autostereoscopic images 553
DBR 284
DBR mirror 321
DCPBH 274
de-duplexed demodulation 452
DE-MSW DFB lasers 281
de-multiplexed demodulation 453
degenerate four-wave mixing 537
delayed nonlinear response 100
delta-function-like optical coherence function 563
density of states 263, 304
depletion region 404
design of a training set 512
detection of the amplitude diffrence 114
detection of the phase change 113
dielectric reflector 294
difference-frequency generation 86
differential gain 264, 307
diffraction-assisted phase matching 139
diffraction loss 176
digital optical cellular image processor 40
direct transition type semiconductor 403
disc pickup 239
discrete nonlinear Schrödinger equation 435
distributed Bragg reflector (DBR) 284
distributed electrode 278
distributed feedback laser 54
distributed feedback semiconductor laser 272
DMD 556
DOCIP 40
domain inversion 70
double-stage phase-diversity (DSPD) 463
DSPD 463
dual-scale topology optoelectronic processor 41

dynamic soliton 424

EB- and MO-doped polyvinyl alcohol ((EB+MO)/PVA) film 109
EDAC 404
EDFA 3, 17, 60, 437, 456
effect of physical injuries on the ultraweak biophoton emission 574
effective mass 263
effective nonlinear coefficient 118
EL emission spectra 299
electron beam exposure (EBX) direct writing 341
electron beam spatial light modulator 502
electrooptic phase modulator 17
electrooptic sampling 17, 61
emission pattern 299
energy function 518, 538
EO deflector 64, 69
EO modulator 64, 69
epoxy sample plate 597
equivalent thickness of the core 189
erbium-doped fiber amplifier (EDFA) 3, 17, 60, 437, 456
erythrosin-B-doped polyvinyl alcohol (EB/PVA) film 109
evanescent waves 126
evolution strategy 554
external input 518
extinction ratio 404
extremely highly-sensitive, vacuum-isolated photon counting system 577

f-domain control 72
Fabry-Perot EO modulator 66
Fabry-Perot microcavity 293
far-field pattern 203
Faraday effect 227
FDM 47, 461
ferroelectric liquid crystal 376
fiber 99
fiber-optic phase modulator 167
finete element method 198
first cladding layer 185
flipping of 3D image 550
fluorescence 211
FM mode-locking 168
FMCW 47

Index

focusing grating coupler 239
forward phase-conjugate 162
four-wave mixing 157, 442
Fourier transform 58, 582
Fourier transform limit 59
frequency comb generation 18
frequency-conversion demodulation 449
frequency-division multiplexing (FDM) 47, 461
frequency down-shift 102
frequency grid 85
frequency multiplexed modulation 451
frequency noise 86
frequency reference 88
frequency stability 91
fundamental standing wave 137
fundamental travelling wave 136

GaAs 406
GaAs/Si 283
gain coefficient 333
gain compression factor 56
gain saturation 220
gain switching 54
GaInAs/GaInAsP/InP quantum wire laser 341
GaInAs/InP quantum wire laser 340
Galerkin-Urabe method 436
garnet crystal film 211
generalized plane strain (GPS) model 467
germinating soybean 573
GI 153
Gordon-Haus limit 5
Gordon-Haus noise limitation 17
graded-index (GI) 153
grating coupler 238
Green's function formalism 141
group delay dispersion 65
group-veloity dispersion 437
guest-host LC-EO device 378

H-OPALS 478
harmonic bandwidth 17
Helmholtz's reciprocity theorem 537
heterojunction phototransistor (HPT) 249
hidden layer 500
hierarchical processing system 487
high-frequency demodulation 448
higher-order diffraction 160
holographic component 111

HPT 249
hybrid computing 496
hybrid-waveguide 152

image 463
image associative memory 520
in-plane magnetization 230
incoherent associative memory 36
index dispersion 142
index grating 139
information retrieval system from image content 523
input layer 500
integrated computing 518
integrated optic device 238
integration of optical interferometers 244
intelligent associative memory 519
intensity modulator 402
interaction length 143
interference cladding 185
interferogram 583
intraband relaxation time 305
intrafiber phase modulation 170
I/O bottleneck 487
ion beam implantation 316
ion beam milling 316
isolator 210

J-learning 511
J-OGNN 511
J-optimally generalizing neural network (J-OGNN) 511
J-over-learning 513
joint transform correlator 526
J_P-OGNN 511
J_P-optimally generalizing neural network (J_P-OGNN) 511

K factor 270
Kerr effect 423, 437
Kramer-Kronig relation 402
Krawczyk operator 436

laminated polarization splitter 178
large-scale substrate 24
laser 210
laser diode (LD) 219, 249, 299
laser Doppler veloimeter 25
laser-scanning confocal microscope 126
lateral excitation connection (LEC) 414
lateral inhibition 414

LD   249
learning criteiron   511
learning problem   510
LEC   414
LEDF   459
lens-plate method   549
lenticular TV display   552
light amplification   145
light-controlled optical bistable function   257
light-controlled optical thresholding function   257
light modulator   13
$LiNbO_3$ 118, 211
linear compression   58
linear frequency chirping   65
linear grating   140
liquid crystal   371, 586
liquid crystal switch   182
liquid phase epitaxy   274
liquid phase epitaxy growth   230
$LiTaO_3$ 118
lithium niobate (LN)   18
lithium niobate crystal   131
LN   18
LNP crystal   223
logic soliton   429
logical neighborhood operation   478
logitudinal spatial hole burning   278
low-dimensional quantum-well   330
low-dimensional quantum well structures   340
low-erbium-doped fiber (LEDF)   459
LPE growth   230
luminescent   217

Mach-Zehnder interferometric modulator   446
Mach-Zehnder structure   25
magnetic field   319
Maker fringe   142
memorization criterion   511
MESFET   289
metalorganic chemical vapor deposition (MOCVD)   284, 316
metalorganic vapor phase deposition   144
methyl-orange-doped polyvinyl alcohol (MO/PVA) film   109
MFP   66
micro-bonding   391

micro cavity   29, 303
micro-channel plate spatial light modulator   501
micro-fabrication technique   26
microlens   389
micromirror   389
microwave phase shifter   357
midpoint isolator   459
Miller's rule   142
MOCVD   284, 316
mode-locked fiber laser   167
mode-locking   12, 63
modified Fabry-Perot (MFP)   66
modulated stripe width   278
modulation characteristics   266
modulation polarity reversal   351
modulator   210
molecular beam epitaxy   144
monochromator   219
MSM photodiode   61
multi-dimensional information processing   46
multi-layer feedforward neural networks   510
multi-layer structure   404
multi-stage interconnection network   37
nano-structure   304

Nd_3+-doped single-mode fiber   168
near-field pattern   203
neodymium   213
neural network   500, 517
new polymer optical amplifier   153
NF   329, 456
noise   329
noise figure (NF)   329, 456
nonlinear gain   310
nonlinear grating   140
nonlinear Kerr effect   459
nonlinear organic fiber   19
nonlinear response   530
nonlinear Schrödinger equation   59, 99, 434
nonlinear Schrödinger wave equation   424
nonlinear susceptibility   97
nonreciprocal phase shift   229
notch-shaped coherence function   563
novelty filtering   115

O-CLIP   40
OAL   37, 478

Index

OEIC 283
OEID 249
off-diagonal term 467
OFSG 85
OPALS 39, 478
optical absorption 211
optical adaptive device 412
optical amplification function 252
optical amplifier 210
optical analog computing systems 35
optical analog/electronic hybrid computing systems 35
optical array logic (OAL) 37, 478
optical bistable function 253
optical Bloch equation 158
optical cellular logic image processor (O-CLIP) 40
optical coherence domain reflectometry by synthesis of the coherence function 567
optical coherence function 561
optical communications 264
optical computing 28
optical computing systems 33
optical correlator 525
optical coupling flip-flop 394
optical coupling sense amplifier 394
optical digital computer 36
optical digital computing 477
optical disk pickup head 21
optical feedback 250
optical frequency comb generator 92
optical frequency sweep generator (OFSG) 85
optical frequency synthesizer 18
optical gain 267
optical heterodyne interferometer 593
optical integrated circuits 22
optical interconnect 27
optical interconnection 290, 386
optical interconnection device 416
optical interconnection system 36
optical interferometry 44
optical joint transform correlator 531
optical memory 125
optical microcavity 293
optical modulator array 400
optical neural chip 30
optical neural computing systems 33

optical nonlinear materials 74
optical parallel array logic system (OPALS) 39, 478
optical parallel digital computing system 36
optical parallel processing 369
optical path length 295
optical phase conjugator 438
optical phase shift 168
optical pumping 213
optical soliton 434
optical switching 250
optical threshold function 253
optical tristable function 257
optical waveguide 386
optical waveguide device 210
optoelectronic integrated device (OEID) 249
optoelectronics integrated circuit (OEIC) 283
optoelectronics technology 45
ORAM-bus memory 394
organic materials 94
orthogonal polarization configuration 110
oscillating fields 538
outer emission angle 297
output layer 500

p-NA 150
P-OPALS 480
PANDA fiber 466
PANDA profile 465
para-nitroaniline (p-NA) 150
parallel pickup 240
parallel polarization configuration 110
parametric amplification 87
periodic domain inversion 69
periodic domain reversal 18
periodic structure 18
permittivity tensor 467
perpendicular magnetization 230
PGNN 511
phase compensation 101
phase-conjugate mirror 539
phase conjugate (PC) wave 109
phase conjugation 538
phase constant 201
phase matching 135, 229
phase modulator 402
photo diode 219

photo refractive crystal 526
photodetector 290
photodiode 61
photon counting distribution 577
photon density 214
photon lifetime 269
photon number states 579
photon recycling 29
photopolymer 127
photorefractive crystal 131
photorefractive effect 131
planar microlens 361
planar optics 22
plastic fiber amplifier 19
PMF 465
POEM 41
POFA 146
polarization control 364
polarization crosstalk 471
polarization-independent optical isolator 181
polarization-interferometer 586
polarization-maintaining optical fiber (PMF) 465
poled polymers 243
polymer optical fiber amplifier (POFA) 146
polyvinylalchohol (PVA) 150
population inversion 214
population inversion parametry 333
positive feedback 250
power coupling coefficient 471
periodic domain structure 117
processing element 489
projection criterion 511
projection generalizing neural network (PGNN) 511
propagation loss 217
proper J-learning 511
pulse compression 65, 95
pulse synthesizer 72
pump-and-probe absorption measurement 80
pump-probe spectroscopy 89
pumping power 215
pumping rate 215
put-in micro-connector 368
PVA 150

Q-switching 12
QCSE 402
QPM 18, 117, 136
quantum biophoton statistics 572
quantum box 330
quantum box laser 303
quantum confined Stark effect (QCSE) 402
quantum film 330
quantum-mechanical information channel 579
quantum statistical properties of the biophoton 575
quantum wire 330
quantum wire laser 303, 336
quasi-phase matching (QMP) 18, 117, 136
quasi-velociy matching 69

Rabi's frequency 159
rapid degradation 289
rare-earth-doped crystal 224
rate equation 54, 214
reciprocal phase shift 229
recombination lifetime 305
reconfigurable optical interconnection 492
red shift chirp 56
reflection type 401
refractive index 402
relaxation resonant frequency 307
repeater-free distance 4
reproducing kernel 510
resolution-point 550
resonance condition 159
response time 100
RFsputtering 212
Rhodamine B 155
ridge-type channel waveguide 146

S-OAL 481
S-SEED device 39
SAP 465
saturation-spectroscopy 89
SBS 460
scalar FEM 199
scanning optical microscope 241
scattering of a 3D image point 551
Schawlow-Townes spectral linewidth 308
screening effect 408
second cladding layer 185
second-harmonic generation (SHG) 87, 135, 145

Index

second-order intensity correlation function 577
selective growth 316, 336
self-organization 418
self-organizing optics 31
self-phase modulation (SPM) 19, 95
self-steepening 102
Sellmeier equation 124
semiconductor laser 84, 266
semiconductor laser amplifier (SLA) 328
semiconductor laser pulse source 53
semiconductor light modulator 400
semileaky waveguide isolator 229
SHG 87, 135, 145
sideband 64
signal light 214
signal sampler/multiplier 355
signal-to-noise ratio (SNR) 329, 532
single-input optical array logic (S-OAL) 481
slab waveguide 220
SLM 379, 400, 531
SNR 329, 532
soliton compression 17
soliton effect 59
soliton equation 17
soliton propagation 59
solitonics 423
SOP 591
space-variant logic operation 496
space-variant parallel logic operation 496
spatial light modulator (SLM) 379, 400, 531
spatial method 499
spatial resolution 11
spatio-temporal carrier frequency 597
spatio-temporal optical information 45
spatio-temporal orthogonal phase 592
spatio-temporal polarization analysis 598
spatio-temporal principal stress 592
spectral imaging 581
spectral resolution 583
spin flip transition 18
spin relaxation 77
SPM 19, 95
spontaneous emission 214, 299, 330
spontaneous emission control 29
spontaneous emission factor 56
spot size 201

spread-spectrum (SS) 460
sputtering target 217
square modulation 446
squeezed states 579
SS (spread-spectrum) 460
SS (symbolic substitution) 37
SSB modulator/frequency shifter 355
SSPS 416
stacked configuration of optical interconnect 190
standing-wave QPM 137
stimulated Brillouin scattering (SBS) 460
stimulated emission 214, 299, 330
strained quantum well 261, 336
strained quantum well structures 343
strained quantum wire 311
strained short-period superlattices (SSPS) 416
streak camera 57
stripe lateral confinement structure 190
strongly directed emission 299
subcarrier multiplexing 451
sum-frequency generation 86
supersoliton 425
surface emitting laser 28
surface-emitting function 249
surface-emitting SHG device 142
switch 210
switching energy 250
switching soliton 426
symbolic substitution (SS) 37
synaptic connections 411
synaptic connections circuit 411
synaptic weight 413, 538
synthesis of the coherent function 561

TDM 46
TE-TM mode conversion 227
TEC fiber 175
temperature dependence 346
temporal coding 496
temporal method 499
temporal resolution 11, 594
ternary-pulse-code switching 80
therminic emission 267
thick hologram 528
thin-film electroluminescent (EL) diode 292
thin-film waveguide integration 22
third-order phase compensation 103

threading dislocation 290, 416
three paractical conditions
    for ultrafast optical switching 76
three-dimensional growth 287
three-dimensional intergration 391
three-dimensional LSI 392
three-dimensional optical memory 126
three-dimensional stacked configuration of
    ARROWs 194
three-electrode coplanar waveguide 350
threshold current 304
Ti:sapphire laser 324
time multi/demultiplexer 355
titled-etched facet 274
training data 510
trans-cis isomerization 111
transfer matrix technique 141
transmission type 401
transparent distance 4
travelling-type laser-diode amplifier
    (TW-LDA) 6
travelling-wave mode of operation 350
travelling-wave type EO modulator 68
tristable flip-flop function 258
tunable birefringence mode 372
tunable laser 273
tuning range 275
tunneling time 305
TW-LDA 6
twisted nematic 375
two-dimensional array 30
two-dimensional Fourier transform
    processor 35

two-dimensional imaging of
    ultraweak biophoton emission 572
two-dimensional optical correlator 35
type II quantum well structure 19, 76

ultrafast optical electronics 10
ultrafast optical switch 74
ultrafast optoelectronics 16
ultraweak photon imagery systems 572

vacuum-sublimed dye film 292
validation of the simulation 435
VCSEL 286
vector FEM 198
velocity matching 68
velocity mismatch 351
vertical cavity surface-emitting laser 361
vertical photonics 175
view-point 550
visibility 562
volume hologram 528

wafer bonding 391
walk-off 87
waveguide isolator 227
wavelength chirping 278
wavelength tuning 367
WDM 47
WDM solition 424
Wiener criterion 511
winner-take-all (WTA) circuit 414

zero-method measurement 80
ΓX scattering process 77